FLORA ZAMBESIACA

Flora terrarum Zambesii aquis conjunctarum

VOLUME TEN: PART THREE

T0132982

FLORA ZAMBESIACA

MOZAMBIQUE

MALAWI, ZAMBIA, ZIMBABWE

BOTSWANA

VOLUME TEN: PART THREE

Edited by
E. LAUNERT & G. V. POPE

on behalf of the Editorial Board:

G. Ll. LUCAS
Royal Botanic Gardens, Kew

E. LAUNERT
British Museum (Natural History)

M. L. GONÇALVES
*Centro de Botânica, Instituto de Investigação
Cientifíca Tropical, Lisboa*

G. V. POPE
Royal Botanic Gardens, Kew

Published by the Managing Committee on behalf of
the contributors to Flora Zambesiaca
1989

Typeset at the Royal Botanic Gardens, Kew, by
Pam Arnold, Christine Beard, Brenda Carey,
Margaret Newman, Helen O'Brien
and Pam Rosen

ISBN 0 950 7682 6 X

Printed in Great Britain by
Whitstable Litho Printers Ltd., Whitstable, Kent

Correction of key to the species of *Amaranthus*

1. Leaf axils with paired spines - - - - - - - - - - - - - - - - - - - 4. *spinosus*
– Leaf axils without paired spines - 2
2. Inflorescence consisting entirely of axillary, cymose clusters, no terminal leafless spike or panicle present - 3
– Inflorescence not entirely of axillary, cymose clusters — a terminal, leafless spike or panicle present - 8
3. Female flowers with 4–5 perianth segments; leaves narrowly oblanceolate to narrowly elliptic-oblong; female perianth segments long-aristate   8 *praetermissus*
– Female flowers with 3 perianth segments; if leaves so narrow, then female perianth-segments very shortly mucronate - 4
4. Female perianth-segments distinctly shorter than the fruit - - - - - - - - - - 5
– Female perianth-segments equalling or exceeding the fruit - - - - - - - - - - 6
5. Leaves generally broadly and conspicuously emarginate; fruit strongly compressed, indehiscent - 10. *lividus*
– Leaves not, or narrowly and inconspicuously emarginate; fruit not strongly compressed, circumcissile or more rarely indehiscent - - - - - - - - - - - - 9. *graecizans*
6. Leaves of the main stem broadest distinctly below the middle   5. *tricolor*
– Leaves of the main stem broadest at or above the middle - - - - - - - - - - 7
7. Stems furnished above with long, crisped hairs; female perianth-segments with a long (0.75–1.75 mm.), slender, flexuose or divergent, generally colourless arista - 6. *thunbergii*
– Stems glabrous (sometimes puberulous above especially about the nodes), without elongate crisped hairs, frequently papillose-scabrid; female perianth-segments with a short 0.1–0.5(0.75) mm. erect or divergent mucro - 7. *dinteri*
8. Perianth segments 4–5 - 9
– Perianth segments 3 - 13
9. Capsule indehiscent - 10
– Capsule circumcissile - 11
10. Capsule globular and extremely muricate, not or scarcely exceeding the perianth - 11. *viridis*
– Capsule compressed, not muricate, exceeding the perianth - - - - - - - - 10. *lividus*
11. Female perianth-segments spathulate to narrowly oblong-spathulate, greenish along the midrib, which ceases below the apex or becomes colourless and faint; plant densely clothed with floccose, multicellular hairs - - - - - - - - - - - - - - - - - 1. *retroflexus*
– Female perianth-segments lanceolate to oblong, shortly aristate to mucronate with the excurrent midrib; plant glabrous to moderately pilose, usually with shorter hairs - - - - - - 12
12. Male flowers confined to a normally quite short length at the apex of each spike, rarely mixed with the females; female perianth-segments obtuse, narrowly spathulate to narrowly oblong; lid of capsule strongly wrinkled, with a swollen neck formed by the inflated style-bases - 3. *dubius*
– Male flowers mixed with the females; female perianth-segments acute, or if obtuse then the capsule not strongly wrinkled, with a short, smooth, firm beak, and the style-bases not inflated - 2. *hybridus*
13. Capsule circumcissile; bracteoles and perianth segments with a long, fine, colourless awn - 5. *tricolor*
– Capsule indehiscent; bracteoles and perianth segments mucronate or shortly aristate only - 14
14. Capsule more or less globular, extremely muricate, not or scarcely exceeding the perianth segments; seeds with shallow, scurfy verrucae on the reticulate pattern of the testa - 11. *viridis*
– Capsule ellipsoid or distinctly compressed, not muricate, distinctly exceeding the perianth segments; seeds with no shallow verrucae - - - - - - - - - - - - - - - 15
15. Capsule ellipsoid, scarcely compressed, seed also ellipsoid; leaves rarely (feebly) retuse - 12. *deflexus*
– Capsule compressed, lenticular or shortly pyriform, seed lenticular; leaves usually broadly and distinctly emarginate, rarely broadly truncate - - - - - - - - - - - 10. *lividus*

CONTENTS

LIST OF FAMILIES INCLUDED IN
VOLUME X, PART 3

200. Gramineae (Paniceae, Isachneae and Arundinelleae)*

*The tribal arrangement, as set out in Volume **10**, part 1, on page 7 and in the key on **page** 13, has been modified in the light of more recent research. Paniceae becomes tribe **number** XXIV, Isachneae tribe XXV and Arundinelleae tribe XXVI.

200. GRAMINEAE

XXIV. PANICEAE R.Br.

By W.D. Clayton*

Paniceae R.Br. in Flinders, Voy. Terra Austral. **2**: 582 (1814).

Inflorescence an open to spiciform panicle, or of unilateral racemes arranged digitately or racemosely, usually terminal, sometimes the spikelets subtended or surrounded by reduced bristle-like branches. Spikelets all alike, single or inconspicuously paired, mostly dorsally compressed, 2-flowered without rhachilla extension, falling entire, usually awnless. Glumes 2, membranous or herbaceous, the superior often as long as spikelet, the inferior usually shorter and sometimes rudimentary. Inferior floret ♂ or barren, the lemma similar to the superior glume. Superior floret bisexual, the lemma typically coriaceous to crustaceous with narrow inrolled margins clasping the edge of the indurated palea, but sometimes cartilaginous with flat margins or rarely chartaceous. Lodicules 2, fleshy. Stamens 3. Stigmas 2. Caryopsis ± ellipsoid with large embryo; hilum punctiform, rarely linear. Chromosomes small, basic number usually 9, sometimes 10.

A tribe comprising 101 genera. Distributed throughout the tropics and extending into warm temperate regions.

The distinction between paniculate and racemose inflorescences is usually quite clear but there are a few ambiguities, as when a panicle becomes contracted around its primary branches, or a raceme bears its spikelets in clusters or in secondary racemelets.

Diagnosis also makes use of the orientation of the spikelet in relation to the rhachis — a spikelet is said to be adaxial when its inferior glume is adjacent to the rhachis, and abaxial when turned away from it. However, the criterion is only reliable when the spikelets are borne singly.

1. Spikelets not subtended by bristles or scales - - - - - - - - 2
 – Spikelets, or some of them, subtended by 1 to many bristles, spines or scales - - 30
2. Leaf surface raised into sinuous lamellae; plant aquatic - - - 6. **Hydrothauma**
 – Leaf surface not lamellate - - - - - - - - - - - - 3
3. Spikelets supported on a globular bead-like callus - - - - 15. **Eriochloa**
 – Spikelets without a basal bead, though sometimes with a short cylindrical stipe - - - - - - - - - - - - - - - 4
4. Inflorescence a panicle, occasionally ± condensed about the primary branches - - - - - - - - - - - - - - - - 5
 – Inflorescence consisting of 1-sided racemes, these either digitate or scattered along a central axis, rarely solitary; the racemes sometimes with short secondary branchlets (especially *Echinochloa crus-pavonis* & *Brachiaria malacodes*) or with the spikelets long-pedicelled and distant (*Brachiaria deflexa*) - - - - - - - - - - - - 11
5. Panicle cylindrical, spiciform - - - - - - - - 5. **Sacciolepis**
 – Panicle open or contracted - - - - - - - - - - - 6
6. Spikelets dorsally compressed, rarely slightly gibbous - - - - - 7
 – Spikelets laterally compressed; inferior glume absent or minute (up to 0.5 mm. in *Sacciolepis*) - - - - - - - - - - - - - - 9
7. Inferior lemma awnless (if superior lemma tipped with a dark green spot or crest see *Acroceras*) - - - - - - - - - - - 4. **Panicum**
 – Inferior lemma awned - - - - - - - - - - - - 8
8. Awn over 5 mm. long - - - - - - - - - - 10. **Oryzidium**
 – Awn up to 1 mm. long - - - - - - - - - - 23. **Hylebates**
9. Superior lemma laterally compressed - - - - - - - 22. **Melinis**
 – Superior lemma dorsally compressed - - - - - - - - - 10

---

* *Digitaria* by P. Goetghebeur & P. Van der Veken, *Melinis* & *Tricholaena* by G. Zizka and *Panicum* by S.A. Renvoize.

1

10. Superior glume gibbous, distinctly ribbed - - - - - - - 5. **Sacciolepis**
 – Superior glume neither gibbous nor ribbed - - - - - 21. **Tricholaena**
11. Racemes very short, of 1–8 spikelets, ± impressed in a thickened
 axis - - - - - - - - - - - - - 20. **Stenotaphrum**
 – Racemes free, appressed or divergent - - - - - - - - 12
12. Superior lemma with a tiny green crest at the tip; superior glume and inferior lemma ± thickened
 at apex - - - - - - - - - - - - - - 8. **Acroceras**
 – Superior lemma not crested (sometimes mucronate or awned) - - - - 13
13. Spikelets laterally compressed or the inferior glume awned - - - - 14
 – Spikelets dorsally compressed, the inferior glume at most with a brief
 awn-point - - - - - - - - - - - - - - 16
14. Superior lemma dorsally compressed; inferior glume awned - - - 3. **Oplismenus**
 – Superior lemma laterally compressed; inferior giume usually awnless - - - 15
15. Superior glume neither gibbous nor armed with hooks; inferior glume shorter than
 spikelet - - - - - - - - - - - - 1. **Poecilostachys**
 – Superior glume gibbous, armed with hooks at maturity; inferior glume c. as long as
 spikelet - - - - - - - - - - - - 2. **Pseudechinolaena**
16. Superior lemma coriaceous to crustaceous, with narrow inrolled margins clasping only edges of
 palea; sometimes thinner but then margins inrolled - - - - - - - 17
 – Superior lemma cartilaginous, with flat thin margins covering most of the palea and often
 overlapping - - - - - - - - - - - - - - 27
17. Inferior glume absent (rarely represented by a tiny triangular scale on some of the
 spikelets) - - - - - - - - - - - - - - 18
 – Inferior glume present - - - - - - - - - - - - 19
18. Back of superior lemma facing rhachis; spikelets strongly plano-convex and often
 orbicular - - - - - - - - - - - - - 16. **Paspalum**
 – Back of superior lemma turned away from rhachis - - - - - 17. **Axonopus**
19. Superior lemma hairy - - - - - - - - - - 7. **Entolasia**
 – Superior lemma glabrous - - - - - - - - - - - 20
20. Superior lemma awned, crisply chartaceous - - - - - 11. **Alloteropsis**
 – Superior lemma at most mucronate, coriaceous to crustaceous - - - - 21
21. Spikelets paired or in racemelets - - - - - - - - - 22
 – Spikelets borne singly - - - - - - - - - - - 24
22. Tip of superior palea reflexed and slightly protruberant; spikelets cuspidate to awned, usually in
 4 rows - - - - - - - - - - - - - 9. **Echinochloa**
 – Tip of superior palea not reflexed - - - - - - - - - 23
23. Superior lemma as long as spikelet, acute, obtuse or rarely mucronulate; spikelets plump, obtuse
 to acute - - - - - - - - - - - - 12. **Brachiaria**
 – Superior lemma shorter than spikelet, broadly obtuse with a mucro c. 0.5–1 mm. long; spikelets
 plano-convex, cuspidate - - - - - - - - - - 14. **Urochloa**
24. Inferior glume turned towards rhachis - - - - - - - - 25
 – Inferior glume turned away from rhachis - - - - - - - - 26
25. Superior floret sessile - - - - - - - - - - 12. **Brachiaria**
 – Superior floret stipitate; superior glume and inferior lemma resembling woven
 fabric - - - - - - - - - - - - 13. **Eccoptocarpha**
26. Superior lemma shorter than spikelet, broadly obtuse, nearly always
 mucronate - - - - - - - - - - - - 14. **Urochloa**
 – Superior lemma as long as spikelet, acute - - - - - - 19. **Paspalidium**
27. Spikelets awned - - - - - - - - - - - - 28
 – Spikelets awnless - - - - - - - - - - - - 29
28. Racemes digitate - - - - - - - - - 24. **Stereochlaena**
 – Raceme single - - - - - - - - - - 25. **Baptorhachis**
29. Inferior glume as long as spikelet - - - - - - 26. **Megaloprotachne**
 – Inferior glume shorter than spikelet or suppressed - - - - 27. **Digitaria**
30. Bristles persisting on the axis after spikelets have fallen - - - 18. **Setaria**
 – Bristles or scales falling with the spikelets - - - - - - - 31
31. Spikelet subtended by a single bristle and supported on a slender
 stipe - - - - - - - - - - - - 30. **Paratheria**
 – Spikelet surrounded by an involucre of bristles or scales, or bristle single and spikelet
 sessile - - - - - - - - - - - - - - 32
32. Involucre composed of glumaceous bracts - - - - - 31. **Anthephora**
 – Involucre composed of bristles or spines - - - - - - - 33
33. Bristles free throughout, ± filiform - - - - - - 28. **Pennisetum**
 – Bristles flattened and connate below, often forming a spiny cup - - 29. **Cenchrus**

1. POECILOSTACHYS Hackel

Poecilostachys Hackel in Sitz.-Ber. Math.-Nat. Akad. Wiss. Wien **89**: 131 (1884). *Chloachne* Stapf in Ic. Pl. (Hook.) **31**: t.3072 (1916).

Inflorescence of lax racemes along a central axis, the spikelets in pairs or little clusters. Spikelets laterally compressed. Glumes ⅓–¾ length of spikelet, the inferior often awned. Superior lemma strongly laterally compressed, membranous to cartilaginous with flat or involute margins.

A genus of c. 20 species, all but one confined to Madagascar.

Poecilostachys oplismenoides (Hackel) Clayton in Kew Bull. **42**: 403 (1987). TAB. 1. Type: Mozambique, Gorongosa, *Carvalho* (W, holotype).
　　Panicum oplismenoides Hackel in Bol. Soc. Brot. **6**: 141 (1888). Type as above.
　　Poecilostachys flaccidula Rendle in Journ. Linn. Soc., Bot. **40**: 231 (1911). Type: Zimbabwe, Chipete Forest, *Swynnerton* 409a (BM, holotype).
　　Chloachne secunda Stapf in Ic. Pl. (Hook.) **31**: t. 3072 (1916); in F.T.A. **9**: 489 (1919). Type from Cameroon.
　　Oplismenus anomalus Peter, Fl. Deutsch Ost-Afr. **1**: 220 (1931). Type from Tanzania.
　　Chloachne oplismenoides (Hackel) Robyns in Bull. Jard. Bot. Brux. **9**: 173 (1932). —Sturgeon in Rhod. Agric. Journ. **50**: 420 (1953). —Jackson & Wiehe, Annot. Check List Nyasal. Grass.: 54 (1958). —Clayton in F.W.T.A. **3**: 436 (1972). —Clayton & Renvoize in F.T.E.A., Gramineae: 545 (1982). Type as for *Panicum oplismenoides*.

Trailing perennial with lanceolate leaf laminae and often with aerial roots at the nodes. Culms 30–100 cm. high. Spikelets 6–8 mm. long, lanceolate. Inferior glume ⅓–⅔ length of spikelet, acute to acuminate. Superior glume and inferior lemma armed with stiff bristles based in yellowish tubercles. Superior lemma cartilaginous.

Zambia. E: Nyika Plateau, 21.x.1958, *Robson* 217 (K). **Zimbabwe**. E: Bunga Forest, 30.iv.1973, *Crook* 1084 (K). **Malawi**. N: Nyika Plateau, 29.vi.1952, *Jackson* 887 (K). S: Great Ruo Gorge, Mulanje, 17.vi.1962, *Robinson* 5373 (K). **Mozambique**. MS: Gorongosa, 6.v.1964, *Torre & Paiva* 12301 (K). Northwards to Nigeria and Ethiopia. Shady places in evergreen forest.

2. PSEUDECHINOLAENA Stapf

Pseudechinolaena Stapf in F.T.A. **9**: 494 (1919). —Bosser in Adansonia **15**: 121–137 (1975).

Inflorescence of slender racemes along a central axis, the spikelets in distant pairs but the sessile of each pair often much reduced. Spikelets laterally compressed. Glumes ¾ to as long as spikelet, the inferior acute to awned, the superior gibbous, eventually armed with tubercle-based hooks and sometimes with wings, rarely awned. Superior lemma laterally compressed, cartilaginous to coriaceous with flat or involute margins.

A genus of 6 species, all but one confined to Madagascar.

Pseudechinolaena polystachya (Kunth) Stapf in F.T.A. **9**: 495 (1919). —Sturgeon in Rhod. Agric. Journ. **50**: 420 (1953). —Chippindall in Meredith, Grasses & Pastures of S. Afr.: 365 (1955). —Jackson & Wiehe, Annot. Check List Nyasal. Grass.: 55 (1958). —Bor, Grasses of B.C.I. & P.: 462 (1960). —Clayton in F.W.T.A. **3**: 436 (1972). —Clayton & Renvoize in F.T.E.A. Gramineae: 547 (1982). TAB. **2**. Type from Colombia.
　　Echinolaena polystachya Kunth in Humb. & Bonpl., Nov. Gen. Sp. **1**: 119 (1816). Type as above.
　　Panicum uncinatum Raddi, Agrost. Bras.: 41 (1823). Type from Brasil.
　　Panicum polystachyum (Kunth) K. Schum., Pflanzenw. Ost-Afr. **C**: 103 (1895) non L. (1759). Type as for *Pseudechinolaena polystachya*.
　　Panicum heterochlamys Peter, Fl. Deutsch Ost-Afr. **1**, Anh.: 23, fig. 8/2 (1930). Type from Tanzania.
　　Loxostachys uncinata (Raddi) Peter, Fl. Deutsch. Ost-Afr. **1**: 204 (1930). Type as for *Panicum uncinatum*.
　　Echinochloa polystachya (Kunth) Roberty in Bull. Inst. Fond. Afr. Noire sér. A, **17**: 64 (1955) non (Kunth) Hitchc. (1920). Type as for *Pseudechinolaena polystachya*.

Trailing annual with lanceolate leaf laminae, and often with aerial roots at the nodes. Culms 10–30 cm. high. Spikelets 3.5–5.7 mm. long, obliquely ovoid. Glumes acute to

4

Tab. 1. POECILOSTACHYS OPLISMENOIDES. 1, habit (× ½), *Faden* 70/71; 2, raceme (× ½); 3, spikelet (× 4); 4, superior lemma (× 10), 2–4 from *Greenway* 3562. From F.T.E.A.

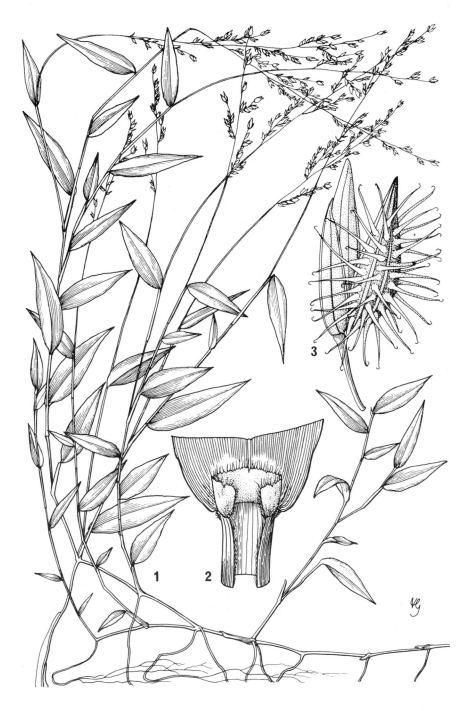

Tab. 2. PSEUDECHINOLAENA POLYSTACHYA. 1, habit (× ⅔), from *Baldwin* 13471; 2, ligule (× 6), from *Morton* 9784; 3, spikelet (× 4), from *Baldwin* 13471. From F.T.E.A.

acuminate, wingless, the inferior almost as long as spikelet. Superior lemma subcoriaceous.

Zambia. N: Mbala, 11.vii.1960, *Robinson* 3778 (K). **Zimbabwe**. E: Lusitu R., Forest, 22.iv.1973, *Ngoni* 212 (K; SRGH). **Malawi**. S: Mulanje, 27.xi.1950, *Wiehe* N/711 (K). **Mozambique**. Z: Serra Morrumbala, 13.xii.1971, *Müller & Pope* 2029 (K). MS: 20 km. W. of Dombe, 23.iv.1974, *Pope & Müller* 1267 (K).

Throughout the tropics. In forest shade; 200–1500 m.

The appearance of the inflorescence varies, partly because the hooks only develop after fertilization, and partly because the flowering period is often extended by delayed maturation of the sessile spikelets. Lucas (Isleyea 1: 115–139, 1979) regards these as adaptations to animal dispersal in a forest environment.

3. OPLISMENUS Beauv.

Oplismenus Beauv., Fl. Owar. **2**: 14 (1810). —Davey & Clayton in Kew Bull. **33**: 147–157 (1978); U. Scholz, Monograph Oplis. (1981).

Inflorescence of short racemes along a central axis, the spikelets in pairs (but the inferior often reduced). Spikelets laterally compressed, lanceolate. Glumes $\frac{1}{2}$–$\frac{3}{4}$ length of spikelet, the inferior of both awned. Superior lemma dorsally compressed, coriaceous with involute margins.

A genus of 5 species, occurring throughout the tropics.

1. Awns scaberulous - - - - - - - - - - - 1. *burmannii*
- Awns smooth and sticky - - - - - - - - - - - 2
2. Spikelets distant, adjacent pairs on the lower racemes 4–12 mm. apart; lowest raceme 2.5–10 cm. long - - - - - - - - - - - 2. *compositus*
- Spikelets contiguous, adjacent pairs on the lower racemes 0.5–4 mm. apart; lowest raceme 0.5–3 cm. long - - - - - - - - - - - - - 3
3. Racemes elongated, or the upper reduced to fascicles (rarely all fascicles but then the awns less than 7 mm. long) - - - - - - - - - - - 3. *hirtellus*
- Racemes reduced to cuneate fascicles of 2–6 spikelets, rarely the lowest forming a short raceme; awns 7–14 mm. long - - - - - - - - - 4. *undulatifolius*

1. **Oplismenus burmannii** (Retz.) Beauv., Ess. Agrost.: 54, 169 (1812). —Stapf in F.T.A. **9**: 636 (1920). —Sturgeon in Rhod. Agric. Journ. **50**: 430 (1953). —Jackson & Wiehe, Annot. Check List Nyasal. Grass.: 49 (1958). —Bor, Grasses of B.C.I. & P.: 317 (1960). —Launert in Merxm., Prodr. Fl. SW. Afr. **160**: 133 (1970). —Clayton in F.W.T.A. **3**: 437 (1972). —Clayton & Renvoize in F.T.E.A., Gramineae: 542 (1982). Type from India.
   Panicum burmannii Retz., Obs. Bot. **3**: 10 (1783). Type as above.

Prostrate or trailing annual with delicate lanceolate to narrowly ovate leaf laminae. Culms 10–60 cm. long. Racemes up to 2.5 cm. long, the spikelets contiguous. Spikelets 2.5–3.5 mm. long, usually pubescent; awns antrorsely scaberulous, the inferior 2.5–20 mm. long.

Botswana. N: Moanachira, 14.v.1973, *Smith* 583 (K; SRGH). **Zambia**. N: Chisonga, 3.iv.1952, *Richards* 1285 (K). C: Mulungushi R., 17.iii.1973, *Kornaś* 3501 (K). E: Lukusuzi, 12.iv.1971, *Sayer* 1160 (K; SRGH). S: Kalomo, 19.iii.1963, *Astle* 2276 (K). **Zimbabwe**. N: Mtoko, 15.iii.1960, *Phipps* 2533 (K). W: Victoria Falls, 23.iv.1970, *Simon & Hill* 2141 (K; SRGH). C: Shurugwi, 2.iv.1967, *Biegel* 2042 (K; SRGH). E: Mutare, 13.ii.1974, *Davidse* 6596 (K; MO). S: Kyle Nat. Park, 10.iv.1971, *Basera* 320 (K; SRGH). **Malawi** N: Rumphi, 2.vi.1976, *Pawek* 11348 (K). C: Chipata Mt., 4.v.1963, *Verboom* 979 (K). S: Lengwe Game Reserve, 9.iii.1970, *Brummitt & Hall-Martin* 8985 (K). **Mozambique**. N: Malema, 16.iii.1964, *Torre & Paiva* 11189 (K).

Throughout the tropics. Shady places; 100–1600 m.

2. **Oplismenus compositus** (L.) Beauv., Ess. Agrost.: 54, 169 (1812). —Jackson & Wiehe, Annot. Check List Nyasal. Grass.: 49 (1958). —Bor, Grasses of B.C.I. & P.: 317 (1960). —Clayton & Renvoize in F.T.E.A., Gramineae: 542 (1982). TAB. **3**. Type from Sri Lanka.
   Panicum compositum L., Sp. Pl.: 57 (1753). Type as above.

Rambling perennial with lanceolate to narrowly ovate leaf laminae. Culms 15–150 cm. long. Racemes up to 11 cm. long, the spikelet pairs distant. Spikelets 2.5–4 mm. long, glabrous to pubescent; awns smooth and sticky, the longest in each pair of spikelets 3–10 mm.

Tab. 3. OPLISMENUS COMPOSITUS. 1, habit (× ½); 2, portion of raceme (× 4); 3, spikelet (× 5), all from *Newbould & Jefford* 2746. From F.T.E.A.

Zambia. N: Katibunga, 30.iii.1961, *Angus* 2584 (K; FHO). **Zimbabwe**. C: Sharugwi 8.xii.1953, *Wild* 4290 (K; SRGH). E: Nyachohwa Falls, 30.iv.1973, *Crook* 1086 (K). S: Mt. Buchwa, 1.v.1973, *Simon, Pope & Biegel* 2409 (K; SRGH). **Malawi**. N: Karonga, 27.vi.1951, *Jackson* 554 (K). C: Ntchisi Forest Reserve, 27.iii.1970, *Brummitt* 9455 (K). S: Zomba Plateau, 6.iii.1977, *Brummitt & Seyani* 14812 (K). **Mozambique**. MS: Chimoio, 5.iv.1958, *Chase* 6874 (K; SRGH).

Mainly tropical Asia, but extending from eastern Africa to Polynesia; rare in S. America. Forest shade; 300–2000 m.

O. compositus, O. hirtellus and *O. undulatifolius* differ in appearance and in geographical distribution. However, variation between them is quite continuous and intermediates cannot be assigned with certainty (see Davey & Clayton who provide a discriminant function as an aid to identification). It seems prudent to maintain them until the cytogenetic basis of their variation has been investigated, but the case for their retention as separate species is tenuous.

3. **Oplismenus hirtellus** (L.) Beauv., Ess. Agrost.: 54, 170 (1812). —Stapf in F.T.A. **9**: 631 (1920). —Sturgeon in Rhod. Agric. Journ. **50**: 430 (1953). —Chippindall in Meredith, Grasses & Pastures of S. Afr.: 362 (1955). —Jackson & Wiehe, Annot. Check List Nyasal. Grass.: **49** (1958). —Clayton in F.W.T.A. **3**: 437 (1972). —Clayton & Renvoize in F.T.E.A., Gramineae: 542 (1982). Type from Jamaica.
 Panicum hirtellum L., Syst. Nat. ed. 10, **2**: 870 (1759). Type as above.
 Oplismenus africanus Beauv., Fl. Owar. **2**: 15 (1810). Type from Nigeria.

Rambling perennial with lanceolate to narrowly ovate leaf laminae. Culms 15–100 cm. long or more. Racemes with a distinct rhachis up to 3 cm. long, the spikelet pairs contiguous; rarely the racemes all ± cuneate, but then awns also short. Spikelets 2–4 mm. long, glabrous to sparsely pubescent; awns smooth and sticky, the longest in each pair of spikelets 3–14 mm.

Botswana. N: Khianadiandavhu R., 16.v.1979, *Smith* 2747 (K; SRGH). **Zambia**. N: Mbala, 11,v.1951, *Bullock* 3864 (K). W: Ndola, 28.iii.1974, *Chisumpa* 140 (K: NDO). C: Lusaka, 16.iv.1962, *Angus* 3120 (K; FHO). E: Nyika Plateau, 27.x.1958, *Robson* 415 (K). S: Victoria Falls, 25.iv.1932, *St. Clair-Thompson* 1340 (K). **Zimbabwe**. N: Mazowe (Mazoe), 14.iii.1965, *Bingham* 1424 (K; SRGH). W: Victoria Falls, 23.iv.1970, *Simon & Hill* 2140 (K; SRGH). C: Marondera, 21.v.1968, *Simon, Rushworth & Mavi* 1825 (K; SRGH). E: Chirinda Forest, iv.1961, *Goldsmith* 19/61 (K; SRGH). S: Chibi, 3.v.1962, *Drummond* 7895 (K; SRGH). **Malawi**. N: Nyika Plateau, 14.vi.1951, *Jackson* 496 (K). S: Zomba Forest Reserve, 28.iii.1981, *Chapman* 5583 (K).

Throughout the tropics (except Asia where it is mainly replaced by *O. compositus*). Forest shade; 1000–1800 m.

Disjunct segregates of *O. hirtellus* can be recognized in Australia and South America, but the differences between them are bridged by continuous variation in the African population whose leaf and inflorescence characters are extremely variable.

4. **Oplismenus undulatifolius** (Ard.) Roem. & Schult., Syst. Veg. ed. 15, **2**: 482 (1817) non Beauv (1812) nom. nud. —Bor, Grasses of B.C.I. & P.: 318 (1960). —Clayton & Renvoize in F.T.E.A., Gramineae: 544 (1982). Type from Italy.
 Panicum undulatifolium Ard., Animad. Spec. Alt.: 14 (1754). Type as above.
 Oplismenus capensis Hochst. in Flora **29**: 114 (1846). Type from S. Africa.
 Panicum kraussii Steud., Syn. Pl. Glum. **1**: 45 (1854) non *P. capense* Roem. & Schult. (1817). Type as for *Oplismenus capensis*.
 Oplismenus simplex K. Schum. in Engl., Glied. Veg. Usambara: 48 (1894) nom. nud.
 Oplismenus africanus var. *capensis* (Hochst.) Stapf in Fl. Cap. **7**: 417 (1899). Type as for *O. capensis*.
 Oplismenus africanus var. *simplex* Stapf in Fl. Cap. **7**: 417 (1899). Type from Tanzania.
 Oplismenus hirtellus subsp. *capensis* (Hochst.) U. Scholz, Monograph Oplis.: 113 (1981). Type as for *O. capensis*.

Trailing perennial with narrowly lanceolate to narrowly ovate leaf laminae. Culms 15–50 cm. long. Racemes reduced to cuneate fascicles without evident rhachis, or rarely the lowermost longer. Spikelets 2.5–4 mm. long, glabrous to sparsely pubescent; awns smooth and sticky, the longest in each fascicle 7–14 mm.

Zambia. C: Kundalila Falls, iii.1969, *Williamson* 1567 (K; SRGH). **Zimbabwe**. E: Kasipiti, 1.vi.1966, *Loveridge* 1559 (K; SRGH). **Malawi**. N: Nyika Plateau, 18.v.1970, *Brummitt* 10886 (K). S: Mulanje, 17.vi.1962, *Robinson* 5370 (K). **Mozambique**. MS: Beira to Inhaminga, 28.iv.1942, *Torre* 4018 (K).

Warm temperate regions of the northern hemisphere, extending southward on the uplands of Africa and India. Forest shade; 800–2400 m.

Essentially a temperate variant, but intergrading with depauperate specimens of *O. hirtellus*.

4. PANICUM L.

By S.A. Renvoize

Panicum L., Sp. Pl.: 55 (1753); Gen. Pl., ed. 5: 29 (1754).

Spikelets spherical, ovate, lanceolate or oblong. Glumes hyaline to membranous, usually the inferior shorter than and the superior as long as the spikelet. Inferior floret male or barren, its lemma usually resembling the superior glume, with or without a palea. Superior lemma as long as or shorter than the spikelet, crustaceous, the margins involute and clasping only the edges of the palea. Caryopsis more or less ellipsoid, dorsally compressed. Annuals or perennials. Leaf laminae involute or flat, linear to ovate. Inflorescence a panicle, usually much branched but occasionally contracted about the primary branches.

A large genus of 470 species found throughout the tropical, subtropical and temperate regions of the world.

Panicum antidotale Retz., a native of India, has been introduced as a drought resistant fodder grass at Marondera in Zimbabwe. It may key out with *P. subalbidum* or *P. coloratum*, from which it is distinguished by its stout, knotty, pubescent rhizome and its long lower glume, which is $\frac{1}{2}$–$\frac{2}{3}$ the length of the spikelet.

Panicum miliaceum L., a cultivated species commonly known as 'proso', is occasionally found as an escape. It is distinguished by its dense, often drooping panicle, its ovate to ovate-oblong spikelets, (4)4.5–5.5 mm. long, clustered towards the upper parts of the branches and by its short lower palea which is reduced to a small scale. Recorded from Malawi, Zimbabwe and Botswana.

1. Whole plant or just the panicle clavellate hairy - - - - - - - - 2
- Whole plant glabrous or hairy but not with clavellate hairs - - - - - 15
2. Glumes and inferior lemma pectinate - - - - - - - - - - 3
- Glumes and inferior lemma entire - - - - - - - - - - 5
3. Spikelets 1.5–2 mm. long - - - - - - - - - 1. *pectinellum*
- Spikelets 2.5–3.5 mm. long - - - - - - - - - - - 4
4. Culms 60–150 cm. high; leaf laminae 18–37(48) cm. × 10–18 mm.   2. *lukwangulense*
- Culms 15–80 cm. high; leaf laminae 5–18(27) cm. × 1–9 mm. - - - - 3. *ecklonii*
5. Inferior glume 3–9-nerved; plant annual or perennial - - - - - - 6
- Inferior glume 0–1-nerved; plant annual - - - - - - - - 12
6. Superior lemma puberulous at the apex - - - - - - - - 7
- Superior lemma glabrous at the apex - - - - - - - - - 9
7. Plant annual; spikelets yellowish-green - - - - - - 4. *flacciflorum*
- Plant perennial; spikelets pale green - - - - - - - - - 8
8. Culms unbranched; leaf laminae lanceolate - - - - - - 5. *claytonii*
- Culms pseudodichotomously branched; leaf-laminae narrowly ovate - - - 6. *peteri*
9. Inferior glume $\frac{1}{2}$–$\frac{2}{3}$ the length of the spikelet - - - - - - - 10
- Inferior glume $\frac{1}{4}$–$\frac{1}{2}$ the length of the spikelet; inferior lemma male   10. *adenophorum*
10. Rhizomatous perennial; culms erect 20–40 cm. high - - - - - 7. *bullockii*
- Tufted or rhizomatous perennial; culms erect, 70–250 cm. high or scrambling and 45–240 cm. long - - - - - - - - - - - - - - 11
11. Leaf laminae 15–48 cm. × 5–35(40) mm.; panicle 10–40 cm. long; spikelets 3.5–5(5.5) mm. long, inferior lemma male - - - - - - - - - - - - 8. *deustum*
- Leaf laminae 4–12 cm. × 6–20 mm.; panicle 7–13 cm. long; spikelets 2.5–3.5 mm. long, inferior lemma sterile - - - - - - - - - - - - 9. *sadinii*
12. Apex of pedicels bearing long hairs - - - - - - - 11. *habrothrix*
- Apex of pedicels glabrous - - - - - - - - - - - 13
13. Spikelets 1.2–1.5 mm. long - - - - - - - 12. *nigromarginatum*
- Spikelets 2–2.5(3) mm. long - - - - - - - - - - - 14
14. Panicle moderately branched, ovate; pedicels flexuous and spreading   13. *pole-evansii*
-. Panicle sparsely branched, oblong; pedicels stiff and often appressed to the branches - - - - - - - - - - - 14. *hymeniochilum*
15. Superior lemma rugose - - - - - - - - - - - - 16
- Superior lemma smooth, tessellate, minutely scaberulous, papillose or verruculose   18
16. Spikelets sulcate on the dorsal side - - - - - - - 15. *infestum*
- Spikelets rounded on the dorsal side - - - - - - - - - 17
17. Plant with pubescent rhizomes, perennial; culms 40–100 cm. high - - 16. *sabiense*
- Plant tufted, annual or perennial; culms (25)75–200 cm. high - - - 17. *maximum*
18. Superior lemma verruculose; spikelets globose or ovate; leaf laminae linear or narrowly lanceolate, if leaf laminae ovate then see 64. *pseudoracemosum* & 65. *heterostachyum*   19
- Superior lemma smooth, tessellate, minutely scaberulous or papillose - - - 24

19. Plant annual - - - - - - - - - - - - - - 20
 – Plant perennial - - - - - - - - - - - - - - 21
20. Spikelets 1–1.5 mm. long; culms 10–35 cm. high; leaf lamina linear or
    filiform - - - - - - - - - - - - 21. *lindleyanum*
 – Spikelets (1.8)2–2.5 mm. long; culms 25–70 cm. high; leaf laminae narrowly
   lanceolate - - - - - - - - - - - 22. *gracilicaule*
21. Culms weak, scrambling, 30–130 cm. long - - - - - - 23. *nervatum*
 – Culms erect from a caespitose or tufted base - - - - - - - 22
22. Spikelets 1.2–1.7 mm. long, pubescent, rarely glabrous - - - 24. *brazzavillense*
 – Spikelets 2–3 mm. long, glabrous - - - - - - - - - 23
23. Culms wiry at the base; inferior glume $\frac{2}{3}$–$\frac{3}{4}$ the length of the spikelet   25. *natalense*
 – Culms tough but not wiry; inferior glume $\frac{3}{4}$ to as long as the spikelet   26. *margaritiferum*
24. Spikelets ovate, the tips of the glumes and inferior lemma acuminate, usually recurved   25
 – Spikelets orbicular, oblong or ovate, the tips of the glumes obtuse, acute or acuminate, not
   recurved - - - - - - - - - - - - - - 36
25. Plant annual - - - - - - - - - - - - - - 26
 – Plant perennial - - - - - - - - - - - - - - 30
26. Robust plant; culms 60–180 cm. high, 4–7 mm. in diam. at the base   27. *hanningtonii*
 – Slender plant; culms (15)20–100(150) cm. high, 1–2 mm. in diam. at the base - - 27
27. Spikelets 2.5–3 mm. long. - - - - - - - - - - - 28
 – Spikelets 1.5–2 mm. long; panicle moderately to much branched - - 31. *walense*
28. Inferior floret male; panicle much branched - - - - - 28. *massaiense*
 – Inferior floret sterile; panicle sparsely to moderately branched - - - - 29
29. Culms erect; panicles mostly terminal - - - - - - 29. *zambesiense*
 – Culms decumbent or ascending; panicles terminal and axillary - - - 30. *kasumense*
30. Plant caespitose; basal sheaths pubescent - - - - - - 32. *dregeanum*
 – Plant tufted or shortly rhizomatous; basal sheaths pilose, hispid or glabrous - - 31
31. Panicle pilose - - - - - - - - - - - - - 33. *pilgeri*
 – Panicle glabrous - - - - - - - - - - - - - 32
32. Superior glume and inferior lemma 7(9)-nerved - - - - - - 33
 – Superior glume and inferior lemma 5(7)-nerved - - - - - - 34
33. Culms 30–90 cm. high; panicle 6–12(16) cm. long - - - - 34. *poaeoides*
 – Culms 100–200 cm. high; panicle (20)30–60 cm. long - - - - 35. *phragmitoides*
34. Panicle moderately and irregularly branched, the spikelets somewhat clustered on contracted
    secondary branches; plant 30–150 cm. high - - - - 36. *graniflorum*
 – Panicle much branched, the branching uniform with the spikelets evenly distributed on
   spreading or ascending branches - - - - - - - - - 35
35. Panicle oblong, the branches ascending - - - - - - 37. *fluviicola*
 – Panicle ovate-oblong, the branches spreading - - - - - 38. *graciliflorum*
36. Superior lemma and palea glossy black or brown - - - - - - 37
 – Superior lemma and palea glossy or dull, pallid - - - - - - 41
37. Plant a tufted perennial - - - - - - - - 39. *carneovaginatum*
 – Plant annual - - - - - - - - - - - - - 38
38. Inferior glume $\frac{2}{3}$ the length of the spikelet; panicle much
    branched - - - - - - - - - - 40. *atrosanguineum*
 – Inferior glume $\frac{1}{3}$–$\frac{1}{2}$ the length of the spikelet - - - - - - - 39
39. Spikelets densely pilose; inferior glume nerveless - - - - 41. *phippsii*
 – Spikelets glabrous; inferior glume 3–7-nerved - - - - - - 40
40. Plant hispid, culms geniculately ascending; panicle moderately to
    much-branched - - - - - - - - - - 42. *arcurameum*
 – Plant glabrous or sparsely pilose, culms slender, erect; panicle sparsely to moderately
   branched - - - - - - - - - - 43. *haplocaulos*
41. Inferior glume up to $\frac{1}{3}$ the length of the spikelet, clasping or cuff-like, if $\frac{1}{3}$–$\frac{2}{3}$ the length of the
    spikelet and the base of the plant lanate-hairy see 60. *lanipes* - - - - 42
 – Inferior glume $\frac{1}{2}$ to equal the length of the spikelet, if less then scale-like or narrowly ovate, not
   clasping or cuff-like - - - - - - - - - - - - 57
42. Plant annual, if nodes sharply demarcated and culms (30)60–200 cm. high see
    53. *subalbidum* - - - - - - - - - - - - 43
 – Plant perennial - - - - - - - - - - - - - 49
43. Spikelets oblong, obtuse; leaf laminae lanceolate - - - - - 44
 – Spikelets ovate-oblong, ovate or lanceolate; leaf laminae linear - - - 45
44. Spikelets 1.8–2.5 mm. long; leaf laminae narrowed at the base; panicle sparsely
    branched - - - - - - - - - - - 44. *comorense*
 – Spikelets 2.5–3.7 mm. long; leaf laminae cordate; panicle much
   branched - - - - - - - - - - 45. *madipirense*
45. Spikelets 3.5–5.5 mm. long, acuminate; panicle branches appressed - - - 46
 – Spikelets 2.2–3.5(4) mm. long, obtuse or acute; panicle branches appressed or
   spreading - - - - - - - - - - - - - - 47

46. Culms soft, spongy, 30–200 cm. high; leaf laminae 10–30 cm. × 5–12 mm. 46. *pilgerianum*
– Culms slender, herbaceous, 25–30 cm. high; leaf laminae 5–8 cm. × 1–2.5 mm. - - - - - - - - - - - - - - - 47. *perangustatum*
47. Panicles scarcely exserted from the uppermost leaf sheaths, sparsely branched, the branches appressed - - - - - - - - - - - - - - - - - 48. *gilvum*
– Panicles fully exserted, moderately to much branched - - - - - - - 48
48. Inferior glume $\frac{1}{3}$ the length of the spikelet, 3-5-nerved - - - - - 49. *mlahiense*
– Inferior glume $\frac{1}{6}$–$\frac{1}{5}$ the length of the spikelet, 1-nerved - - - - - - 50. *schinzii*
49. Plant rhizomatous, inferior glume hyaline - - - - - - - - - - - - 50
– Plant caespitose, loosely tufted or very shortly rhizomatous; inferior glume membranous or herbaceous - - - - - - - - - - - - - - - - - - - 51
50. Culms 30–100 cm. high; leaf laminae convolute, pungent - - - - - - 51. *repens*
– Culms 20–60 cm. high; leaf laminae flat, soft - - - - - - - - 52. *repentellum*
51. Spikelets acuminate, 3–3.5 mm. long - - - - - - - - - - - - - 52
– Spikelets obtuse or acute - - - - - - - - - - - - - - - - 53
52. Nodes sharply demarcated - - - - - - - - - - - - - 53. *subalbidum*
– Nodes not sharply demarcated, pigment often spreading above and below - - - - - - - - - - - - - - - - - 54. *porphyrrhizos*
53. Nodes conspicuously bearded or pubescent - - - - - - - - 55. *trichonode*
– Nodes glabrous or hairy but not conspicuously bearded - - - - - - - - 54
54. Plant hispid; culms 60–120 cm. high from a knotty pubescent base - - - - - - - - - - - - - - - - - 56. *bechuanense*
– Plant glabrous or hairy, if hispid then base not knotty - - - - - - - - 55
55. Culms 20–50 cm. high; panicle 2.5–6 cm. long; inferior glume saccate - - - - - - - - - - - - - - - - - 78. *subflabellatum*
– Culms 50–200 cm. high; panicle (10)25–45 cm. long - - - - - - - - - 56
56. Culms 50–100 cm. high; plant glabrous or pilose - - - - - - 57. *coloratum*
– Culms 100–200 cm. high; plant generally hispid - - - - - - - 58. *merkeri*
57. Inferior glume clasping the base of the spikelet, $\frac{1}{2}$–$\frac{2}{3}$ the length of the spikelet - - 58
– Inferior glume ovate, oblong or scale-like, not clasping - - - - - - - - 62
58. Plant perennial - - - - - - - - - - - - - - - - - - 59
– Plant annual - - - - - - - - - - - - - - - - - - - 60
59. Culms 100–200 cm. high, pubescent at the base; spikelets 3–4 mm. long - - - - - - - - - - - - - - - - - 59. *kalaharense*
– Culms 20–90 cm. high, lanate hairy at the base; spikelets (2)2.5–3 mm. long - - - - - - - - - - - - - - - - - 60. *lanipes*
60. Spikelets paired; panicle 15–40 cm. long - - - - - - - - 61. *pansum*
– Spikelets not paired - - - - - - - - - - - - - - - - 61
61. Plant hispid; culms branched; panicle 10–20 cm. long, much branched ovate - - - - - - - - - - - - - - - - - 62. *novemnerve*
– Plant glabrous; culms unbranched; panicle 1.5–8 cm. long, very sparsely branched, narrowly oblong - - - - - - - - - - - - - - - 63. *ephemerum*
62. Panicle with glandular patches on the branches - - - - - - - - - 63
– Panicle without glandular patches on the branches - - - - - - - - - 65
63. Panicle with secondary branches poorly developed or absent, the spikelets appearing to be scattered and racemose - - - - - - - - - - - 64. *pseudoracemosum*
– Panicle with secondary branches well developed - - - - - - - - - 64
64. Panicles much branched the branches fine and tangled; spikelets 1.5 mm. long - - - - - - - - - - - - - - - - - 65. *heterostachyum*
– Panicle moderately branched; spikelets 1.5–2 mm. long - - 66. *glandulopaniculatum*
65. Spikelets oblong, obtuse - - - - - - - - - - - - - - - 66
– Spikelets ovate or ovate-oblong, acute, acuminate or with an awnlet - - - - 71
66. Leaf laminae lanceolate; culms decumbent or scrambling - - - - - - 67
– Leaf laminae linear; culms erect, tufted - - - - - - - - - - - 69
67. Inferior glume $\frac{1}{8}$–$\frac{1}{4}$ the length of the spikelet - - - - - - 18. *trichocladum*
– Inferior glume $\frac{1}{2}$–$\frac{2}{3}$ the length of the spikelet - - - - - - - - - - 68
68. Plant annual - - - - - - - - - - - - - - - - 14. *hymeniochilum*
– Plant perennial - - - - - - - - - - - - - - - - 7. *bullockii*
69. Culms soft, spongy; plant aquatic - - - - - - - - - - - 19. *funaense*
– Culms hard; plant of dry or damp situations - - - - - - - - - - 70
70. Inferior glume $\frac{1}{3}$ the length of the spikelet, 3-nerved; culms 45–90 cm. high - - - - - - - - - - - - - - - - - - 20. *chambeshii*
– Inferior glume $\frac{1}{2}$–$\frac{2}{3}$ the length of the spikelet, 5–7-nerved; culms 70–250 cm. high - - - - - - - - - - - - - - - - - - 8. *deustum*
71. Inferior glume separated from the superior glume by a small internode - - - - 72
– Inferior glume not separated by an internode - - - - - - - - - - 76
72. Inferior glume as long as the spikelet - - - - - - - - - 67. *brevifolium*
– Inferior glume much shorter than the spikelet - - - - - - - - - - 73

73. Inferior glume 0–1-nerved - - - - - - - - - - - - - - 74
- Inferior glume 3–5-nerved; if panicle 3–10 cm. long with fine ascending branches see 68. *mueense* - - - - - - - - - - - - - - - - - 75
74. Panicle 3–10 cm. long, moderately branched, the branches fine and somewhat appressed - - - - - - - - - - - - - - 68. *mueense*
- Panicle 5–15(25) cm. long, much branched, the branches fine and spreading - - - - - - - - - - - - - - 69. *trichoides*
75. Panicle 4–10 cm. long, much branched, dense, the branches ascending or contracted and appressed - - - - - - - - - - - - - - 70. *pleianthum*
- Panicle 10–25 cm. long, moderately to much branched, the branches fine and spreading - - - - - - - - - - - - - - 71. *laticomum*
76. Leaf laminae glaucous, narrowly ovate to lanceolate; plant of swampy places - - - - - - - - - - - - - - 72. *parvifolium*
- Leaf laminae green, not glaucous - - - - - - - - - - - 77
77. Leaf laminae linear or lanceolate; plant of open situations, if plant annual with spikelets 2–3 mm. long and crowded in short 1-sided racemes in a branching panicle see *Brachiaria malacodes* - - - - - - - - - - - - - - - - - 78
- Leaf laminae lanceolate or ovate; plant of forest shade - - - - - - - 82
78. Plant annual or ephemeral; culms slender; spikelets 1.5–2 mm. long - - - - 79
- Plant perennial, if annual then spikelets 3–4 mm. long - - - - - - - 80
79. Plant annual, tufted; culms branched; inferior glume $\frac{2}{3}$–$\frac{3}{4}$ the length of the spikelet - - - - - - - - - - - - - - 73. *pusillum*
- Plant ephemeral with erect unbranched culms; inferior glume $\frac{1}{2}$ the length of the spikelet - - - - - - - - - - - - - - 74. *nymphoides*
80. Spikelets 3–4 mm. long; panicle 4–8(20) cm. long - - - - - 75. *aequinerve*
- Spikelets 2–3 mm. long; panicle 0.7–7 cm. long - - - - - - - - - 81
81. Panicle 0.7–3.5 cm. long, sparsely branched, the branches often somewhat contracted - - - - - - - - - - - - - - 76. *eickii*
- Panicle 1.5–7 cm. long, sparsely branched, the branches spreading - - - - - - - - - - - - - - 77. *inaequilatum*
82. Inferior glume a nerveless, hyaline scale $\frac{1}{5}$–$\frac{1}{4}$ the length of the spikelet; plant annual; spikelets acuminate - - - - - - - - - - - 79. *delicatulum*
- Inferior glume 0–3-nerved, $\frac{1}{4}$ to as long as the spikelet; plant annual or perennial; spikelets acute or acuminate - - - - - - - - - - - - - - - - 83
83. Plant perennial; panicle sparsely to moderately branched, secondary branches often poorly developed - - - - - - - - - - - - - - 80. *monticola*
- Plant annual; panicle sparsely to much branched, the branches spreading - - 84
84. Inferior glume $\frac{3}{4}$ to as long as the spikelet - - - - - - - - 81. *chionachne*
- Inferior glume $\frac{1}{4}$–$\frac{1}{3}$ the length of the spikelet - - - - - - - - 82. *wiehei*

1. **Panicum pectinellum** Stapf in F.T.A. **9**: 720 (1920). Type from Zaire (Katanga).

Perennial with slender erect culms 20–60(90) cm. high from a knotty, pubescent base. Leaves mostly cauline, laminae 3–9 × 0.2–0.6 cm., linear, flat, sharply acute. Panicle (3)6–13 cm. long, ovate or elliptic, usually much branched, the branches fine and flexuous, glandular hairy. Spikelets 1.5–2 mm. long, ovate-elliptic; inferior glume $\frac{1}{6}$–$\frac{1}{4}$ the length of the spikelet, 0–3-nerved; superior glume $\frac{1}{2}$ the length of the spikelet, 5–7-nerved; inferior lemma $\frac{2}{3}$–$\frac{3}{4}$ the length of the spikelet, 7-nerved, its palea absent, sterile; superior lemma and palea pallid glossy, puberulous at the apex.

Zambia. N: Chishinga Ranch, 1.ii.1962, *Astle* 1357 (K). W: Mwinilunga Distr., Kalenda Plain, 30.i.1938, *Milne-Redhead* 4409 (K). C: Serenje Distr., 15 km. NE. of Kanona on Great North Road, 6.iv.1961, *Phipps & Vesey-FitzGerald* 2973 (K; SRGH). **Malawi**. N: Chitipa, Chisanga Falls, 13.ii.1968, *Simon, Williamson & Ball* 1798 (K; SRGH).

Also in Zaire (Katanga) and Angola. Damp places in rocky grassland or woodland; 1300–2200 m.

2. **Panicum lukwangulense** Pilg. in Not. Bot. Gart. Berlin **12**: 380 (1935). —Clayton & Renvoize in F.T.E.A., Gramineae: 467 (1982). Type from Tanzania.

Caespitose perennial; culms 60–150 cm. high, often densely mixed hispid and glandular hairy. Leaf laminae 18–37(48) × 1–1.8 cm., lanceolate, often densely hairy with both hispid and glandular hairs, subulate at the apex. Panicle 10–44 cm. long, oblong to broadly ovate, much branched, the branches fine and spreading or ascending, often densely glandular hairy. Spikelets 2.5–3.5 mm. long, oblong, pilose at the apex, otherwise glabrous; inferior glume $\frac{1}{4}$–$\frac{1}{2}$ the length of the spikelet, 3-nerved, pectinate or acute and entire; superior glume $\frac{3}{4}$ to as long as the spikelet, 3–5-nerved, pectinate; inferior lemma as long as the spikelet, 7-nerved, pectinate, sterile, its palea absent; superior lemma pallid.

Malawi. N: Nyika Plateau, Lake Kaulime, 5.ii.1968, *Simon, Williamson & Ball* 1647 (K; SRGH).
Also in southern Tanzania. Submontane grassland, in damp situations; 2300–3000 m.
This species is very similar to *Panicum ecklonii*, being distinguished by its greater height and especially by its broader and generally longer leaf laminae.

3. **Panicum ecklonii** Nees, Fl. Afr. Austr.: 43 (1841). —Clayton in F.W.T.A. **3**: 429 (1972). —Clayton & Renvoize in F.T.E.A. Gramineae: 466 (1982). Type from S. Africa.
 Panicum pectinatum Rendle in Trans. Linn. Soc. Bot. **4**: 54 (1894). —Stapf in F.T.A. **9**: 719 (1920). —Jackson & Wiehe, Annot. Check List. Nyasal. Grass.: 52 (1958). Type: Malawi, Mt. Mulange (Milanje), *Whyte* (K, isotype).
 Panicum catangense Chiov. in Ann. Bot. Roma **13**: 44 (1914). Type from Zaire (Shaba, Katanga).
 Panicum katentaniense Robyns in Mém. Inst. Col. Belge, Sec. Sci. Nat. & Med. **1**: 57, pl. 5, f. A-F (1932). Type from Katanga.

Caespitose perennial, usually densely pubescent at the base. Culms 15–80 cm. high, erect, simple. Leaves mostly basal, pilose with both simple and gland-tipped hairs; laminae 5–18(27) × 0.1–0.9 cm., linear to lanceolate, flat, bluntly to setaceously acute. Panicle 6–12(18) cm. long, oblong or ovate, glabrous or clavellate-hairy, moderately to much-branched, the branches ascending or spreading usually fine and flexuous. Spikelets 2.5–3.5 mm. long, oblong, glabrous except for the puberulous apex; inferior glume $\frac{1}{5}$–$\frac{1}{2}$ the length of the spikelet, 3-nerved, pectinate; superior glume $\frac{1}{2}$–$\frac{3}{4}$ the length of the spikelet, 5-nerved, pectinate; inferior lemma $\frac{3}{4}$ to as long as the spikelet, 7-nerved, pectinate; sterile; superior lemma and palea glossy, puberulous at the apex.

Zambia. N: Mbala, 11.x.1966, "Comp. herb." 2248 (K). W: Mwinilunga Distr., SW. of Dobeka Bridge, 4.ix.1937, *Milne-Redhead* 3088 (K). **Zimbabwe**. E: Nyanga (Inyanga) Nat. Park, 9.vii.1975, *Cleghorn* 3060 (K; SRGH). **Malawi**. N: Nyika Plateau, 23.xii.1977, *Pawek* 13342 (K; SRGH). S: Mulanje, 6.xi.1972, *Leach* 14965 (K; SRGH). **Mozambique**. MS: Beira Distr., vii.1971, *Tinley* 2133 (K; SRGH).
W. Africa eastwards to Zaire and southern Tanzania, southwards to S. Africa. Grassland, in rocky places; 1200–2700 m.
A wide range of leaf lamina size has been admitted to this species. Very narrow forms were previously separated as *Panicum pectinatum* but the range of variation appears to be continuous and in the absence of any other distinguishing features this name is here included under *Panicum ecklonii*. *Panicum omega* Renv. from southern Tanzania is closely related to *Panicum ecklonii*, it is distinguished by its smaller spikelets 1.5–2 mm. long and the total absence of clavellate hairs.

4. **Panicum flacciflorum** Stapf in F.T.A. **9**: 720 (1920). —Clayton & Renvoize in F.T.E.A., Gramineae: 467 (1982). Type from Tanzania.
 Panicum microcephalum Peter, Fl. Deutsch Ost-Afr. 1, Anh.: 44 (1930). Type from Burundi.

Annual with erect culms 10–60(90) cm. high. Leaf laminae 3.5–10.5 × 0.4–1.1 cm., narrowly lanceolate, cordate, pilose with both tapering and clavellate hairs, acute. Panicle 3–9(18) cm. long, oblong, clavellate-hairy. Spikelets 3–4 mm. long, oblong, yellowish green; inferior glume broadly ovate, $\frac{1}{3}$–$\frac{1}{2}$ the length of the spikelet, 9-nerved; superior glume 9-nerved; inferior lemma 9-nerved, male, its palea well developed; superior lemma and palea puberulous at the apex, dull.

Zambia. N: Mporokoso Distr., 2.5 km. N. of Mugambwe, W. side of Mweru-Wantipa, 1200 m., 16.iv.1961; *Phipps & Vesey-FitzGerald* 3229 (K; SRGH).
Also in Zaire, Burundi, Tanzania and Uganda. Open places on poor soils with impeded drainage.

5. **Panicum claytonii** Renvoize in Kew Bull. **22**: 484 (1968). —Clayton & Renvoize in F.T.E.A., Gramineae: 467 (1982). Type: Malawi, Nyika Plateau, *Robson* 332 (K, holotype).

Perennial; culms (15)40–75 cm. high, erect or ascending from a pubescent, rhizomatous or knotty base. Leaf laminae 4–11.5 cm. long, 5–16 mm. wide, lanceolate, acute. Panicle 3–10 cm. long, oblong, or ovate, sparsely branched, the secondary branches very short or absent, densely clavellate-hairy; pedicels slender, 3–25 mm. long. Spikelets 3.5–5.5 mm. long, oblong, pale green often tinged purple, obtuse; inferior glume broadly ovate; $\frac{1}{3}$–$\frac{1}{2}$ the length of the spikelet, 7-nerved; superior glume almost as long as the spikelet, 9-nerved, acute; inferior lemma 9-nerved, male, its palea well developed; superior lemma puberulous at the apex, dull.

Zambia. N: Mbala, 1800 m., 23.x.1967, *Simon, Williamson, Richards & Vesey-FitzGerald* 1182 (K; SRGH). W: Mwinilunga, 22.xi.1937, *Milne-Redhead* 3341 (K). E: Lundazi Distr., Nyika Plateau,

ix.1968, *Williamson* 1014 (K; SRGH). **Malawi**. N: Nyika Plateau, Chelinda, 2500 m., 2.ii.1978, *Pawek* 13731 (K; SRGH; MO; MA). S: Zomba Plateau, 30.xi.1971, *Banda* 1161 (K; SRGH).
Also in Zaire (Shaba) and southern Tanzania. In woodland or grassland, on red soils.

6. **Panicum peteri** Pilg. in Engl. & Prantl, Pflanz., ed. 2, **14e**: 20 (1940). —Clayton & Renvoize in F.T.E.A. Gramineae: 468 (1982). Type from Tanzania.
   Polyneura squarrosa Peter, Fl. Deutsch Ost-Afr. **1**: 203 & Anh.: 53, t. 30/1 (1930) non *Panicum squarrosum* Retz. Type as for species.

Perennial with branching culms 50–150 cm. high. Leaf sheaths clavellate hairy; laminae 4–8 cm. long and 12–20 mm. wide, narrowly ovate, narrowed and asymmetric at the base, clavellate hairy on the margins, acuminate. Panicle 4–6 cm. long, scarcely exserted from the uppermost leaf sheath, oblong, sparsely branched. Spikelets oblong, 4–5 mm. long, glabrous; inferior glume ovate, ½ the length of the spikelet, 7-nerved; superior glume as long as the spikelet, 11-nerved; inferior lemma 13-nerved, sterile, its palea well developed; superior lemma and palea dull, minutely puberulous at the apex.

Zimbabwe. E: Makurupini Forest, southern limit of Chimanimani Nat. Park, 400 m., 25.xi.1967, *Simon & Ngoni* 1296 (K; SRGH). **Mozambique**. MS: 20 km. W. of Dombe, SE. end of Chimanimani Mts., 24.iv.1974, *Pope & Müller* 1290 (K; SRGH).
Also in Tanzania. Forest shade.

7. **Panicum bullockii** Renvoize in Kew Bull. **44**, 3: 543 (1989). Type: Zambia, Chisimba Falls, *Drummond & Williamson* 10092 (SRGH, holotype; K, isotype).

Rhizomatous perennial; culms 20–40 cm. long, loosely tufted, decumbent or ascending, branching. Leaf laminae 2.5–7 cm. long and 4–10 mm. wide, lanceolate, flat, acute or finely pointed. Panicles 5–7.5 cm. long ovate-oblong, moderately to sparsely branched, pilose, with or without clavellate hairs. Spikelets 2.5–3 mm. long, oblong, obtuse; inferior glume ½ the length of the spikelet, ovate, 5–7-nerved; superior glume 9-nerved; inferior lemma 9-nerved, its palea moderately developed, sterile; superior lemma and palea glossy, pallid.

Zambia. N: Chisimba Falls, 28.11.1970, *Drummond & Williamson* 10092 (K; SRGH).
Only known from northern Zambia. Woodland or mountain slopes, on arenaceous soils; 1750–2200 m.

8. **Panicum deustum** Thunb., Prodr. Pl. Cap.: 19 (1794). —Stapf in F.T.A. **9**: 651 (1920). —Sturgeon in Rhod. Agric. Journ. **50**: 433 (1954). —Chippindall in Meredith, Grasses & Pastures of S. Afr.: 328 (1955). —Clayton in F.W.T.A. **3**: 435 (1972). —Clayton & Renvoize in F.T.E.A., Gramineae: 468 (1982). Type from S. Africa.
   Panicum leptocaulon Trin. in Mém. Acad. St. Pétersb., sér. 6, Sci. Nat. **3**: 275 (1834). Type from East Africa.
   Panicum unguiculatum Trin. loc. cit.: 275. Type from S. Africa.
   Panicum corymbiferum Steud., Syn. Pl. Glum. **1**: 76 (1854). Type: Mozambique, Delagoa Bay, *Forbes* (K, isotype).
   Panicum arundinifolium Schweinf. in Bull. Herb. Boiss. **2**, App. 2: 22 (1894). Type from Ethiopia.
   Panicum pubivaginatum K. Schum. in Pflanzenw. Ost-Afr. **C**: 102 (1895). Type from Tanzania.
   Panicum menyharthii Hack. in Schinz in Bull. Herb. Boiss., Sér. 2, **1**: 766 (1901). Type: Mozambique, Boroma, *Menyharth* 902 (Z, holotype).

Tufted or shortly rhizomatous perennial; culms slender or robust, 70–250 cm. high. Leaf laminae 15–48 × 0.5–3.5(4) cm., linear to narrowly lanceolate, cordate or amplexicaul at the base, acuminate. Panicle 10–40 cm. long, ovate to oblong, the primary branches stiff and usually ascending, rarely spreading, secondary branches usually very short, antrorsely scaberulous, pilose and clavellate-hairy or glabrous. Spikelets 3.5–5(5.5) mm. long, oblong, glabrous, obtuse. Inferior glume broadly ovate, ½–⅔ the length of the spikelet, 5–7-nerved; superior glume 7-nerved, as long as the spikelet. Inferior lemma 5-nerved, male, its palea well developed. Superior lemma and palea dull or shining.

Zambia. C: Chilanga, Quien Sabe, 9.ix.1929, *Sandwith* 25A (K). S: banks of Kafue R., 29.vi.1963, *van Rensburg* 1875 (K). **Zimbabwe**. N: Makonde (Lomagundi) Distr., Mhangura (Mangula) Area, 1300 m., 9.ii.1969, *Jacobsen* 3671 (K; SRGH). C: Harare, 1600 m., 25.i.1931, *Brain* 4946 (K; SRGH). E: E. of Umvumvumu Bridge, 1100 m., 6.iv.1969, *Crook* 864 (K; SRGH). S: Beitbridge, Tuli Circle, bank of Shashi R., 24.iv.1972, *Cleghorn* 2582 (K; SRGH). **Mozambique**. N: Marrupa, between Montanhas Mirenge and Mucuwango, 20.ii.1981, *Nuvunga* 650 (K; LMU). GI: Margins of Limpopo

R., between Mapai and Guija, 7.v.1944, *Torre* 6589 (K). M: Maputo, between Changalane and Catuane, 22.xii.1952, *Myre & Carvalho* 1436 **(K).**

Also recorded from Sudan and Ethiopa to S. Africa, introduced sporadically throughout the tropics. In forest, deciduous bushland and grassland on argillaceous or arenaceous soils.

9. **Panicum sadinii** (Vanderyst) Renvoize in Kew Bull. **22**: 485 (1968). —Clayton in F.W.T.A. **3**: 429 (1972). Type from Zaire.

    Brachiaria sadinii Vanderyst in Bull. Agr. Congo Belge **16**: 665 (1925). Type as above.

    Panicum acuminatifolium Robyns in Mém. Inst. Col. Belge, Sec. Sci. Nat. & Méd. **1**(6): 27, pl. 2, A–F (1932). Type from Zaire.

Perennial with scrambling, branched culms 90–150 cm. long. Leaf laminae 4–12 cm. long and 6–20 mm. wide, lanceolate, cordate or amplexicaul, acuminate. Panicle 7–13 cm. long, ovate, sparsely to moderately branched, sparsely to densely clavellate hairy. Spikelets 2.5–3.5 mm. long, oblong, glabrous; inferior glume broadly ovate $\frac{1}{3}-\frac{2}{3}$ the length of the spikelet, 5-nerved, superior glume 7–9-nerved; inferior lemma 7–9-nerved, sterile, its palea well developed; superior lemma and palea shining.

    Zambia. N: Mbala Distr., Lunzua R., 14. vi.1964, *Vesey-FitzGerald* 4284 (K). W: Mwinilunga, 1.5 km. S. of Salujinga, 1400 m., 24.xii.1969, *Simon & Williamson* 1982 (K; SRGH).

Also in West Africa, Zaire and Angola. Forest shade.

10. **Panicum adenophorum** K. Schum. in Engl., Pflanzenw. Ost-Afr. **C**: 103 (1895). —Stapf in F.T.A. **9**: 654 (1920). —Clayton & Renvoize in F.T.E.A., Gramineae: 469 (1982). Type from Uganda.

Annual or short-lived perennial with scrambling, branched culms 45–240 cm. long. Leaf laminae 4–11 cm. long and 4–13 mm. wide, lanceolate, cordate or amplexicaul, acuminate. Panicle 5–15 cm. long, ovate, moderately to much-branched, sparsely to densely clavellate hairy. Spikelets 2.5–3(4) mm. long, oblong, glabrous; inferior glume $\frac{1}{4}-\frac{1}{2}$ the length of the spikelet, broadly ovate, 5–7-nerved, superior glume almost as long as the spikelet, 9–11-nerved; inferior lemma 9–11-nerved, male, its palea well developed; superior lemma dull or shining.

    Zambia. N: 32 km. ESE. of Mporokoso, xii.1968, *Williamson* 1261 (K; SRGH). **Malawi**. N: Nyika Plateau, Chowo Rocks, 2300 m., 6.ii.1968, *Simon, Williamson & Ball* 1659 (K; SRGH).

Also in Uganda, Tanzania, Zaire and Burundi. Forest margins, grassland or scrub.

11. **Panicum habrothrix** Renvoize in Kew Bull. **22**: 486 (1968). Type: Zambia, Mwinilunga, just south of Matonchi Farm, 24.i.1938, *Milne-Redhead* 4305 (K, holotype).

Delicate annual with erect culms 12–60 cm. high. Leaf laminae 1.5–4.5 cm. long and 4–15 mm. wide, lanceolate, cordate, acute. Panicle 3–8 cm. long, elliptic, clavellate-hairy; the tips of the pedicels bearing long hairs which overtop the spikelets. Spikelets 2–2.2 mm. long, oblong; inferior glume $\frac{1}{2}$ the length of the spikelet; superior glume 9-nerved; inferior lemma 9-nerved, male, its palea well developed; superior lemma pilose at the apex.

    Zambia. W: Mwinilunga, 24.i.1938, *Milne-Redhead* 4305 (K).

Also known from Burundi. Woodland shade.

12. **Panicum nigromarginatum** Robyns in Mém. Inst. Roy. Col. Belge, Sec. Sci. Nat. & Méd. **1**, **6**: 24, t. 1/A–F (1932); in Fl. Agrost. Congo Belge **2**: 165 (1934). —Clayton & Renvoize in F.T.E.A., Gramineae: 469 (1982). Type from Zaire.

Aquatic annual with weak culms 30–55 cm. long. Leaf laminae 1.5–3.5(4.5) cm. long and 1–4 mm. wide, narrowly lanceolate, often clavellate-hairy on the margins, acute or acuminate. Panicle 1–6 cm. long, ovate or oblong, clavellate hairy. Spikelets 1.2–1.5 mm. long; oblong; inferior glume ovate, $\frac{1}{3}-\frac{1}{2}$ the length of the spikelet, nerveless; superior glume 7-nerved; inferior lemma 7-nerved, sterile, its palea poorly developed; superior lemma dull.

    Zambia. N: Mbala, 1600 m., 14.iv.1959, *McCallum-Webster* A320 (K). W: Mwekera, near Kitwe, 24.iv.1956, *Mortimer* 137(K; SRGH). S: 4 km. N. of Choma, 1400 m., 22.iii.1957, *Robinson* 2161 (K).

Also in Zaire (Shaba) and Uganda. Swampy places.

13. **Panicum pole-evansii** C.E. Hubbard in Kew Bull. **1934**: 113 (1934). —Clayton & Renvoize in
   F.T.E.A., Gramineae: 470 (1982). Type: Zambia, 14 km. S. of Lake Tanganyika, *Pole Evans* 3039
   (K, holotype).

Annual with decumbent culms 50–110 cm. long. Leaf laminae 4–10 cm. long and 4–12
mm. wide linear-lanceolate. Panicle 7–9 cm. long, ovate-oblong, moderately branched,
clavellate-hairy. Spikelets 2–2.3 mm. long, oblong, purple tinged, glabrous; inferior glume
narrowly ovate, $\frac{1}{3}$–$\frac{1}{2}$ the length of the spikelet, 1-nerved; superior glume almost as long as
the spikelet, 9-nerved; inferior lemma 9-nerved, male, its palea well developed; superior
lemma and palea dull.

   Zambia. N: 25 km. S. of Mbala, 1600 m., 8.iv.1961, *Phipps & Vesey-FitzGerald* 3005 (K; SRGH).
   Also in Tanzania and Zaire (Katanga). Swampy places.

14. **Panicum hymeniochilum** Nees, Fl. Afr. Austr.: 46 (1841). —Chippindall in Meredith, Grasses &
   Pastures of S. Afr.: 324 (1955). —Clayton in F.W.T.A. **3**: 429 (1972). —Clayton & Renvoize in
   F.T.E.A., Gramineae: 470 (1982). Type from S. Africa.
      Panicum filiculme Schinz in Bull. Herb. Boiss. **3**: 377 (1895). Type from S. Africa.
      Panicum schlechteri Schinz in Bull. Herb. Boiss. **7**: 24 (1899). Type from S. Africa.
      Panicum snowdenii C.E. Hubbard in Kew Bull. **1928**: 132 (1928). —Jackson & Wiehe, Annot.
   Check List Nyasal. Grass: 52 (1958). Type from Uganda.
      Panicum kisantuense Robyns in Mém. Inst. Roy. Col. Belge, Sec. Sci. Nat. & Méd. 1, **6**: 25, t.
   1/G–M (1932); in Fl. Agrost. Congo Belge **2**: 166 (1934). Type from Zaire.

Aquatic annual with slender culms 14–120 cm. long. Leaf laminae 1.2–7(8.5) cm. long
and 2–5(10) mm. wide, lanceolate, acute. Panicle 2–10 cm. long, oblong, pilose with both
tapering and clavellate hairs, rarely the clavellate hairs absent. Spikelets 2–2.5(3) mm.
long, oblong; inferior glume $\frac{1}{2}$–$\frac{2}{3}$ the length of the spikelet, 0–1-nerved; superior glume
7–9-nerved; inferior lemma 7–9-nerved, sterile, its palea poorly developed; superior
lemma shiny.

   Botswana. N: Nqaga R., 5.v.1977, *Smith* 1997 (K; SRGH). **Zambia**. N: 20 km. W. of Mporokoso,
   1600 m., 12.iv.1961, *Phipps & Vesey-FitzGerald* 3106 (K; SRGH). **Zimbabwe**. E: Nyanga (Inyanga)
   Distr., Inyangombe Falls, 17.vi.1973, *Simon* 2357 (K; SRGH). S: Masvingo 30.iii.1973, *Chiparawasha*
   667 (K; SRGH). **Malawi**. C: Maperere Mission, 2.xi.1950, *Jackson* 248 (K). **Mozambique**. MS: Dororo
   near Bandula, 17.vii.1948, *Fisher & Schweicherdt* 254 (K; PRE).
   Also from Guinée and Ethiopia southwards to S. Africa, also in Madagascar. River margins and
   swampy places.

15. **Panicum infestum** Anderss. ex Peters, Reise Mossamb., Bot. **2**: 546 (1865). —Stapf in F.T.A. **9**: 658
   (1920). — Sturgeon in Rhod. Agric. Journ. **50**: 434 (1954). —Chippindall in Meredith, Grasses &
   Pastures of S. Afr.: 330 (1955). —Jackson & Wiehe, Annot. Check List Nyasal. Grass.: 51 (1958).
   —Clayton & Renvoize in F.T.E.A., Gramineae: 472 (1982). Type: Mozambique, Quirimba, *Peters*
   (K, isotype).

Tufted perennial with erect culms 50–200 cm. high. Leaf laminae 15–50 cm. long and
2–10 mm. wide, acuminate. Panicle 10–28 cm. long, oblong, secondary branches very
short or absent. Spikelets 2.5–4 mm. long, oblong, glabrous, sulcate on the dorsal side;
inferior glume broadly ovate, $\frac{1}{4}$–$\frac{1}{3}$ the length of the spikelet, 1–3-nerved; superior glume
5-nerved; inferior lemma 5-nerved, male, its palea well developed; superior lemma
transversely rugose.

   Zambia. N: Mbala Distr., Mpulungu, 800 m., 18.ii.1959, *McCallum-Webster* A82 (K). **Zimbabwe**.
   C: Marondera, 1000 m., iii.1955, *Davies* 964 (K; SRGH). E: Mutare, 1100 m., 16.i.1970, *Crook* 901
   (K). S: Masvingo (Fort Victoria), Glyntor, 25.i.1948, *Robinson* 216 (K; SRGH). **Malawi**. N: Igembe
   (Yembe), 23.iii.1953, *Jackson* 1180 (K). **Mozambique**. N: Memba, 300 m., 10.xii.1963, *Torre & Paiva*
   9494 (K). MS: Beira Distr., Gorongosa Nat. Park, v.1972, *Tinley* 2587 (K; SRGH). M: Umbeluzi to
   Catuane, 19.iv.1944, *Torre* 6466 (K).
   Also from Somalia and Zaire to S. Africa. Grassland, bushland or deciduous woodland on
   argillaceous or arenaceous soils.

16. **Panicum sabiense** Renvoize in Kew Bull. **34**: 552 (1979). Type: Zimbabwe, Lower Sabi, 500 m.,
   27.i.1948, *Rattray* 1270 (K, holotype).

Perennial with slender pubescent rhizomes; culms 40–100 cm. high, wiry at the base.
Leaf laminae 9–27 cm. × 2–6 mm., acute. Panicle 8–18 cm. long, oblong. Spikelets 2.4–2.8
mm. long, oblong, glabrous, obtuse; inferior glume broadly ovate $\frac{1}{3}$ the length of the
spikelet, 3-nerved; superior glume 5-nerved; inferior lemma 5-nerved, male, its palea well
developed; superior lemma transversely rugulose.

Zimbabwe. N: Gokwe Distr., Sengwa Research Station, 18.i.1975, *Guy* 2259 (K; SRGH). C: Kwekwe (Que Que) Distr., Sable Park, 5.ii.1977, *Chipunga* G33 (K; SRGH). E: Chipinge Distr.. upper Rupembe, 400 m., 22.i.1957, *Phipps* 69 (K; SRGH). S: Buhera Distr., Sabi Valley, 12.ii.1966, *Wild* 7533 (K; SRGH).

Only known from Zimbabwe. In woodland or bush.

17. **Panicum maximum** Jacq., Ic. Pl. Rar. **1**: 2, t.13 (1781). —Stapf in F.T.A. **9**: 655 (1920). —Sturgeon in Rhod. Agric. Journ. **50**: 433 (1954). —Chippindall in Meredith, Grasses & Pastures of S. Afr.: 329 (1955). —Jackson & Wiehe, Annot. Check List Nyasal. Grass.: 51 (1958). —Bor, Grasses of B.C.I. & P.: 327 (1960). —Launert in Merxm., Prodr. Fl. SW. Afr. **160**: 141 (1970). —Clayton in F.W.T.A. **3**: 429 (1972). —Clayton & Renvoize in F.T.E.A., Gramineae: 471 (1982). Type from the Lesser Antilles.

    Panicum maximum var. *altissimum* Kuntze, Rev. Gen. Pl. **3**: 362 (1898). Type: Mozambique, Beira (B, isotype).

    Panicum tephrosanthum Schinz in Bull. Herb. Boiss., sér. 2, **1**: 766 (1901). Type: Mozambique, Boruma, *Menyharth* 898 (Z, holotype).

    Panicum maximum var. *trichoglume* Robyns in Mém. Inst. Roy. Col. Belge, Sec. Sci. Nat. & Méd. **1**, **6**: 31 (1932). Type from Zaire.

Tufted perennial, occasionally annual, with culms (25)75–200(450) cm. high. Leaf laminae (6)12–40(100) cm. long and (4)12–35 mm. wide, acuminate. Panicle 12–45(60) cm. long, oblong or pyramidal, the lowest branches often arranged in a whorl. Spikelets (2.5)3–4.5(5) mm. long, oblong, glabrous or pubescent, obtuse or acute; inferior glume broadly ovate $\frac{1}{3}-\frac{1}{2}$ the length of the spikelet, 3-nerved; superior glume 5-nerved; inferior lemma 5-nerved, male or sterile, its palea well developed; superior lemma and palea conspicuously transversely rugose.

Botswana. N: 3.5 km. E. of Masoko pan, 16.iii.1973, *Smith* 456B (K; SRGH). SW: 30 km. N. of Kang, 19.ii.1960, *de Winter* 7391 (K; PRE). **Zambia**. C: Luangwa (Feira), 27.ix.1962, *Angus* 3345 (K; FHO). E: Chipata (Fort Jameson), i.1963, *Verboom* 509 (K). S: Kafue Nat. Park, 11.iii.1961, *Mitchell* 6/46 (K). **Zimbabwe**. N: Mazowe (Mazoe) Distr., Mutorashanga Pass, 1350 m., 2.iii.1965, *Simon* 189 (K; SRGH). W: Hwange to Victoria Falls, 5–9.iii.1956, *Rattray* 1785. (K; SRGH). C: Harare, 1400 m. i.1981, *Burrows* 1606 (K; SRGH). E: Mutare, Fairbridge Park, 400 m., 3.iii.1975, *Crook* 2073 (K). S: Sabi-Lundi Junction, 300 m., 6.vi.1950, *Wild* 3368 (K; SRGH). **Malawi**. S: Chikwawa, Lower Mwanza R., 180 m., 6.x.1946, *Brass* 18023 (K; NY). **Mozambique**. N: Nampula, 26.iv.1946, *Morais* 13 (K; PRE). T: Boruma, ii.1891, *Menyharth* 898 (Z). GI: Gaza Distr., Xai-Xai, 24.i.1980, *de Koning* 8104 (K). M: Vila Luiza to Manhica, 12.ii.1949, *Myre* 336 (K).

Tropical to S. Africa and Madagascar, introduced widely throughout the tropics. In shady or open habitats, stable or unstable soils in natural or unnatural situations.

18. **Panicum trichocladum** Hackel ex K. Schum. in Pflanzenw. Ost-Afr. **C**: 103 (1895). —Stapf in F.T.A. **9**: 659 (1920). —Jackson & Wiehe, Annot. Check List Nyasal. Grass.: 52 (1958). —Clayton & Renvoize in F.T.E.A., Gramineae: 473 (1982). Types from Tanzania.

Perennial with slender, scrambling culms 20–300 cm. long. Leaf laminae 5–15(18) cm. long and 4–18 mm. wide, lanceolate, acuminate. Panicle 6–20 cm. long, ovate, densely hairy on the main axis immediately below the lowest branches, rarely glabrous. Spikelets (2.2)2.5–3 mm. long, oblong, overtopped by long cilia from the apex of the pedicel; inferior glume $\frac{1}{8}-\frac{1}{4}$ the length of the spikelet, scale-like, nerveless, obtuse or truncate; superior glume 5-nerved; inferior lemma 5–7-nerved, enclosing a male flower, its palea well developed; superior lemma smooth.

Zambia. C: Serenje, 23.ii.1956, *Hinds* 313 (K). **Zimbabwe**. E: Chimanimani (Melsetter), 28.ix.1967, *Simon & Ngoni* 1369 (K; SRGH). **Malawi**. N: 2.5 km. W. of Mzuzu, 1400 m., 3.vi.1976, *Phillips* 1823 (K; MO). S: Nsanje, Malawe Hill, 800 m., 23.iii.1960, *Phipps* 2661 (K; SRGH). **Mozambique**. N: Malema, 27 km. from Mutuáli towards Nova Freixo, 550 m. 13.ii.1964 *Torre & Paiva* 10563 (K). Z: 10 km. N. of Quelimane, 10.viii.1962, *Wild* 5876 (K; SRGH). MS: Cheringoma, Lower Chiniziua, 45 m., 13.vii.1972, *Ward* 7878 (K).

Northwards to Zaire and Ethiopia. Bush or forest shade, on stony or arenaceous soils.

19. **Panicum funaense** Vanderyst in Bull. Agr. Congo Belge **10**: 248 (1919). TAB **4**. Type from Zaire.

    Panicum spongiosum Stapf in F.T.A. **9**: 661 (1920). Type from Zaire.

Perennial with soft spongy culms 40–200 cm. long. Leaf laminae 15–35 cm. long, linear, acute. Panicle 15–25 cm. long, narrowly oblong, the secondary branches generally short. Spikelets 2.5–3 mm. long, oblong; inferior glume $\frac{1}{4}-\frac{1}{3}$ the length of the spikelet, 3-nerved; superior glume 9-nerved; inferior lemma 9-nerved enclosing a male flower, its palea well

Tab. 4. PANICUM FUNAENSE. 1, habit (× ⅔); 2, spikelet (× 16), both from *Smith* 2758.

developed; superior lemma smooth, shiny, puberulous at the apex.

Botswana. N: Okovango R., Seronga, 20.ii.1983, *Smith* 4120 (K; SRGH). **Zambia**. B: Kataba Valley, Senanga, 19.xii.1964, *Verboom* 1557 (K). N: Chambeshi Flats, 1300 m., 27.i.1962, *Astle* 1297 (K). Also in Zaire. Marshy or aquatic situations.

20. **Panicum chambeshii** Renvoize in Kew Bull. **34**: 553 (1979). —Clayton & Renvoize in F.T.E.A, Gramineae: 474 (1982). Type: Zambia, Isoka Distr., Chambeshi R., *Vesey-FitzGerald* 1428 (K, holotype).

Tufted perennial; culms erect, 45–90 cm. high. Leaf laminae 7–20 cm. long, 3–5 mm. wide, linear. Panicle 7–14 cm. long, ovate, sparsely branched. Spikelets 3–4.5 mm. long, oblong, blunt or acute; inferior glume narrowly ovate, $\frac{1}{3}$ the length of the spikelet, 3-nerved; superior glume ovate, 9-nerved, acute; inferior lemma ovate, 9-nerved, enclosing a male flower, its palea well developed; superior lemma shiny and minutely scaberulous at the apex.

Zambia. N: Chambeshi R., Mpika to Kasama, 1200 m., 29.xii.1967, *Simon & Williamson* 1596 (K; SRGH).
Also in southern Tanzania. Grassland beside rivers, on loamy soil.

21. **Panicum lindleyanum** Nees ex Steud., Syn. Pl. Glum. **1**: 91 (1854). —Stapf in F.T.A. **9**: 673 (1920). —Clayton in F.W.T.A. **3**: 431 (1972). Type from Sierra Leone.

Tufted annual with slender erect culms 10–35 cm. high. Leaf laminae 2.5–9 cm. long, linear to filiform, flat or involute. Panicle 3–11 cm. long, ovate or oblong, finely branched. Spikelets 1–1.5 mm. long, orbicular, ovate or elliptic, glabrous or pubescent; inferior glume as long as the spikelet, 3-nerved; superior glume 5-nerved; inferior lemma 5-nerved, enclosing a male flower, its palea well developed; superior lemma and palea verruculose.

Zambia. N: Kawambwa Distr., Ntumbacushi (Timnatushi) Falls, 1260 m., 18.iv.1957, *Richards* 9306 (K). W: Mwinilunga, 1470 m., 14.v.1972, *Kornaś* 1781 (K). C: Serenje Distr., Kundalila Falls, 1500 m., 5.iv.1961, *Phipps & Vesey-FitzGerald* 2958 (K; SRGH). **Mozambique**. Z: Montes do Ile, 2.iv.1943, *Torre* 5053 (K).
This rather variable species is centred in West Africa but extends eastwards to the Sudan and southwards through Zaire to Zambia. Damp places on arenaceous soils.

22. **Panicum gracilicaule** Rendle, Cat. Afr. Pl. Welw. **2**: 179 (1899). —Stapf in F.T.A. **9**: 672 (1920). —Jackson & Wiehe, Annot. Check List Nyasal. Grass.: 50 (1958). —Clayton in F.W.T.A. **3**: 432 (1972). —Clayton & Renvoize in F.T.E.A., Gramineae: 474 (1982). Type from Angola.
Panicum verruciferum Mez in Engl. Bot. Jahrb. **57**: 188 (1921). Type: Malawi, *Buchanan* 247 (B, holotype).

Annual with slender culms 25–70 cm. high. Leaf laminae 4–10 cm. long, narrowly lanceolate. Panicle 5–12 cm. long, ovate or oblong. Spikelets (1.8)2–2.5 mm. long, ovate to elliptic, glabrous; inferior glume $\frac{2}{3}$ the length of the spikelet, 3-nerved; superior glume 5-nerved; inferior lemma 5-nerved, enclosing a male flower, its palea well developed; superior lemma and palea verruculose.

Zambia. N: Kalambo Falls, 1350 m., 16.iv.1966, *Richards* 21436 (K). W: Mwinilunga Distr., Ikelenge, 16.iv.1965, *Robinson* 6600 (K). C: Chakwenga Headwaters, 100–129 km. E. of Lusaka, 27.iii.1965, *Robinson* 6561 (K). E: Kachalola, 500 m., 17.iii.1959, *Robson* 1737 (K). S: Muvuma Hills, Munali Pass, 1200 m., 25.iii.1972, *Kornaś* 1454A (K). **Zimbabwe**. N: Guruve (Sipolilo) Distr., Nyamayanyetsi, 13.v.1985, *Martin* s.n. (K; SRGH). **Malawi**. N: Karonga, Vinthukhutu Forest, 550 m., 26.iv.1975, *Pawek* 9584 (K). C: Mchinje (Fort Manning), 20.iv.1952, *Jackson* 770 (K). S: Chaone, 600 m., 24.iv.1937, *Lawrence* 375 (K). **Mozambique**. N: Cabo Delgado, Mecúfi, 380 m., 6.iv.1964, *Torre & Paiva* 11672 (K; LISC).
Also in Tanzania, Angola and West Africa. Woodland on stony or arenaceous soils.

23. **Panicum nervatum** (Franch.) Stapf in F.T.A. **9**: 669 (1920). —Clayton & Renvoize in F.T.E.A., Gramineae: 475 (1982).
Isachne nervata Franch. in Bull. Soc. Hist. Nat. Autun **8**: 340 (1895). Type from Zaire.
Panicum baumannii K. Schum. in Engl. Bot. Jahrb. **24**: 331 (1898). —Clayton, F.W.T.A. **3**: 431 (1972). Type from Togo.
Panicum subrepandum Rendle, Cat. Afr. Pl. Welw. **2**: 178 (1899). —Jackson & Wiehe, Annot. Check List Nyasal. Grass.: 52 (1958). Type from Angola.

Panicum fulgens Stapf in F.T.A. **9**: 668 (1920). Type from Uganda.

Perennial with weak scrambling culms 30–130 cm. long. Leaf laminae 5–15 cm. long, narrowly lanceolate to linear. Panicle 6–18 cm. long, ovate-oblong. Spikelets 1.5–2.5 mm. long, ovoid to orbicular, glabrous or pubescent; inferior glume $\frac{2}{3}$–$\frac{3}{4}$ the length of the spikelet, 3-nerved; superior glume 5-nerved; inferior lemma 5-nerved, enclosing a male flower, its palea well developed; superior lemma and palea verruculose.

Zambia. N: 60 km. SE. of Mporokoso, 1600 m., 11.x.1967, *Simon & Williamson* 1130 (K; SRGH). C: Kabwe, 26.iii.1961, *Angus* 2504 (K; FHO). S: Mapanza East, 1100 m., 21.iii.1954, *Robinson* 627 (K). **Zimbabwe**. E: Chimanimani foothills, Tarka Forest Reserve, 800 m., x.1969, *Goldsmith* 81/69 (K; SRGH). **Malawi**. N: Mzimba Distr., Mbawa, 19.vi.1952, *Jackson* 844 (K).
Also from West Africa to the Sudan and southwards to Zimbabwe and Angola. Wooded grassland and open hillsides, on arenaceous or stony soils.

24. **Panicum brazzavillense** Franch. in Bull. Soc. Hist. Nat. Autun **8**: 341 (1895). —Stapf in F.T.A. **9**: 662 (1920). —Clayton & Renvoize in F.T.E.A., Gramineae: 476 (1982). Type from Zaire.
    Panicum frederici Rendle, Cat. Afr. Pl. Welw. **2**: 180 (1899). Types from Angola.
    Panicum ianthum Stapf in F.T.A. **9**: 663 (1920). —Sturgeon in Rhod. Agric. Journ. **59**: 434 (1954). Types from Angola and Zimbabwe: Mrewa, i.1911, *Appleton* 6 (K, syntype).

A caespitose perennial with slender erect culms (20)30–60(85) cm. high. Leaf laminae 10–20(25) cm. long, linear, acuminate. Panicle 3–15(20) cm. long, ovate or oblong. Spikelets 1.2–1.7 mm. long, ovate or orbicular, sparsely to densely pubescent, rarely glabrous; inferior glume $\frac{2}{3}$–$\frac{3}{4}$ the length of the spikelet, 3-nerved; superior glume 5-nerved; inferior lemma 5-nerved, enclosing a male flower, its palea well developed; superior lemma and palea verruculose.

Zambia. B: Lake Lutende, 20 km. East of Mongu, 6.xi.1965, *Robinson* 6700 (K). N: Mpika Distr., Danger Hill, 1500 m., 18.xii.1967, *Simon & Williamson* 1446 (K; SRGH). W: Mwinilunga Distr., 18 km. E. of Kalene Hill, 16.xii.1963, *Robinson* 6110 (K). C: Serenje Distr., Kundalila Falls, 1500 m., 17.xii.1967, *Simon & Williamson* 1415A (K; SRGH). **Zimbabwe**. C: Marondera, 8.i.1949, *Corby* 352 (K; SRGH). E: Chimanimani (Melsetter), Stonehenge Plateau, 1800 m., 1.ii.1957, *Phipps* 359 (K; SRGH). **Mozambique**. N: Malema, 650 m., 3.ii.1967, *Torre & Paiva* 10417 (K).
Also in Angola and northwards to Uganda and West Africa.

25. **Panicum natalense** Hochst. in Flora **29**: 113 (1846). —Chippindall in Meredith, Grasses & Pastures of S. Afr.: 333 (1955). Type from S. Africa.
    Panicum juncifolium Stapf in F.T.A. **9**: 664 (1920). Type from Angola.

Perennial with erect, wiry culms 25–120 cm. high. Leaf laminae 12–28 cm. long, linear. Panicle 7–15 cm. long, ovate or broadly oblong. Spikelets 2–2.5 mm. long, orbicular oblong, glabrous; inferior glume $\frac{2}{3}$–$\frac{3}{4}$ the length of the spikelet, 3–5-nerved; superior glume 5-nerved; inferior lemma 5-nerved, male, its palea well developed; upper lemma and palea verruculose.

Zambia. B: Loma Pan, 2.iv.1964, *Verboom* 1175 (K). W: Mwinilunga Distr., 7 km. S. of Ikelenge, 1420 m., 21.i.1975, *Brummitt, Chisumpa & Polhill* 13939 (K). **Zimbabwe**. N: Guruve (Sipolilo), 14.ii.1979, *Nyariri* 684 (K; SRGH). C: Nyanga (Inyanga) to Rusape, 1840 m., 26.ii.1985, *Keller-Grein* 7-79-3. (K). E: Chimanimani Mts., 1200 m., 23.ix.1966, *Simon* 894 (K; SRGH). **Malawi**. N: Mbawa, near Mzimba, 13.iii.1953, *Jackson* 1145. **Mozambique**. MS: Chimanimani Mts., 1400 m., 27.ix.1966, *Simon* 870 (K; SRGH).
Also from Zaire to S. Africa. Grassland or savanna.

26. **Panicum margaritiferum** (Chiov.) Robyns in Mém. Inst. Col. Belge Sec. Sci. Nat. & Med. **1**: 36 (1932). Type from Zaire.
    Isachne margaritifera Chiov. in Nuov. Giorn. Bot. Ital. n.s. **26**: 65, 67 (1919). —Stapf in F.T.A. **9**: 1098 (1934).

Caespitose perennial with erect culms 60–150 cm. high. Leaf laminae 18–30 cm. long, linear, pungent. Panicle 10–20 cm. long, ovate, the branches rather stiff. Spikelets 2–3 mm. long, globose, glabrous; inferior glume $\frac{3}{4}$ to as long as the spikelet, 5-nerved; superior glume 5-nerved; inferior lemma 5-nerved, male, its palea well developed; superior lemma and palea verruculose.

Zambia. N: Kawambwa, 1400 m., 23.vi.1957, *Robinson* 2371 (K; SRGH). W: Mwinilunga Distr., Matonchi Farm, 11.ii.1938, *Milne-Redhead* 4530 (K).
Also in Angola and Zaire (Katanga). Wooded grassland on arenaceous soils.

27. **Panicum hanningtonii** Stapf in F.T.A. **9**: 676 (1920). —Clayton & Renvoize in F.T.E.A., Gramineae: 480 (1982). Type from Tanzania.

Robust annual; culms 60–180 cm. high, erect, rarely decumbent and rooting at the lower nodes. Leaf-sheaths glabrous or the upper one hispid; laminae 15–45 cm. long and 8–18 mm. wide, linear, amplexicaul, acuminate. Panicle 20–35 cm. long, oblong, much branched, dense. Spikelets (2)2.5–3 mm. long, ovate, often purple-tinged; inferior glume $\frac{1}{2}$–$\frac{2}{3}$ the length of the spikelet, broadly ovate, 1–3-nerved, recurved acuminate; superior glume ovate, (5)7-nerved, recurved acuminate; inferior lemma 9-nerved, its palea well developed, male; superior lemma pale and glossy.

Zambia. E: Chipata, 24.iii.1963, *Verboom* 928 (K). **Malawi**. N: Karonga, 25.iii.1953, *Jackson* 1186 (K). **Mozambique**. MS: Mazzaro, 21.iii.1860, *Kirk* (K).
Also in Tanzania. Disturbed open habitats; 400–1700 m.

28. **Panicum massaiense** Mez in Engl., Bot. Jahrb. **34**: 144 (1904). —Stapf in F.T.A. **9**: 680 (1920). —Jackson & Wiehe, Annot. Check List Nyasal. Grass.: 51 (1958). —Clayton & Renvoize in F.T.E.A., Gramineae: 480 (1982). Type from Tanzania.
Panicum manicatum Stapf in F.T.A. **9**: 693 (1920). Type: Mozambique, Zambezi delta, *Scott* (K, holotype).

Tufted annual with culms 25–100 cm. high. Leaf laminae 10–25 cm. long and 3–8 mm. wide, linear, straight at the base or cordate, acute or acuminate. Panicle 7–20 cm. long, ovate or ovate oblong, much branched, the branches fine and often flexuous. Spikelets 2.5–3 mm. long, ovate; inferior glume $\frac{1}{2}$–$\frac{2}{3}$ the length of the spikelet, broadly ovate, 1(5)-nerved, acuminate; superior glume ovate, 5–7-nerved, recurved acuminate; inferior lemma 5–7-nerved, its palea well developed, male; superior lemma pale and glossy.

Zambia. C: Luangwa South Game Reserve, 7.iv.1972, *Abel* 567 (K; SRGH). E: Chipata to Nsefu, 11.v.1963, *van Rensburg* 2122 (K; SRGH). **Malawi**. C: Salima, 26.iv.1951, *Jackson* 477 (K). **Mozambique**. Z: Mopeia, ix.1974, *Bond* W562 (K; SRGH).
Also in Uganda and Tanzania. Damp places in savanna; 300–1700 m.

29. **Panicum zambesiense** Renvoize in Kew Bull. **44**, 3: 544 (1989). Type: Zimbabwe, Hwange, *Gonde* 289 (SRGH, holotype).

Annual; culms erect, 25–60 cm. high. Leaf laminae 6–16 cm. long and 4–8 mm. wide, linear or linear lanceolate, cordate, acute or finely pointed. Panicle 6–18 cm. long, oblong, moderately branched, the branches flexuous, ascending or appressed to the axis. Spikelets 2.5–3 mm. long, lanceolate, glabrous; inferior glume $\frac{1}{2}$ the length of the spikelet, ovate, 1-nerved, acuminate; superior glume 7-nerved, acuminate; inferior lemma 7-nerved, acuminate, its palea poorly developed, sterile; superior lemma and palea glossy, pale or dark brown.

Zambia. C: Luangwa R., 14.iv.1972, *Abel* 586 (K; SRGH). S: Kalomo Distr., Senkobo, 14.i.1964, *Astle* 2863 (K). **Zimbabwe**. N: Gokwe, 27.ii.1984, *Mahlangu* 923 (K; SRGH). W: Hwange, 1.v.1978, *Gonde* 189 (K; SRGH).
Not known elsewhere. Woodland on argillaceous or arenaceous soils; 1100–1300 m.

30. **Panicum kasumense** Renvoize in Kew Bull. **44**, 3: 546 (1989). Type: Zimbabwe, Hwange, Kazuma Range, *Simon* 2187 (K, holotype).

Annual with weakly erect or decumbent culms 10–35 cm. long. Leaf laminae 3–4.5 cm. long and 2–5 mm. wide, linear, sharply acute. Panicle axillary and terminal, 3–8 cm. long, ovate, sparsely to moderately branched, the secondary branches short and appressed. Spikelets 2.5–3 mm. long, narrowly ovate, glabrous; inferior glume $\frac{1}{2}$ the length of the spikelet, broadly ovate, 1-nerved, acuminate; superior glume 7-nerved, acuminate; inferior lemma 5-nerved, acuminate, its palea poorly developed, sterile; superior lemma and palea pallid.

Zimbabwe. W: Hwange, Kazuma Range, 10.v.1972, *Simon* 2187 (K; SRGH).
Only known from the type specimen. Grassland; 1000 m.

31. **Panicum walense** Mez in Engl., Bot. Jahrb. **34**: 146 (1904), as '*watense*'. —Clayton, F.W.T.A. **3**: 433 (1972). —Clayton & Renvoize in F.T.E.A. Gramineae: 480 (1982). Type from Senegal.

Slender annual with branching culms (15)20–60 cm. high. Leaf laminae 7–20 cm. long ovate or ovate oblong, moderately to much branched, the branches often fine and tangled. Spikelets 1.5–2 mm. long, ovate; inferior glume $\frac{1}{2}$–$\frac{3}{4}$ the length of the spikelet, (1)3-nerved, acuminate or with an awn point; superior glume 3(5)-nerved,acuminate; inferior lemma 3(5)-nerved, sterile; superior lemma pale and glossy.

Zambia. N: Mbala Distr., Mpulungu, 12.iv.1959, *McCallum-Webster* A308 (K). E: Chipata, 28.vi.1963, *Verboom* 111 (K). S: Kafue Nat. Park, 20.iv.1961, *Mitchell* 7/53 (K).
Also from Senegal to Sudan and Tanzania; also in India, China and Malaysia. Damp places in grassland; 775–1600 m.

32. **Panicum dregeanum** Nees, Fl. Afr. Austr.: 42 (1841). —Stapf in F.T.A. **9**: 684 (1920). —Sturgeon in Rhod. Agric. Journ. **50**: 434 (1954). —Chippindall in Meredith, Grasses & Pastures of S. Afr.: 332 (1955). —Jackson & Wiehe, Annot. Check List Nyasal. Grass.: 50 (1958). —Clayton in F.W.T.A. **3**: 432 (1972). —Clayton & Renvoize in F.T.E.A., Gramineae: 478 (1982). TAB. **5**. Type from S. Africa.

Densely caespitose perennial with culms (30)50–130 cm. high. Leaves mostly basal; sheaths pubescent at the base; laminae (14)18–35(50) cm. long and 1.5–5 mm. wide, linear, acute or acuminate. Panicles (8)12–20(30) cm. long, narrowly oblong to ovate, much branched. Spikelets 2–2.5(3) mm. long, ovate, glabrous, often purple tinged; inferior glume $\frac{1}{2}$–$\frac{2}{3}$ as long as the spikelet, (3)5-nerved, recurved acute to acuminate; superior glume 5-nerved, recurved acute to acuminate; inferior lemma 5-nerved, its palea well developed, male; superior lemma smooth and glossy.

Zambia. N: Mbala, 10.i.1965, *Richards* 19451 (K). W: Mwinilunga Distr., Dobeka Bridge, 11.xii.1937, *Milne-Redhead* 3610 (K). C: Mkushi, 16.xii.1967, *Simon & Williamson* 1398 (K; SRGH). E: Chipata to Katete, 9.i.1959, *Robson* 1118 (K). S: Mpanza, 28.i.1956, *Robinson* 1335 (K). **Zimbabwe**. N: Sengwa Research Station, 19.i.1975, *Guy* 2265 (K; SRGH). W: Kazuma depression, 21.ii.1973, *Chiparawasha* 565 (K; SRGH). C: Mazowe (Mazoe), i.1981, *Burrows* 1609 (K; SRGH). E: Chimanimani, 5.ii.1965, *Simon* 107 (K; SRGH). **Malawi**. N: Katoto, 10.ii.1974, *Pawek* 8072 (K). C: Mchinje, 9.x.1951, *Jackson* 613 (K). S: Thyolo, 30.i.1951, *Wiehe* 744 (K). **Mozambique**. MS: Nyamaruza Dambo v.1973, *Tinley* 2911 (K; SRGH).
Throughout tropical Africa. Damp grassland; 330–2000 m.

33. **Panicum pilgeri** Mez in Engl., Bot. Jahrb. **34**: 146 (1904). —Stapf in F.T.A. **9**: 686 (1920). —Clayton in F.W.T.A. **3**: 432 (1972). Type from Nigeria.

Tufted, pilose perennial with slender erect culms 30–120 cm. high. Leaf laminae 20–40 cm. long and 2–4 mm. wide, linear, acuminate. Panicle 10–27 cm. long, oblong, moderately to much branched, the branches loose or contracted. Spikelets 2–3 mm. long, ovate, glabrous; inferior glume $\frac{1}{2}$ to as long as the spikelet, broadly ovate, acuminate or with the mid nerve produced into an awnlet, 3–5-nerved; superior glume ovate, 5–7-nerved; inferior lemma 5-nerved, acuminate, its palea well developed, enclosing a male flower; superior lemma glossy.

Mozambique. N: Lake Chilwa, Mesanhelas, 11.v.1971, *Trevor-Jones* UG3 (K; SRGH).
Also from Guinea eastwards to Chad. Damp or swampy grassland; 300–1300 m.

34. **Panicum poaeoides** Stapf in F.T.A. **9**: 681 (1920). —Clayton & Renvoize in F.T.E.A., Gramineae: 477 (1982). Types from Kenya.
Panicum acutissimum Peter, Fl. Deutsch Ost-Afr. **1**, Anh.: 48, t.13/4 (1930). Type from Tanzania.
Panicum graciliculme Napper in Kirkia **3**: 127 (1963). Type from Tanzania.

Tufted perennial; culms 30–80 cm. high from a knotty base. Leaf laminae 6–8 cm. long and 3–8 mm. wide, linear, acuminate. Panicle 6–12(16) cm. long, ovate, moderately to much branched, usually dense and compact. Spikelets 2.3–3.2 mm. long, ovate; inferior glume $\frac{1}{3}$–$\frac{2}{3}$ the length of the spikelet, broadly ovate, 1-nerved, straight or recurved acuminate; superior glume ovate, 7-nerved, recurved acute or acuminate; inferior lemma ovate, 7-nerved, acute, its palea well developed, enclosing a male flower; superior lemma glossy.

Zimbabwe. W: Gwampa For. Res., i.1955, *Goldsmith* 125/55 (K; SRGH). **Mozambique**. T: Sisitso, 8.vii.1950, *Chase* 2668 (K; SRGH).
Also from Ethiopia to Zaire. Savanna on various soils, often in damp ground; 300–2100 m.

Tab. 5. PANICUM DREGEANUM. 1, habit (× $\frac{2}{3}$); 2, spikelet (× 16), both from *Simon & Williamson* 1868.

35. **Panicum phragmitoides** Stapf in F.T.A. **9**: 677 (1920). —Sturgeon in Rhod. Agric. Journ. **50**: 434 (1954). —Jackson & Wiehe, Annot. Check List Nyasal. Grass.:52 (1958). —Launert in Merxm., Prodr. Fl. SW. Afr. **160**: 142 (1970). —Clayton & Renvoize in F.T.E.A., Gramineae: 477 (1982). Types from Central African Republic.
  Panicum trypheron var. *giganteum* Rendle, Cat. Afr. Pl. Welw. **2**: 181 (1899). Type from Angola.
  Panicum phragmitoides var. *lasioneuron* Stapf in F.T.A. **9**: 678 (1920). —Sturgeon in Rhod. Agric. Journ. **50**: 434 (1954). Type: Mozambique, Messalo (Msalu) R., *Allen* 131 (K, holotype).

Robust, tufted or shortly rhizomatous perennial with erect culms 100–200 cm. high. Leaf laminae 50–75 cm. long and 5–15(20) mm. wide, linear, acuminate. Panicle (20)30–60 cm. long, ovate to oblong, moderately branched. Spikelets 3–4 mm. long, ovate; inferior glume $\frac{1}{2}$–$\frac{2}{3}$ the length of the spikelet, ovate, 5-nerved, straight or recurved acuminate; superior glume 7-nerved, acuminate; inferior lemma 7–9-nerved, its palea well developed, enclosing a male flower, superior lemma pale and glossy.

Zambia. N: Isoka, 28.i.1958, *Vesey-FitzGerald* 1418 (K; SRGH). **Zimbabwe**. W: Hwange, 5.ii.1970, *Rushworth* 2357A (K; SRGH). E: Chipinge, 10.ii.1977 *Izzett* s.n. (K; SRGH). **Malawi**. N: Karonga, 26.iv.1975, *Pawek* 9585 (K). C: Nkhota Kota, 19.ii.1953, *Jackson* 1064 (K). S: Nankupu Stream, iii.1929, *Barker* D15 (K). **Mozambique**. N: Marrupa, 11.ii.1981, *Nuvunga* 502 (K; LM).
  Also from Tanzania westwards to Guinée and Angola. Wooded grassland on sandy or red soils; 450–1800 m.

36. **Panicum graniflorum** Stapf in F.T.A. **9**: 681 (1920). —Jackson & Wiehe, Annot. Check List Nyasal. Grass.: 51 (1958). Type: Mozambique, Kongone, *Kirk* (K, holotype).

Tufted or shortly rhizomatous perennial; culms 30–150 cm. high. Leaf laminae 10–25 cm. long and 2–5 mm. wide, linear, often glaucous, acute. Panicle (7)12–25(35) cm. long, oblong, moderately branched with the secondary branches long and ascending. Spikelets 2–2.5 mm. long, ovate, shortly pedicelled and clustered on short tertiary branches; inferior glume $\frac{1}{3}$–$\frac{2}{3}$ the length of the spikelet, broadly ovate, 1–3-nerved, recurved acuminate; superior glume ovate, 5-nerved, recurved acute to acuminate; inferior lemma ovate, 5-nerved, acute, its palea well developed, enclosing a male flower, superior lemma pale and glossy.

Botswana. N: Kwando, 24.iv.1975, *Williamson* 45 (K; SRGH). **Zambia**. B: Senanga, 2.viii.1952, *Codd* 7352 (K; PRE). W: Mwinilunga, 22.ix.1972, *Strid* 2598 (K). S: Katambora, 3.iv.1956, *Robinson* 1390 (K). **Zimbabwe**. N: Gokwe, 19.i.1981, *Mahlangu* 7A (K; SRGH). W: Victoria Falls, 23.iv.1970, *Simon & Hill* 2144 (K; SRGH). **Malawi**. N: Mbawa Research Station, 28.ix.1950, *Jackson* 179 (K). C: Nkhota Kota, 1.x.1950, *Jackson* 200 (K). **Mozambique**. Z: Kongone, i.1861, *Kirk* (K).
  Known only from the Flora Zambesiaca area. Grassland on sandy, often damp soil; 400–1600 m.

37. **Panicum fluviicola** Steud., Syn. Pl. Glum. **1**: 89 (1854). —Stapf in F.T.A. **9**: 689 (1920). —Clayton in F.W.T.A. **3**: 432 (1972). —Clayton & Renvoize in F.T.E.A., Gramineae: 478 (1982). Type from Gabon.
  Panicum aphanoneurum Stapf in F.T.A. **9**: 687 (1920). —Launert in Merxm., Prodr. Fl. SW. Afr. **160**: 136 (1970). Type from Nigeria.
  Panicum purpurascens Mez in Engl., Bot. Jahrb. **57**: 188 (1921). Types from Togo, Nigeria and Tanzania.

Tufted perennial with robust unbranched or sparsely branched culms 60–230 cm. high. Leaf laminae 25–50 cm. long and 3–12 mm. wide, linear, tapering to a pungent apex. Panicle 15–45 cm. long, oblong, moderately to much branched, the branches appressed or ascending, loose or contracted. Spikelets mostly shortly pedicelled, 2–2.5 mm. long, ovate, glabrous; inferior glume $\frac{1}{2}$–$\frac{2}{3}$ the length of the spikelet, broadly ovate, (1)3–5-nerved, recurved acuminate or produced into an awn point; superior glume 5(7)-nerved, ovate, recurved acute or acuminate; inferior lemma 5-nerved, ovate, acute to acuminate, its palea well developed, enclosing a male flower, superior lemma smooth and glossy.

Caprivi Strip. Mpilila Isl., 12.i.1959, *Killick & Leistner* 3328 (K; PRE). **Botswana**. N: Gwetshoa Isl., 12.iv.1973, *Smith* 524 (K; SRGH). **Zambia**. N: Chibutabuta Forest Area, 7.iv.1961, *Phipps & Vesey-FitzGerald* 2993 (K; SRGH). W: Kabompo R. Bridge, 19.xii.1969, *Simon & Williamson* 1874 (K; SRGH). E: Chipata, Mpangwe Hills, 10.vi.1963, *Verboom* 996 (K). **Zimbabwe**. N: Sengwa Res. Sta., 19.i.1975, *Guy* 2268 (K; SRGH). W: Victoria Falls, 9.ii.1912, *Rogers* 5717 (K). **Malawi**. C: Nkhota Kota, 16.vi.1970, *Brummitt* 11445 (K). S: Zomba, 1936, *Cormack* 316 (K). **Mozambique**. N: Montepuez, 6.iv.1964, *Torre & Paiva* 11684 (K; LISC). M: Bilene, 14.xii.1940, *Torre* 2417 (K; LISC).
  Distributed throughout tropical Africa in wet grassland; 350–1630 m.
  This species is only distinguished with difficulty from *Panicum genuflexum* Stapf which occurs in Zaire, Angola, Kenya, Tanzania, Zanzibar and Pemba and has slender culms 30–75 cm. high,

branching upwards and bearing lax, sparsely branched panicles with flexuous branches. *Panicum graniflorum* is also very similar but may be distinguished by its moderately branched panicle with somewhat congested spikelets on short tertiary branches.

38. **Panicum graciliflorum** Rendle, Cat. Afr. Pl. Welw. **2**: 181 (1899). —Clayton & Renvoize in F.T.E.A., Gramineae: 479 (1982). Type from Angola.
   Panicum vagiliflorum Stapf in F.T.A. **9**: 683 (1920). Type: Mozambique, Messalo (Msalu) R., *Allen* 115 (K, holotype).

Loosely tufted perennial with slender culms 70–130 cm. high. Leaf laminae 14–35 cm. long and 3–5 mm. wide, linear, acuminate. Panicle 7–30 cm. long, ovate-oblong, much branched, the branches fine and spreading with the spikelets evenly distributed. Spikelets 2–2.5 mm. long, ovate; inferior glume $\frac{1}{2}-\frac{3}{4}$ the length of the spikelet, 3(5)-nerved, acuminate; superior glume 5-nerved, recurved acuminate; inferior lemma 5-nerved, palea well developed, male; superior lemma glossy.

   Zambia. C: Chakwenga headwaters, 27.iii.1965, *Robinson* 6483 (K). E: Chipata, 11.v.1963, *van Rensburg* 2121 (K). **Mozambique**. N: Messalo (Msalu) R., 20.iii.1912, *Allen* 115 (K).
   Also in Angola. Damp or marshy ground; 1200 m.
   In F.T.E.A. *Panicum graciliflorum* was included under *Panicum fluviicola* but a reappraisal of the material suggests that the panicle shape, along with its delicate spreading branches, distinguishes it sufficiently from *Panicum fluviicola* to justify recognition as a separate species. *Panicum vagiliflorum* was placed under *Panicum genuflexum* Stapf in F.T.E.A. but is now considered more appropriately placed under *Panicum graciliflorum*.

39. **Panicum carneovaginatum** Renvoize in Kew Bull. **34**: 555 (1979); Clayton & Renvoize in F.T.E.A., Gramineae: 489 (1982). Type: Zambia, Isoka, *Vesey-FitzGerald* 1405 (K, holotype).

Tufted perennial with culms 45–145 cm. high, the lower leaf-sheaths carinate and usually tinged pink. Leaf laminae 20–55 cm. long, linear, glabrous, tapering to a firm, abrupt apex. Panicle 10–30 cm. long, oblong, the branches ascending, usually bare on the lower part. Spikelets 2.2–2.8 mm. long, ovate-oblong, glabrous, yellow or orange, often purple tinged; inferior glume broadly ovate, $\frac{1}{3}-\frac{1}{2}$ the length of the spikelet, 3-nerved, acute; superior glume ovate, 7-nerved acute; inferior lemma similar to the superior glume sterile, its palea absent; superior lemma and palea dark, glossy.

   Zambia. N: Mbala. Kawimbe, 26.iii.1959, *McCallum-Webster* A246 (K). W: Kasempa Distr., Manyinga to Solwezi, 27.xii.1969, *Simon & Williamson* 2064 (K; SRGH). S: Choma, i.1979, *Heery* 28 (K). **Malawi**. N: Mbawa, Mzimba, 20.i.1951, *Jackson* 375 (K). **Mozambique**. MS: Beira Distr., Cheringoma, v.1973, *Tinley* 2855 (K; SRGH).
   Also recorded Northwards to Sudan. Grassland, often in waterlogged soil; 1100–2000 m.

40. **Panicum atrosanguineum** A. Rich., Tent. Fl. Abyss. **2**: 375 (1851). —Stapf in F.T.A. **9**: 703 (1920). —Sturgeon in Rhod. Agric. Journ. **50**: 433 (1954). —Bor, Grasses of B.C.I. & P.: 322 (1960). —Clayton & Renvoize in F.T.E.A., Gramineae: 488 (1982). Type from Ethiopia.

Tufted, pilose annual with culms 10–40 cm. high. Leaf laminae 6–14 cm. long, narrowly lanceolate, acute. Panicle 4–20 cm. long, ovate or oblong, much branched, the branches fine and spreading or ascending. Spikelets 1.8–2 mm. long, ovate oblong, glabrous, purple tinged; inferior glume broadly ovate $\frac{2}{3}$ the length of the spikelet, 3–5-nerved, acuminate; superior glume ovate, 5–7-nerved, acuminate; inferior lemma similar to the superior glume, sterile, its palea poorly developed; superior lemma and palea dark, glossy.

   Botswana. N: Maun, 8.iii.1978, *Smith* 2366 (K; SRGH). **Zambia**. N: Mbala Distr., Mpulungu, 10.iv.1961, *Vesey-FitzGerald* 3024 (K; SRGH). S: Pemba, 8.ii.1963, *Astle* 2067 (K). **Zimbabwe**. N: Gokwe Distr., Sengwa Res. Sta., 17.i.1975, *Guy* 2226 (K; SRGH). W: Nyamandhlovu, Edwaleni Farm, ii.1972, *Keogh* 28 (K; SRGH). S: Mberengwa Distr. (Belingwe), Buhwa Mine, 5.v.1973, *Simon, Pope & Biegel* 2461 (K; SRGH).
   Also occurring Northwards into Zaire and Ethiopia, also in NW. India. Disturbed places in farmland and bushland; 100–2200 m.

41. **Panicum phippsii** Renvoize in Kew Bull. **44**, 3: 546 (1989). Type: Zambia, Mporokoso, *Phipps & Vesey-FitzGerald* 3238 (K, holotype)

Annual; culms 30–50 cm. long, erect or decumbent and rooting at the lower nodes. Leaf laminae 3.5–6 cm. long and 5–10 mm. wide, lanceolate, thin, acuminate. Panicle 6–10 cm.

long, ovate, sparsely to moderately branched, the branches slender and flexuous. Spikelets 3 mm. long, lanceolate, densely pilose; inferior glume a small nerveless scale $\frac{1}{6}$ the length of the spikelet; superior glume 5-nerved; inferior lemma 5-nerved, its palea poorly developed, sterile; superior lemma and palea glossy, dark brown.

Zambia. N: Mporokoso Distr., Mweru–Wantipa, 16.iv.1961, *Phipps & Vesey-FitzGerald* 3238 (K; SRGH).
Only known from the type collection. Dense scrub, in shade; 1350 m.

42. **Panicum arcurameum** Stapf in F.T.A. **9**: 704 (1920). Type from S. Africa (Transvaal).

Hispid annual with geniculately ascending, branching culms 15–40 cm. high. Leaf laminae 5–15 cm. long and 5–15 mm. wide, linear or linear-lanceolate, acute. Panicle 4–13 cm. long, ovate, moderately to much branched, usually scarcely exserted from the uppermost leaf-sheath. Spikelets 2–2.5 mm. long, ovate, glabrous, apiculate; inferior glume $\frac{1}{2}$ the length of the spikelet, broadly ovate, 5–7-nerved; superior glume as long as the spikelet, 9-nerved; inferior lemma similar to the superior glume, sterile, its palea poorly developed; superior lemma and palea dark, glossy.

Botswana. SE: Mahalapye, ii.1957, *de Beer* 5 (K; SRGH). **Zimbabwe**. N: Makonde (Lomagundi) Distr., Kildonan, 2.iii.1965, *Simon* 183 (K; SRGH). W: Shangani, iii.1918, *Eyles* 947 (K). C: Gweru Distr., c. 29 km. SSE. of Kwekwe, 4.iv.1966, *Biegel* 1075 (K; SRGH). E: Mutare, 5.vii.1970, *Crook* 933 (K; SRGH). S: Masvingo Kyle Nat. Park, 16.iv.1971, *Basera* 330 (K; SRGH).
Also in S. Africa. Disturbed ground on sandy or clay soils; 1200–1700 m.

43. **Panicum haplocaulos** Pilg. in Engl. Jahrb. **33**: 49 (1902). —Stapf in F.T.A. **9**: 705 (1920). —Clayton & Renvoize in F.T.E.A., Gramineae: 487 (1982). Type from Sudan.

Slender, glabrous or sparsely pilose annual with erect culms 25–60 cm. high. Leaf laminae 7–25 cm. long and 2–6 mm. wide, linear, acuminate. Panicle 5–20 cm. long, ovate, sparsely to moderately branched, the branches delicate, ascending or spreading. Spikelets 2–2.5 mm. long, ovate-elliptic, glabrous, acute or apiculate; inferior glume $\frac{1}{3}$–$\frac{1}{2}$ the length of the spikelet, broadly ovate, 3-nerved, acute; superior glume ovate, 7-nerved, acute; inferior lemma ovate, 7-nerved, sterile, its palea absent; superior lemma and palea dark, glossy.

Zambia. N: Mbala, 21.ii.1959, *McCallum-Webster* A115 (K). S: Mapanza, 11.ii.1956, *Robinson* 1343 (K). **Malawi**. N: Rumphi, Cilungiro, 26.ii.1978, *Pawek* 13919 (K; SRGH). C: Nkhota Kota Distr., Kanyenda Dambo, 24.ii.1953, *Jackson* 1107 (K).
Also from Senegal to Sudan and southwards to Zambia and Malawi. In dambos and flooded grassland; 1100–1950 m.

44. **Panicum comorense** Mez in Engl. Jahrb. **57**: 185 (1921) —Jackson & Wiehe, Annot. Check List Nyasal. Grass.: 50 (1958) —Clayton in F.W.T.A. **3**: 434 (1972) —Clayton & Renvoize in F.T.E.A., Gramineae: 492 (1982). Types from Tanzania and Comoro Islands.
Panicum microlemma Pilg. in Not. Bot. Gard. Berlin **15**: 707 (1942). Type from Tanzania.

Annual; culms 30–110 cm. long, erect or trailing. Leaf laminae 6–15 cm. long and 10–15 mm. wide, lanceolate, membranous, flat, cross-veins present, acuminate. Panicle 12–30 cm. long, oblong, moderately branched, the secondary branches usually appressed. Spikelets 1.8–2.2 mm. long, oblong, glabrous, obtuse; inferior glume $\frac{1}{4}$–$\frac{1}{3}$ the length of the spikelet, ovate, cuff-like, hyaline, 1-nerved; superior glume 3-nerved, membranous, obtuse; inferior lemma 5-nerved, its palea absent, sterile; superior lemma and palea pallid, glossy, minutely scaberulous, the lemma with a green spot at the apex.

Zambia. E: Chipata, 24.iii.1963, *Verboom* 938 (K). **Malawi**. C: Citala R., 27.iv.1951, *Jackson* 487 (K). S: Nsanje, 23.iii.1960, *Phipps* 2651 (K; SRGH). **Mozambique**. N: Nampula, 2.iv.1964, *Torre & Paiva* 11583 (K; LISC). Z: Metalola, 6.vi.1972, *Bowbrick* JWIOB (K; SRGH). MS: Chimoio, 11.iii.1962, *Chase* 7658 (K; SRGH). M: Maputo, 20.iv.1949, *Myre & Balsinhas* 682 (K).
Throughout tropical Africa also in the Comoro Islands and Madagascar. Forest shade; 200–800 m.

45. **Panicum madipirense** Mez in Engl., Bot. Jahrb. **57**: 189 (1921). —Clayton & Renvoize in F.T.E.A., Gramineae: 474 (1982). Type from Tanzania.

Annual with erect culms 25–120 cm. high. Leaf laminae 8–24 cm. long and 6–20 mm. wide, linear-lanceolate, cordate, acute or acuminate. Panicle 10–20(30) cm. long, oblong, finely branched. Spikelets 2.5–3.7 mm. long, oblong, glabrous; inferior glume $\frac{1}{4}$ the length of the spikelet, cuff-like, membranous 1–sub-7-nerved; superior glume 9-nerved; inferior lemma 9-nerved, enclosing a male flower, its palea well developed; superior lemma smooth or scaberulous towards the apex.

Botswana. SE: Eastern Bamangwato Territory, v.1883, *Holub* (K). **Zambia**. B: Mongu, 20.i.1966, *Robinson* 6817 (K). N: Luangwa Valley Game Reserve, south of Katete R., 17.ii.1967, *Prince* 239 (K). C: Mupamadzi R., 600 m., 14.i.1966, *Astle* 4390 (K). E: Lugomo, 600 m., 4.ii.1958, *Stewart* 115 (K). S: Kafue Nat. Park, Nkala-Ngoma, 11.iii.1961, *Mitchell* 6/45 (K). **Zimbabwe**. N: Hurungwe Distr., (Urungwe), Mensa Pan, 460 m., 29.i.1958, *Drummond* 5320 (K; SRGH). **Malawi**. S: Nsanje Distr., Chiromo, 80 m., 22.iii.1960, *Phipps* 2608 (K; SRGH). **Mozambique**. MS: Beira Distr., Gorongosa Nat. Park, 4.ii.1971, *Tinley* 2016 (K; SRGH).

Also in Kenya and Tanzania. Swamps and damp areas in bushland or wooded grassland.

46. **Panicum pilgerianum** (Schweickerdt) Clayton in Kew Bull. **42**, 2: 402 (1987). Type from Namibia.
   Acroceras pilgerianum Schweickerdt in Not. Bot. Gart. Berlin **14**: 199 (1938) —Chippindall in Meredith, Grasses & Pastures of S. Afr.: 386 (1955). Type as above.
   Psilochloa pilgeriana (Schweickerdt) Launert in Mitt. Bot. Staatss. München **8**: 156 (1970); in Merxm., Prodr. Fl. SW. Afr. **160**: 157 (1970). Type as above.

Glabrous annual; culms 30–200 cm. high erect, thick and soft. Leaf laminae 10–30 cm. long and 5–12 mm. wide, linear, flat, finely acute. Panicle 6–36 cm. long, narrowly oblong or linear, the primary branches long, ascending or appressed to the axis, secondary branches short, appressed. Spikelets appressed, 4–5.5 mm. long, lanceolate, acuminate; inferior glume $\frac{1}{6}-\frac{1}{5}$ the length of the spikelet, broadly ovate, 0–3-nerved; superior glume as long as the spikelet, 7–9-nerved; inferior lemma 5–7-nerved, male; superior lemma and palea oblong, apiculate.

Botswana. N: Samedupe Bridge, 5.ii.1977, *Smith* 1899 (K; SRGH). SE: Content Farm, 10.iii.1978, *Hansen* 3370 (K; PRE; SRGH).

Also in Namibia. Seasonally flooded areas, growing in water; 1050 m.

47. **Panicum perangustatum** Renvoize in Kew Bull. **44**, 3: 545 (1989). Type: Zambia, Misamfu, *Robinson* 4293 (K, holotype).

Tufted annual with erect or geniculately ascending, slender culms 20–30 cm. high. Leaf laminae 3–8 cm. long and 1–3 mm. wide, flat or folded, acute or finely pointed, ascending or appressed to the culms. Panicle 2–5 cm. long, linear, sparsely branched, the branches appressed to the axis, scarcely exserted from the uppermost leaf-sheath. Spikelets 3.5–4.5 mm. long, lanceolate, glabrous; inferior glume $\frac{1}{4}$ the length of the spikelet, broadly ovate, clasping, 3-nerved; superior glume 9–11-nerved, acuminate; inferior lemma 9–11-nerved, acuminate, its palea poorly developed, sterile; superior lemma and palea 2.5 mm. long, pallid, beaked.

Zambia. N: Kasama, Misamfu, 22.i.1961, *Robinson* 4293 (K).
Only known from the type collection. Seasonally wet places.

48. **Panicum gilvum** Launert in Mitt. Bot. Staatss. München **8**: 153 (1970); in Merxm., Prodr. Fl. SW. Afr. **160**: 141 (1970). Type from Namibia.

Annual with culms 20–70 cm. high, erect or geniculately ascending, branched, soft. Leaf laminae 7–15 cm. long and 3–6 mm. wide, linear, flat, sharply acute. Panicles 4–10 cm. long, narrowly oblong, seldom fully exserted from the uppermost leaf-sheath, sparsely branched, the branches appressed. Spikelets 2.5–4 mm. long, ovate-oblong, acute; inferior glume $\frac{1}{5}-\frac{1}{4}$ the length of the spikelet, membranous, cuff-like, obtuse or bluntly acute 0–1-nerved; superior glume as long as the spikelet 11-nerved; inferior lemma 9-nerved; its palea absent, sterile; superior lemma and palea glossy.

Botswana. N: 20°43'S. & 21°05'E. 23.iii.1980, *Smith* 3300 (K; SRGH).
Also in Namibia. Seasonal rainwater pans.
This species may be confused with *P. impeditum* Launert from Namibia and S. Africa, which differs in its exserted ovate panicles, acuminate spikelets clustered somewhat towards the distal half of the primary branches and by its generally shorter leaf laminae 5–10 cm. long. *P. subalbidum* Kunth is distinguished by its greater stature, 60–200 cm. high and larger, exserted panicles 20–50 cm. long. *P. schinzii* Hack., which is an extremely variable, but mainly S. African species, is distinguished by its bluish-green foliage, exserted panicles, elliptic oblong, obtuse spikelets with the inferior floret male.

49. **Panicum mlahiense** Renvoize in Kew Bull. **34**: 554 (1979). —Clayton & Renvoize in F.T.E.A. Gramineae: 485 (1982). Type from Tanzania.

Slender annual with erect or geniculately ascending culms 30–65 cm. high. Leaf laminae 8–14 cm. long and 3–9 mm. wide, linear, flat, acute or acuminate. Panicles 10–18 cm. long, ovate, moderately branched, terminal and axillary. Spikelets 2.2–2.5 mm. long, narrowly ovate, narrowly acute; inferior glume ⅓ the length of the spikelet, broadly ovate, 3–5-nerved; superior glume 9–11-nerved; inferior lemma 9-nerved, sterile, its palea absent; superior lemma and palea pallid and glossy.

Zambia. N: Mporokoso Distr., Mugombwe, 15.iv.1961, *Phipps & Vesey-FitzGerald* 3220 (K; SRGH). Also in southern Tanzania. Riverine thicket and laterite pans; 275–1800 m.

50. **Panicum schinzii** Hack. in Verh. Bot. Ver. Brand. **30**: 142 (1888). —Stapf in F.T.A. **9**: 715 (1920). —Chippindall in Meredith, Grasses & Pastures of S. Afr.: 334 (1955). —Launert in Merxm., Prodr. Fl. SW. Afr. **160**: 142 (1970). Type from Namibia.
    Panicum laevifolium Hack. in Bull. Herb. Boiss. **3**: 378 (1895) —Sturgeon in Rhod. Agric. Journ. **50**: 433 (1954). —Chippindall in Meredith, Grasses & Pastures of S. Afr.: 334 (1955). —Bor, Grasses of B.C.I. & P.: 327 (1960). Types from S. Africa (Transvaal).

Glabrous annual with erect or ascending culms 30–120 cm. high. Leaf laminae 10–40 cm. long and 5–15 mm. wide, linear, typically bluish green. Panicle 10–35 cm. long, oblong or ovate oblong, moderately branched, the branches fine and flexuous, axillary panicles often present. Spikelets 2.3–2.8 mm. long, elliptic oblong, obtuse or bluntly acute, glabrous, typically bluish-green; inferior glume ⅛–⅕ the length of the spikelet, cuff-like, 1-nerved, lateral nerves usually obscure, obtuse or bluntly acute; superior glume as long as the spikelet, ovate, 7–9-nerved; inferior lemma similar to the superior glume, enclosing a male flower; superior lemma and palea pale brown, shiny.

Botswana. SE: Content Farm, 27.i.1978, *Hansen* 3341 (K). **Zimbabwe**. C: Hunyani, 3.ii.1932, *Stent* 5557 (K; SRGH). E: Chimanimani (Melsetter) 12.ii.1950, *Williams* 61 (K; SRGH).
    The following two specimens, N: Binga, 16.i.1968, *Thomson* 53 (K); S: Masvingo (Fort Victoria), 21.i.1971, *Chiparawasha* 297 (K) are provisionally included here.
    Also widespread in Namibia and S. Africa. Seasonally flooded pans and weedy places in open situations on a variety of soils; 520–2000 m.
    This species exhibits a wide range of different forms which tempt recognition of taxonomic and nomenclatural identities, however, discreet taxa are hard to delimit so the wide but continuous variation is here recognised in a single species. The typical plant is 20–30 cm. high, bluish-green overall with both terminal and axillary, oblong, moderately branched panicles of oblong, obtuse spikelets. Variation of the culms extends to thick succulent types and very slender delicate types. Although panicles are usually multiple, single terminal types are frequent. Spikelets are usually oblong and obtuse, occasionally ovate-oblong and acute.
    Panicum schinzii may be confused with several other species, which are listed below with their distinguishing features:-
    Panicum madipirense Mez — leaves yellow-green, panicle much branched, the branches fine and flexuous, spikelets ovate, superior lemma apiculate.
    Panicum subalbidum Kunth — spikelets acuminate.
    Panicum novemnerve Stapf — plant hispid, inferior glume ovate.
    The typical form described above matches the type of *Panicum laevifolium* Hack., however, although not typical for this taxon, the type of *Panicum schinzii* Hack. cannot be segregated and since the name is earlier it must take precedence over *Panicum laevifolium*.

51. **Panicum repens** L., Sp. Pl., ed 2: 87 (1762). —Stapf in F.T.A. **9**: 708 (1920). —Sturgeon in Rhod. Agric. Journ. **50**: 435 (1954). —Chippindall in Meredith, Grasses & Pastures of S. Afr.: 333 (1955). —Jackson & Wiehe, Annot. Check List Nyasal. Grass.: 52 (1958). —Bor, Grasses of B.C.I. & P.: 330 (1960). —Launert in Merxm., Prodr. Fl. SW. Afr. **160**: 142 (1970). —Clayton in F.W.T.A. **3**: 434 (1972). —Clayton & Renvoize in F.T.E.A., Gramineae: 481 (1982). Type from Spain.

Perennial with long rhizomes and occasionally with surface stolons; culms 30–100 cm. high, erect or decumbent, tough, often arising from a knotty base. Leaf laminae 7–25 cm.long and 2–8 mm. wide, linear, flat or involute, often strongly distichous and ascending close to the stem, usually stiff and pungent. Panicle 5–20 cm. long, narrowly oblong, sparsely to moderately branched, the branches usually ascending. Spikelets 2.5–3 mm. long, ovate-elliptic or ovate, acute, inferior glume ⅓ the length of the spikelet, broadly ovate or cuff-like, membranous, 0–3-nerved, clasping the base of the spikelet, obtuse or acute; superior glume 7–9-nerved; inferior lemma similar to the superior glume, male, its palea well developed; superior lemma and palea pallid, glossy.

Botswana. N: Gwetshoa Island, 21.i.1973, *Smith* 359 (K; SRGH). **Zambia**. B: Zambezi R., Barotse Flood Plain, 7.x.1964, *Verboom* 1159 (K). N: Mbala, 23.ii.1959, *McCallum-Webster* A151 (K). W: Ndola,

2.v.1953, *Hinds* 151 (K). S: Mazabuka Distr., Zongwe Estuary, 8.iv.1960, *Phipps* 2803 (K; SRGH). **Zimbabwe**. N: Gokwe Distr., Sengwa Camp, 20.v.1983, *Mahlangu* 730 (K; SRGH). W: Matopos, 26.ii.1954, *Rattray* 1663 (K; SRGH). C: Chegutu, 31.i.1944, *Hornby* 2332 (K; SRGH). S: Masvingo, Great Zimbabwe, Nat. Park, 11.iii.1971, *Chiparawasha* 366 (K; SRGH). **Malawi**. C: Salima, Mpatsanjoka Dambo, 7.i.1953, *Jackson* 1019 (K). S: Lake Chilwa, Mapira Dock, 13.v.1952, *Jackson* 820 (K). **Mozambique**. N: Mogincual, 30.iii.1964, *Torre & Paiva* 11486 (K; LISC). Z: Massingire, Aguas Guentes, 15.v.1943, *Torre* 5325 (K; LISC). MS: Beira, Estoril Camp, 24.iv.1968, *Crook* 824 (K).
Throughout the tropics and subtropics. Marshy places; 150–2000 m.

52. **Panicum repentellum** Napper in Kirkia **3**: 127 (1963). —Clayton & Renvoize in F.T.E.A., Gramineae: 482 (1982). Type from Tanzania.

Rhizomatous aquatic perennial, often also with creeping stolons; culms 20–60(90) cm. high, usually soft, slender, erect or decumbent. Leaf laminae 5–12 cm. long and (1)3–5 mm. wide, linear, flat, soft, pilose or glabrous, usually bluntly acute, occasionally acuminate and slightly pungent. Panicle 6–14 cm. long, ovate to narrowly oblong, moderately to sparsely branched, the branches ascending or appressed. Spikelets 2.5–3 mm. long, ovate, pallid, acuminate; inferior glume ¼ the length of the spikelet, hyaline, 0(3)-nerved, clasping the base of the spikelet, obtuse or bluntly acute; superior glume as long as the spikelet, membranous, 7-nerved, acuminate; inferior lemma similar to the superior glume, male, its palea well developed; superior lemma and palea pallid, glossy.

Botswana. N: Moremi Wildlife Reserve, 21.iii.1977, *Smith* 1952 (K; SRGH). **Zambia**. C: Munali, 10 km. E. of Lusaka, 15.vi.1955, *Robinson* 1302 (K). S: Kafue Flats near Mazabuka, 16.i.1954, *Hinds* 178 (K). **Zimbabwe**. N: Kariba, 10.iv.1966, *Jarman* A6 (K; SRGH). C: Marondera, Chikokorana Pan, 29.iv.1972, *Gibbs-Russell* 2002 (K; SRGH). **Malawi**. N: Lake Kasuni, 22.vi.1951, *Jackson* 545 (K). S: Lake Chilwa, 3.x.1963, *Vesey-FitzGerald* 4194 (K).
Also recorded from Ethiopia to S. Africa (Transvaal). Marshes and lakes; 200–1800 m.

53. **Panicum subalbidum** Kunth, Rév. Gram. **2**: 397 (1831). —Clayton in F.W.T.A. **3**: 434 (1972). —Clayton & Renvoize in F.T.E.A. in Gramineae: 484 (1982). Type from Senegal.
    Panicum glabrescens Steud., Syn. Pl. Glum. **1**: 71 (1854). —Sturgeon in Rhod. Agric. Journ. **50**: 436 (1954). —Chippindall in Meredith, Grasses & Pastures of S. Afr.: 333 (1955). —Jackson & Wiehe, Annot. Check List Nyasal. Grass.: 50 (1958). —Launert in Merxm., Prodr. Fl. SW. Afr. **160**: 141 (1970). Type from Senegal.
    Panicum proliferum var. *longijubatum* Stapf in F.C. **7**: 406 (1899). Types from S. Africa.
    Panicum longijubatum (Stapf) Stapf in F.T.A. **9**: 718 (1920).

Robust annual or short-lived perennial with soft herbaceous culms 60–200 cm. high, erect or decumbent and rooting at the lower nodes, the upper nodes sharply demarcated. Leaf laminae 20–50 cm. long and 7–15 mm. wide, linear, flat, acute or acuminate. Panicle 20–50 cm. long, ovate or oblong, sparsely to moderately branched, the tertiary and sometimes the secondary branches appressed at maturity. Spikelets (2.7)3–3.5 mm. long, narrowly ovate, acuminate; inferior glume ¼–⅓ the length of the spikelet, broadly ovate, 1–3-nerved, obtuse or acute; superior glume 7–9-nerved; inferior lemma similar to the superior glume, sterile, its palea poorly developed or absent; superior lemma and palea pallid or dark, glossy.

Botswana. N: Shamatoka Pan, 15.ii.1983, *Smith* 4075 (K; SRGH). **Zambia**. B: Zambezi R., Barotse Flood Plain, 19.iii.1964, *Verboom* 1157 (K). N: Mbala, 26.ii.1959, *McCallum-Webster* A165 (K). E: Petauke Distr., Chilongozi, ii.1963, *Verboom* 908 (K). S: Kafue Nat. Park, Ngoma, 11.iii.1961, *Mitchell* 6/47 (K). **Zimbabwe**. N: Makonde (Lomagundi), 9.ii.1961, *Phipps* 2852 (K; SRGH). W: Shangani, i.1955, *Goldsmith* 145/55 (K; SRGH). C: Chegutu, 24.ii.1969, *Mavi* 950 (K; SRGH). E: Chimanimani (Melsetter), Tarka Dam, xii.1968, *Goldsmith* 173/68 (K; SRGH). S: Beitbridge, 25.iv.1972, *Cleghorn* 2590 (K; SRGH). **Malawi**. C: Lilongwe, 1.ii.1951, *Jackson* 388 (K). S: Zomba, 23.i.1950, *Wiehe* N/416 (K). **Mozambique**. MS: Beira Distr., Gorongosa, i.1972 *Tinley* 2340 (K; SRGH). M: Namaacha, 22.xi.1944, *Torre* 6941 (K; LISC).
Occurring throughout tropical Africa. Margins of rivers, lakes and swamps; 200–3400 m.

54. **Panicum porphyrrhizos** Steud., Syn. Pl. Glum. **1**: 72 (1854). —Stapf in F.T.A. **9**: 712 (1920). —Clayton in F.W.T.A. **3**: 434 (1972). —Clayton & Renvoize in F.T.E.A., Gramineae: 484 (1982). Type from Ethiopia.

Perennial with tough, robust culms 50–200 cm. high from a shortly rhizomatous or tufted base; the upper nodes not sharply demarcated, the pigment diffusing above and below. Leaf laminae 15–30 cm. long and 5–10 mm. wide, linear, flat, acuminate. Panicle (20)30–50 cm. long, ovate or oblong, moderately to much branched, the tertiary and sometimes the secondary branches appressed. Spikelets 3–3.5 mm. long, narrowly ovate, acuminate; inferior glume $\frac{1}{4}$–$\frac{1}{3}$ the length of the spikelet, broadly ovate, 1–3-nerved; superior glume 7–9-nerved; inferior lemma similar to the superior glume, sterile, its palea poorly developed; superior lemma and palea pallid, glossy.

Botswana. N: Mababe Depression, 15.vi.1978, *Smith* 2455 (K; SRGH). **Zambia**. C: Lusaka Distr., Cheta R., 20.iii.1963, *Vesey-FitzGerald* 3988 (K). E: Chipata Distr., Ngoni land, i.1963, *Verboom* 501 (K). S: Kafue Flats, 13.v.1957, *Angus* 1588 (K). **Zimbabwe**. N: Gokwe Distr., Sengwa, 13.ii.1977, *Guy* 2471 (K; SRGH). W: Hwange, iii.1960, *West* in GHS 107817 (K; SRGH). C: Chegutu, Avondale Farm, 25.ii.1969, *Mavi* 1010 (K; SRGH). **Malawi**. C: Kasungu, 14.i.1959, *Robson* 1172 (K). **Mozambique**. N: Lake Chilwa, 11.v.1971, *Bowbrick* L4 (K; SRGH).
Also occurring Northwards to West Africa and Ethiopia. Damp ground near swamps and rivers; 400–2250 m.

55. **Panicum trichonode** Launert & Renvoize in Merxm., Prodr. Fl. SW. Afr. **160**: 226 (1970). —Launert in Merxm. loc. cit.: 143. Type: Zambia, *Robinson* 6124 (K, holotype).

Caespitose or shortly rhizomatous, glabrous or pilose perennial with erect or ascending culms 35–120 cm. high from a pubescent base; nodes conspicuously bearded or pubescent with white hairs. Leaf laminae 20–30 × 3–5 mm., linear, flat, sharply acute. Panicle 10–22 cm. long, elliptic or oblong, moderately branched. Spikelets 2.5–3 mm. long, ovate-elliptic, glabrous, acute; inferior glume $\frac{1}{8}$–$\frac{1}{4}$ the length of the spikelet, broadly ovate, membranous, 0–1-nerved; superior glume herbaceous, 7–9-nerved; inferior lemma similar to the superior glume, male, its palea well developed; superior lemma and palea pallid, shiny.

Botswana. N: Okovango River, 14.ii.1979, *Smith* 2640 (K; SRGH). **Zambia**. B: Mongu, 20.i.1966, *Robinson* 6816 (K). W: 80 km. W. of Chingola, 18.xii.1963, *Robinson* 6124 (K). S: Namwala, 10.xii.1962, *van Rensburg* 1051 (K). **Zimbabwe**. N: Guruve, 3.i.1979, *Nyariri* 614 (K; SRGH). W: Hwange, 6.iii.1985, *Martin* 110 (K; SRGH). C: Gweru, 30.xii.1965, *Simon* 593 (K; SRGH). S: Makoholi, 2.iii.1984, *Sibanda* 4 (K; SRGH).
Also in Namibia. Damp grassland on clay soils; 1000–1900 m.

56. **Panicum bechuanense** Brem. & Oberm. in Ann. Transvaal Mus. **16**: 403 (1935). —Chippindall in Meredith, Grasses & Pastures of S. Africa: 336 (1955). Type: Botswana, Kuke Pan, *van Son* in Herb. Trans. Mus. 28611 (K, isotype).

Tufted, hispid perennial with culms 60–120 cm. high from a knotty, pubescent base. Leaf laminae 15–25 cm. long and 2–6 mm. wide, linear, flat, acuminate. Panicle 17–26 cm. long, elliptic-oblong, moderately branched. Spikelets 2.5–3(3.5) mm. long, ovate; inferior glume $\frac{1}{3}$ the length of the spikelet, broadly ovate, 1(3)-nerved; superior glume 5–7-nerved; inferior lemma similar to the superior glume, male, its palea well developed; superior lemma and palea pallid, glossy.

Botswana. N: Kumaga, 10.iv.1966, *McKay* 1982 (K; SRGH). SW: Kuke Pan, 24.iii.1930, *van Son* in Herb. Trans. Mus. 28611 (K). SE: Khutsa, 22.iv.1972, *Coleman* 79 (K). **Zimbabwe**. W: Gwaai to Dhlemeni, ii.1949, *Davies* D120 (K; SRGH).
Also known from Namibia and S. Africa (northern Cape Province). Grassy pans on Kalahari Sand; 1000–1100 m.

57. **Panicum coloratum** L., Mant. Pl. **1**: 30 (1767). —Stapf in F.T.A. **9**: 713 (1920). —Sturgeon in Rhod. Agric. Journ. **50**: 435 (1954). —Chippindall in Meredith, Grasses & Pastures of S. Afr.: 335 (1955). —Jackson & Wiehe, Annot. Check List Nyasal. Grass.: 50 (1958). —Bor, Grass. of B.C.I. & P.: 325 (1960). —Launert in Merxm., Prodr. Fl. SW. Afr. **160**: 140 (1970). —Clayton in F.W.T.A. **3**: 434 (1972). —Clayton & Renvoize in F.T.E.A., Gramineae: 485 (1982). Type from Egypt.

Var. **coloratum**
      Panicum swynnertonii Rendle in Journ. Linn. Soc. **40**: 230 (1911). — Stapf in F.T.A. **9**: 317 (1920). Type: Zimbabwe, *Swynnerton* 1702a (BM, holotype).
      Panicum crassipes Mez in Engl., Bot. Jahrb. **57**: 187 (1921). Type from Namibia.

Tufted perennial with erect or ascending culms 50–100 cm. high. Leaf laminae 7–20(30) cm. long and 3–10 mm. wide, linear, flat, acute or acuminate. Panicle 10–26(40) cm. long, ovate, moderately to much branched, contracted or spreading, the spikelets somewhat clustered or evenly dispersed. Spikelets 2–3 mm. long, ovate-elliptic or ovate; inferior glume $\frac{1}{4}$–$\frac{1}{3}$ the length of the spikelet, 1–3-nerved, broadly ovate, cuff-like; superior glume 7–9-nerved; inferior lemma similar to the superior glume, male, its palea well developed; superior lemma and palea pallid, shiny.

Botswana. N: Samedupe Bridge to Tolankwe, 27.ii.1977, *Smith* 1924 (K; PRE). SW: Takatshwane to Lehututu, 21.ii.1960, *de Winter* 7421 (K; PRE). SE: Gaborone, 7.i.1978, *Hansen* 3328 (K; PRE; SRGH). **Zambia**. N: Chishinga Ranch, 6.vii.1961, *Astle* 522 (K). S: Kafue R. Rail., Bridge, 12.ii.1954, *Hinds* 195 (K). **Zimbabwe**. N: Hurungwe (Urungwe) Distr., Mensa Pan, 29.i.1958, *Drummond* 5316 (K; SRGH). W: Hwange Distr., Matetsi Safari Area, 14.i.1980, *Gonde* 273 (K; SRGH). C: Kadoma, 18.ii.1976, *Dye* 341, (K; SRGH). E: Northern Chimanimani, iv.1907, *Swynnerton* 1702A (BM). S: Mwenezi, 23.ii.1967, *Cleghorn* 1417 (K; SRGH). **Malawi**. S: Chikwawa, 24.iii.1960, *Phipps* 2673 (K; SRGH). **Mozambique**. MS: Beira, Urema Plains, iii.1969, *Tinley* 1803 (K; SRGH). GI: Gaza Distr., Banhine Nat. Park, ix.1973, *Tinley* 2983 (K: SRGH). M: Umbeluzi, 12.x.1940, *Torre* 1770 (K). Tropical and subtropical Africa, introduced elsewhere. Bushland and grassland; 50–2300 m.

Var. **makarikariense** Goosens in Kew Bull., **1934**: 195 (1934). —Chippindall in Meredith, Grasses & Pastures of S. Afr.: 336 (1955). —Clayton & Renvoize in F.T.E.A., Gramineae: 486 (1982). Type: Botswana, Makarikari Pan, xii.1929, *Pentz* in Nat. Herb. Pre. 8416 & *Phillips & Goosens* in Nat. Herb. Pre. 8787 (both PRE, syntypes).

Glaucous robust perennial with ascending or erect culms 120–140 cm. high. Leaf laminae 15–35 cm. long and 5–12 mm. wide, linear, cordate, sharply acute or acuminate.

Botswana. N: Makarikari Pan xii.1929, *Pentz* in Nat. Herb. Pre. 8416 (K; PRE). SE: Gaborone, *Parker* in G.H.S. 263423 (K; SRGH).
This variety is only known outside Botswana as an introduction. As a pasture grass it has proved to have considerable advantages over the traditional cultivars selected from *P. coloratum* var. *coloratum*; it has increased resistance to flooding and frost and tolerance of heavy clay soils; it is more robust than var. *coloratum* and has a high herbage yield. From this variety several cultivars have been selected which are not only cultivated elsewhere in Africa but also in Australia (see Bogdan, Tropical Pasture and Fodder Plants: 178 (1977)).
Panicum stapfianum Fourc. which occurs in S. Africa (the southern Transvaal, Orange Free State and eastern and southern Cape Province), is very similar to *P. coloratum*. It is distinguished by its caespitose habit, short, erect culms 22–50(90) cm. high and smaller panicle 7–10(18) cm. long. Apart from the basal differences the only other difference is its small size, in spikelet characteristics it is similar to *P. coloratum*.

58. **Panicum merkeri** Mez in Engl., Bot. Jahrb. **34**: 144 (1904). —Jackson & Wiehe, Annot. Check List Nyasal. Grasses: 51 (1958). Types from Tanzania and Malawi, *Whyte* s.n. (K, isosyntype).
Panicum radula Mez in Engl., Bot. Jahrb. **57**: 189 (1921). Type from Namibia.

Robust, often hispid tufted perennial with culms 100–200 cm. high. Leaf laminae 20–35 cm. long and 7–15 mm. wide, flat, acuminate. Panicle (20)25–35 cm. long, elliptic-oblong, moderately to much branched, the branches ascending. Spikelets (2)2.5–3 mm. long, ovate-elliptic, sharply acute; inferior glume $\frac{1}{4}$–$\frac{1}{3}$ the length of the spikelet, ovate 1(3)-nerved; superior glume 9-nerved; inferior lemma similar to the superior glume, male.

Zambia. C: Mumbwa, 9.i.1962, *Mitchell* 12/37 (K). S: Mazabuka, 20.i.1964, *van Rensburg* 2806 (K). **Zimbabwe**. W: cult. Queensland, C.S.I.R.O. Res. Stat. Samford, xi.1963, originally from Bulawayo, (K; BRI). **Malawi**. C: Mehenzi, 14.i.1953, *Jackson* 1026 (K). S: Zomba, 5.i.1979, *Banda et al.* 1350 (K; SRGH). **Mozambique**. MS: Shupanga, i.1961, *Kirk* s.n. (K).
Also in Namibia, and northwards to Uganda and Kenya. Swampy and seasonally damp places on heavy clay soils; 90–1900 m.
This species intergrades with *P. coloratum*, being distinguished principally by its more robust habit, taller culms and larger panicles.

59. **Panicum kalaharense** Mez in Engl., Bot. Jahrb. **57**: 187 (1921). —Chippindall in Meredith, Grasses & Pastures of S. Afr.: 338 (1955). —Launert in Merxm., Prodr. Fl. SW. Afr. **160**: 141 (1970). TAB **6**. Type from Namibia.

Robust perennial of caespitose or shortly rhizomatous habit; culms 100–200 cm. tall, arising from a pubescent base. Leaf laminae 20–50 cm. long and 2–7 mm. wide, tough, linear, flat or involute, acuminate. Panicle 19–27 cm. long, broadly ovate, moderately to much branched. Spikelets 3–4 mm. long ovate, pinched at the apex; inferior glume $\frac{1}{2}$–$\frac{2}{3}$ the length of the spikelet, broadly ovate, 5-nerved, acute or acuminate; superior glume 5–7-nerved; inferior lemma 7-nerved, male, its palea well developed; superior lemma and palea pallid or dark, glossy.

Botswana. N: Mababe to Seronga, 12.iv.1982, *Smith* 3842 (K; SRGH). SW: Boso Bogolo (Boshobogolo) to Mpathutlwa, 11.iii.1976, *Ellis* 2632 (K; PRE). SE: Mahalapye, Lephepe, i.1965, *Yalala* 489 (K; SRGH). **Zimbabwe**. W: Hwange, 9.v.1967, *Cleghorn* 1674 (K; SRGH). S: Mwenezi, 4.vi.1971, *Ngoni* 154 (K; SRGH). **Mozambique**. M: Bela Vista, 7.xii.1961, *Lemos & Balsinhas* 251 (K).

Tab. 6. PANICUM KALAHARENSE. 1, habit (×⅔); 2, spikelet (× 10), both from *Skarpe* S-269.

Also in Namibia and S. Africa. Grassland and savanna on sandy soils; 500–1200 m.

60. **Panicum lanipes** Mez in Engl., Bot. Jahrb. **57**: 187 (1921). —Chippindall in Meredith, Grasses & Pastures of S. Africa: 337 (1955). —Launert in Merxm., Prodr. Fl. SW. Afr. **160**: 141 (1970). Type from S. Africa.

Tufted perennial with densely lanate hairy base; culms 20–90 cm. high. Leaf laminae 7–20 cm. long and 2–3(5) mm. wide, linear, flat, glabrous to densely hirsute with tubercle-based hairs, acuminate. Panicle 7–20 cm. long, ovate, moderately to densely branched. Spikelets (2)2.5–3 mm., ovate, acute; inferior glume $\frac{1}{3}$–$\frac{2}{3}$ the length of the spikelet, broadly ovate, 0–3-nerved; superior glume 5-nerved; inferior lemma 5–7-nerved, male, its palea well developed; superior lemma and palea pallid, glossy.

Botswana. SW: Zanye Pan, 13.iii.1976, *Ellis* 2659 (K; PRE). SE: 64 km. W. of Lutlhe, 17.ii.1960, *de Winter* 7331 (K; PRE).
Also in South Africa and Namibia. Dry sandy soils of river beds or in pans; 2000 m.
Only one other species of *Panicum* in southern Africa has a lanate hairy base and that is *P. pearsonii* Bolus which is distinguished by its more robust habit and leaf laminae 3–7 mm. wide.

61. **Panicum pansum** Rendle, Cat. Afr. Pl. Welw. **2**: 177 (1899). —Stapf in F.T.A. **9**: 700 (1920). —Launert in Merxm., Prodr. Fl. SW. Afr. **160**: 142 (1970). —Clayton in F.W.T.A. **3**: 434 (1972). —Clayton & Renvoize in F.T.E.A., Gramineae: 487 (1982). Type from Angola.

Annual with slender erect culms 20–100 cm. high. Leaf laminae 8–30 cm. long, linear, pilose or hispid, tapering to the apex. Panicle 15–40 cm. long, oblong or ovate, much branched, the branches long, fine, flexuous and spreading at maturity, the spikelets borne in pairs at their apex. Spikelets 2.5–3 mm. long, ovate, glabrous, acuminate; inferior glume broadly ovate, $\frac{1}{2}$–$\frac{2}{3}$ as long as the spikelet, 3–5-nerved, acute to acuminate; superior glume broadly ovate, 5–7-nerved, acute; inferior lemma 7–9-nerved, sterile, its palea poorly developed; superior lemma and palea pale and glossy.

Zambia. N: Mpulungu, 900 m., 12.iv.1959, *McCallum-Webster* A307 (K). **Malawi**. N: Kilwa, 500 m., 18.vi.1973, *Pawek* 6907 (K).
Also in West Africa, Sudan to Angola. Wooded grassland on arenaceous soils; 275–930 m.

62. **Panicum novemnerve** Stapf in F.T.A. **9**: 702 (1920). —Sturgeon in Rhod. Agric. Journ. **50**: 435 (1954). —Chippindall in Meredith, Grasses & Pastures of S. Afr.: 327 (1955). —Launert in Merxm., Prodr. Fl. SW. Afr. **160**: 142 (1970). Type: Zimbabwe, Harare, *Craster* 27 and several other specimens (K, syntypes).

Hispid annual with branching culms 30–60 cm. high. Leaf laminae 6–20 cm. long, linear to narrowly lanceolate, acuminate or tapering to a finely pointed apex. Panicle 10–20 cm. long, ovate, much branched, the branches spreading or ascending, flexuous. Spikelets, 2.2–2.5 mm. long; narrowly ovate, inferior glume broadly ovate, $\frac{1}{2}$ the length of the spikelet, acute or acuminate; superior glume as long as the spikelet, 7–9-nerved, acute; inferior lemma similar to the superior glume, its palea poorly developed, sterile; superior lemma and palea glossy.

Zambia. C: Lusaka Distr., Chilanga, 31.i.1968 *Simon & Anton-Smith* 1627 (K; SRGH). **Zimbabwe**. N: Makonde (Lomagundi) Distr., Great Dyke, 31.iii.1975, *Wild* 7991 (K; SRGH). W: Bulawayo, Waterford, 6.ii.1974, *Norrgrann* 501 (K; SRGH). C: Chegutu Distr., Norton township, 23.i.1975, *Campbell* s.n. (K; SRGH). E: Mutare, Hospital Hill, 21.i.1974, *Crook* 2011 (K). S: Masvingo, Glenglivet Hotel, 15.ii.1974, *Davidse* 6645 (K; SRGH; MO). **Malawi**. S: Chipata, Ngoniland, ii.1963, *Verboom* 914 (K).
Also in S. Africa (Transvaal) and Namibia. Weedy and disturbed open habitats; 1080–1400 m.

63. **Panicum ephemerum** Renvoize in Kew Bull. **34**: 551 (1979). Type: Zambia, Mwinilunga, *Milne-Redhead* 3926 (K, holotype).

Annual with very slender, unbranched, erect culms 25–35 cm high. Leaf laminae 9–16 cm. long, filiform, involute, glabrous, tapering to an abrupt point. Panicle 1.5–8 cm. long, narrowly oblong, very sparsely branched, the branches contracted, ascending or appressed to the axis, spikelets numbering 2–12 on each panicle. Spikelets 3–3.5 mm. long, oblong or ovate oblong, glabrous, purple-tinged; inferior glume ovate, $\frac{1}{2}$ the length of the spikelet, 3-nerved, acute; superior glume ovate-oblong, 7-nerved, bluntly acuminate, longer than the superior floret; inferior lemma similar to the superior glume, sterile, its palea absent; superior lemma and palea glossy, apiculate.

Zambia. W: Mwinilunga Distr., Matonchi Farm, 2.ii.1938, *Milne-Redhead* 3926 (K). Growing in 15 cm. of water on shallow soil over laterite rock.

64. **Panicum pseudoracemosum** Renvoize in Kew Bull. **44**, 3: 544 (1989). Type: Zambia, Mwinilunga, *Milne-Redhead* 3680 (K, holotype).

Annual; culms erect, 30–70 cm. high. Leaf laminae 3–6 cm. long and 8–17 mm. wide, narrowly ovate, amplexicaul, pilose, sharply acute. Panicle 6–10 cm. long, broadly ovate or oblong, secondary branches usually reduced or absent, the spikelets appearing to be pseudoracemosely arranged on the ascending primary branches. Spikelets 1.5–2 mm. long, obliquely ovate, sparsely to densely pilose; inferior glume as long as the spikelet, 3-nerved; superior glume 5-nerved; inferior lemma 5-nerved, its palea well developed, male; superior lemma and palea pallid, densely papillose.

Zambia. N: Mbala, Lunzua R., 5.iv.1959, *McCallum-Webster* A277 (K). W: Mwinilunga, Kaomba R., 15.xii.1937, *Milne-Redhead* 3680 (K).
Not known from elsewhere. Damp places in shade; 1350–1650 m.

65. **Panicum heterostachyum** Hackel in Oest. Bot. Zeitschr. **51**: 430 (1901). —Stapf in F.T.A. **9**: 733 (1920). —Sturgeon in Rhod. Agric. Journ. **50**: 18 (1954). —Chippindall in Meredith, Grasses & Pastures of S. Afr.: 327 (1955). —Clayton in F.W.T.A. **3**: 429 (1972). —Clayton & Renvoize in F.T.E.A., Gramineae: 496 (1982). Type from Ethiopia.

Annual with culms 20–80 cm. long, erect or ascending. Leaf laminae 8–12 cm. long and 10–25(40) mm. wide, ovate or narrowly ovate, amplexicaul, thin, acute or acuminate. Panicle 4–15 cm. long, much-branched, the branches rather stiff and bearing conspicuous glandular patches. Spikelets 1.5 mm. long, asymmetrically ovate, sparsely to densely pubescent or pilose; inferior glume as long as the spikelet, 3-nerved; superior glume 5-nerved; inferior lemma hyaline, 5-nerved, its palea moderately developed, sterile; superior lemma and palea membranous, not glossy, scaberulous.

Botswana. N: Gubatsaa Hills, 21.v.1977, *Smith* 2077 (K; SRGH). **Zambia**. B: Kabompo, 24.iii.1961, *Drummond & Rutherford-Smith* 7275 (K; SRGH). N: Mbala Distr., Lunzua R., 24.v.1967, *Richards* 22257 (K). W: Solwezi, 20.iii.1971, *Drummond & Rutherford-Smith* 7113 (K; SRGH). C: Mumbwa, 20.iii.1963, *van Rensburg* 1751 (K). E: Luangwa R., 17.iii.1959, *Robson* 1752 (K). S: Livingstone, 3.iii.1963, *Mitchell* 17/92 (K; SRGH). **Zimbabwe**. N: Hurungwe (Urungwe), 15.iii.1966, *Simon* 699 (K; SRGH). W: Hwange, 12.iii.1981, *Gonde* 357 (K; SRGH). C: Harare, 6.iii.1966, *Crook* 798 (K). E: Chimanimani, 29.v.1969, *Müller & Kelly* 1170 (K; SRGH). S: Mwenezi, Clarendon Cliffs, 29.iv.1962, *Drummond* 7797 (K; SRGH).
Throughout tropical Africa. Bushland or savanna; 30–1700 m.

66. **Panicum glandulopaniculatum** Renvoize in Kew Bull. **44**, 3: 544 (1989). Type: Malawi, Nkhata Bay, Bandawe, *Jackson* 910 (K, holotype).

Annual with erect or decumbent branching culms 30–100 cm. long, often rooting at the lower nodes. Leaf laminae 2.5–8.5 cm. long and 10–20 mm. wide, lanceolate or narrowly ovate, amplexicaul, acute or acuminate. Panicle 6–9(14) cm. long, ovate, moderately branched, the branches stiff and bearing glandular patches. Spikelets 1.5–2 mm. long, asymmetrically ovate, sparsely to densely pilose; inferior glume as long as the spikelet, 3-nerved, separated from the superior glume by a short internode; superior glume 5-nerved; inferior lemma 5-nerved, its palea well developed, male; superior lemma and palea pallid, scaberulous.

Zambia. N: Samfya, 10.vi.1962, *Symoens* 9636 (K). **Zimbabwe**. E: Nyanga (Inyanga), Nyamingura R., 21.iv.1958, *Phipps* 1171 (K; SRGH). **Malawi**. N: Nkhata Bay, Kabunduli, 3.vi.1973, *Pawek* 6791 (K; MA; MO). **Mozambique**. Z: Gúruè, 7.iv.1943, *Torre* 5091 (K; LISC). MS: Beira Distr., Gorongosa Nat. Park, 7.vii.1972, *Ward* 7775 (K).
Also in S. Africa (Natal). Forest or woodland shade; 150–1300 m.

67. **Panicum brevifolium** L., Sp. Pl.: 59 (1753). —Stapf in F.T.A. **9**: 731 (1920). —Jackson & Wiehe, Annot. Check List Nyasal. Grass.: 50 (1958). —Bor, Grasses of B.C.I. & P.: 324 (1960). —Launert in Merxm., Prodr. Fl. SW. Afr. **160**: 140 (1970). —Clayton in F.W.T.A. **3**: 429 (1972). —Clayton & Renvoize in F.T.E.A., Gramineae: 496 (1982). Type from India.

Annual; culms 15–100 cm. long, rambling, often rooting at the lower nodes. Leaf laminae 5–10 cm. long and 10–25 mm. wide, thin amplexicaul, glabrous or pilose, cross-veins present, finely acute. Panicle 5–15 cm. long, ovate, much-branched, the branches delicate and often appearing somewhat tangled. Spikelets 1.5–2 mm. long, ovate or ovate-elliptic, sparsely puberulous to pilose; inferior glume as long as the spikelet, 3-nerved; superior glume 5-nerved; inferior lemma 5-nerved, sterile, its palea present; superior lemma and palea pallid, glossy.

Zambia. S: Kalomo, 19.iii.1963, *Astle* 2272 (K). **Zimbabwe**. E: Chimanimani, 29.xi.1967, *Simon & Ngoni* 1375 (K; SRGH). **Malawi**. S: Nsanje, 23.ii.1960, *Phipps* 2619 (K; SRGH). **Mozambique**. MS: Chimanimani Mts., 27.v.1969, *Müller & Kelly* 1067 (K; SRGH).
Throughout the Old World tropics. Forest shade; 0–1700 m.

68. **Panicum mueense** Vanderyst in Bull. Agric. Congo Belge **10**: 248 (1919). —Clayton in F.W.T.A. **3**: 433 (1972). —Clayton & Renvoize in F.T.E.A., Gramineae: 497 (1982). Types from Zaire.

Annual, culms 15–60 cm. long, decumbent. Leaf laminae 3–8 cm. long and 5–10 mm. wide, lanceolate, finely acute. Panicle 3–10 cm. long, narrowly oblong or ovate, moderately branched, the branches fine and ascending. Spikelets 1.7–2.3 mm. long, elliptic-oblong, sparsely to densely pilose; inferior glume ½ the length of the spikelet, 0–1(3)-nerved, separated from the superior glume by a short internode; superior glume 3(5)-nerved; inferior lemma 3(5)-nerved, its palea absent, sterile; superior lemma and palea pallid, glossy.

Mozambique. MS: Beira Distr., Cheringoma, *Tinley* 2670 (K; SRGH).
Also in Nigeria, Cameroon and Zaire. Forest shade; 360–470 m.
This species is very similar to *Panicum pleianthum*.

69. **Panicum trichoides** Sw., Prodr. Veg. Ind. Occ.: 24 (1788). —Stapf in F.T.A. **9**: 730 (1920). —Clayton in F.W.T.A. **3**: 429 (1972). —Clayton & Renvoize in F.T.E.A., Gramineae: 497 (1982). Type from Jamaica.

Delicate annual with decumbent culms 15–100 cm. long. Leaf laminae 3–7 cm. long and 5–20 mm. wide, lanceolate or narrowly ovate, thin, pilose, acuminate. Panicle 5–15(25) cm. long, broadly ovate to ovate-oblong, finely branched, the branching regular. Spikelets 1–1.5 mm. long, asymmetrically ovate, puberulous, bluntly acute; inferior glume up to ½ the length of the spikelet, 1-nerved, separated from the superior glume by a short internode; superior glume 3–5-nerved; inferior lemma 3–5-nerved, its palea poorly developed, sterile; superior lemma and palea usually pallid, granulose.

Zambia. S: Gwembe Distr., Chirundu Bridge, 6.ii.1958, *Drummond* 5507 (K; SRGH). **Malawi**. S: Chikwawa Distr., Lengwe, 15.iv.1970, *Hall-Martin* 632 (K). **Mozambique**. N: Cabo Delgado Distr., Montepuez, 8.iv.1964, *Torre & Paiva* 11752 (K; LISC). MS: Beira Distr., Chitengo, iii.1972, *Tinley* 2498 (K; SRGH).
Native in tropical America, probably introduced elsewhere. Riverine forest; 0–900 m.

70. **Panicum pleianthum** Peter, Fl. Deutsch Ost.-Afr. **1**, Anh.: 47 (1930). —Clayton & Renvoize in F.T.E.A., Gramineae: 498 (1982). TAB **7**. Type from Tanzania.

Annual or short-lived perennial. Culms 20–60 cm. long, decumbent. Leaf sheaths ciliately fringed, laminae 5–10 cm. long and 6–10 mm. wide, lanceolate, pilose, acuminate. Panicle 4–10 cm. long, oblong or obovate, much branched, dense, the branches ascending or contracted and appressed. Spikelets 1.7–2.2 mm. long, obovate, glabrous or sparsely pilose, apiculate; inferior glume ½ as long as the spikelet, 3–5-nerved, separated from the superior glume by a short internode; superior glume 5-nerved; inferior lemma 5-nerved, its palea poorly developed, sterile; superior lemma and palea glossy.

Mozambique. MS: Amatongas Forest, iv.1952, *Schweickerdt* 2352 (K). GI: Nhacoongo, 4.iv.1959, *Barbosa & Lemos* 8487 (K).
Also in Kenya and Tanzania. Coastal Forest; 50–550 m.

71. **Panicum laticomum** Nees, Fl. Afr. Austr.: 43 (1841). —Chippindall in Meredith, Grasses & Pastures of S. Afr.: 325 (1955). —Clayton & Renvoize in F.T.E.A., Gramineae: 498 (1982). Type from S. Africa (Natal).

Annual with scrambling or geniculately ascending culms 30–200 cm. long. Leaf laminae 6–10 cm. long and 10–28 mm. wide, lanceolate, acuminate. Panicle 10–25 cm. long, ovate, much to moderately branched, the branches fine and spreading. Spikelets 1.5–2 mm. long, asymmetrically narrowly ovate, glabrous or pilose; inferior glume ½ the length of the spikelet, 3-nerved, separated from the superior glume by a short internode; superior glume ⅔–¾ the length of the spikelet, 5-nerved; inferior lemma 5–7-nerved, its

Tab. 7. PANICUM PLEIANTHUM. 1, habit (×⅔); 2, spikelet (× 16), both from *Barbosa & Lemos* 8487.

palea poorly developed, sterile; superior lemma and palea coriaceous, granulose, pallid, the apiculate apex generally exposed.

Zimbabwe. E: Chipinge, 30.i.1975, *Gibbs-Russell* 2689 (K; SRGH). S: Masvingo, 23.i.1974, *Vernon* 103 (K; SRGH). **Malawi**. S: Nsanje, 23.iii.1960, *Phipps* 2659 (K; SRGH). **Mozambique**. N: Ribáuè, 23.i.1964, *Torre & Paiva* 10158 (K; LISC). MS: Dombe, 22.iv.1974, *Pope & Müller* 1241 (K; SRGH).
Extends from Kenya to Natal. Coastal forest; 1–1100 m.

72. **Panicum parvifolium** Lam., Tab. Encycl. Méth. Bot. **1**: 173 (1791). —Stapf in F.T.A. **9**: 726 (1920). —Chippindall in Meredith, Grasses & Pastures of S. Afr.: 325 (1955). —Clayton in F.W.T.A. **3**: 433 (1972). —Clayton & Renvoize in F.T.E.A., Gramineae: 490 (1982). Type from S. America.
   Panicum beccabunga Rendle, Cat. Afr. Pl. Welw. **2**: 179 (1899). —Stapf in F.T.A. **9**: 727 (1920). Type from Angola.

Perennial; culms 8–1500 cm. long, slender, wiry, branching, decumbent, ascending or scrambling. Leaf laminae 1.5–3 cm. long and 2–7 mm. wide, narrowly ovate to lanceolate, cordate, flat, glaucous, cross-veins present, acute; spreading to reflexed at maturity. Panicle 1–3 cm. long, ovate, moderately branched, the branches spreading or reflexed at maturity. Spikelets 1–2 mm. long, ovate-oblong, glabrous; inferior glume $\frac{1}{2}$–$\frac{2}{3}$ the length of the spikelets, 3-nerved; superior glume 5-nerved; inferior lemma 5-nerved, sterile; superior lemma and palea pallid, glossy.

Botswana. N: Moanchira R., 16.i.1980, *Smith* 2985 (K; SRGH). **Zambia**. B: Mongu Boma, 24.iii.1964, *Verboom* 1331 (K). N: Mbala, 21.x.1967, *Simon & Williamson* 1142 (K; SRGH). C: Serenje, 5.iv.1961, *Phipps & Vesey-FitzGerald* 2955 (K; SRGH). **Malawi**. N: Mzimba, Mzuzu, 30.i.1975, *Pawek* 9025 (K; SRGH). **Mozambique**. N: Ribáuè, 28.i.1964, *Torre & Paiva* 10306 (K). MS: Chimanimani Mts., Dombe, 25.iv.1974, *Pope & Müller* 1296 (K; SRGH). M: Maputo, 4.i.1980, *de Koning* 7883 (K).
Throughout tropical Africa, Madagascar and tropical America. Swamps; 0–2300 m.
This species is similar to *Panicum pusillum* but may be distinguished by the presence of cross-veins in the bluish green or glaucous leaf laminae.

73. **Panicum pusillum** Hook.f. in Journ. Linn. Soc., Bot. **7**: 227 (1864). —Clayton in F.W.T.A. **3**: 433 (1972). —Clayton & Renvoize in F.T.E.A., Gramineae: 490 (1982). Type from Cameroon.

Delicate annual; culms 3–20(60) cm. long, slender, erect or more often decumbent. Leaf laminae 1–4 cm. long and 1.5–6 mm. wide, lanceolate, flat, pilose, acute or acuminate, cross veins absent. Panicle 1–2(5) cm. long, ovate, sparsely to moderately branched, the branches spreading to deflexed at maturity. Spikelets 1.5–2 mm. long, narrowly ovate, pilose; inferior glume $\frac{2}{3}$–$\frac{3}{4}$ the length of the spikelet, 3-nerved, acuminate; superior glume 5–7-nerved; inferior lemma 5-nerved; its palea poorly developed, sterile; superior lemma and palea pallid, glossy.

Malawi. N: Nyika Plateau, Rumphi Bridge, 16.iv.1975, *Pawek* 9263 (K; MO; SRGH; MAL).
Also occurring Northwards to Ethiopia and Sierra Leone. Mountain grassland and bushland; 1350–3300 m.

74. **Panicum nymphoides** Renvoize in Kew Bull. **44**, 4: 545 (1989). Type: Malawi, Mt. Mulanje, *Hilliard & Burtt* 6396 (K, holotype).

Ephemeral; culms 9–10 cm. high, erect, delicate. Leaf laminae 0.5–1.5 cm. long and 1–3 mm. wide, lanceolate, thin, acute. Panicle 0.5–1 cm. long, reduced to 2–3 spikelets on short slender pedicels. Spikelets 2 mm. long, oblong, sparsely pilose; inferior glume $\frac{1}{2}$ the length of the spikelet, a narrow, 1-nerved scale; superior glume 7-nerved; inferior lemma 5-nerved, its palea well developed, male; superior lemma and palea pallid, glossy.

Malawi. S: Mt. Mulanje, 13.vii.1956, *Jackson* 1892 (K).
Only known from the two collections cited. Growing amongst boulders, moss and other grasses; 2100 m.

75. **Panicum aequinerve** Nees, Fl. Afr. Austr.: 40 (1841). —Chippindall in Meredith, Grasses & Pastures of S. Afr.: 324 (1955). —Clayton & Renvoize in F.T.E.A., Gramineae: 495 (1982). Type from S. Africa.

Annual or short-lived perennial with delicate trailing culms 15–70 cm. long, from a slender rhizome. Leaf laminae 2.5–8 cm. long and 2–9 mm. wide, narrowly lanceolate, flat, acuminate. Panicle 4–8(20) cm. long, broadly ovate, glabrous or pilose, sparsely branched, the branches spreading at maturity. Spikelets 3–4 mm. long, ovate, pilose or glabrous, acuminate; inferior glume as long as the spikelet, 3(5)-nerved; superior glume 7-nerved; inferior lemma 5-nerved, its palea poorly developed, sterile; superior lemma and palea pallid.

Malawi. S: Mulanje Mts., Litchenya Plateau, 8.vi.1962, *Robinson* 5310 (K).
A South African species which extends northwards to Ethiopia. Forest margins and rocky mountain grasslands; 1000–2700 m.

76. **Panicum eickii** Mez in Engl., Bot. Jahrb. **57**: 185 (1921). —Jackson & Wiehe, Annot. Check List Nyasal. Grass.: 50 (1958). —Clayton & Renvoize in F.T.E.A., Gramineae: 491 (1982). Type from Tanzania.

Small perennial with culms 10–30 cm. long straggling or forming tufts from slender wiry rhizomes. Leaf laminae 2–5 cm. long and 1–5 mm. wide, linear-lanceolate, flat, sharply acute. Panicle small 0.7–3.5 cm. long, narrowly oblong very sparsely branched, the branches often somewhat contracted. Spikelets 2–3 mm. long, ovate-elliptic, glabrous; inferior glume as long as the spikelet, (3)5-nerved; superior glume 7-nerved; inferior lemma 5-nerved its palea well developed, male; superior lemma and palea pallid, glossy.

Zimbabwe. E: Chimanimani Mts., 29.xii.1957, *Goodier & Phipps* 189 (K; SRGH). **Malawi.** S: Mt. Mulanje, Chambe Plateau, 17.xi.1949, *Wiehe* N344 (K).
Also in Tanzania, Usambara & Uluguru Mountains. Submontane grassland in damp, rocky open situations; 2300–2850 m.
In F.T.E.A. this species was credited with a wider distribution than that described above; the preparation of this present account has provided an opportunity to reappraise the taxonomy of this species and has led to the reinstatement of *P. striatissimum* Hubb. as a separate species distinguished by its appressed leaf laminae.

77. **Panicum inaequilatum** Stapf & Hubbard in Kew Bull. **1927**: 267 (1927). —Sturgeon in Rhod. Agric. Journ. **50**: 436 (1954). Type: Zimbabwe, Mutare, *Dept. of Agriculture*, ? *Eyles* 3375 (K, holotype).

Perennial with delicate, often tufted, erect or decumbent culms from a slender rhizome; culms 10–40(60) cm. long. Leaf laminae 2–6 cm. long and 2–6 mm. wide, flat, acute or acuminate. Panicle 1.5–7 cm. long, ovate, sparsely branched, the branches spreading at maturity. Spikelets 2–3 mm. long, elliptic, sparsely to densely pilose; inferior glume as long as the spikelet, 3-nerved; superior glume 7-nerved; inferior lemma 5-nerved, its palea poorly developed, sterile; superior lemma and palea pallid, glossy.

Zimbabwe. E: Chimanimani, 19.xi.1964, *Cleghorn* 1054 (K; SRGH). **Malawi.** N: Nyika Plateau, Chelinda Camp, 19.v.1970, *Brummitt* 10904; (K; MAL; SRGH). S: Mt. Mulanje, Litchenya Plateau, 6.vi.1962, *Robinson* 5286 (K). **Mozambique.** MS: Beira Distr., Gorongosa Mt., 12.iii.1972, *Tinley* 2418 (K; SRGH).
Also in southern Tanzania (Mbeya). Submontane grassland, in wet places; 1600–2400 m.

78. **Panicum subflabellatum** Stapf in F.T.A. **9**: 711 (1920). —Clayton & Renvoize in F.T.E.A., Gramineae: 481 (1982). Type: Mozambique, W. Luabo, *Kirk* (K, holotype).

Tufted perennial; culms 20–50 cm. high, erect or decumbent, often wiry. Leaf laminae 3–7 cm. long and 2–3 mm. wide, linear, flat and acuminate, or involute and somewhat pungent. Panicle 2.5–6 cm. long, oblong, moderately to sparsely branched, the branches short, ascending or appressed. Spikelets 1.5–1.8 mm. long, broadly ovate or elliptic-oblong, glabrous, obtuse; inferior glume ⅓ the length of the spikelet, broadly ovate, 3–5-nerved, slightly saccate; superior glume ovate, 5-nerved; inferior lemma similar to the superior glume but 5–7-nerved, male, its palea well developed; superior lemma and palea pallid, shining.

Mozambique. MS: Beira, 21.viii.1929, *Hitchcock* 24369 (K). GI: Inhambane, Ponta Zavora, *Barbosa & Lemos* 8491 (K). M: Maputo, 10.iii.1944, *Torre* 6400 (K).
Also in Tanzania and S. Africa (Natal). Coastal sand dunes.

79. **Panicum delicatulum** Fig. & De Not. in Mem. Acad, Sci. Torino, ser. 2, **14**: 351 (1854). —Clayton & Renvoize in F.T.E.A., Gramineae: 491 (1982). Type from Sudan.
   Panicum caudiglume Stapf in F.T.A. **9**: 727 (1920). —Jackson & Wiehe, Annot. Check List Nyasal. Grass.: 50 (1958) non Hack (1901) nom. illegit. Type: Malawi, *Cameron* 16 (K, holotype).
   Panicum nyassense Napper in Kirkia **3**: 130 (1963). Based on *P. caudiglume* Stapf.

Annual; culms 10–30(50) cm. high, weak, erect or ascending. Leaf laminae 3–8 cm. long and 5–12 mm. wide, lanceolate, thin, flat, acuminate. Panicle 3–8 cm. long, ovate, moderately to much branched, the branches very fine. Spikelets 2–2.5 mm. long, ovate, pilose, acuminate; inferior glume $\frac{1}{5}$–$\frac{1}{4}$ the length of the spikelet, a hyaline, obtuse, nerveless scale; superior glume 3-nerved, acuminate; inferior lemma 3-nerved, its palea poorly developed, sterile; superior lemma and palea pallid, glossy.

Zambia. N: Mbala, 14.iv.1959, *McCallum-Webster* A319 (K). **Malawi**. N: Nyika Plateau, 11.iv.1969, *Pawek* 2110 (K). C: Ntchisi Mt., 6.v.1963, *Verboom* 965 (K). S: Zomba Mt., 17.ix.1950, *Wiehe* N648 (K). Northwards to Sudan and Ethiopia. Submontane forest; 1500–2400 m.

80. **Panicum monticola** Hook.f. in Journ. Linn. Soc., Bot. **7**: 226 (1864). —Stapf in F.T.A. **9**: 722 (1920). —Sturgeon in Rhod. Agric. Journ. **50**: 17 (1954). —Chippindall in Meredith, Grasses & Pastures of S. Afr.: 326 (1955). —Jackson & Wiehe, Annot. Check List Nyasal. Grass.: 52 (1958). —Clayton in F.W.T.A. **3**: 433 (1972). —Clayton & Renvoize in F.T.E.A., Gramineae: 494 (1982). Type from Cameroon Mt.

Perennial with slender, decumbent culms 30–100 cm. long. Leaf laminae 7–15 cm. long and 5–25 mm. wide, lanceolate to narrowly ovate, acuminate. Panicle 10–25 cm. long, ovate to oblong, sparsely to moderately branched, the secondary branches often poorly developed. Spikelets 2.2–3.5 mm. long, ovate to ovate lanceolate, acuminate; inferior glume $\frac{1}{4}$–$\frac{1}{2}(\frac{2}{3})$ the length of the spikelet, 0–1-nerved; superior glume 5-nerved; inferior lemma 5-nerved, sterile, its palea absent or rudimentary; superior lemma and palea pallid, glossy.

Zimbabwe. E: Mutare Distr., Vumba Mts., 30.iv.1973, *Crook* 1085 (K; SRGH). **Malawi**. N: Viphya, 9.vii.1952, *Jackson* 952 (K). C: Ntchisi Mt., 6.v.1963, *Verboom* 962 (K). S: Litchenya Plateau, 8.vi.1962, *Robinson* 5309 (K).
Throughout tropical Africa. Submontane forest; 600–2600 m.

81. **Panicum chionachne** Mez in Engl., Bot. Jahrb. **57**: 185 (1921). —Jackson & Wiehe, Annot. Check List Nyasal. Grass.: 50 (1958). —Clayton & Renvoize in F.T.E.A., Gramineae: 495 (1982). TAB. **8**. Type from Tanzania.

Annual or short-lived perennial with slender, decumbent or trailing culms 35–100(300) cm. long. Leaf laminae 5–13 cm. long and 5–16 mm. wide, flat, lanceolate, acuminate. Panicle 6–22 cm. long, broadly ovate, sparsely to moderately branched, the branches fine, spreading at maturity. Spikelets 2–3 mm. long, ovate, glabrous, acute or acuminate; inferior glume $\frac{3}{4}$ to as long as the spikelet, 3-nerved; superior glume 5–7-nerved; inferior lemma 5-nerved, male or its palea poorly developed and sterile; superior lemma and palea pallid, glossy.

Zambia. N: Mbala, 1.iv.1958, *Vesey-FitzGerald* 1579 (K; SRGH). W: Solwezi, 10.iv.1960, *Robinson* 3526 (K). **Malawi**. N: Mzimba, Mzuzu, 29.iv.1974, *Pawek* 8556 (K). C: Ntchisi Mt., 6.v.1969, *Verboom* 974a (K). S: Zomba, Mulungusi Stream, 16.viii.1950, *Wiehe* N/611 (K).
Extending northwards to Zaire and Sudan. Forest shade, or mountain grassland, in damp places; 1000–2750 m.

82. **Panicum wiehei** Renvoize in Kew Bull. **34**: 554 (1979). —Clayton & Renvoize in F.T.E.A., Gramineae: 499 (1982). Type: Malawi, Dedza Mt., *Wiehe* N/469 (K, holotype).

Annual with scrambling culms 30–100 cm. long. Leaf laminae 6–16 cm. long and 8–18 mm. wide, lanceolate, acuminate. Panicle 7–15 cm. long, ovate, moderately to much branched. Spikelets 2–2.5 mm. long, ovate, glabrous or sparsely pubescent; inferior glume $\frac{1}{4}$–$\frac{1}{3}$ the length of the spikelet, 0–1-nerved; superior glume 5-nerved; inferior lemma 5-nerved, male, its palea well developed; superior lemma and palea pallid, glossy.

Zimbabwe. E: Mutare, Murahwa's Hill, 22.iii.1970, *Simon* 2118 (K; SRGH). **Malawi**. N: Mzimba, Viphya, 17.iii.1951, *Jackson* 429 (K). C: Mwera Hill Res. Sta., 8.v.1951, *Jackson* 491. S: Zomba Plateau, 15.iii.1970, *Brummitt* 9096 (K; LISC; MAL; SRGH).
Extending northwards to Sudan. Evergreen forest; 900–2200 m.

40

Tab. 8. PANICUM CHIONACHNE. 1, habit (× ⅔); 2, spikelet (× 15), both from *Robinson* 3526.

5. SACCIOLEPIS Nash

Sacciolepis Nash in Britton, Man. Fl. North. States: 89 (1901). —Simon in Kew Bull. **27**: 387–406 (1972).

Inflorescence a dense spiciform panicle, rarely open. Spikelets laterally (*S. africana, S. interrupta* dorsally) compressed, asymmetrical. Glumes prominently ribbed, the inferior $\frac{1}{3}-\frac{3}{4}$ length of spikelet (shorter in species 1–3), the superior equalling spikelet and gibbous. Inferior lemma resembling superior glume but less gibbous. Superior lemma dorsally compressed, thinly coriaceous to cartilaginous, with flat or involute but never hyaline margins, the floret readily deciduous.

A genus of 30 species, occurring throughout the tropics but best represented in Africa.

1. Panicle open or somewhat contracted   -   -   -   -   -   -   -   -   1. *curvata*
- Panicle dense, spike-like   -   -   -   -   -   -   -   -   -   -   -   - 2
2. Spikelets slightly dorsally compressed, light green   -   -   -   -   -   - 3
- Spikelets laterally compressed   -   -   -   -   -   -   -   -   -   - 4
3. Spikelets 2.5–3.5 mm. long, obtuse to subacute   -   -   -   -   - 2. *interrupta*
- Spikelets 3.8–5 mm. long, acute to acuminate   -   -   -   -   - 3. *africana*
4. Plants annual; palea of inferior floret usually short, up to c. $\frac{1}{2}$ length of lemma   -   -   -   -   -   -   -   -   -   -   -   - 5
- Plants perennial; palea of inferior floret $\frac{2}{3}$ to almost as long as lemma   -   -   - 7
5. Leaf laminae not papillose; spikelets 1.5–3 mm. long   -   -   -   -   - 6. *indica*
- Leaf laminae papillose above   -   -   -   -   -   -   -   -   -   - 6
6. Spikelets 0.7–1 mm. long, glabrous   -   -   -   -   -   -   - 4. *micrococca*
- Spikelets 1–1.7(2) mm. long, glabrous or hispidulous   -   -   -   - 5. *spiciformis*
7. Inferior lemma with a prominent transverse fringe of hairs above the middle; spikelets 3–4 mm. long   -   -   -   -   -   -   -   -   - 10. *transbarbata*
- Inferior lemma glabrous or uniformly pubescent   -   -   -   -   -   - 8
8. Spikelets 1.5–2.5 mm. long   -   -   -   -   -   -   -   -   -   - 9
- Spikelets 3–5 mm. long   -   -   -   -   -   -   -   -   -   -   - 10
9. Leaf laminae tightly rolled; culms hard, the inferior sheaths seldom with cross-nerves   -   -   -   -   -   -   -   -   -   -   - 7. *chevalieri*
- Leaf laminae flat or folded; culms spongy, the inferior sheaths usually with cross-nerves   -   -   -   -   -   -   -   -   -   - 8. *typhura*
10. Leaf sheaths with auricles 1–4 mm. long   -   -   -   -   -   - 9. *rigens*
- Leaf sheaths without auricles   -   -   -   -   -   -   -   -   - 11
11. Panicle cylindrical, dense; rhizomes slender   -   -   -   - 11. *seslerioides*
- Panicle interrupted, scanty; rhizomes short and knotty   -   -   - 12. *catumbensis*

1. **Sacciolepis curvata** (L.) Chase in Proc. Biol. Soc. Wash. **21**: 8 (1908). —Stapf in F.T.A. **9**: 766 (1920). —Chippindall in Meredith, Grasses & Pastures of S. Afr.: 357 (1955). —Bor, Grasses of B.C.I. & P.: 357 (1960). —Clayton & Renvoize in F.T.E.A., Gramineae: 455 (1982). Type from India.
Panicum curvatum L., Syst. Nat. ed. 12, **2**: 732 (1767). Type as above.

Perennial, geniculately ascending from a slender rhizome. Culms weak, 20–75 cm. high. Panicle open and ovate to contracted and linear, 3–11 cm. long, the branches ± wavy and pedicels distinct. Spikelets 2–3 mm. long, laterally compressed, obliquely oblong in profile, soon gaping, glabrous or with a few scurfy hairs. Inferior glume a tiny ovate scale, 0.2–0.5 mm. long. Superior glume saccate at the base, herbaceous, prominently ribbed.

Zimbabwe. E: Chimanimani (Melsetter), 11.i.1969, *Biegel* 2793 (K; SRGH). S: Tshiturapadsi, 19.iii.1967, *Drummond* 9050 (K; SRGH). **Mozambique**. Z: Maganja, 28.ix.1949, *Barbosa & Carvalho* 4239 (K). MS: Chimoio (Vila Pery), 21.vii.1948, *Fischer & Schweickerdt* 280 (K; PRE). GI: Mocucuni Is., 26.xi.1958, *Mogg* 29295 (K). M: Marracuene (Vila Luiza), 1.x.1957, *Barbosa & Lemos* 7884 (K).
Eastern Africa from Kenya to Natal; also in Madagascar, India and Sri Lanka. A weedy species of damp shady places subject to disturbance; 0–600 m.
A deceptive species whose loose panicle could be mistaken for *Panicum*, but whose ribbed gibbous spikelets clearly place it in *Sacciolepis*.

2. **Sacciolepis africana** C.E. Hubbard & Snowden in Kew Bull. **1936**: 294 (1936). —Sturgeon in Rhod. Agric. Journ. **50**: 418 (1953). —Chippindall in Meredith, Grasses & Pastures of S. Afr.: 357 (1955). —Jackson & Wiehe, Annot. Check List Nyasal. Grass.: 56 (1958). —Launert in Merxm., Prodr. Fl. SW. Afr. **160**: 164 (1970). —Clayton in F.W.T.A. **3**: 425 (1972). —Clayton & Renvoize in F.T.E.A., Gramineae: 455 (1982). TAB. **9**. Type from Nigeria.

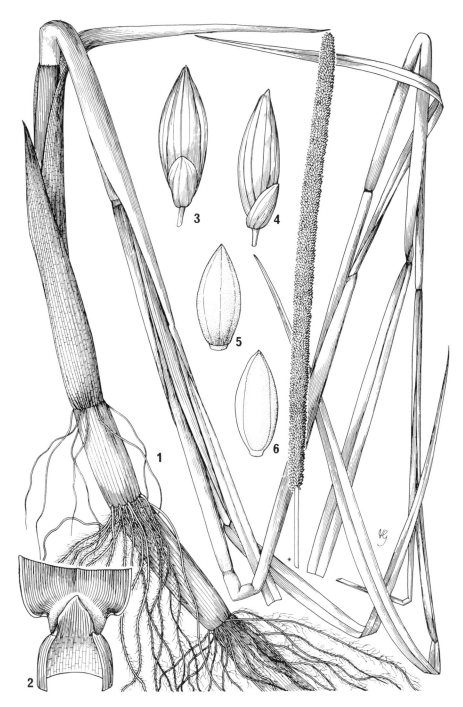

Tab. 9. SACCIOLEPIS AFRICANA. 1, habit (×½); 2, ligule (×½), 1–2 from *Hall* 369; 3, spikelet (×12); 4, lateral view of spikelet (×12); 5, superior lemma (×12); 6, superior floret showing palea (×12), 3–6 from *Ankrah* GC 20440. From F.T.E.A.

Rhizomatous perennial. Culms 30–180 cm. high, thick, spongy, decumbent and rooting below. Panicle 4–30 cm. long, cylindrical. Spikelets dorsally compressed, 2.5–3.5 mm. long, plumply elliptic, light green, glabrous, obtuse to subacute. Inferior glume subrotund, c. ¼ length of spikelet.

Botswana. N: Moanachira R., 17.v.1973, *Smith* 598 (K; SRGH). **Zambia**. B: Kaoma (Mankoya), 3.iv.1969, *Verboom* 1169 (K). N: Kawambwa, 18.iv.1957, *Richards* 9244 (K). C: Mumbwa, 22.iii.1963, *van Rensburg* 1776 (K). S: Kabulamwanda, 21.iv.1955, *Robinson* 1242 (K). **Zimbabwe**. N: Gokwe, 25.v.1974, *Guy* 2164 (K; SRGH). **Malawi**. C: Lilongwe, *Jackson* 56 (K). S: Lukulezi swamps, 20.vi.1962, *Robinson* 5395 (K). **Mozambique**. MS: Gorongosa, v.1972, *Tinley* 2552 (K; SRGH).
Throughout tropical Africa. Shallow water and flood plain grassland; 650–1500 m.

3. **Sacciolepis interrupta** (Willd.) Stapf in F.T.A. **9**: 757 (1920). —Bor, Grasses of B.C.I. & P.: 358 (1960). —Clayton & Renvoize in F.T.E.A., Gramineae: 456 (1982). Type from India.
　　Panicum interruptum Willd., Sp. Pl. **1**: 341 (1797). Type as above.

Like *S. africana*, but spikelets 3.8–5 mm. long and acute to acuminate.

Botswana. N: 18°10'S, 23°19'E, 24.vi.1975, *Smith* 1401 (K; SRGH). **Zambia**. N: Luangwa Valley Game Reserve, 23.iii.1967, *Prince* 413 (K). E: Chipata (Fort Jameson), 11.v.1963, *van Rensburg* 2123 (K). S: Namwala, 13.ii.1962, *Mitchell* 12/85 (K). **Zimbabwe**. N: Hurungwe (Urungwe), 1.iv.1970, *Guy* 465 (K; SRGH). W: Shangani, *West* 3166 (K). C: Chegutu, 21.iv.1983, *Mavi* 1604 (K; SRGH).
India to SE. Asia; also Tanzania. In water up to 1 m. deep.
S. interrupta is an Asiatic species which extends into eastern Africa. It resembles *S. africana* except for its larger pointed spikelets, and intermediate specimens can be difficult to assign. The continued recognition of two species finds some support from the reported chromosome counts: *S. africana* 2n=18, *S. interrupta* 2n=36.

4. **Sacciolepis micrococca** Mez in Fedde, Repert. **15**: 122 (1918). —Stapf in F.T.A. **9**: 753 (1920). —Jackson & Wiehe, Annot. Check List Nyasal. Grass.: 57 (1958). —Clayton in F.W.T.A. **3**: 425 (1972). —Clayton & Renvoize in F.T.E.A., Gramineae: 458 (1982). Types from Senegal, Nigeria & Sudan.
　　Sacciolepis nana Stapf in F.T.A. **9**: 753 (1920). Type from Central African Republic.
　　Panicum micrococcum (Mez) Peter, Fl. Deutsch Ost-Afr. **1**, Anh.: 52 (1930). Types as for *S. micrococca*.

Annual, growing in clumps. Culms 15–70 cm. high. Leaf laminae filiform, folded, papillose on the nerves above. Panicle 2–15 cm. long, cylindrical, slender. Spikelets 0.7–1 mm. long, elliptic, slightly laterally compressed, glabrous.

Zambia. E: Luangwa Valley, Chief Nsefu area, *Verboom* 960 (K). **Zimbabwe**. N: Gokwe, 25.iii.1963, *Bingham* 581 (K; SRGH). **Malawi**. N: Karonga, 26.vi.1951, *Jackson* 548 (K). C: Nkhota Kota, 16.vi.1970, *Brummitt* 11448 (K).
Tropical Africa, but mainly in the west. Swampy places; 500–1500 m.
S. micrococca and *S. spiciformis* are closely related but can be distinguished by spikelet length; there is also partial geographical separation.

5. **Sacciolepis spiciformis** (A. Rich.) Stapf in F.T.A. **9**: 756 (1920). Type from Ethiopia.
　　Panicum spiciforme A. Rich., Tent. Fl. Abyss. **2**: 359 (1851). Type as above.
　　Panicum huillense Rendle, Cat. Afr. Pl. Welw. **2**: 174 (1899). Type from Angola.
　　Sacciolepis huillensis (Rendle) Stapf in F.T.A. **9**: 755 (1920). —Sturgeon in Rhod. Agric. Journ. **50**: 418 (1953). — Chippindall in Meredith, Grasses & Pastures of S. Afr.: 358 (1955). —Jackson & Wiehe, Annot. Check List Nyasal. Grass.: 57 (1958). —Clayton & Renvoize in F.T.E.A., Gramineae: 458 (1982). Type as above.
　　Sacciolepis lebrunii Robyns in Bull. Jard. Bot. Brux. **9**: 185 (1932). Type from Zaire.
　　Sacciolepis geniculata Simon in Kirkia **8**: 86 (1971). Type: Zambia, Mweru-Wantipa, *Phipps & Vesey-FitzGerald* 3295 (SRGH, holotype; K, isotype).
　　Sacciolepis luciae Simon in Kirkia **8**: 85 (1971). Type: Zimbabwe, Nyanga (Inyanga), *Crook* 789 (SRGH, holotype; K, isotype).

Annual, growing in clumps. Culms 5–80 cm. high. Leaf laminae linear, papillose on raised nerves above. Panicle 1–20 cm. long, cylindrical, slender, borne above the foliage or rarely nestling amongst it. Spikelets 1–1.7(2) mm. long, ovate, glabrous or hispidulous.

Zambia. N: Mbala, Uningi Pans, 24.vi.1956, *Robinson* 1769 (K). W: Kalenda dambo, Mwinilunga, 16.iv.1960, *Robinson* 3611 (K). C: Chakwenga Headwaters, 27.iii.1965, *Robinson* 6553 (K). S: Victoria Falls, 17.viii.1947, *Brenan & Trapnell* 7727 (K). **Zimbabwe**. W: Matobo, iv.1955, *Miller* 2763 (K; SRGH). C: Chindamora Reserve, 25.iii.1952, *Wild* 3789 (K; SRGH). E: Chimanimani, 18.iv.1957, *Goodier & Phipps* 18 (K; SRGH). S: Bikita, 5.v.1969, *Biegel* 3016 (K; SRGH). **Malawi**. N: Nyika Plateau,

5.ii.1968, *Simon, Williamson & Ball* 1650 (K; SRGH). S: Citembo dambo, Mposa, 13.v.1952, *Jackson* 817 (K). **Mozambique**. N: Ribáuè, 31.i.1964, *Torre & Paiva* 10357 (K). Z: Mologué R., 8.vii.1943, *Torre* 5688 (K). MS: Dororo, 13.vii.1948, *Fischer & Schweickerdt* 7480 (K; PRE). GI: Massangena to Machaila, 3.ix.1955, *Myre & Carvalho* 2237 (K).

Cameroon and Ethiopia to Natal; also in Madagascar. Damp or swampy soils; pools on rock and ironstone outcrops; 0–2300 m.

The types of *S. spiciformis* and *S. luciae* are plants of small stature with large (up to 2 mm.) spikelets, but seem to represent mountain forms at the limit of normal variation rather than distinct taxa. *S. geniculata* has geniculate basal nodes and panicles nestling amongst the foliage, but the distinction is obscured by intermediates and it seems to be no more than a rock pool ecotype.

6. **Sacciolepis indica** (L.) Chase in Proc. Biol. Soc. Wash. **21**: 8 (1908). —Jackson & Wiehe, Annot. Check List Nyasal. Grass.: 57 (1958). —Bor, Grasses of B.C.I. & P.: 357 (1960). —Clayton & Renvoize in F.T.E.A., Gramineae: 458 (1982). Type from India.

   Aira indica L., Sp. Pl.: 63 & errata (1753) *non A. spicata* op. cit.: 64. Type as above.
   Panicum indicum (L.) Mant. Pl. Alt. **2**: 184 (1771) *non P. indicum* Mill. (1768). Type as above.
   Sacciolepis auriculata Stapf in F.T.A. **9**: 762 (1920). —Clayton in F.W.T.A. **3**: 425 (1922). — Launert in Merxm., Prodr. Fl. SW. Afr. **160**: 164 (1970). Type from Nigeria.
   Sacciolepis pergracilis Chiov., Pl. Nov. Min. Not. Aeth.: 24 (1928). Type from Ethiopia.
   Panicum glaucidulum Peter, Fl. Deutsch Ost-Afr. **1**, Anh.: 52 (1930). Type from Tanzania.
   Sacciolepis gracilis Stent & Rattray in Proc. Trans. Rhod. Sci. Ass. **32**: 31 (1933). —Sturgeon in Rhod. Agric. Journ. **50**: 418 (1953). —Jackson & Wiehe, Annot. Check List Nyasal. Grass.: 57 (1958). Type: Zimbabwe, Harare, *Eyles* 2264 (K, isotype).
   Sacciolepis claviformis Simon in Kirkia **8**: 85 (1917). Type: Zambia, Choma, *Robinson* 1205 (K, isotype).

Decumbent or ascending annual. Culms 10–100 cm. high, often with aerial roots from the lower nodes. Leaf laminae broadly linear, not papillose nor the nerves raised, with or without sheath auricles up to 2 mm. long. Panicle 1–13 cm. long, cylindrical to oblong. Spikelets 1.5–3 mm. long, narrowly ovate, glabrous or pubescent.

Zambia. B: Mankoya Distr., Kasempa Rd., 3.iv.1964, *Verboom* 1181 (K). N: Sansia Falls, 7.v.1961, *Vesey-FitzGerald* 3387 (K). W: Solwezi, 9.iv.1960, *Robinson* 3472 (K). E: Lundazi, 1.iv.1954, *Robinson* 806 (K). S: Choma, 9.v.1963, *Astle* 2387 (K). **Zimbabwe**. C: Chindamora Reserve, 15.iv.1922, *Eyles* 3387 (K). **Malawi**. C: Lilongwe, 23,iv.1951, *Jackson* 466 (K). S: Limbe, 17.viii.1950, *Jackson* 102 (K). **Mozambique**. N: Marrupa to Nungo, 5.viii.1981, *Jansen, de Koning & de Wilde* 60 (K). MS: Gorongosa, iv.1972, *Tinley* 2532 (K; SRGH).

Old World tropics. Streamsides and marshy places; 750–1700 m.

S. indica is very similar to *S. spiciformis*; it can usually be distinguished by spikelet length and habit, but absence of leaf papillae is the most reliable character though it requires use of a microscope. African specimens often have little auricles and have been separated as *S. auriculata*, but they intergrade completely with the commoner form lacking auricles.

7. **Sacciolepis chevalieri** Stapf in F.T.A. **9**: 754 (1920). —Jackson & Wiehe, Annot. Check List Nyasal. Grass.: 57 (1958). —Clayton in F.W.T.A. **3**: 425 (1972). —Clayton & Renvoize in F.T.E.A., Gramineae: 459 (1982). Types from Mali and Central African Republic.

   Sacciolepis brevifolia Stapf in F.T.A. **9**: 755 (1920). —Sturgeon in Rhod. Agric. Journ. **50**: 418 (1953). Type from Angola.
   Sacciolepis incurva Stapf in F.T.A. **9**: 761 (1920). Type from Tanzania.
   Panicum incurvum (Stapf) Peter, Fl. Deutsch Ost-Afr. **1**: 203 (1930). Type as above.
   Sacciolepis strictula Pilger in Notizbl. Bot. Gart. Berl. **11**: 651 (1932). Type from Angola.
   Sacciolepis palustris Napper in Kirkia **3**: 128 (1963). Type from Tanzania.

Shortly rhizomatous perennial. Culms 20–100 cm. high, hard, usually without cross-nerves in the basal sheaths. Leaf laminae 1–2 mm. wide, tightly rolled, strongly ribbed, papillose or not. Panicle 2–16 cm. long, cylindrical, often somewhat interrupted. Spikelets 1.5–2 mm. long, ovate, mostly pubescent.

Zambia. B: Senanga, 3.viii.1952, *Codd* 7363 (K; PRE). N: Shiwa Ngandu, 19.i.1938, *Greenway & Trapnell* 5720 (K). W: South Mwinilunga, ix.1934, *Trapnell* 1612 (K). C: Kundalila Falls, 15.x.1967, *Simon & Williamson* 1000 (K; SRGH). S: Choma, 21.viii.1963, *Astle* 2626 (K). **Zimbabwe**. N: Nyamunyeche Estate, 18.x.1978, *Nyariri* 425 (K; SRGH). W: Matobo, ix.1956, *Miller* 3664 (K; SRGH). C: Harare, 12.ix.1932, *Eyles* 7185 (K; SRGH). E: Chimanimani, 13.x.1950, *Crook* 183 (K; SRGH). **Malawi**. N: Mzuzu, 24.x.1969, *Pawek* 2918 (K). C: Tembwe, 8.x.1951, *Jackson* 624 (K). S: Kamfaloma, 16.xi.1950, *Jackson* 285 (K). **Mozambique**. N: Marrupa to Lichinga, 8.viii.1981, *Jansen, de Koning de Wilde* 155 (K).

Tropical and S. Africa; Madagascar. Swamps and near water; 500–2000 m.

8. **Sacciolepis typhura** (Stapf) Stapf in F.T.A. **9**: 760 (1920). —Sturgeon in Rhod. Agric. Journ. **50**: 418 (1953). —Chippindall in Meredith, Grasses & Pastures of S. Afr.: 358 (1955). —Jackson & Wiehe, Annot. Check List Nyasal. Grass.: 57 (1958). —Launert in Merxm., Prodr. Fl. SW. Afr. **160**: 164 (1970). —Clayton & Renvoize in F.T.E.A., Gramineae: 460 (1982). Type from S. Africa.
  Panicum typhurum Stapf in Fl. Cap. **7**: 414 (1899). Type as above.
  Panicum cinereo-vestitum Pilger in Wiss. Ergebn. Schwed. Rhod.-Kongo-Exped. **1**: 201 (1915). Type: Zambia, Bangweulu, *Fries* 897 (UPS, holotype).
  Sacciolepis scirpoides Stapf in F.T.A. **9**: 759 (1920). —Jackson & Wiehe, Annot. Check list Nyasal. Grass.: 57 (1958). Type from Angola.
  Sacciolepis glaucescens Stapf in F.T.A. **9**: 759 (1920). —Sturgeon in Rhod. Agric. Journ. **50**: 418 (1953). —Chippindall in Meredith, Grasses & Pastures of S. Afr.: 358 (1955). Types: Zimbabwe, Charter District, *Mundy & Dept. Agric.* 2102 (K, syntypes).
  Sacciolepis wombaliensis Vanderyst in Bull. Agric. Congo Belge **16**: 679 (1925). Type from Zaire.
  Sacciolepis kimpasaensis Vanderyst in Bull. Agric. Congo Belge **16**: 679 (1925). Type from Zaire.
  Sacciolepis wittei Robyns in Bull. Jard. Bot. Brux. **9**: 188 (1932). Type from Zaire.
  Sacciolepis cinereo-vestita (Pilger) C.E. Hubbard in Kew Bull. **1934**: 110 (1934). Type as for *P. cinereo-vestitum.*
  Sacciolepis trollii Pilger in Not. Bot. Gart. Berl. **12**: 381 (1935). Type from Tanzania.
  Sacciolepis velutina Napper in Kirkia **3**: 129 (1963). Type from Kenya.

Shortly rhizomatous perennial. Culms 20–150 cm. high, usually spongy at the base with cross-nerves in the papery sheaths. Leaf laminae 2–10 mm. wide, flat or folded, strongly ribbed, papillose or not. Panicle 3–30 cm. long, cylindrical, dense. Spikelets 1.7–2.5 mm. long, ovate, mostly glabrous.

Caprivi Strip. Andara, 17.i.1956, *de Winter* 4314 (K; PRE). **Botswana**. N: Kakanaga Is., 18.iv.1977, *Smith* 1983 (K; SRGH). **Zambia**. B: Mongu, 8.iv.1964, *Verboom* 1182 (K). N: Mwamba R., 18.iii.1966, *Astle* 4659 (K). E: Mchinje (Fort Manning) Rd., iii.1962, *Verboom* 584 (K). C: Mkushi R., 26.iii.1946, *Grassl* 46–39 (K). **Zimbabwe**. N: Chirisa, 12.ii.1981, *Mahlangu* 438 (K; SRGH). W: Victoria Falls, 7.iii.1977, *Elias* 32 (K; SRGH). C: Lake McIlwaine Nat. Park, 14.iii.1965, *Simon* 205 (K; SRGH). E: Rusape, 28.x.1972, *Crook* 1059 (K). S: Makohole Expt. Farm, 16.xii.1947, *Robinson* 97 (K; SRGH). **Malawi**. N: Mbara, 20.i.1951, *Jackson* 372 (K). S: Nyambi, 22.iv.1955, *Jackson* 1631 (K). **Mozambique**. Z: Milange to Quelimane, iv.1972, *Bowbrick* J3C (K; SRGH).
Tropical and S. Africa. Swampy and marshy places; 500–2500 m.
S. typhura is distinguished from *S. chevalieri* by its more luxuriant habit, but variation is continuous and the dividing line arbitrary. It is possible that they are merely habitat forms of a single species. The spikelets are prone to proliferation, assuming a superficially eragrostoid appearance but composed of disorganized barren scales.

9. **Sacciolepis rigens** (Mez) A. Chev. in Rev. Bot. Appliq. **14**: 29 (1934). —Clayton in F.W.T.A. **3**: 425 (1972). —Clayton & Renvoize in F.T.E.A., Gramineae: 460 (1982). Type from Togo.
  Panicum rigens Mez in Bot. Jahrb. **34**: 141 (1904). —Stapf in F.T.A. **9**: 767 (1920). Type as above.
  Sacciolepis leptorhachis Stapf in F.T.A. **9**: 763 (1920). Types from Zaire & Angola.
  Sacciolepis johnstonii C.E. Hubbard & Snowden in Kew Bull. **1936**: 314 (1936). Type from Uganda.

Shortly rhizomatous perennial. Culms 60–200 cm. high. Leaf sheaths with auricles 1–4 mm. long. Panicle 6–20 cm. long, cylindrical. Spikelets 3–4.5 mm. long, narrowly ovate, glabrous or sometimes pubescent. Superior glume membranous to subcoriaceous.

Caprivi Strip. Andara, 17.i.1956, *de Winter* 4310 (K; PRE). **Zambia**. N: Mporokoso, 14.iv.1961, *Phipps & Vesey-FitzGerald* 3177 (K; SRGH). W: Lusongwa, 26.xii.1969, *Simon & Williamson* 2050 (K; SRGH).
Tropical Africa, but nowhere frequent. Swamps; 1100–1300 m.
Not easy to distinguish from *S. indica* without seeing the base.

10. **Sacciolepis transbarbata** Stapf in F.T.A. **9**: 761 (1920). —Jackson & Wiehe, Annot. Check List Nyasal. Grass.: 57 (1958). —Clayton & Renvoize in F.T.E.A., Gramineae: 456 (1982). Syntypes: Zambia, Chibemba stream, *Kassner* 2087 & 2157 (both K, isosyntypes).
  Sacciolepis barbiglandularis Mez in Bot. Jahrb. **57**: 194 (1921). Type from Tanzania.
  Sacciolepis circumciliata Mez in Bot. Jabrb. **57**: 194 (1921). Type from Tanzania.
  Sacciolepis incana Mez in Bot. Jahrb. **57**: 194 (1921) nom superfl pro *S. transbarbata.*

Tufted perennial from a short knotty rhizome. Culms 30–100 cm. high. Panicle 2–10 cm. long, cylindrical, dense or sometimes interrupted. Spikelets 3–4 mm. long, elliptic, silky pillose all over. Superior glume herbaceous to papyraceous. Inferior lemma bearing a transverse fringe of long hairs across the back.

Zambia. N: Chilongowelo, 5.iii.1952, *Richards* 871 (K). W: Dobeka Bridge, 14.xii.1937, *Milne-Redhead* 3668 (K). C: Kanona, 17.xii.1967, *Simon & Williamson* 1413 (K; SRGH). S: Pemba, 7.ii.1963, *Astle* 2061 (K). **Zimbabwe**. N: Sengwa Research St., 12.i.1975, *Guy* 2237 (K; SRGH). **Malawi**. N: Mzuzu, 2.xii.1973, *Pawek* 7565 (K). C: Nkhota Kota, 28.ii.1953, *Jackson* 1135 (K). Also in Tanzania. *Brachystegia* woodland; 1200–1900 m.

11. **Sacciolepis seslerioides** (Rendle) Stapf in F.T.A. **9**: 764 (1920). —Clayton & Renvoize in F.T.E.A., Gramineae: 461 (1982). Type from Angola.
    Panicum seslerioides Rendle, Cat. Afr. Pl. Welw. **2**: 174 (1899). Type as above.
    Sacciolepis albida Stapf in F.T.A. **9**: 765 (1920). Type from Zaire.

Perennial with slender rhizomes. Leaf laminae flat. Culms 30–100 cm. high. Panicle 2–10 cm. long, cylindrical to oblong, dense. Spikelets 3–5 mm. long, narrowly ovate, glabrous, pallid. Superior glume subcoriaceous.

Zambia. N: Mbala, 11.iv.1961, *Phipps & Vesey-FitzGerald* 3060 (K; SRGH). W: Mwinilunga, 23.xi.1937, *Milne-Redhead* 3349 (K).
Also in Zaire, Tanzania and Angola. *Brachystegia* woodland; 1200–1700 m.

12. **Sacciolepis catumbensis** (Rendle) Stapf in F.T.A. **9**: 764 (1920). Type from Angola.
    Panicum catumbense Rendle, Cat. Afr. Pl. Welw. **2**: 175 (1899). Type as above.
    Panicum squamigerum Pilger in Wiss. Ergebn. Schwed. Rhod.-Kongo-Exped. **1**: 203 (1915). Type: Zambia, Kalungwishi R., *Fries* 1140 (UPS, holotype).
    Sacciolepis squamigera (Pilger) C.E. Hubbard in Kew Bull. **1934**: 110 (1934). Type as above.

Tufted perennial from a short knotty rhizome. Leaf laminae narrow, usually involute. Culms 15–70 cm. high. Panicle 2–5 cm. long, much interrupted, scanty. Spikelets 3–4.5 mm. long, narrowly ovate, glabrous. Superior glume chartaceous.

Zambia. N: Mpika, 4.i.1962, *Astle* 1153 (K). W: Mwinilunga Distr., W. of Kabompo, 11.ix.1930, *Milne-Redhead* 1180 (K).
Also in Angola. *Brachystegia* woodland; 1500 m.
Distinguished from *S. seslerioides* mainly by its habit, and perhaps no more than a depauperate form.

6. HYDROTHAUMA C.E. Hubbard

Hydrothauma C.E. Hubbard in Ic. Pl. (Hook.) **35**: t. 3458 (1947).

Inflorescence a slender spiciform panicle, sometimes reduced to a unilateral raceme with paired spikelets. Spikelets dorsally compressed. Inferior glume $\frac{1}{2}$ length of spikelet, hyaline, truncate, the superior equalling spikelet and gibbous. Superior lemma cartilaginous with flat margins.
A monotypic genus from Zambia and Zaire, related to *Sacciolepis*.

Hydrothauma manicatum C.E. Hubbard in Ic. Pl. (Hook.) **35**: t. 3458 (1947). TAB **10**. Type: Zambia, Mwinilunga Distr., Matonchi Farm, *Milne-Redhead* 4625 (K, holotype).

Aquatic annual. Culms 10–100 cm. long. Leaf laminae floating, the inferior on long false petioles, the superior surface raised into sinuous longitudinal lamellae. Spikelets 1.8–2.5 mm. long, narrowly ovate.

Zambia. N: Kasama, 1.iii.1960, *Richards* 12674 (K). W: Kalene Hill, 16.iv.1965, *Robinson* 6591 (K).
Also in Zaire. Shallow pools on rock and ironstone outcrops; 1200–1400 m.
The strange leaf lamellae presumably improve buoyancy by entrapping air bubbles. The culm and false petiole contain air canals.

7. ENTOLASIA Stapf

Entolasia Stapf in F.T.A. **9**: 739 (1920).

Inflorescence of racemes along a central axis, the spikelets borne singly on the rhachis or in neat appressed secondary racemelets. Spikelets narrowly elliptic. Inferior glume short, triangular, the superior as long as spikelet. Superior lemma coriaceous, pubescent.

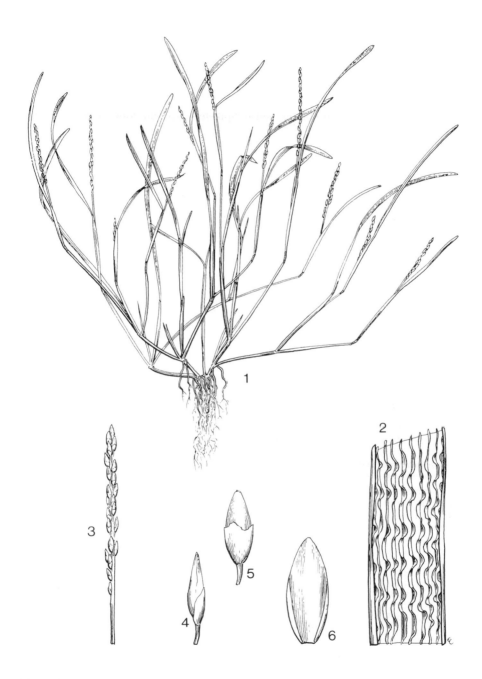

Tab. 10. HYDROTHAUMA MANICATUM. 1, habit (×⅔), from *Richards* 12679; 2, superior surface of leaf lamina (× 10), from *Robinson* 5901; 3, inflorescence (× 2); 4, spikelet, lateral view (× 8); 5, spikelet showing inferior glume (× 8); 6, superior lemma (× 15), 3–6 from *Richards* 12679.

A genus of 5 species. Tropical Africa and Australia.

Entolasia imbricata Stapf in F.T.A. **9**: 739 (1920). —Jackson & Wiehe, Annot. Check List Nyasal. Grass.: 38 (1958). —Clayton & Renvoize in F.T.E.A., Gramineae: 573 (1982). TAB. **11**. Type from Tanzania.
 Panicum endolasion Peter, Fl. Deutsch Ost-Afr. **1**, Anh.: 24 (1930). Type from Tanzania.

Tufted perennial. Culms 60–150 cm. high. Inflorescence 10–45 cm. long, bearing appressed racemes 1–7 cm. long, these sessile and simple, or the inferior stalked and sometimes compound with the spikelets in little appressed racemelets 1–2 cm. long; pedicels strongly flattened. Spikelets 4.5–6.5 mm. long; inferior glume a broadly obtuse cuff up to $\frac{1}{10}$ length of spikelet; superior lemma obtuse, pubescent all over.

Caprivi Strip. Singalamwe, 1.i.1959, *Killick & Leistner* 3240 (K; PRE). **Botswana**. N: Xazaxa Is., 10.iii.1982, *Smith* 3788 (K; SRGH). **Zambia**. N: Mpika to Kasama, 29.xii.1967, *Simon & Williamson* 1608 (K; SRGH). B: Kabompo, 26.xii.1969, *Simon & Williamson* 2041 (K; SRGH). S: Namwala, 15.i.1962, *Mitchell* 12/66 (K). **Malawi**. N: Mbawa, 11.xii.1952, *Jackson* 1005 (K). C: Kasunga, 15.i.1959, *Robson* 1195 (K).
Northwards to Kenya and Uganda. Marshes and streamsides; 1000–1300 m.

8. ACROCERAS Stapf

Acroceras Stapf in F.T.A. **9**: 621 (1920).

Inflorescence of loose racemes along a central axis bearing paired spikelets, but sometimes approaching a panicle due to irregular secondary branching (true panicle in *A. attenuatum*). Spikelets dorsally or weakly laterally compressed. Inferior glume $\frac{1}{2}$–$\frac{3}{4}$ length of spikelet; superior glume and inferior lemma thickened and laterally compressed at apex as if nipped by pincers (*A. attenuatum* scarcely so). Superior lemma dorsally compressed, crustaceous, with a little green crest at apex; palea apex usually reflexed and slightly protruding from lemma, usually bearing 2 green spots.

A genus of 19 species, including 12 Madagascan endemics. Tropics.

1. Glumes and inferior lemma not laterally compressed at apex, the glumes separated by a short internode   -   -   -   -   -   -   -   -   -   -   -   -   - 1. *attenuatum*
- Glumes and inferior lemma compressed at apex, the glumes not separated by an internode.   -   -   -   -   -   -   -   -   -   -   -   -   -   -   - 2
2. Leaf laminae narrowly lanceolate to lanceolate   -   -   -   -   -   - 2. *zizanioides*
- Leaf laminae convolute to broadly linear   -   -   -   -   -   -   - 3. *macrum*

1. **Acroceras attenuatum** Renvoize in Kew Bull. **34**: 556 (1980). —Clayton & Renvoize in F.T.E.A., Gramineae: 567 (1982). Type from Tanzania.

Prostrate annual. Culms 20–40 cm. high, rooting at the nodes below. Leaf laminae broadly linear to narrowly lanceolate, 5–20 mm. wide, with indistinct cross-nerves. Inflorescence a panicle. Spikelets 3–3.5(4) mm. long, narrowly elliptic, turning black at maturity. Glumes separated by a short internode, they and the inferior lemma scarcely thickened at apex. Superior lemma tipped with a dark green spot, this slightly depressed or raised into a shallow crest.

Mozambique MS: Amatongas Forest, iv.1952, *Schweickerdt* 2353 (K; PRE).
Also in Uganda, Kenya and Tanzania. Coastal forest.
An atypical member of the genus, linking it to *Panicum* and *Cyrtococcum*.

2. **Acroceras zizanioides** (Kunth) Dandy in Journ. Bot. **69**: 54 (1931). —Bor, Grasses of B.C.I. & P.: 275 (1960). —Clayton in F.W.T.A. **3**: 435 (1972). —Clayton & Renvoize in F.T.E.A., Gramineae: 565 (1982). Type from Colombia.
 Panicum oryzoides Sw., Prodr. Veg. Ind. Occ.: 23 (1788) non Ard. (1763). Type from Jamaica.
 Panicum zizanioides Kunth in Humb. & Bonpl., Nov. Gen. Sp. **1**: 100 (1816). Type as for *A. zizanioides*.
 Panicum ogowense Franch. in Bull. Soc. Hist. Nat. Autun **8**: 344 (1895). Types from Congo.
 Panicum lutetense K. Schum. in Bot. Jahrb. **24**: 332 (1897). Type from Zaire.
 Acroceras oryzoides Stapf in F.T.A. **9**: 622 (1920). Type as for *P. oryzoides*.
 Echinochloa oryzoides (Stapf) Roberty in Bull. I.F.A.N., Sér. A, **17**: 67 (1955). Type as above.

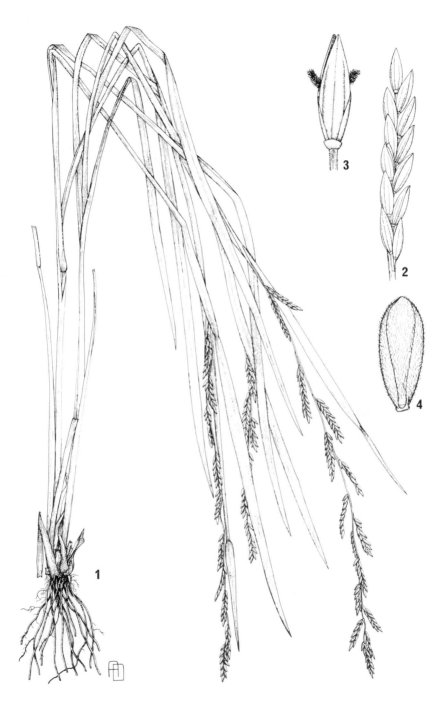

Tab. 11. ENTOLASIA IMBRICATA. 1, habit (× ½); 2, raceme (× 2), 1–2 from *Richards* 25987; 3, spikelet (× 5); 4, superior floret (× 7), 3–4 from *Shabani* s.n. From F.T.E.A.

Rambling perennial. Culms 30–100 cm. high, decumbent below. Leaf laminae narrowly lanceolate to lanceolate, 6–26 mm. wide, faintly cross-nerved, broadly rounded at base. Inflorescence 10–25 cm. long, the racemes 4–12 cm. long, mostly divergent from the central axis, sometimes with secondary branchlets. Spikelets 4.5–6.5 mm. long, lanceolate. Glumes not separated, the superior glume and inferior lemma distinctly nipped at apex.

Mozambique. MS: Nyemesembe, vii.1972, *Tinley* 2676 (K; SRGH).
Tropical Africa, mainly in the west from Guinée to Angola; also in India and tropical America. Damp places or shallow water in shade of coastal forest.

3. **Acroceras macrum** Stapf in F.T.A. **9**: 624 (1920). —Sturgeon in Rhod. Agric. Journ. **50**: 430 (1953). —Chippindall in Meredith, Grasses & Pastures of S. Afr.: 386 (1955). —Jackson & Wiehe, Annot. Check List Nyasal. Grass.: 27 (1958). —Launert in Merxm., Prodr. Fl. SW. Afr. **160**: 19 (1970). —Clayton & Renvoize in F.T.E.A., Gramineae: 565 (1982). TAB **12**. Types: Mozambique, Beira, *Swynnerton* 1596 (BM, syntype) & Zimbabwe, Harare, *Craster* 22 (K, syntype) & *Craster* 81 (K, syntype) & Angola, Cunene, *Pearson* (K, syntype).
  Neohusnotia macra (Stapf) Hsu in Journ. Fac. Sci. Univ. Tokyo, sect. 3, Bot. **9**: 94 (1965).

Rhizomatous perennial. Culms 20–100 cm. high, decumbent below. Leaf laminae linear to broadly linear, flat or convolute, 1–8 mm. wide, glabrous or pubescent, without cross-nerves, the base as wide as the sheath or slightly cordate. Inflorescence 5–25 cm. long, the racemes 2–8 cm. long mostly rather short and appressed to the central axis. Spikelets 4–4.8 mm. long, oblong. Glumes not separated, the superior glume and inferior lemma distinctly nipped at the apex.

Botswana. N: Gwetshaa Is., 22.ii.1973, *Smith* 413 (K; SRGH). **Zambia**. B: Senanga, 30.vii.1952, *Codd* (K; PRE). N: Kawimbe, 26.ii.1959, *McCallum-Webster* A161 (K). W: Lusongwa, 26.xii.1969, *Simon & Williamson* 2058 (K: SRGH). C: Kabwe, 25.iv.1953, *Hinds* 114 (K). S: Kafue Pilot Polder, 5.ii.1963, *van Rensburg* 1314 (K). **Zimbabwe**. W: Kazuma pan, 26.iii.1969, *Rushworth* 1751a (K; SRGH). C: Mlezu, 2.xii.1965, *Biegel* 651 (K; SRGH). S: Mtilikwe, 12.v.1948, *Robinson* 339 (K; SRGH). **Malawi**. N: Litutu Ridge, 14.v.1952, *Jackson* 825 (K). C: Kada, 26.ii.1953, *Jackson* 1114 (K). S: Zomba, 1950, *Jackson* (K). **Mozambique**. Z: Namacurra, 25.iii.1943, *Torre* 4974 (K).
From Ethiopia to S. Africa. Streamsides, pond margins and swampy places; 1000–1800 m.

9. ECHINOCHLOA Beauv.

Echinochloa Beauv., Ess. Agrost.: 53 (1812) nom conserv.

Inflorescence of racemes along a central axis, the spikelets paired or in short secondary racemelets, typically densely packed in 4 rows. Spikelets narrowly elliptic to subrotund, flat on one side gibbous on the other, often hispidulous, the apex usually cuspidate or awned. Glumes acute to acuminate, the inferior usually c. $\frac{1}{3}$, the superior as long as spikelet. Inferior lemma often stiffly awned. Superior lemma crustaceous, terminating in a short, membranous, laterally compressed, incurved beak; superior palea acute, the apex briefly reflexed and slightly protruberant from lemma.

A genus of 30–40 species Tropical and warm temperate regions.
The genus can usually be recognized by its 4-rowed racemes of pointed spikelets. It is sometimes difficult to separate from *Brachiaria* and then the slightly reflexed apex of the superior palea, though requiring careful observation, seems the most reliable discriminatory character. Despite their apparent similarity, *Echinochloa* and *Brachiaria* seem not to be closely related since they possess different forms of leaf anatomy.
It is a difficult genus comprising several complexes of intergrading species, these further subdivided into microtaxa distinguished as much by their phenology and ecology as by their morphology. This diversity apparently springs from a tendency to self-pollination, coupled with fluent adaptability to the wealth of ecological niches associated with ruderal and seasonally wet habitats. The problem is compounded by blurred discriminatory characters, such as ligular hairs which may be suppressed in the superior leaves, and longevity which is not easy to ascertain from culms sprawling in mud; many of the other characters depend on features of shape and size which are difficult to define or imperfectly disjunct.
Consequently there is much uncertainty as to the most appropriate taxonomic treatment. On the whole it seems better, in the present unsatisfactory state of knowledge, to treat species in a wide sense and to accept that intermediates are difficult to apportion than to designate narrower taxa corresponding to subjectively recognizable, but weakly circumscribed, facies differences. Spikelet measurements exclude awns and awn-points.

Tab. 12. ACROCERAS MACRUM. 1, habit (× ½), from *Richards* 10745; 2, raceme (× 2); 3, spikelet (× 5), 2–3 from *Davidse & Handlos* 7192. From F.T.E.A.

1. Ligule absent   -   -   -   -   -   -   -   -   -   -   -   -   -   2
– Ligule represented by a line of hairs, at least in the inferior leaves   -   -   -   8
2. Plants perennial   -   -   -   -   -   -   -   -   -   -   -   8. *haploclada*
– Plants annual   -   -   -   -   -   -   -   -   -   -   -   -   -   3
3. Racemes distinctly compound with short secondary branchlets, the inflorescence untidily ovate;
spikelets 2–3(3.5) mm. long, often with a short curved awn   -   -   -   1. *crus-pavonis*
– Racemes not or inconspicuously compound   -   -   -   -   -   -   -   4
4. Spikelets acuminate to awned, in 2–several irregular rows; longest raceme 2–10 cm. long; leaf
laminae over 4 mm. wide (if less than 3 mm. see 10. *E. stagnina*)   -   -   -   5
– Spikelets acute to cuspidate; racemes seldom over 3 cm., simple   -   -   -   -   6
5. Spikelets mostly 3–4 mm. long; superior floret 2–3 mm. long; lowest raceme with secondary
branchlets   -   -   -   -   -   -   -   -   -   -   -   2. *crus-galli*
– Spikelets 3.8–6.5 mm. long; superior floret 3.5–5 mm. long; racemes all
simple   -   -   -   -   -   -   -   -   -   -   -   3. *oryzoides*
6. Racemes neatly 4-rowed, openly spaced, commonly c. $\frac{1}{2}$ their length apart and appressed to the
axis, but sometimes subverticillate and spreading (rarely forming a lanceolate head but then
inferior floret ♂); spikelets 1.5–3 mm. long   -   -   -   -   -   -   5. *colona*
– Racemes crowded with plump spikelets, congested into a dense lanceolate head; spikelets 2.5–4
mm. long, the inferior floret barren   -   -   -   -   -   -   -   -   7
7. Spikelets tinged with purple   -   -   -   -   -   -   -   -   -   4. *utilis*
– Spikelets pallid   -   -   -   -   -   -   -   -   -   -   6. *frumentacea*
8. Plants perennial   -   -   -   -   -   -   -   -   -   -   -   -   9
– Plants annual   -   -   -   -   -   -   -   -   -   -   -   -   -   11
9. Spikelets seldom over 2.5 mm. long, crowded, commonly subglobose, often
awned   -   -   -   -   -   -   -   -   -   -   -   -   8. *haploclada*
– Spikelets seldom under 3 mm. long, ± elliptic   -   -   -   -   -   -   -   10
10. Spikelets awnless (rarely with a subulate mucro), plump, 2.5–4 mm. long; culms robust,
erect   -   -   -   -   -   -   -   -   -   -   -   -   9. *pyramidalis*
– Spikelets awned, tapering; culms spongy, floating, decumbent or
ascending   -   -   -   -   -   -   -   -   -   -   -   10. *stagnina*
11. Inferior lemma with or without a brief awn-point up to 3 mm. long; spikelets 2.5–3.5 mm.
long   -   -   -   -   -   -   -   -   -   -   -   -   7. *ugandensis*
– Inferior lemma with an awn 3–50 mm. long, rarely shorter and then spikelets 4 mm. long
or more   -   -   -   -   -   -   -   -   -   -   -   -   -   12
12. Inflorescence loose, the branches secund; spikelets narrowly ovate, 3.5–6 mm.
long   -   -   -   -   -   -   -   -   -   -   -   -   10. *stagnina*
– Inflorescence dense; spikelets narrowly elliptic, 3–4 mm. long   -   -   -   11. *jubata*

1. **Echinochloa crus-pavonis** (Kunth) Schult., Syst. Veg. Mant. **2**: 269 (1824). —Stapf in F.T.A. **9**: 612
(1920). —Chippindall in Meredith, Grasses & Pastures of S. Afr.: 361 (1955). —Jackson & Wiehe,
Annot. Check List Nyasal. Grass.: 37 (1958). —Bor, Grasses of B.C.I. & P. : 310 (1960).
—Clayton in F.W.T.A. **3**: 439 (1972). —Clayton & Renvoize in F.T.E.A., Gramineae : 556 (1982).
Type from Venezuela.
    Oplismenus crus-pavonis Kunth in Humb. & Bonpl., Nov. Gen. & Sp. **1**: 108 (1815). Type as
    above.
    Panicum crus-pavonis (Kunth) Nees, Agrost. Bras.: 259 (1829). Type as above.
    Panicum crus-pavonis var. *rostratum* Stapf in Fl. Cap. **7**: 396 (1898). Type from S. Africa.
    Echinochloa crus-galli var. *crus-pavonis* (Kunth) Hitchc. in Contrib. U.S. Nat. Herb. **22**: 149
    (1920). —Pfitscher & Barreto in An. Tec. Inst. Peq. Zootec. Francisco Osorio **3**: 272 (1976). Type
    as for *Oplismenus crus-pavonis*.

Robust annual. Culms 50–200 cm. high, stout, often decumbent and rooting in mud at
lower nodes; ligule absent. Inflorescence large, 10–30 cm. long, loose, untidily ovate;
racemes typically compound with short secondary branchlets but superior racemes of
weaker plants simple; lowest raceme 3–15 cm. long. Spikelets 2–3(3.5) mm. long, elliptic,
hispid. Inferior lemma acute to acuminate, or with a short curved awn 1–3(7) mm. long.
Superior lemma 2–2.5(3) mm. long.

Botswana. N: Okavango R., 8.v.1977, *Smith* 2004 (K; SRGH). **Zambia**. N: Kawimbe, Lumi R.,
8.ii.1957, *Richards* 8106 (K). W: Mufulira, 28.v.1934, *Eyles* 8410 (K; SRGH). **Malawi**. N: Umbaka R.,
30.v.1888, *Scott* (K). S: Mulanje, Mimosa, 4.xii.1950, *Jackson* 322 (K).
    Tropical and S. Africa; tropical America. Shallow water, swamps and streamsides; 1400–1900 m.
    The species is related to *E. crus-gallii*, being distinguished by smaller spikelets and a large
inflorescence whose racemes bear copious secondary branchlets. It is the characteristic species of
tropical Africa, where *E. crus-galli* occurs only as a rare adventive; but in S. America the two species
are sympatric and the boundary between them is indistinct.

2. **Echinochloa crus-galli** (L.) Beauv., Ess. Agrost.: 53, 161 (1812). —Stapf in F.T.A. **9**: 610 (1920).

—Chippindall in Meredith, Grasses & Pastures of S. Afr.: 361 (1955). —Bor, Grasses of B.C.I. & P.: 310 (1960). —Launert in Merxm., Prodr. Fl. SW. Afr. **160**: 71 (1970). —Clayton & Renvoize in F.T.E.A., Gramineae: 557 (1982). Type from Europe.

Panicum crus-galli L., Sp. Pl.: 56 (1753). Type as above.

Echinochloa glabrescens Kossenko in Bot. Mat. Gerb. Bot. Inst. Komarov **11**: 40 (1949). —Bor, Grasses of B.C.I. & P.: 311 (1960). Type from India.

Coarse annual. Culms 25–100 cm. high, erect or ascending; ligule absent; leaf laminae 62–20 mm. wide. Inflorescence 6–22 cm. long, linear to ovate; racemes untidily 2–several-rowed, the longest 2–10 cm. long usually with short secondary branchlets at the base. Spikelets mostly 3–4 mm. long, ovate-elliptic, hispid. Inferior lemma acuminate or with an awn up to 5 cm. long. Superior lemma 2–3 mm. long, including the short herbaceous apex.

Malawi. S: Zomba, Domas Irrigation Rice Scheme, 28.iii.1980, *Blackmore* 1152 (K).

Warm temperate and sub-tropical regions of the world; adventive in the Flora Zambesiaca area. A weed of inhabited, often damp, places.

It is a polymorphic species whose numerous intergrading races are apparently the consequence of cleistogamous self-pollination. There is much uncertainty as to which segregates are worth recognizing as species (see, for example, Gould, Fairbrothers & Ali in Amer. Midl. Nat. **87**: 36–59 (1972) for N. America; and Vickery in Flora New South Wales, Gram.: 189–211 (1975) for Australia).

3. **Echinochloa oryzoides** (Ard.) Fritsch in Verh. Zool. —Bot. Ges. Wien **41**: 742 (1891). —Yabuno in Cytologia **49**: 673–678 (1984). Type from Italy.

Panicum oryzoides Ard., Animad. Spec. Alt.: 16 (1764). Type as above.

Like *E. crus-galli* but lowest raceme simple, spikelets 3.8–6.5 mm. long, superior lemma 3.5–5 mm. long.

Zimbabwe. W: Bulawayo & Matopo Hills, *Appleton* 11 (K).

Apparently originating in the Mediterranean region and Middle East, but now widespread. A crop mimic infesting rice fields.

4. **Echinochloa utilis** Ohwi & Yabuno in Acta Phytotax. Geobot. **20**: 50 (1962). Type from Japan.

Echinochloa crus-galli var. *utilis* (Ohwi & Yabuno) Kitamura in Acta Phytotax. Geobot. **36**: 93 (1985). Type as above.

Like *E. frumentacea*, but spikelets tinged with red or purple, the inferior lemma shortly acuminate to cuspidate and firmer in texture.

Zimbabwe. N: Chinhoyi Distr., Stonybyres, 24.vi.1955, *Steele* in GHS 53688 (K; SRGH).

Japan and China; introduced to most warm temperate countries; adventive in the Flora Zambesiaca area. A minor grain crop, often employed in birdseed mixtures which are probably responsible for adventive occurrences.

The species is a domesticated derivative of *E. crus-galli*.

5. **Echinochloa colona** (L.) Link, Hort. Berol. **2**: 209 (1833). —Stapf in F.T.A. **9**: 607 (1920). — Sturgeon in Rhod. Agric. Journ. **50**: 429 (1953). —Chippindall in Meredith, Grasses & Pastures of S. Afr.: 361 (1955). —Jackson & Wiehe, Annot. Check List Nyasal. Grass.: 36 (1958). —Bor, Grasses of B.C.I. & P.: 308 (1960). —Launert in Merxm., Prodr. Fl. SW. Afr. **160**: 70 (1970). —Clayton in F.W.T.A. **3**: 439 (1972). —Clayton & Renvoize in F.T.E.A., Gramineae: 557 (1982). TAB. **13**. Type from Jamaica.

Panicum colonum L., Syst. Nat., ed. 10, **2**: 870 (1759). Type as above.

Echinochloa divaricata Anderss. in Peters, Reise Mossmb., Bot. **2**: 549 (1863). Type: Mozambique, Tete, *Peters* (B, holotype).

Panicum echinochloa Th. Dur. & Schinz, Consp. Fl. Afr. **5**: 748 (1894). Based on *Echinochloa divaricata* Anderss. non *Panicum divaricatum* L. (1759).

Panicum colonum var. *atroviolaceum* Hack. in Viert. Nat. Ges. Zürich **56**: 71 (1911). Type from Namibia.

Echinochloa crus-galli subsp. *colona* (L.) Honda in Bot. Mag. Tokyo **37**: 122 (1923). Type as for *Panicum colonum*.

Panicum colonum var. *angustatum* Peter, Fl. Deutsch Ost-Afr. **1**, Anh.: 33 (1930). Type from Tanzania.

Echinochloa subverticillata Pilger in Notizbl. Bot. Gart. Berl. **15**: 451 (1941). Type from Namibia.

Brachiaria longifolia Gilli in Ann. Nat. Mus. Wien **69**: 39 (1966). Type from Tanzania.

Annual. Culms 10–150 cm. high, erect or ascending; ligule absent. Inflorescence 1–20 cm. long, typically linear; racemes up to 3 (rarely 5) cm. long, neatly 4-rowed, simple,

D.E.

Tab. 13. ECHINOCHLOA COLONA. 1, habit (×⅔); 2 & 3, portion of raceme (× 4); 4, 5 & 6, spikelet (× 4); 7, inferior glume (× 4); 8, superior glume (× 4); 9, inferior lemma (× 4); 10, inferior palea (× 4); 11, superior lemma (× 4); 12, superior palea (× 4); 13, flower (× 6); 14 & 15, caryopsis (× 6); 16, ligule (× 2), all from *Haddad* 14752. From F.T.E.A.

commonly c. ½ their length apart and appressed to the axis but sometimes subverticillate and spreading, rarely congested. Spikelets 1.5–3(3.5) mm. long, subglobose to ovate-elliptic, pubescent to hispidulous, often purple tinged. Inferior floret ♂ or barren, its lemma acute to cuspidate (rarely mucronate). Superior lemma 2–3 mm. long.

Botswana. N: Qangwa R., 19.iv.1981, *Smith* 3602 (K; SRGH). SW: Ghanse Pan Farm, 11.iii.1970, *Brown* 8789 (K; PRE). SE: Nata R. delta, 12.iv.1976, *Ngoni* 477 (K; SRGH). **Zambia.** N: Chilongowelo, Plain of Death, 9.ii.1957, *Richards* 8131 (K). C: Feira, 4.vii.1972, *Kornaś* 1927 (K). E: Petauke Distr., Cholongozi, ii.1963, *Verboom* 906 (K). S: Monze, 10.i.1964, *Astle* 2818 (K). **Zimbabwe.** N: Hurungwe (Urungwe) Nat. Park, Chirundu, 14.ii.1981, *Philcox, Leppard & Dini* 8538 (K; SRGH). W: Hwange Distr., Kazangula Border Post, 8.iii.1981, *Gonde* 327 (K; SRGH). C: Chegutu, Poole Farm, 29.i.1944, *Hornby* 2301 (K; SRGH). E: Chipinge Distr., Muhenye, 24.i.1957, *Phipps* 146 (K; SRGH). S: Mwenezi (Nuanetsi) Distr., Malipate, 24.iv.1961, *Simon* 16 (K; SRGH). **Malawi.** N: Mzimba, Kasitu Valley, 30.i.1938, *Fenner* 241 (K). C: Salima to Citala Rd. junction, 27.iv.1951, *Jackson* 485 (K). S: Mulanje, Mimosa, 27.ix.1950, *Wiehe* N/710 (K). **Mozambique.** N: Lugenda R., N. of bridge on Rd. from Marrupa to Mecula, 11.viii.1981, *Jansen, Koning & de Wilde* 227 (K; L). Z: Mocuba, 12.iii.1943, *Torre* 4934 (K). T: Dique, 21.ix.1948, *Wild* 2576 (K; SRGH). MS: Gorongosa, 15.xi.1963, *Torre & Paiva* 9235 (K; COI). GI: Vila Alferes Chamusca to Vila Pinto Teixeira, 30.vi.1971, *Correia & Marques* 1276 (K). M: Marracuene, 10.iv.1946, *Gomes e Sousa* 3420 (K).

Tropics and subtropics. A weedy species of muddy or swampy places; sea level to 2000 m.

E. colona is always annual without a ligule and can usually be recognized by its short separate neatly 4-rowed racemes; the typical form also has little round awnless spikelets with soft indumentum. This form is predominant in Asia, but in Africa the species is markedly polymorphic. Forms with a mucro up to 2 mm. long occur from Senegal to Ethiopia, their racemes distinguishing them from *E. crus-galli*; forms in Tanzania and Zambia with a lanceolate inflorescence of subverticillate spreading racemes imitate *E. haploclada*; and, particularly in Zambia, unusually robust forms with dense purple inflorescences closely approach *E. frumentacea*.

The epithet is sometimes treated as a noun (colonum = of the farmers), but there is no reason to reject the adjectival form sanctioned by lexicographers of Linnaeus' time.

6. **Echinochloa frumentacea** Link, Hort. Reg. Bot. Berol. **1**: 204 (1827). —Jackson & Wiehe, Annot. Checklist Nyasal. Grass.: 37 (1959). —Bor, Grasses of B.C.I. & P.: 311 (1960). —Clayton & Renvoize in F.T.E.A., Gramineae: 559 (1982). Type from India.
Echinochloa colona var. *frumentacea* (Link) Ridl., Fl. Malay Penin. **5**: 223 (1925). Type as above.

Robust annual. Culms 30–150 cm. high, erect; ligule absent. Inflorescence 6–20 cm. long, lanceolate, congested; racemes 1–3 cm. long, several-rowed with crowded spikelets, simple, densely overlapping. Spikelets 2.5–3.5 mm. long, plump and often gaping to expose superior floret, ± tardily deciduous, yellowish or pallid, pubescent to hispid. Inferior floret barren, its lemma shortly acute. Superior lemma 2–3 mm. long.

Zambia. N: Mulila, Munkanya, 27.ii.1968, *Phiri* 64 (K; SRGH). **Zimbabwe.** W: Nyamandhlovu Pasture Station, iv.1953, *Plowes* 1624 (K; SRGH). C: Makwiro, Msengezi Exp. Farm, 22.i.1954, *Conradie* in GHS 45126 (K; SRGH). E: Chipinge Distr., Eastleigh, 15.v.1948, *Dixon* in GHS 20482 (K; SRGH). S: Masvingo Distr., Glyntor, 17.i.1948, *Robinson* 201 (K; SRGH). **Malawi.** S: Zomba, Makwapala, 26.iii.1934, *Lawrence* 164 (K). **Mozambique.** N: Erati, Estação Experimental de Namapa, 2.iii.1960, *Lemos & Macúaca* 4 (K). M: Umbeluzi, 26.v.1949, *Myre & Balsinhas* 721 (K).

Also from Tropical Asia. Cultivated as a minor grain crop, or occurring as an escape in inhabited places.

Yabuno (in Cytologia **27**: 296–305 (1962) & in Jap. Journ. Bot. **19**: 277–323 (1966)) discusses the cytology of *E. frumentacea* and regards it as a derivative of *E. colona*. The two species are distinct enough in Asia, but approach one another closely in central Africa, where wild forms of *E. colona* with a dense lanceolate inflorescence (but the spikelets purplish and the inferior floret male) are often gathered for food during the dry season.

7. **Echinochloa ugandensis** Snowden & Hubbard in Kew Bull. **1936**: 315 (1936). —Clayton & Renvoize in F.T.E.A., Gramineae: 561 (1982). Type from Uganda.

Annual. Culms 25–100 cm. high; ligule a line of hairs. Inflorescence 5–20 cm. long, linear; racemes up to 4 cm. long, like *E. colona*. Spikelets 2.5–3.5 mm. long, elliptic, pubescent. Inferior floret ♂ or barren, its lemma acute or with an awn point up to 3 mm. long. Superior lemma 2–2.5 mm. long.

Botswana. N: Mababe Depression, 2 km. N. of Segxebe Pan, 15.vi.1978, *Smith* 2454 (K; SRGH). **Zimbabwe.** N: Mutoko Distr., Matudza Dam, iii.1972, *Davies* 3165 (K; SRGH). C: Gweru Distr., Mlezu Government Agric. School farm, 3.xii.1965, *Biegel* 6750 (K; SRGH). E: 56 km. N. of Mutare, 11.i.1967, *Biegel* 2067 (K; SRGH). S: Masvingo Distr., L. Kyle, 11.iii.1971, *Chiparawasha* 368 (K; SRGH). **Mozambique.** N: Pemba (Porto Amélia), 19.xii.1963, *Torre & Paiva* 9609 (COI; K). M: Goba, 5.xi.1960, *Balsinhas* 203 (K).

Also from Kenya to S. Africa. In or near shallow pools and other wet places; sea level to 1500 m.
E. ugandensis closely resembles *E. colona* in its inflorescence, but has hairy ligules and a tendency
to awned spikelets. It can be confused with depauperate or short-awned forms of *E. stagnina*, but
these have laxer racemes and longer spikelets (4 mm. or more).

8. **Echinochloa haploclada** (Stapf) Stapf in F.T.A. **9**: 613 (1920). —Jackson & Wiehe, Annot. Check
   List Nyasal. Grass.: 37 (1958). —Clayton & Renvoize in F.T.E.A., Gramineae: 560 (1982). Types
   from Kenya and Tanzania.
   Panicum haplocladum Stapf in Kew Bull. **1908**: 58 (1908). Type as above.
   Panicum aristiferum Peter, Fl. Deutsch Ost-Afr. 1, Anh.: 32 (1930). Type from Tanzania.
   Echinochloa aristifera (Peter) Robyns & Tournay, Fl. Parc Nat. Alb. **3**: 93 (1955). Type as above.

Tussocky perennial. Culms 30–300 cm. high, arising from a short oblique rhizome,
often wiry; ligule absent or a line of hairs; leaf laminae occasionally marked with purple
bars. Inflorescence 7–25 cm. long, lanceolate (occasionally linear); racemes 1–5 cm. long,
densely crowded with appressed spikelets. Spikelets small, 1.5–2.5(3) mm. long,
subglobose to elliptic, ± hispid. Inferior floret ♂, its lemma acute or with a short curved
awn up to 5(15) mm. long. Superior lemma 1.5–2.3 mm. long.

Zambia. B: Mongu Distr., in flood plain, 5.xii.1964, *Verboom* 1148 (K). N: Mbesuma, Chambeshi
flood plain, 6.i.1962, *Astle* 1195 (K). W: Luanshya Distr., Baluba stream on Fisenge to Kitwe Rd.,
18.xii.1969, *Simon & Williamson* 2075 (K; SRGH). C: Serenje, 23.xii.1963, *Symoens* 10687
(K). E: Petauke Distr., Cholongozi, ii.1963, *Verboom* 904 (K). S: Sinazongwe, 18.iv.1965, *Astle* 3099
(K). **Zimbabwe**. W: Gwai, ii.1949, *Davies* 142 (K; SRGH). S: Lower Sabi, 2.ii.1948, *Rattray* 1325 (K;
SRGH). **Malawi**. C: Nkhota Khota, Kanyenda dambo, 24.ii.1953, *Jackson* 1099 (K). S: Nsanje,
between Muona and Shire R., 20.iii.1960, *Phipps* 2590 (K; SRGH). **Mozambique**. N: Erati, Namapa,
8.iii.1960, *Lemos & Macúaca* 19 (K). Z: Megaza, 26.x.1971, *Bowbrick* R35E (K; SRGH). MS: Gorongosa
Nat. Park, Xivulo, 6.vii.1972, *Ward* 7757 (K). M: Magude to Guijá, 5.v.1944, *Torre* 6575 (K).
Northwards to Zaire, Sudan and Ethiopia. Stream banks, dry river beds, alluvial flood plains and
black clays; sea level to 1500 m.
Echinochloa haploclada characteristically has a lanceolate inflorescence, whose spreading racemes
are densely crowded with little plump spikelets which are often shortly awned. The presence or
absence of a ligule seems not to be correlated with any other character. The species displays
considerable variation, and is not easily separated from *E. colona* with which it hybridizes (Yabuno in
Cytologia **38**: 131–135 (1973)). It can usually be distinguished from *E. pyramidalis* by the smaller,
more densely packed spikelets and by the less robust habit; and from *E. jubata* by its small round
spikelets.

9. **Echinochloa pyramidalis** (Lam.) Hitchc. & Chase in Contr. U.S. Nat. Herb. **18**: 345 (1917). —Stapf
   in F.T.A. **9**: 615 (1920). —Sturgeon in Rhod. Agric. Journ. **50**: 429 (1953). —Chippindall in
   Meredith, Grasses & Pastures of S. Afr.: 361 (1955). —Jackson & Wiehe, Annot. Check List.
   Nyasal. Grass.: 37 (1958). —Bor, Grasses of B.C.I. & P.: 311 (1960). —Launert in Merxm., Prodr.
   Fl. SW. Afr. **160**: 71 (1970). —Clayton in F.W.T.A. **3**: 439 (1972). —Clayton & Renvoize in
   F.T.E.A., Gramineae: 561 (1982). Type from Senegal.
   Panicum pyramidale Lam., Tab. Encycl. Méth. Bot. **1**: 171 (1791). Type as above.
   Panicum holubii Stapf in Fl. Cap. **7**: 394 (1899). Types from S. Africa.
   Panicum pyramidale var. *hebetatum* Stapf in Fl. Cap. **7**: 396 (1899). Types from S. Africa.
   Echinochloa holubii (Stapf) Stapf in F.T.A. **9**: 606 (1920). —Chippindall in Meredith, Grasses &
   Pastures of S. Afr.: 361 (1955). —Launert in Merxm., Prodr. Fl. SW. Afr. **160**: 71 (1970). Types
   from S. Africa.
   Panicum pyramidale var. *spadiceum* Peter, Fl. Deutsch Ost-Afr. 1, Anh.: 33 (1930). Type from
   Tanzania.
   Panicum crus-galli var. *molle* Peter, Fl. Deutsch Ost-Afr. **1**: 179 (1930). Type from Tanzania.

Reed-like rhizomatous perennial. Culms robust, 1–4 m. high, firm, erect; ligule a line of
hairs. Inflorescence 8–40 cm. long, ovate to narrowly lanceolate with numerous
overlapping racemes or linear with 4–10 ± distant appressed racemes; racemes 2–20 cm.
long, simple or compound, straight, ascending, coarsely spiculate. Spikelets 2.5–3.5(4)
mm. long, narrowly ovate to broadly elliptic, plump, glabrous to hispid. Inferior lemma
acute to acuminate, rarely with a subulate point up to 2 mm. long. Superior lemma 2–3
mm. long.

Caprivi Strip. 32 km. from Katima on Rd. to Linyanti, 26.xii.1958, *Killick & Leistner* 3124 (K; PRE).
Botswana. N: Nqoga-Moanachira cross-channel, 16.i.1980, *Smith* 2984 (K; SRGH). SE: Gaborone,
18.xii.1974, *Mott* 479 (K; SRGH). **Zambia**. B: Shangombo, 8.viii.1952, *Codd* 7438 (K; PRE). N:
Chambeshi Flats, 27.i.1962, *Astle* 1296 (K). W: Ndola, L. Kashiba, 14.ii.1975, *Williamson & Gassner*
2386 (K; SRGH). C: Lusaka, *Lawton* 1201 (K). E: Petauke, Kaula Dam, 3.xii.1958, *Robson* 809 (K;
SRGH). S: Victoria Falls, 4.iii.1962, *Mitchell* 13/24 (K). **Zimbabwe**. N: Hurungwe (Urungwe) Distr.,

Nyamnyetsi Dam, 5.ix.1978, *Nyariri* 325 (K; SRGH). W: Shangani Distr., Gwampa Forest Reserve, i.1955, *Goldsmith* 76/55 (K; SRGH). C: Harare, Newmarsh's Farm, 10.i.1966, *Simon* 632 (K; SRGH). **Malawi**. N: Kondowe to Karonga, *Whyte* 355 (K). C: Lilongwe Agric. Res. St., 1.ii.1951, *Jackson* 391 (K). S: Mangochi, 15.xii.1950, *Jackson* 334 (K). **Mozambique**. N: Amaramba, 21.ii.1964, *Torre &* *Paiva* 10717 (COI; K). Z: Megaza, 26.x.1971, *Bowbrick* R35L (K; SRGH). T: Zumbo, 20.iv.1972, *Macêdo* 5229 (K). MS: Gorongosa Nat. Park, i.1972, *Tinley* 2364 (K; SRGH). GI: Xai-Xai (Vila de João Belo), 17.vi.1960, *Lemos & Balsinhas* 123 (K). M: Maputo, 25.xi.1946, *Hornby* 2498 (K; SRGH).

Also from Tropical and S. Africa. Swamps and riversides, usually standing in water; sea level to 1800 m.

E. pyramidalis is a variable species with some tendency to intergrade with its neighbours, though it can usually be recognized by its robust stature and plump awnless spikelets. In particular the spikelets are larger and more coarsely disposed along the raceme than in *E. haploclada*; and the inferior lemma, when mucronate, is abruptly subulate rather than gradually tapering as in *E. stagnina.*

The form with a linear inflorescence (*E. holubii*) is sympatric and intergrading with the large waterside form. It is more frequent towards the margins of the species' geographical range and in drier habitats, suggesting that it is a response to stressful conditions rather than a distinct taxon.

Yabuno (Jap. Journ. Genet. **45**: 189–192 (1970)) reports that whereas most members of the genus are fully self-fertile, *E. pyramidalis* is partially self-incompatible.

10. **Echinochloa stagnina** (Retz.) Beauv., Ess. Agrost.: 161 (1812). —Stapf in F.T.A. **9**: 617 (1920). —Sturgeon in Rhod. Agric. Journ. **50**: 429 (1953). —Chippindall in Meredith, Grasses & Pastures of S. Afr.: 361 (1955). —Jackson & Wiehe, Annot. Check List Nyasal. Grass.: 37 (1958). —Bor, Grasses of B.C.I. & P. : 311 (1960). —Launert in Merxm., Prodr. Fl. SW. Afr. **160**: 72 (1970). —Clayton in F.W.T.A. **3**: 439 (1972). Type from India.

Panicum stagninum Retz., Obs. Bot. **5**: 17 (1789). Type as above.

Panicum scabrum Lam., Tab. Encycl. Méth. Bot. **1**: 171 (1791). Type from Senegal.

Echinochloa scabra (Lam.) Roem. & Schult., Syst. Veg. **2**: 479 (1817). —Clayton & Renvoize in F.T.E.A., Gramineae: 562 (1982). Type as above.

Panicum crus-galli var. *maximum* Franch. in Bull. Soc. Hist. Nat. Autun **8**: 347 (1895). Types from Congo (Brazzaville).

Panicum crus-galli var. *submuticum* Franch. in Bull. Soc. Hist. Nat. Autun **8**: 347 (1895). Types from Congo (Brazzaville) & Zaire.

Panicum crus-galli var. *leiostachyum* Franch. in Bull. Soc. Hist. Nat. Autun **8**: 348 (1895). Type from Congo (Brazzaville).

Panicum subaristatum Peter, Fl. Deutsch Ost-Afr. **1**, Anh.: 31 (1930). Type from Tanzania.

Annual, but apparently sometimes a rhizomatous perennial. Culms 20–200 cm. high, spongy, decumbent and rooting at nodes; ligule a line of hairs. Inflorescence 6–25 cm. long, ovate to narrowly lanceolate, typically loose with the racemes secund, flexuous and ± nodding, but displaying much variation; racemes 2–10 cm. long, simple, coarsely spiculate. Spikelets 3.5–6 mm. long, narrowly ovate, often tuberculate hispid. Inferior glume $\frac{1}{4}$–$\frac{3}{4}$ length of spikelet; superior glume awnless or with an awn up to 4 mm. long. Inferior lemma tapering to an awn (1)3–20(50) mm. long. Superior lemma 3–5 mm. long.

Botswana. N: Khwai to Selinda Rd., 27.ii.1983, *Smith* 4148 (K; SRGH). SE: 3 km. E. of Mahalapye, 3.v.1978, *Hansen* 3418 (C; K). **Zambia**. B: Barotse flood plain, 19.iii.1964, *Verboom* 1151 (K). N: Chambeshi Flats, Mbesuma, 29.vi.1963, *Symoens* 10543 (BR; K). C: Chilanga, 11.ii.1958, *Verboom* 302 (K; SRGH). E: Ngoni area, i.1962, *Verboom* 549 (K). S: Kalomo, Siantambo, 28.ii.1962, *Mitchell* 13/38 (K). **Zimbabwe**. N: Banket, 8.i.1971, *Crook* 944 (K; SRGH). W: Hwange, Kazuma Depression, 13.ii.1985, *Gonde* 429 (K; SRGH). C: 15 km. SE. of Gweru, 14.iv.1967, *Biegel* 2074 (K; SRGH). S: Masvingo, 14.i.1948, *Robinson* 193 (K; SRGH). **Malawi**. N: Mzimba, L. Kazuni, 20.v.1970, *Brummitt* 10940 (K). S: Ncheu, Kamfalmoma, 9.iii.1951, *Jackson* 420 (K).

Also from Tropical Africa, Madagascar, Assam to Indo-China. Streamsides and pond margins, growing in water; sometimes floating in deep water, or collapsed in a tangle on mud after recession of flood; 600–1600 m.

E. stagnina is a floating species with soft spongy culms and long narrow spikelets tapering towards an awn; the inferior glume (not over $\frac{1}{2}$ the spikelet in other species, except *E. jubata*) and tuberculate spikelets are helpful but inconstant supporting characters.

Its longevity in deep water and mud is uncertain, but in shallow water it is clearly annual, and in ephemeral pools it can produce depauperate forms approaching *E. ugandensis*. A few slender narrow-leaved specimens with no ligule from shallow pans on laterite in northern Zambia seem to be extreme examples of the depauperate form.

The species is replaced in India and SE. Asia by *E. picta* (Koenig) Michael, a plant with similar habit but plump rounded spikelets.

11. **Echinochloa jubata** Stapf in F.T.A. **9**: 619 (1920). —Jackson & Wiehe, Annot. Check List Nyasal. Grass: 37 (1958). —Clayton & Renvoize in F.T.E.A., Gramineae: 563 (1982). Type: Malawi, Mwaremba, *McClounie* 20 (K, holotype).

Annual. Culms 50–200 cm. high, soft, rambling or ascending; ligule a line of hairs. Inflorescence 8–20 cm. long, linear to lanceolate, typically dense with the racemes overlapping to form an oblong head, but sometimes interrupted; racemes 2–4 cm. long, simple, the spikelets closely packed. Spikelets 3–3.5(4) mm. long, narrowly elliptic. Inferior glume $\frac{1}{4}$–$\frac{3}{4}$ length of spikelet; superior glume distinctly awned (1–6 mm.). Inferior lemma tapering to a slender awn 3–25 mm. long. Superior lemma 2.5–3 mm. long.

Botswana. N: L. Ngami, Dijengs Fishing Camp, 22.ii.1977, *Smith* 1911 (K; SRGH). SW: Rietfontein, 14.ii.1928, *Blenkiron* 724 (K). SE: Palapye, Malede Sponge, 17.i.1958, *de Beer* 562 (K; SRGH). **Zambia**. B: Sesheke Distr., Moshi R. fringe, 30.x.1964, *Verboom* 1140 (K). S: Kafue Rd. Bridge, 20.i.1964, *van Rensburg* 2805 (K). **Zimbabwe**. N: L. Kariba, Kessessi Bay, 26.viii.1983, *Denny* 1279 (K; SRGH). W: Bulawayo, Hillside Dams Reserve, ii.1962, *Miller* 8181 (K; PRE). C: Gweru, Senka, Fletcher Dam, 7.ii.1967, *Biegel* 1906 (K; SRGH). S: Masvingo Distr., Great Zimbabwe, Muzero Farm, 2.ii.1971, *Chiparawasha* 330 (K; SRGH). **Malawi**. N: Nyika Plateau, Mwaremba, *McClounie* 20 (K). **Mozambique**. Z: Luabo to Chinde, 13.x.1941, *Torre* 3632 (COI; K). MS: Gorongosa Nat. Park, iii.1969, *Tinley* 1736 (K; SRGH). GI: Inhambane, Chongola, 12.ii.1954, *Bettencourt* 360 (K). M: Delagoa Bay, 1822, *Forbes* (K).

Also from Zaire to S. Africa. Lake and river margins; sea level to 1500 m.

E. jubata has the narrow tapering spikelets of *E. stagnina*, a feature which distinguishes it from *E. haploclada*. It differs from *E. stagnina* in the smaller, narrower, slenderly awned, densely packed spikelets on racemes which are usually congested into a head. However, the distinction is not sharp and *E. jubata* may be no more than a southern variant of *E. stagnina*.

10. ORYZIDIUM Hubbard & Schweick.

Oryzidium Hubbard & Schweick. in Kew Bull. **1936**: 326 (1936).

Inflorescence a panicle. Spikelets lanceolate, falling with pedicel attached. Inferior glume small, the superior as long as spikelet and stiffly awned. Inferior lemma acuminate, separated from superior floret by an internode c. 1.5 mm. long. Superior lemma softly coriaceous.

A monotypic genus. Southern Africa.

Oryzidium barnardii Hubbard & Schweick. in Kew Bull. **1936**: 328 (1936). —Chippindall in Meredith, Grasses & Pastures of S. Afr.: 425 (1955). —Launert in Merxm., Prodr. Fl. SW. Afr. **160**: 135 (1970). TAB. **14**. Type from Namibia.

Perennial. Culms 80–150 cm. long, spongy, rooting at the nodes. Panicle 10–25 cm. long, narrowly ovate. Spikelets 8–10 mm. long. Inferior glume usually 1–2 mm. long and obtuse, rarely longer and acute; superior glume scabrid on lateral nerves, tipped by an awn 6–18 mm. long. Superior floret 4–5 mm. long, acuminate.

Botswana. N: Moremi Wildlife Reserve, Mboma Is. Rd., 20.iii.1977, *Smith* 1944 (K; SRGH). **Zambia**. S: Kafue Nat. Park, 3 km. S. of Shimulimbo Hill, 21.iv.1961, *Mitchell* 7/46 (K; SRGH). **Zimbabwe**. W: Hwange Nat. Park, 11.iv.1972, *Chiparawasha* 436 (K; SRGH). Also from Namibia. Floating in water; 1000 m.

11. ALLOTEROPSIS Presl

Alloteropsis Presl, Rel. Haenk. **1**: 343 (1830). —Butzin in Willdenowia 5: 123–143 (1968).

Inflorescence of irregular racemes, these digitate or in whorls on a short common axis, each with a slender weakly unilateral rhachis which is sometimes bare at the base, bearing the spikelets in pairs or clusters. Spikelets narrowly ovate to elliptic. Glumes acute to briefly awned, the superior as long as spikelet and ciliate on margins, the inferior c. $\frac{1}{2}$ as long. Superior lemma crisply chartaceous with involute margins, shortly awned.

A genus of 5 species. Old World tropics.

A genus of nondescript appearance, prone to misidentification because the untidy racemes of awned spikelets are not obviously recognisable as panicoid.

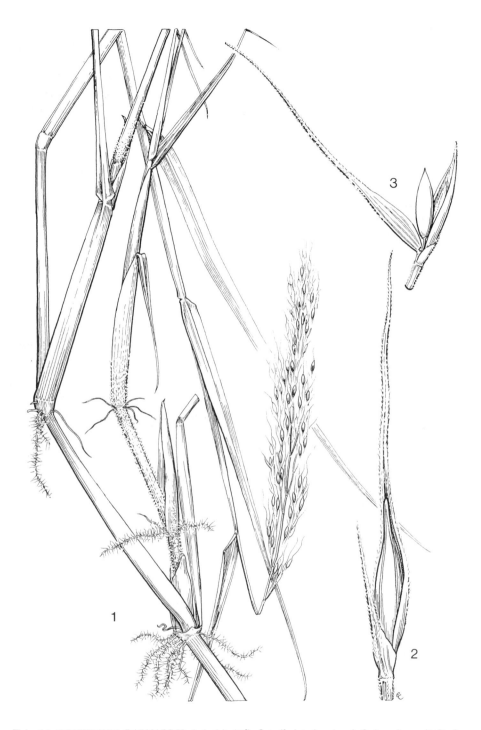

Tab. 14. ORYZIDIUM BARNARDII. 1, habit ($\times \frac{2}{3}$); 2, spikelet showing inferior glume ($\times$ 6); 3, spikelet dissected to show superior floret ($\times$ 4), all from *Ellis* 2917.

1. Plants perennial; leaf laminae broadly linear to convolute - - - - - - 2
- Plants annual; leaf laminae narrowly lanceolate - - - - - - - - 4
2. Culms slender, straggling, not basally thickened, the lowermost sheaths at most
   pubescent - - - - - - - - - - - - - - 1. *angusta*
- Culms erect or ascending, ± enlarged at the base, the lowermost sheaths silky hairy to
   tomentose - - - - - - - - - - - - - - - 3
3. Palea of superior floret glabrous or sparsely pubescent - - - - 2. *semialata*
- Palea of superior floret papillose with globular warts - - - - - 3. *papillosa*
4. Racemes bare at the base for at least 1.5 cm.; culm nodes hairy - - - 4. *cimicina*
- Racemes bearing spikelets almost or quite to the base; culm nodes
   glabrous - - - - - - - - - - - - - 5. *paniculata*

1. **Alloteropsis angusta** Stapf in F.T.A. **9**: 485 (1919). —Clayton in F.W.T.A. **3**: 448 (1972). —Clayton &
   Renvoize in F.T.E.A., Gramineae: 615 (1982). Type from Angola.

Straggling perennial, the basal sheaths at most pubescent and not bulbously thickened.
Culms 50–100 cm. high, slender, wiry. Leaf laminae 1–2 mm. wide, mostly convolute.
Inflorescence of 2–3 racemes, digitate; racemes 4–13 cm. long, slender, the spikelets
somewhat distant, often bare at the base. Spikelets 2.5–4 mm. long. Superior glume
membranous, appressed ciliate on the margins, acute. Superior lemma with an awn 0.5–1
mm. long, its palea sparsely pubescent with turgid hairs.

Zambia. W: Mwinilunga, Dobeka-Kamwezhi confluence, 17.xii.1937, *Milne-Redhead* 3711 (K;
SRGH). C: Chakwenga Headwaters, 100–129 km. E. of Lusaka, 10.i.1964, *Robinson* 6202 (K).
Sparsely distributed from Nigeria to Angola. Swamps; 1300–1500 m.
Perhaps no more than an impoverished form of *A. semialata*.

2. **Alloteropsis semialata** (R. Br.) Hitchc. in Contr. U.S. Nat. Herb. **12**: 210 (1909). —Stapf in F.T.A. **9**:
   483 (1919). —Sturgeon in Rhod. Agric. Journ. **50**: 285 (1953). —Chippindall in Meredith,
   Grasses & Pastures of S. Afr.: 423 (1955). —Jackson & Wiehe, Annot. Check List Nyasal. Grass.:
   28 (1958). —Bor, Grasses of B.C.I. & P.: 276 (1960). —Clayton in F.W.T.A. **3**: 448 (1972).
   —Clayton & Renvoize in F.T.E.A., Gramineae: 616 (1972). Type from Australia.
   Panicum semialatum R. Br., Prodr. Fl. Nov. Holl.: 192 (1810). Type as above.
   Urochloa semialata (R. Br.) Kunth, Rév. Gram. **1**: 31 (1829). Type as above.
   Oplismenus semialatus (R. Br.) Desv., Opusc.: 81 (1831). Type as above.
   Coridochloa semialata (R. Br.) Nees in Edinb. New Phil. Journ. **15**: 381 (1833). Type as above.
   Axonopus semialatus (R. Br.) Hook.f., Fl. Brit. India **7**: 64 (1896). Type as above.
   Pterochlaena catangensis Chiov. in Ann. Bot. Roma **13**: 47 (1914). Type from Zaire.
   Paspalum semialatum (R. Br.) Eyles in Trans. Roy. Soc. S. Afr. **5**: 299 (1916). Type as for
   Panicum semialatum.
   Alloteropsis homblei Robyns in Bull. Jard. Bot. Brux. **9**: 172 (1932). Type from Zaire.
   Alloteropsis gwebiensis Stent & Rattray in Proc. Trans. Rhod. Sci. Ass. **32**: 21 (1933). —Sturgeon
   in Rhod. Agric. Journ. **50**: 285 (1953). Type: Zimbabwe, Gwebi, *Rattray* 428 (K, isotype).
   [*Alloteropsis semialata* var. *ecklonii* sensu Jackson & Wiehe, Annot. Check List Nyasal. Grass.:
   28 (1958) non Stapf].

Tufted perennial, the basal sheaths silky pubescent to tomentose and often bulbously
thickened. Culms 20–150 cm. high. Leaf laminae 1–12 mm. wide, broadly linear to
convolute. Inflorescence of 2–4(8) racemes, digitate; racemes 2–25 cm. long, usually with
spikelets all along but sometimes bare towards the base. Spikelets 4–7.5 mm. long,
sometimes with transverse purple bars. Superior glume membranous, appressed ciliate
on the margins and occasionally expanded laterally into membranous wings, acuminate.
Superior lemma with an awn 1.5–3 mm. long, its palea glabrous or sparsely pubescent
with turgid hairs.

A. *semialata* is a polymorphic species unique in possessing both kranz and non-kranz leaf
anatomy, the latter generally of an unusual intermediate type (Ellis in S. Afr. Journ. Sci. **70**: 169–173
(1974) & in Bothalia **11**: 273–275 (1974)). It is therefore of interest for studies on the origin of
anatomical variants and the associated differences in photosynthetic pathway.
   The bulk of the species has kranz anatomy. Non-kranz specimens occur in southern Africa, and
Gibbs-Russell (Bothalia **14**: 205–213 (1983)) has shown that they can be distinguished
morphologically with reasonable certainty, though further study is needed to determine the full
range of morphological and anatomical variation.
   Frean & Marks (Bot. Journ. Linn. Soc. **97**: 255–259, (1988)) report chromosome numbers of 2n = 18
for A. *semialata* subsp. *eckloniana* and 2n = 54 for subsp. *semialata*.
   An unusual feature is the occasional presence of a minute rhachilla extension above the superior
floret.

Leaf laminae narrow, rolled upwards, tapering gradually to the apex; nerves of basal sheaths usually
   enlarged (but see note)   -   -   -   -   -   -   -   -   -   -   subsp. *semialata*
Leaf laminae broad, flat, narrowing abruptly at apex; nerves of basal
   sheaths slender   -   -   -   -   -   -   -   -   -   -   subsp. *eckloniana*

Subsp. **semialata**

Base bulbous, the outer sheaths with broad flat ribs (0.5–1.1 mm. across). Leaf laminae
1–5 mm. wide, rolled upwards, sparsely hairy, tapering gradually to apex. Racemes often
longer than 8 cm. and loosely packed with light coloured spikelets. Leaf anatomy kranz.

Zambia. B: Mongu Distr., flood plain, 5.xii.1964, *Verboom* 1147 (K). N: Mbesuma, Chambeshi R.,
29.i.1958, *Vesey-FitzGerald* 1434 (K; SRGH). W: Mwinilunga, 7 km. towards Kalene Hill, 20.xi.1972,
Strid 2534 (K). E: Chipata, Senegallia Farm, i.1954, *Grout* 121 (K; SRGH). S: Choma, 13.xii.1962,
Astle 1800 (K). **Zimbabwe**. N: Guruve Distr., Nyamanyetsi Estate, 13.xii.1978, *Nyariri* 568 (K;
SRGH). W: Gwampa Forest Reserve, 17.xii.1954, *Goldsmith* 77/55 (K; SRGH). C: Chegutu Distr.,
Msengezi Expt. Farm, 15.xi.1955, *Conradie* 7 (K; SRGH). E: Mutare, Kelly's Park, 26.iv.1949, *Chase*
1146 (K; SRGH). S: Masvingo, Makaholi Expt. St., 7.i.1948, *Newton & Juliasi* 5 (K; SRGH).
Malawi. N: Mzuzu, 16.xii.1973, *Pawek* 7609 (UC; K; MO; SRGH; MA). C: Dedza, 2.i.1950, *Wiehe* 396
(K). S: Kirk Range, Dovikos, 14.xi.1950, *Jackson* 266 (K). **Mozambique**. N: Mogincual, 27.xi.1963,
Torre & Paiva 9340 (COI; K). Z: Pebane to Mocubela, 25.xi.1942, *Torre* 4686 (COI; K). MS: 10 km. S.
of Senga, 16.xii.1971, *Müller & Pope* 2053 (K; SRGH).
Old World tropics. Dambos and open places in savanna woodland; 20–1600 m.
The outer sheaths of subsp. *semialata* have at least some of the nerves broadened, but they are
sometimes stripped off during collection to reveal the inner sheaths which always have slender
nerves.
A few upland specimens from Tanzania, Zambia and Malawi combine narrow leaves with
slenderly nerved outer sheaths. Preliminary anatomical investigations indicate that they should be
included in subsp. *semialata*, but their status needs clarification.

Subsp. **eckloniana** (Nees) Gibbs-Russell in Bothalia **14**: 211 (1983). Types from S. Africa.
   Bluffia eckloniana Nees in Lehm., Ind. Sem. Hort. Hamb. **1834**: 8 (1834); in Fl. Afr. Austr.: 61
   (1841). Types as above
   Panicum semialatum var. *ecklonianum* (Nees) Th. Dur. & Schinz, Consp. Fl. Afr. **5**: 764 (1894).
   Types as above.
   Axonopus semialatus var. *ecklonii* Stapf in Fl. Cap. **7**: 418 (1899) nom. superfl. Types as above.
   Axonopus ecklonianum (Nees) Chiov. in Ann. Bot. Roma **13**: 47 (1914). Types as above.
   Alloteropsis eckloniana (Nees) Hitchc. in Proc. Biol. Soc. Wash. **29**: 128 (1916). Types as above.
   Paspalum semialatum var. *ecklonii* (Stapf) Eyles in Trans. Roy. Soc. S. Afr. **5**: 299 (1916) nom.
   superfl. Types as above.
   Alloteropsis semialata var. *ecklonii* (Stapf) Stapf in F.T.A. **9**: 485 (1919) nom. superfl. —
   Chippindall in Meredith, Grasses & Pastures of S. Afr.: 423 (1955). Types as above.
   Axonopus semialatus var. *ecklonianus* (Nees) Peter, Fl. Deutsch Ost-Afr. **1**: 165 (1929). Types as
   above.
   Alloteropsis semialata var. *eckloniana* (Nees) C.E. Hubbard in Bor, Grasses of B.C.I. & P.: 277
   (1960). Types as above.

Base somewhat flabellate, the sheaths with slender nerves (up to 0.3 mm. across). Leaf
laminae 3–12 mm. wide, flat, usually densely hairy, narrowing abruptly at the apex.
Racemes usually less than 8 cm. long and tightly packed with dark purplish spikelets. Leaf
anatomy non-kranz.

Zambia. W: Mwinilunga, slope E. of R. Lunga, 23.xi.1937, *Milne-Redhead* 3371 (K; SRGH).
Zimbabwe. E: Nyanga (Inyanga), Erin Forest Reserve, 18.xi.1975, *Crook* 2097 (K; SRGH).
Mozambique. MS: Serra de Choa, Catandica (Vila Gouveia), 7.ix.1943, *Torre* 5845 (COI; K).
Also from Tanzania to S. Africa. Upland grassland; 1300–2000 m.

3. **Alloteropsis papillosa** Clayton in Kew Bull. **33**: 21 (1978). —Clayton & Renvoize in F.T.E.A.,
   Gramineae: 617 (1982). Type from Kenya.

Tufted perennial, the basal sheaths silky pubescent and ± bulbously thickened. Culms
40–70 cm. high. Leaf laminae 2–8(10) mm. wide, linear. Inflorescence of 4–10 racemes,
usually in 1 whorl; racemes 3–12(20) cm. long, often bare at the base. Spikelets 2.5–5 mm.
long. Superior glume thinly cartilaginous, ciliate on the margins, acute. Superior lemma
with an awn 0.5–4 mm. long, its palea papillose with globular-tipped hairs.

Mozambique. MS: Cheringoma section, 4 km. W. of Safrique Hunting Camp, 12.vii.1972, *Ward*
7852 (K). M: Marracuene to Manhiça, 9.xii.1940, *Torre* 2254 (COI; K).

Extends from Kenya to Natal. Savanna woodland; sea level to 100 m.

A. papillosa combines the vegetative habit of *A. semialata* subsp. *semialata* with an inflorescence reminiscent of *A. cimicina*.

4. **Alloteropsis cimicina** (L.) Stapf in F.T.A. **9**: 487 (1919). —Sturgeon in Rhod. Agric. Journ. **50**: 285 (1953). —Jackson & Wiehe, Annot. Check List Nyasal. Grass.: 28 (1958). —Bor, Grasses of B.C.I. & P.: 276 (1960). —Clayton in F.W.T.A. **3**: 449 (1972). —Clayton & Renvoize in F.T.E.A., Gramineae: 617 (1982). TAB. **15**. Type from India.

    Milium cimicinum L., Mant. Pl. Alt.: 184 (1771). Type as above.
    Panicum cimicinum (L.) Retz., Obs. Bot. **3**: 9 (1783). Type as above.
    Axonopus cimicinus (L.) Beauv., Ess. Agrost.: 154, 167 (1812). Type as above.
    Urochloa cimicina (L.) Kunth, Rév. Gram. **1**: 31 (1829). Type as above.
    Coridochloa cimicina (L.) Nees in Edinb. New Phil. Journ. **15**: 381 (1833). Type as above.
    Urochloa quintasii Mez in Bot. Jahrb. **57**: 195 (1921). Types: Mozambique, Maputo, *Quintas* 86 (K, isosyntype) and 2 syntypes from Tanzania.
    Axonopus latifolius Peter, Fl. Deutsch Ost-Afr. **1**, Anh.: 21 (1930). Types from Tanzania.
    Alloteropsis latifolia (Peter) Pilger in Not. Bot. Gart. Berlin **12**: 382 (1935). Type as above.
    Alloteropsis quintasii (Mez) Pilger in Not. Bot. Gart. Berlin **13**: 263 (1936). Type as for *Urochloa quintasii*.

Tufted annual. Culms 30–150 cm. high, erect or ascending; nodes hairy. Leaf laminae 5–20 mm. wide, narrowly lanceolate to narrowly ovate, ± amplexicaul, tuberculate-ciliate on the margins. Inflorescence of 4–11 racemes, usually in 1 whorl; racemes 7–25 cm. long, the lowest $\frac{1}{5}$ to $\frac{1}{2}$ of their length bare. Spikelets 3.5–5.5 mm. long. Superior glume cartilaginous, smooth and shining, subulate-acuminate. Superior lemma with an awn 2–4.5 mm. long, its palea papillose with globular-tipped hairs.

Caprivi Strip. 5 km. E. of Katima Mulilo, 13.ii.1969, *de Winter* 9141 (K; PRE). **Botswana**. N: 26.i.1978, *Smith* 2252 (K; SRGH). **Zambia**. C: Lusaka, 3.iv.1946, *Grassl* 46–51 (K). E: Chipata Distr., i.1963, *Verboom* 596 (K). S: Namwala, 16.i.1964, *van Rensburg* 2776 (K). **Zimbabwe**. N: Gokwe Distr., Sengwa Rd., 9.ii.1984, *Mahlangu* 861 (K; SRGH). W: Hwange Nat. Park, Livingi Pan, 25.i.1981, *Crook* 2365 (K; SRGH). C: Chegutu, Poole Farm, 11.ii.1945, *Hornby* 2370 (K; SRGH). S: Buhera, Mrore R. bridge, 28.ii.1967, *Cleghorn* 1476 (K; SRGH). **Malawi**. N: Rumphi Distr., 21 km. NE. of Rukuru R. bridge, 12.iv.1976, *Pawek* 10965a (K; MO; MA; UC). S: Machinga Distr., Sitola, 2.ii.1978, *Seyani* 792 (K; SRGH). **Mozambique**. N: Ribáuè, 37 km. towards Lalaua, 22.i.1964, *Torre & Paiva* 10125 (COI; K). M: Maputo, *Quintas* 86 (K).

Old World tropics. Open savanna woodland, often bordering damp tracts; sea level to 1300 m.

5. **Alloteropsis paniculata** (Benth.) Stapf in F.T.A. **9**: 486 (1919). —Clayton in F.W.T.A. **3**: 449 (1972). —Clayton & Renvoize in F.T.E.A., Gramineae: 619 (1982). Type from Nigeria.

    Urochloa paniculata Benth. in Hook., Niger Fl.: 558 (1849). Type as above.
    Panicum aubertii Mez in Bot. Jahrb. **34**: 134 (1904). Type from Madagascar.
    Axonopus paniculatus (Benth.) A. Chev., Sudania **1**: 22 (1911). Type as for *Urochloa paniculata*.
    Echinochloa paniculata (Benth.) Roberty in Bull. I.F.A.N. sér. A, **17**: 67 (1955). Type as above.
    Mezochloa aubertii (Mez) Butzin in Willdenowia **4**: 211 (1966) & op. cit. **5**: 124 (1968). Type as for *Panicum aubertii*.

Annual. Culms 60–150 cm. high, often decumbent; nodes glabrous. Leaf laminae 8–20 mm. wide, narrowly lanceolate, ± amplexicaul, glabrous on the margins. Inflorescence of 8–25 racemes, usually in 2 or more whorls; racemes 9–20 cm. long, bearing spikelets almost or quite to the base. Spikelets 3.5–4.5 mm. long. Superior glume membranous, acute to acuminate. Superior lemma with an awn 2.5–7 mm. long, its palea smooth.

Mozambique. Z: Morrumbala, on the banks of the Shire R., 15.i.1863, *Kirk* (K).
Also from Mali to Angola; also Mauritius and Madagascar. Damp soils; sea level.

12. BRACHIARIA (Trin.) Griseb.

Brachiaria (Trin.) Griseb. in Ledeb., Fl. Ross. **4**: 469 (1853).
Panicum subtaxon *Brachiaria* Trin. in Mém. Acad. Sci. Pétersb., sér. 6, **3**: 194 (1834).
Pseudobrachiaria Launert in Mitt. Bot. Staatss. München **8**: 158 (1970).

Inflorescence of racemes along a central axis (this branched in *B. malacodes*), the spikelets single and adaxial, paired, or rarely in fascicles. Spikelets ovate to oblong, plump, obtuse to acute, sometimes the lowest internode elongated and then often accrescent to the sheathing base of the inferior glume to form a short stipe. Inferior

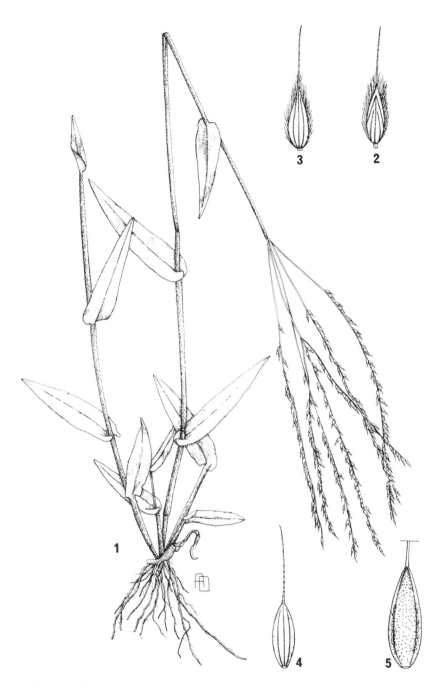

Tab. 15. ALLOTEROPSIS CIMICINA. 1, habit (× ½); 2, spikelet showing inferior glume (× 4); 3, spikelet showing superior glume (× 4); 4, superior lemma (× 5); 5, superior floret showing palea (× 8), all from *Polhill & Paulo* 1182. From F.T.E.A.

glume mostly shorter than spikelet. Inferior lemma awnless, rarely (*B. nigropedata*, *B. serrata*) with an awn-point up to 1 mm. long. Superior lemma filling the spikelet, coriaceous to crustaceous, obtuse to acute, rarely mucronate.

A genus of c. 100 species. Tropics, mainly Old World.

Species with single spikelets are easily recognized by the adaxial orientation of the inferior glume. This character fails when the spikelets are paired, and the distinction then rests upon less clearcut differences in spikelet shape.

The racemose inflorescence is a useful supporting character, but not easily perceived when the spikelets are distant (*B. deflexa*), condensed into oblong masses (*B. glomerata*) or borne on secondary branches (*B. malacodes*).

1. Inferior glume over $\frac{2}{3}$ length of spikelet; superior glume with cross-nerves - - - 2
 – Inferior glume up to $\frac{1}{2}$ (rarely $\frac{2}{3}$) length of spikelet - - - - - - - 8
2. Plants annual (if spikelets have a swollen basal callus see
    Eriochloa rovumensis) - - - - - - - - - - - 2. *brevispicata*
 – Plants perennial - - - - - - - - - - - - - 3
3. Rhachis of racemes flat, ribbon-like - - - - - - - - - 4
 – Rhachis of racemes triquetrous, sometimes narrowly winged - - - - 5
4. Margins of rhachis glabrous - - - - - - - - - 1. *platynota*
 – Margins of rhachis ciliate with long yellow hairs - - - - - - 3. *jubata*
5. Spikelets 3–4.5 mm. long; inferior glume not separated from the superior - - 6
 – Spikelets 4–6 mm. long; glumes separated by a short internode - - - - 7
6. Leaf laminae flat or convolute - - - - - - - - - 4. *bovonei*
 – Leaf laminae filiform, subterete - - - - - - - - 5. *subulifolia*
7. Plants stoloniferous; inflorescence of 2–3 racemes - - - - - 6. *humidicola*
 – Plants densely tufted; inflorescence of 3–8 racemes - - - - - 7. *dictyoneura*
8. Rhachis of racemes flat, ± ribbon-like, narrowly to broadly winged;
    spikelets single - - - - - - - - - - - - - 9
 – Rhachis of racemes solid, ± triangular or crescentic in section, sometimes with very
    narrow wings - - - - - - - - - - - - - 12
9. Glumes not separated; perennials - - - - - - - - 10
 – Glumes separated by an internode; annuals - - - - - - - 11
10. Rhachis margins scabrid - - - - - - - - - 8. *arrecta*
 – Rhachis margins tuberculate-ciliate - - - - - - - 9. *rugulosa*
11. Rhachis margins minutely ciliolate - - - - - - 10. *oligobrachiata*
 – Rhachis margins tuberculate-ciliate - - - - - - - 11. *eminii*
12. Spikelets borne singly on pedicels as long as themselves, 1.5–2 mm.
    long - - - - - - - - - - - - - - 22. *umbellata*
 – Spikelets subsessile or paired - - - - - - - - - 13
13. Plants perennial; spikelets borne singly - - - - - - - 14
 – Plants annual - - - - - - - - - - - - 19
14. Spikelets bearing appressed clavellate hairs; inferior lemma soft - - 14. *clavipila*
 – Spikelet hairs, if any, not clavellate - - - - - - - - 15
15. Racemes 1-rowed; spikelets glabrous to sparsely pubescent, 3.5–6 mm long - - 16
 – Racemes 2-rowed, or the spikelets villous - - - - - - - 17
16. Leaf laminae flat, 3–20 mm. wide - - - - - - - 12. *brizantha*
 – Leaf laminae convolute, 1–2 mm. wide - - - - - - - 13. *dura*
17. Spikelets without a stipe - - - - - - - - 15. *serrata*
 – Spikelets supported on a stipe c. 0.5 mm. long - - - - - - 18
18. Stipe cylindrical, black; superior glume acuminate, with or without an indistinct transverse fringe
    of hairs - - - - - - - - - - - - 16. *nigropedata*
 – Stipe cuneate, light brown; superior glume obtuse, with a prominent
    transverse fringe - - - - - - - - - - - 17. *pungipes*
19. Superior lemma smooth, shining, obtuse, readily deciduous - - - - 20
 – Superior lemma granulose to rugose, not deciduous - - - - - 21
20. Inflorescence racemose, the racemes occasionally with a sessile secondary racemelet at
    the base - - - - - - - - - - - - 18. *eruciformis*
 – Inflorescence paniculate, the primary branches bearing short pedunculate
    racemes - - - - - - - - - - - - 19. *malacodes*
21. Racemes globose to oblong, the spikelets tightly packed in an irregular
    mass - - - - - - - - - - - - - 20. *glomerata*
 – Racemes linear, the spikelets in 1–2 regular rows or distant from one
    another - - - - - - - - - - - - - - 22
22. Glumes separated by a distinct internode; superior lemma coarsely
    rugose - - - - - - - - - - - - 25. *leersioides*
 – Glumes adjacent (or if slightly separated, the superior lemma not coarsely
    rugose) - - - - - - - - - - - - - - 23
23. Spikelets 1.5–2.2 mm. long, glabrous or uniformly pubescent, without a stipe - - 24

-   Spikelets 2.5–5 mm. long, or if smaller then with little tufts of hair above - - - 25
24. Superior lemma granulose to striate, acute - - - - - - - 23. *scalaris*
-   Superior lemma rugose, mucronate - - - - - - - 24. *reptans*
25. Culms wiry, scandent or creeping; leaf laminae narrowly
    lanceolate - - - - - - - - - - 26. *chusqueoides*
-   Culms herbaceous, erect or ascending - - - - - - - - 26
26. Spikelets mostly single, neatly 1–2-rowed, hispidulous - - - - - 27
-   Spikelets mostly paired - - - - - - - - - - 28
27. Spikelets 2.7–4 mm. long, hispidulous, acute to cuspidate - - - 29. *xantholeuca*
-   Spikelets 4–5 mm. long, pilose above with hairs forming loose tufts on either side of mid-nerve,
    caudate-acuminate - - - - - - - - - 30. *leucacrantha*
28. Spikelets 2–2.5 mm. long, with little tufts of hair above, in compact racemes; superior lemma
    striate - - - - - - - - - - - 21. *marlothii*
-   Spikelets glabrous or uniformly pubescent, ± loosely or irregularly arranged; superior lemma
    rugose - - - - - - - - - - - 29
29. Superior lemma coarsely rugose; spikelets 3–5 mm. long, obovate-elliptic - - 20
-   Superior lemma rugose; spikelets 2.5–3.5 mm. long, elliptic to broadly
    elliptic - - - - - - - - - - - 31
30. Leaf margins scaberulous - - - - - - - 27. *grossa*
-   Leaf margins serrately crinkled - - - - - - 28. *serrifolia*
31. Pedicels shorter than spikelet; spikelets contiguous and appressed to rhachis, this simple or
    occasionally the longest with branchlets at base - - - - 31. *ramosa*
-   Pedicels, or some of them, longer than spikelet; spikelets distant, spreading, the racemes often
    compound - - - - - - - - 32. *deflexa*

1. **Brachiaria platynota** (K. Schum.) Robyns in Bull. Jard. Bot. Brux. **9**: 174 (1932). —Clayton &
    Renvoize in F.T.E.A., Gramineae: 579 (1982). Type from Tanzania.
    Panicum platynotum K. Schum. in Pflanzenw. Ost-Afr. **C**: 101 (1895). Type as above.
    Panicum geometra Chiov. in Nuov. Giorn. Bot. Ital. n.s. **26**: 64 (1919). Type from Zaire.
    Urochloa platynota (K. Schum.) Pilger in Pflanzenfam. ed. 2, **14e**: 35 (1940). Type as for
    Panicum platynotum.

Tufted perennial, the basal sheaths silky hairy. Culms 45–100 cm. high. Leaf laminae
linear. Inflorescence of (1)2–3(5) racemes, these 3–13 cm. long, bearing spikelets singly
on a ribbon-like (up to 4 mm. wide) rhachis with scaberulous margins. Spikelets 3.5–4.5
mm. long, glabrous or pubescent, cuspidate. Inferior glume as long as spikelet, 7–9-
nerved. Superior glume and inferior lemma reticulately nerved. Superior lemma
papillose, with a mucro 0.2–0.5 mm. long.

Zambia. W: Ndola, 26.xi.1947, *Brenan* 8373 (FHO; K).
Northwards to Kenya and Uganda. Savanna woodland; 1400 m.
The spikelet resembles *Urochloa* but with adaxial orientation, thus providing a link between the
two genera.

2. **Brachiaria brevispicata** (Rendle) Stapf in F.T.A. **9**: 521 (1919). Type from Angola.
    Panicum brevispicatum Rendle, Cat. Afr. Pl. Welw. **2**: 168 (1899). Type as above.

Annual. Culms 15–45 cm. high. Leaf laminae linear. Inflorescence of 1–4 racemes,
these 1–3 cm. long, bearing spikelets singly on a narrow (up to 0.5 mm.) rhachis, the
margins ciliate though sometimes sparsely so. Spikelets 3–4 mm. long, obtuse. Inferior
glume almost as long as spikelet, 9–15-nerved. Superior glume and inferior lemma
reticulately nerved, usually pubescent. Superior lemma smooth or rugulose.

Zambia. W: Mwinilunga Distr., Matonchi Farm, 24.i.1938, *Milne-Redhead* 4312 (K; SRGH).
Also from Zaire to Angola. Moist soil overlying ironstone; 1500 m.
Intergrades with the West African *B. stigmatisata* (Mez) Stapf, which has glabrous spikelets with an
inferior glume c. $\frac{2}{3}$ their length.

3. **Brachiaria jubata** (Fig. & De Not.) Stapf in F.T.A. **9**: 563 (1919). —Jackson & Wiehe, Annot. Check
    List Nyasal. Grass.: 31 (1958). —Clayton in F.W.T.A. **3**: 443 (1972). —Clayton & Renvoize in
    F.T.E.A., Gramineae: 580 (1982). Type from Sudan.
    Panicum jubatum Fig. & De Not. in Mem. Acad. Sci. Torino, ser. 2, **14**: 331 (1854). Type as
    above.

Tufted perennial. Culms 25–120 cm. high. Leaf laminae 3–17 mm. wide, broadly linear.
Inflorescence of (3)5–10(15) racemes, these 1–6 cm. long, bearing spikelets singly on a
ribbon-like (1–2 mm. wide) rhachis, its margins ciliate with bright yellow tubercle-based
hairs up to 6 mm. long. Spikelets 2.5–4 mm. long, pubescent, obtuse. Inferior glume $\frac{2}{3}-\frac{3}{4}$ as

long as spikelet, 7–11-nerved. Superior lemma finely striate or rugulose.

Zambia. N: Luangwa Valley, Mupamadzi R., 17.i.1966, *Astle* 4426 (K). C: Chilanga, Longridge Dam, 1.ii.1960, *Angus* 2134 (K; SRGH). S: Mapanza, 3.i.1954, *Robinson* 421 (K). **Zimbabwe**. N: Gokwe Distr., Sengwa Research St., 15.i.1976, *Guy* 2373 (K; SRGH). W: Shangani Distr., Gwampa Forest Reserve, i.1956, *Goldsmith* 5/56 (K; SRGH). **Malawi**. C: Nkhota Kota, Kanyenda Dambo, 24.ii.1953, *Jackson* 1098 (K).

Tropical Africa. Watersides and damp depressions; 700–1400 m.

B. jubata is widespread in east and west tropical Africa, being largely replaced by *B. bovonei* in the Flora Zambesiaca area.

4. **Brachiaria bovonei** (Chiov.) Robyns in Bull. Jard. Bot. Brux. **9**: 174 (1932). —Clayton & Renvoize in F.T.E.A., Gramineae: 582 (1982). Type from Zaire.
    Panicum bovonei Chiov. in Ann. Bot. Roma **13**: 42 (1914). Type as above.
    Brachiaria hians Stapf in F.T.A. **9**: 514 (1919). Type from Zaire.
    Brachiaria viridula Stapf in F.T.A. **9**: 515 (1919). —Sturgeon in Rhod. Agric. Journ. **50**: 422 (1953). —Jackson & Wiehe, Annot. Check List Nyasal. Grass.: 32 (1958). Types: Zimbabwe, Harare, *Nobbs* 674 (K, syntype) & *Craster* 19 (K, syntype); Malawi, Mt. Malosa, *Whyte* (K, syntype); other syntypes from Zaire and east Africa.

Densely tufted perennial. Culms 20–100 cm. high. Leaf laminae 2–6 mm. wide, flat or some of them convolute. Inflorescence of (1)2–5 racemes, these 1–5 cm. long, bearing spikelets singly on a triquetrous rhachis c. 0.5 mm. wide (sometimes narrowly winged), its margins sparsely ciliate with white or yellow hairs up to 2 mm. long. Spikelets 3–4(4.5) mm. long, pubescent, subacute. Inferior glume ⅔ to almost as long as spikelet, 5–11-nerved. Superior lemma obscurely rugulose.

Zambia. B: Kaoma (Mankoya), 1.x.1957, *West* 3475 (K; SRGH). N: Mbala, Safu Dambo N. of Chilwa School, 23.x.1967, *Simon, Williamson & Richards* 1166 (K; SRGH). W: Mwinilunga Distr., Matonchi Farm, 3.xi.1937, *Milne-Redhead* 3077 (K; SRGH). C: Kabwe, 23.ix.1947, *Brenan & Greenway* 7933 (FHO; K; SRGH). E: Chipata Distr., Sinde Missale, xii.1962, *Verboom* 505 (K). S: Choma, 5.xii.1962, *Astle* 1763 (K). **Zimbabwe**. N: Hurungwe (Urungwe) Distr., Nyamanyetsi Estate, 4.ix.1978, *Nyariri* 322 (K; SRGH). W: Hwange Distr., Matetsi Safari Area, 8.iii.1985, *Keller-Grein* Z-146 (K; SRGH). C: 12 km. SE. from Gweru, 29.xii.1966, *Biegel* 1593 (K; SRGH). E: 56 km. N. of Mutare, 11.i.1967, *Biegel* 2063 (K; SRGH). S: Makaholi Expt. Farm, 8.xii.1947, *Robinson* 27 (K; SRGH). **Malawi**. N: Nyika Plateau, Chitipa, *Salubeni* 374 (K). C: Mchinji, Tembwe, 9.x.1951, *Jackson* 606 (K). S: Thyolo, 22.xii.1949, *Wiehe* N/391 (K). **Mozambique**. T: 30 km. from Zimbabwe frontier, 18.xii.1941, *Hornby* 3308 (K).

Also from Zaire and Kenya to S. Africa (Transvaal). Damp soils in dambos and by water; 1000–1800 m.

Forms with a narrowly winged, but usually ± wavy, raceme rhachis closely approach *B. jubata*. Sometimes most of the leaves are convolute, and they are then easily mistaken for the subterete laminae of *B. subulifolia*.

5. **Brachiaria subulifolia** (Mez) Clayton in Kew Bull. **34**: 559 (1980). —Clayton & Renvoize in F.T.E.A., Gramineae: 582 (1982). TAB. **16**. Type from S. Africa.
    Panicum subulifolium Mez in Bot. Jahrb. **34**: 135 (1904). Type as above.
    Brachiaria filifolia Stapf in F.T.A. **9**: 516 (1919). —Sturgeon in Rhod. Agric. Journ. **50**: 422 (1953). —Chippindall in Meredith, Grasses & Pastures of S. Afr.: 373 (1955). —Jackson & Wiehe, Annot. Check List Nyasal. Grass.: 31 (1958). Types: Zimbabwe, Charter Distr., *Mundy* (K, syntype) & *Col. Herb.* 2107 (K, syntype); Angola, *Gossweiler* 2001 (K, syntype).

Like *B. bovonei* but leaf laminae up to 1 mm. diam., subterete, filiform.

Zambia. N: Kasama Distr., Chishimba Falls, 15.x.1960, *Robinson* 3987 (K). W: Mwinilunga Distr., Matonchi Farm, 1.ix.1930, *Milne-Redhead* 1005 (K). C: Chakwenga headwaters, 27.x.1963, *Robinson* 5779 (K). **Zimbabwe**. C: Harare, 1.x.1974, *Crook* 2032 (K; SRGH). E: Nyanga (Inyanga), Gairesi Ranch, 9 km. N. of Troutbeck, 13.ix.1957, *Robinson* 1885 (K; NRGH; SRGH). **Malawi**. N: Mzimba Distr., Mzuzu, 18.xii.1973, *Pawek* 7614 (K; MO; SRGH; UC). C: Dedza, Chongoni Forest Reserve, 20.vii.1967, *Salubeni* 774 (K; SRGH). S: Kirk Range, Davikos, 14.xi.1950, *Jackson* 265 (K).

Also from Zaire and Tanzania to S. Africa (Transvaal). Margins of lakes and dambos, often in peaty bogs; 1400–1900 m.

6. **Brachiaria humidicola** (Rendle) Schweick. in Kew Bull. **1936**: 297 (1936). —Sturgeon in Rhod. Agric. Journ. **50**: 422 (1953). —Chippindall in Meredith, Grasses & Pastures of S. Afr.: 372 (1955). —Jackson & Wiehe, Annot. Check List Nyasal. Grass.: 31 (1958). —Launert in Merxm., Prodr. Fl. SW. Afr. **160**: 43 (1970). —Clayton in F.W.T.A. **3**: 443 (1972). —Clayton & Renvoize in F.T.E.A., Gramineae: 583 (1982). Type from Angola.

Tab. 16. BRACHIARIA SUBULIFOLIA. 1, habit ($×\frac{2}{3}$); 2, portion of raceme (× 4); 3, spikelet showing inferior glume (× 8), all from *Crook* 2032.

Panicum humidicola Rendle, Cat. Afr. Pl. Welw. **2**: 169 (1899). Type as above.
Panicum rautanenii Hack. in Bull. Herb. Boiss. sér. 2, **2**: 935 (1902). Type from Namibia.
Panicum golae Chiov. in Ann. Bot. Roma. **14**: 43 (1914). Type from Zaire.
Brachiaria rautanenii (Hack.) Stapf in F.T.A. **9**: 513 (1919). Type as for *Panicum rautanenii*.
Panicum vexillare Peter, Fl. Deutsch Ost-Afr. **1**, Anh.: 22 (1930). Type from Tanzania.

Stoloniferous perennial. Culms 40–150 cm. high, often geniculate and rooting at the nodes. Leaf laminae 2–10 mm. wide, linear to narrowly lanceolate. Inflorescence of 2–3(4) racemes, these 2–7 cm. long, bearing spikelets singly on a triquetrous rhachis. Spikelets 4–6 mm. long, pubescent or very rarely glabrous, subacute. Inferior glume $\frac{3}{4}$ to as long as spikelet, 11-nerved, separated from the superior by a very brief internode. Superior lemma finely rugulose.

Caprivi Strip. Mpilila Is., 12.i.1959, *Killick & Leistner* 3330 (K; PRE). **Botswana**. N: Nxabega Is., 9.iii.1982, *Smith* 3773 (K; SRGH). **Zambia**. B: Mongu, 2.i.1966, *Robinson* 6761 (K). N: Mbala Distr., Ndundu, 13.ii.1959, *McCallum-Webster* A8 (K; SRGH). W: Mwinilunga Distr., Luao R., 10.ii.1938, *Milne-Redhead* 4527 (K; SRGH). C: Lusaka to Kasisi, 27.i.1973, *Kornaś* 3103 (K). S: Mapanza, 15.i.1956, *Robinson* 1321 (K). **Zimbabwe**. N: Gokwe Distr., Sengwa Wild Life Res. Inst., 11.ii.1983, *Mahlangu* 687 (K; SRGH). W: Hwange Distr., Livingi Pan, 30.i.1969, *Rushworth* 1481 (K; SRGH). C: Gweru Distr., Mlezu Govt. Agric. School Farm, 28.i.1966, *Biegel* 846 (K; SRGH). E: Nyanga (Inyanga) Nat. Park, 12.ii.1974, *Davidse, Simon & Pope* 6535 (K; MO; SRGH). S: Great Zimbabwe, 5.iv.1973, *Chiparawasha* 675 (K; SRGH). **Malawi**. N: Mzimba Distr., Mbawa, 11.xii.1952, *Jackson* 1008 (K). C: Lilongwe, 1.ii.1951, *Jackson* 386 (K). **Mozambique**. Z: Nicuadala to Namacurra, 30.vii.1943, *Torre* 5716 (COI; K). M: Montes da Namaacha, 20.xii.1944, *Torre* 6913 (COI; K).

Also from Tropical and S. Africa. Flood plains and damp soils in savanna woodland; 0–1800 m.

The spikelets resemble those of *B. bovonei*. Apart from their larger size they can be distinguished by the internode between the glumes, but this is very short and some dissection may be necessary to observe it.

7. **Brachiaria dictyoneura** (Fig. & De Not.) Stapf in F.T.A. **9**: 512 (1919). —Chippindall in Meredith, Grasses and Pastures of S. Africa: 372 (1955). —Launert in Merxm., Prodr. Fl. SW. Afr. **160**: 42 (1970). —Clayton & Renvoize in F.T.E.A., Gramineae: 582 (1982). Type from Sudan.
   Panicum dictyoneurum Fig. & De Not. in Mem. Accad. Sci. Torino, ser. 2, **14**: 329 (1854). Type as above.

Densely tufted perennial. Culms 40–120 cm. high, never rooting at the nodes. Leaf laminae 3–15(30) mm. wide, linear to lanceolate. Inflorescence of 3–8(12) racemes, these 1–8 cm. long, bearing spikelets singly on a triquetrous rhachis. Spikelets as for *B. humidicola*.

Zambia. E: Chipata Distr., Chipangali, i.1963, *Verboom* 903 (K).

Also from Ethiopia and Sudan to Tanzania; also in S. Africa. Savanna woodland and dambo margins; 1500 m.

The species is a non-stoloniferous segregate from *B. humidicola*, distributed mainly to the north of our area. It also has more numerous racemes, but there is some overlap in this character. Unfortunately the formation of stolons seems, to some extent, to be facultative as in the Botswana specimens treated here as *B. humidicola*.

A further problem is that occasional specimens with unusually broad leaves (including the type and the Zambian records) occur throughout the range, together with intermediates which link them to the common form.

In short the customary division of this complex into two species is likely to be an oversimplification, but the basis for an improved treatment is not yet apparent.

8. **Brachiaria arrecta** (Th. Dur. & Schinz) Stent in Bothalia **1**: 263 (1924). —Chippindall in Meredith, Grasses & Pastures of S. Afr.: 374 (1955). —Clayton & Renvoize in F.T.E.A., Gramineae: 585 (1982). Type from S. Africa.
   Panicum arrectum Th. Dur. & Schinz, Consp. Fl. Afr. **5**: 741 (1894). Type as above.
   Brachiaria latifolia Stapf in F.T.A. **9**: 526 (1919). —Jackson & Wiehe, Annot. Check List Nyasal. Grass.: 31 (1958). Type: Malawi, Shire R., *Kirk* (K, holotype).
   Panicum multifolium Peter, Fl. Deutsch Ost- Afr. **1**, Anh.: 29 (1930). Types from Tanzania.

Sprawling perennial. Culms 50–130 cm. high, prostrate and rooting at the nodes below. Leaf laminae 5–15 mm. wide. Inflorescence of 4–15 racemes, these 1–10 cm. long, bearing spikelets singly in two neat rows on a flat rhachis 0.5–1.5 mm. wide with scabrid margins. Spikelets 3–4.5 mm. long, glabrous, acute. Inferior glume $\frac{1}{3}$–$\frac{1}{2}$ length of spikelet, arising immediately below the superior. Superior lemma rugulose.

Botswana. N: Motsaudi Safari Camp, 17.iii.1982, *Smith* 3809 (K; SRGH). **Zambia**. N: Mbesuma, 28.i.1958, *Vesey-FitzGerald* 1412 (K). S: 10 km. SE. of Choma, i.1979, *Heery* 31 (K). **Zimbabwe**. N: Mazowe (Mazoe) Henderson R. St., 16.v.1950, *Thompson* GHS 27912 (K; SRGH). W: Matobo Distr., Mtsheleli Dam, 16.i.1973, *Simon* 2311 (K; SRGH). C: Harare Distr., Vainona, 28.ii.1974, *Bezuidenhout* 75 (K; SRGH). E: Chipinge, iv.1961, *Goldsmith* 27/61 (K; SRGH). **Malawi**. C: Nkhota Kota, Kalonga R., 20.ii.1953, *Jackson* 1077 (K). S: Blantyre, Chilomoni Dam, 15.iv.1970, *Brummitt & Williams* 9859 (K; SRGH).
Also from Tropical and S. Africa; introduced to tropical America. Swamps, river banks and shallow water; 1000–1500 m.
B. mutica (Forssk.) Stapf is a very similar forage species known as Para grass, which has been introduced to the Agricultural Experiment Station, Harare. It is distinguished by its paired spikelets in 4 untidy rows, but may have single spikelets in the upper part of the raceme. Although pantropical in distribution, it is almost entirely replaced by *B. arrecta* in eastern and southern Africa, hinting that our species may be no more than a geographical race.
The characteristic elliptic shape of the spikelets could be mistaken for *Eriochloa*, but there is no trace of a swelling at their base.

9. **Brachiaria rugulosa** Stapf in F.T.A. **9**: 529 (1919). —Clayton & Renvoize in F.T.E.A., Gramineae: 585 (1982). Types from Kenya.
    Brachiaria umboensis Stent & Rattray in Proc. Rhod. Sci. Ass. **32**: 23 (1933). —Sturgeon in Rhod. Agric. Journ. **50**: 423 (1953). Type: Zimbabwe, *Jack* 6234 (K, isotype).

Loosely tufted perennial. Culms 50–100 cm. high. Leaf laminae 2–10 mm. wide. Inflorescence of 2–12 racemes, these 2–10 cm. long, bearing spikelets singly on a flat rhachis c. 1 mm. wide with tuberculate-ciliate margins. Spikelets 3–3.5 mm. long, glabrous, acute. Inferior glume $\frac{1}{3}$ length of spikelet, arising immediately below the superior. Superior lemma rugose.

Zambia. C: Mumbwa, 27.xii.1963, *van Rensburg* 2641 (K). S: Kafue Nat. Park, Chunga, 31.i.1961, *Mitchell* 5/20 (K). **Zimbabwe**. N: Lomagundi Distr., Umboe Valley, 20.ii.1933, *Jack* 6234 (K; SRGH).
Also from Kenya and Tanzania. Damp soils and streamsides; 1000–1300 m.
Closely resembles *B. arrecta*, but has ciliate rhachis margins.

10. **Brachiaria oligobrachiata** (Pilger) Henr. in Blumea **3**: 436 (1940). Type from Congo (Brazzaville).
    Panicum oligobrachiatum Pilger in Bot. Jahrb. **33**: 50 (1902). Type as above.
    Brachiaria interstipitata Stapf in F.T.A. **9**: 523 (1919). Type from Congo (Brazzaville).
    Brachiaria platytaenia Stapf in F.T.A. **9**: 524 (1919). Type from Zaire.
    Brachiaria vittata Stapf in F.T.A. **9**: 525 (1919). Type from Congo (Brazzaville).

Coarse annual. Culms 60–120 cm. high. Leaf laminae 5–25 mm. wide. Inflorescence of 3–15 racemes, these 2–8 cm. long, bearing spikelets singly on a ribbon-like rhachis 1–3 mm. wide with minutely ciliolate margins. Spikelets 4–6 mm. long, sparsely pubescent above, acute. Inferior glume $\frac{1}{3}$ length of spikelet, acute to acuminate, separated from the superior by a short internode. Superior lemma minutely granulose, with a short stubby mucro.

Zambia. B: Mongu Harbour, 11.v.1964, *Verboom* 1193 (K; SRGH). N: Mporokoso Distr., Mweru-Wantipa, Mofwe Dambo, 14.iv.1961, *Phipps & Vesey FitzGerald* 3181 (K; SRGH). S: Kafue Nat. Park, Kalala Is., 14.iii.1961, *Mitchell* 6/84 (K).
Also from Zaire and Congo (Brazzaville). River banks; 1000 m.
B. plantaginea (Link) Hitchc. is a similar west African species, occurring also in South America. It differs in the obtuse inferior glume and emucronate superior lemma.

11. **Brachiaria eminii** (Mez) Robyns in Bull. Jard. Bot. Brux. **9**: 176 (1932). —Clayton & Renvoize in F.T.E.A., Gramineae: 586 (1982). Types from Tanzania.
    Panicum eminii Mez in Bot. Jahrb. **34**: 135 (1904). Types as above.

Tufted annual. Culms 40–100 mm. high. Leaf laminae 5–18 mm. wide. Inflorescence of 3–10 racemes, these 3–7 cm. long, bearing spikelets singly on a flat rhachis 1–1.5 mm. wide with tuberculate-ciliate margins. Spikelets 3.5–4.5 mm. long, sparsely and stiffly pubescent, sharply acute. Inferior glume $\frac{1}{2}$–$\frac{2}{3}$ length of spikelet, 7–9-nerved, acuminate, separated from the superior by a short internode. Superior lemma rugulose to granulose, subacute, mucronate.

Zambia. N: Mporokoso Distr., Muzombwe, 15.iv.1961, *Phipps & Vesey-FitzGerald* 3201 (K; SRGH). E: Lunzi R., Mkosakwenda, 15.viii.1938, *Greenway & Trapnell* 5614 (EAH; K). S: Kafue Nat. Park, Kalala, 22.xii.1963, *Mitchell* 24/35 (K; SRGH). **Malawi**: N: 6 km. E. of Rumphi, 12.ii.1968, *Simon & Williamson* 1782 (K; SRGH).

Also from Zaire and Tanzania. Weedy places on alluvial soils; 600–1500 m.

Closely similar to and doubtfully distinct from *B. oligobrachiata*.

The cultivated species *B. decumbens* will key out here. It is a stoloniferous perennial with emucronate superior lemma.

12. **Brachiaria brizantha** (A. Rich.) Stapf in F.T.A. **9**: 531 (1919). —Sturgeon in Rhod. Agric. Journ. **50**: 423 (1954). —Chippindall in Meredith, Grasses & Pastures of S. Afr.: 371 (1955). —Jackson & Wiehe, Annot. Check List Nyasal. Grass.: 31 (1958). —Bor, Grasses of B.C.I. & P.: 281 (1960). —Launert in Merxm., Prodr. Fl. SW. Afr. **160**: 42 (1970). —Clayton in F.W.T.A. **3**: 443 (1972). —Clayton & Renvoize in F.T.E.A., Gramineae: 587 (1982). Types from Ethiopia.

    Panicum brizanthum A. Rich., Tent. Fl. Abyss. **2**: 363 (1851). Types as above.

    Brachiaria brizantha var. *angustifolia* Stent & Rattray in Proc. Rhod. Sci. Ass. **32**: 23 (1933). Types: Zimbabwe, Harare, *Eyles* 1921 (K, isosyntype; SRGH, syntype) and many other syntypes.

Tufted perennial. Culms 30–200 cm. high, erect or sometimes geniculately ascending. Leaf laminae 3–20 mm. wide, flat. Inflorescence of (1)2–16 racemes, these mostly 4–20 cm. long, bearing spikelets singly and usually in 1 row; rhachis crescentic in section, c. 1 mm. wide with narrow involute wings, ciliate on the margins. Spikelets 4–6 mm. long, glabrous or sometimes sparsely pubescent, obtuse to subacute with a slight stipe at base. Inferior glume ⅓ length of spikelet, acute or obtuse, separated from the superior by a short internode; superior glume and inferior lemma cartilaginous, dully shining. Superior lemma granulose, subacute.

Zambia. B: Kaoma (Mankoya) Distr., Kafue watershed, 2.iv.1964, *Verboom* 1365 (K). N: Mbala Distr., Ndundu, 13.ii.1959, *McCallum-Webster* A7 (K; SRGH). W: Mwinilunga Distr., Matonchi Farm, 4.i.1938, *Milne-Redhead* 3949 (K; SRGH). C: 12 km. SE. of Lusaka, 14.i.1952, *Best* 17 (K). E: Chipata, Senegallia Farm, 30.i.1953, *Grant* 116 (K; SRGH). S: Mapanza Mission, 7.ii.1953, *Robinson* 75 (K). **Zimbabwe**. N: Gokwe Distr., Sengwa Res. Area, 12.ii.1981, *Mahlangu* 430 (K; SRGH). W: Hwange Distr., Matetsi Safari Area, 16.i.1980, *Gonde* 277 (K; SRGH). C: Gweru Distr., White Waters Dam, 5.ii.1967, *Biegel* 1894 (K; SRGH). E: Mutare Distr., Kukwanisa, 18.i.1968, *Biegel* 2498 (K; SRGH). S: Umzingwane Distr., Balla Balla, Irisvale Ranch, xii.1961, *Poultney* 66 (K; SRGH). **Malawi**. N: Mzimba Distr., Mbawa, 10.xii.1952, *Jackson* 1003 (K). C: Lilongwe, on Salima Rd., 31.xii.1952, *Jackson* 1014 (K). S: Zomba, 6.iv.1949, *Wiehe* N/43 (K). **Mozambique**. N: Malema, 4.ii.1964, *Torre & Paiva* 10442 (COI; K). MS: Karuso, iv.1935, *Gilliland* 1824 (K). M: Namaacha Mts., 20.xii.1944, *Torre* 6922 (COI; K).

Also from Tropical and S. Africa, introduced elsewhere. Savanna woodland and grassland, also commonly grown as a pasture plant under the names *Palisade* or *Signal* grass; 700–2300 m.

B. decumbens is a closely related east African species which has been introduced as a pasture plant throughout the tropics. It is distinguished from *B. brizantha* by its stoloniferous habit, flat raceme rhachis and herbaceous superior glume and inferior lemma, but the two species intergrade and some of the cultivated forms are difficult to assign. Glume and lemma texture seem to be the best and decumbent habit the least reliable diagnostic characters.

13. **Brachiaria dura** Stapf in F.T.A. **9**: 531 (1919). —Launert in Merxm., Prodr. Fl. SW. Afr. **160**: 43 (1970). Type from Angola.

    Brachiaria dura var. *pilosa* J.G. Anderson in Kirkia **1**: 104 (1961). Type from S. Africa.

Densely tufted perennial. Culms 40–100 mm. high. Leaf laminae 1–2 mm. wide (when flattened), convolute, wiry. Inflorescence of 1–2 racemes, these 4–15 cm. long, bearing spikelets singly in 1 row; rhachis crescentic in section, c. 1 mm. wide, with irregularly ciliate margins. Spikelets 3.5–5 mm. long, glabrous or pubescent, obtuse, with a slight stipe at base. Inferior glume ½–⅔ length of spikelet, not separated from the superior. Superior glume and inferior lemma membranous. Superior lemma granulose, subacute.

Caprivi Strip. Katima Mulilo, *de Winter* 9192 (K; PRE). **Botswana**. N: Movombe, 14.ii.1983, *Smith* 4035 (K; SRGH). SE: Morapedi Ranch, 19.i.1978, *Hansen* 3332 (C; GAB; K; PRE; SRGH; WAG). **Zambia**. B: Mongu, 3.viii.1965, *van Rensburg* 3023 (K). N: Lake Bangweulu, 24.iv.1969, *Verboom* 2551 (K). S: Kafue Nat. Park, Siakalongo, 13.i.1962, *Mitchell* 12/62 (K). **Zimbabwe**. W: Hwange Nat. Park, Mbiza, 9.ii.1970, *Rushworth* 2424 (K; SRGH).

Also from Angola and S. Africa. Savanna woodland and grassland on sandy soils; 1000 m.

14. **Brachiaria clavipila** (Chiov.) Robyns in Bull. Jard. Bot. Brux. **9**: 179 (1932). Type from Zaire.

    Panicum clavipilum Chiov. in Ann. Bot. Roma **13**: 43 (1914). Type as above.

Tufted perennial. Culms 30–70 cm. high. Inflorescence of 2–6 racemes, these 1–4 cm. long, bearing spikelets singly in 1–2 rows on a triquetrous rhachis which is clavellate hairy on the back. Spikelets 3–4.5 mm. long, pubescent with appressed clavellate hairs, the nerves often prominent, acute. Inferior glume c. ½ length of spikelet. Superior lemma softly coriaceous, striate, acute.

Zambia. N: Kawambwa Distr., Ntumbachushi Falls, 22.xii.1967, *Simon & Williamson* 1512 (K; SRGH). W: Mwinilunga airfield, 20.i.1975, *Brummitt, Chisumpa & Polhill* 13888 (K; SRGH). Also in Zaire and Angola. Dambos and flood plains; 1000–1400 m.

15. **Brachiaria serrata** (Thunb.) Stapf in F.T.A. **9**: 537 (1919). —Sturgeon in Rhod. Agric. Journ. **50**: 424 (1953). —Chippindall in Meredith, Grasses & Pastures of S. Afr.: 375 (1955). —Jackson & Wiehe, Annot. Check List Nyasal. Grass.: 32 (1958). —Launert in Merxm., Prodr. Fl. SW. Afr. **160**: 44 (1970). —Clayton & Renvoize in F.T.E.A., Gramineae: 588 (1982). Type from S. Africa.
  Holcus serratus Thunb., Prodr. Pl. Cap.: 20 (1794). Type as above.
  Sorghum serratum (Thunb.) Roem. & Schult., Syst. Veg. **2**: 839 (1817). Type as above.
  Panicum serratum (Thunb.) Spreng., Syst. Veg. **1**: 309 (1824). Type as above.
  Panicum scopuliferum Trin., Sp. Gram. Ic. **2**: t.165 (1828). Type from S. Africa.
  Panicum gossypinum A. Rich., Tent. Fl. Abyss. **2**: 366 (1851). Types from Ethiopia.
  Panicum serratum var. *gossypinum* (A. Rich.) Th. Dur. & Schinz, Consp. Fl. Afr. **5**: 765 (1894). Types as above.
  Panicum serratum var. *holosericeum* Th. Dur. & Schinz, Consp. Fl. Afr. **5**: 765 (1894). Type from S. Africa.
  Panicum serratum var. *hirtum* Kuntze, Rev. Gen. Pl. **3**: 364 (1898). Type from S. Africa.
  Panicum andongense Rendle in Cat. Afr. Pl. Welw. **2**: 167 (1899). Type from Angola.
  Panicum nigropedatum var. *basipiliferum* Chiov. in Ann. Bot. Roma **13**: 42 (1914). Type from Zaire.
  Brachiaria serrata var. *gossypina* (A. Rich.) Stapf in F.T.A. **9**: 538 (1919). Types as for *Panicum gossypinum*.
  Brachiaria brachylopha Stapf in F.T.A. **9**: 539 (1919). —Clayton in F.W.T.A. **3**: 443 (1972). Types from Mali, Ivory Coast and Nigeria.
  Brachiaria andongensis (Rendle) Stapf in F.T.A. **9**: 560 (1919). Type as for *Panicum andongense*.
  Panicum serratum var. *brachylophum* (Stapf) A. Chev., Expl. Bot. Afr. Occ. Fr. **1**: 730 (1920). Types as for *Brachiaria brachylopha*.

Tufted perennial, the basal sheaths silky tomentose. Culms 20–100 cm. high, erect or ascending. Leaf laminae glabrous to softly pilose, the margins ± cartilaginous. Inflorescence of 5–15 racemes, these 0.5–2 cm. long, short and compact, bearing the spikelets in 2 rows on a triquetrous rhachis beset with a few bristles. Spikelets 2–4.5 mm. long, ovate-elliptic, with a transverse fringe of pink or silver hairs above (rarely the fringe missing) otherwise pubescent, cuspidate to acuminate (rarely obtuse), without a stipe. Inferior glume ½ length of spikelet, subacute; superior glume obtuse to acuminate. Inferior lemma hollowed along the mid-nerve, sometimes obtuse in our area but usually tipped by a subulate awn-point up to 1 mm. long. Superior lemma striate, acute.

Zambia. B: 50 km. W. of Kabompo, 26.xii.1969, *Simon & Williamson* 2038 (K; SRGH). N: Mbala, 22.xii.1949, *Bullock* 2121 (K; SRGH). W: Ndola Distr., 32 km. N. of Kapiri Mposhi, 1.xii.1952, *Angus* (FHO; K). C: 8 km. W. of Serenje, 16.xii.1967, *Simon & Williamson* (K; SRGH). S: Choma, 19.xii.1962, *Astle* 1833 (K). **Zimbabwe**. N: Guruve Distr., Nyamanyetsi Estate, 13.xii.1978, *Nyariri* 571 (K; SRGH). W: Shangani Distr., Gwampa Forest Reserve, i.1956, *Goldsmith* 6/56 (K; SRGH). C: Chegutu Distr., Poole Farm, 2.iii.1944, *Hornby* 2283 (K; SRGH). E: Chimanimani Distr., Mutare, Rd. to Lookout Point, 13.xi.1975, *Crook* 2091 (K; SRGH). S: Masvingo Distr., Makaholi Expt. Farm, 1.xii.1947, *Robinson* 119 (K; SRGH). **Malawi**. N: Mzimba Distr., Mzuzu, Marymount, 2.xii.1973, *Pawek* 7563 (K; MA; MO; SRGH; UC). C: Kasungu Nat. Park, 21.xii.1970, *Hall-Martin* 1281 (K; SRGH). S: Thyolo, Tung St., 19.xii.1949, *Wiehe* N/387 (K). **Mozambique**. MS: Chimoio, Missão de Amatongas, 10.xi.1941, *Torre* 3802 (COI; K).
  Also from Tropical and S. Africa. Savanna woodland and grassland; 600–1800 m.
  Specimens from west Africa tend to have smaller spikelets (*B. brachylopha*, *B. serrata* var. *gossypina*) than those from the south. Likewise Zambian specimens tend to obtuse puberulous spikelets lacking transverse fringe or awn-point on the inferior lemma (*B. andongensis*). However there are no discernible boundaries between these intergrading forms, and for the present it seems better to treat them as a single species subject to clinal variation.

16. **Brachiaria nigropedata** (Ficalho & Hiern) Stapf in F.T.A. **9**: 535 (1919). —Sturgeon in Rhod. Agric. Journ. **50**: 423 (1953). —Chippindall in Meredith, Grasses & Pastures of S. Afr.: 374 (1955). —Launert in Merxm., Prodr. Fl. SW. Afr. **160**: 43 (1970). —Clayton & Renvoize in F.T.E.A., Gramineae: 587 (1982). TAB. **17**. Types from S. Africa.

Tab. 17. BRACHIARIA NIGROPEDATA. 1, habit (×⅔); 2, portion of raceme (× 4); 3, spikelet (× 8), all from *Rattray* 48742.

Panicum nigropedatum Ficalho & Hiern in Trans. Linn. Soc. Bot., ser. 2, **2**: 29 (1881). Type as above.
Panicum melanotylum Hack. in Bot. Jahrb. **11**: 398 (1890). Type from S. Africa.
Brachiaria melanotyla (Hack.) Henr. in Blumea **3**: 436 (1940). Type as above.

Densely tufted perennial, the base ± bulbous and clad in silky tomentose sheaths. Culms 25–100 cm. high. Leaf laminae pubescent to villous. Inflorescence of 5–16 racemes, these 1–6 cm. long, bearing spikelets singly in 2 rows on a villous triquetrous rhachis. Spikelets 3.5–5 mm. long, ovate-elliptic, silky villous with off-white hairs these becoming longer above and often gathered into a loose transverse fringe, acuminate to a subulate point, supported by a cylindrical black stipe c. 0.5 mm. long. Inferior glume $\frac{1}{3}$–$\frac{1}{2}$ ($\frac{2}{3}$) length of spikelet, acute to acuminate; superior glume and inferior lemma acuminate to a subulate point up to 1 mm. long. Superior lemma obscurely rugulose, acute to mucronate.

Caprivi Strip. Katima Mulilo, 22.xii.1958, *Killick & Leistner* 3037 (K; PRE). **Botswana**. N: Tsantsarra Pan, 26.xii.1976, *Smith* 1860 (K; SRGH). SW: Tsabong, 2.iii.1977, *Mott* 1104 (K; SRGH). SE: Gaborone, 29.xi.1973, *Mott* 28 (K; UBLS). **Zambia**. B: Kalabo airport, 8.xii.1964, *Verboom* 1551 (K). S: Choma, 24.xii.1962, *Astle* 1864 (K). **Zimbabwe**. N: Gokwe Distr., Sengwa Research St., 20.i.1976, *Guy* 2378 (K; SRGH). W: Shangani Distr., Gwampa Forest Reserve, ii.1955, *Goldsmith* 91/55 (K; SRGH). C: Marondera Distr., Digglefold, 3.xii.1948, *Corby* 281 (K; SRGH). E: Mutare Distr., Kukwanisa, 22.xii.1967, *Biegel* 2405 (K; SRGH). S: Mwenzi, 16.i.1976, *Kelly* 555 (K; SRGH). **Mozambique**. GI: Namaacha to Moamba, 9.ii.1952, *Myre & Carvalho* 1102 (K). M: Goba Fronteira, 11.i.1980, *de Koning* 8002 (K).
Also from Kenya, Tanzania, Namibia and S. Africa. Savanna woodland and bushland on dry sandy or stony soils; 500–1500 m.

17. **Brachiaria pungipes** Clayton in Kew Bull. **34**: 558 (1980). Type: Zambia, Mwinilunga, Dobeka Bridge, *Milne-Redhead* 3617 (holotype, K).

Densely tufted perennial, the basal sheaths silky pubescent. Culms 50–80 cm. high. Leaf laminae glabrous to sparsely hirsute, the margins cartilaginous. Inflorescence of 5–10 racemes, these 1–2 cm. long, bearing the spikelets singly in 1–2 rows on a villous triquetrous rhachis. Spikelets 2.5–3 mm. long, turbinate, pubescent with a transverse fringe of hairs above, acute, tapering below into a light brown cuneate stipe c. 0.5 mm. long. Inferior glume $\frac{1}{3}$ length of spikelet, obtuse; superior glume obtuse. Superior lemma striate, apiculate.

Zambia. W: Solwezi Distr., 8 km. E. of Kabompo R., 19.xii.1969, *Simon & Williamson* 1865 (K; SRGH).
Not known elsewhere. Grassland on sandy soils; 1500 m.
This is the perennial counterpart of *B. turbinata* Van der Veken, an annual species from Zaire. For discussion of the latter and its allies see Bull. Jard. Bot. Brux. **28**: 77–81 (1958).

18. **Brachiaria eruciformis** (J.E. Smith) Griseb. in Ledeb., Fl. Ross. **4**: 469 (1853). —Sturgeon in Rhod. Agric. Journ. **50**: 424 (1953). —Chippindall in Meredith, Grasses & Pastures of S. Afr.: 376 (1955). —Jackson & Wiehe, Annot. Check List Nyasal. Grass.: 31 (1958). —Bor, Grasses of B.C.I. & P.: 281 (1960). —Launert in Merxm., Prodr. Fl. SW. Afr. **160**: 43 (1970). —Clayton & Renvoize in F.T.E.A., Gramineae: 590 (1982). Type from Greece.
Panicum eruciforme J.E. Smith in Sibth. & Sm., Fl. Graeca **1**: 44 (1806). Type as above.
Panicum wightii Nees, Fl. Afr. Austr.: 29 (1841). Type from S. Africa.

Loosely tufted annual. Culms 10–60 cm. high, geniculately ascending. Inflorescence of 3–14 racemes, these 0.5–2.5 cm. long, bearing single spikelets on a triquetrous rhachis, sometimes with a secondary racemlet at the base. Spikelets 1.7–3 mm. long, pubescent (very rarely glabrous), subacute. Inferior glume a tiny scale up to $\frac{1}{5}$ length of spikelet. Superior floret readily deciduous, its lemma smooth, shiny and obtuse.

Caprivi Strip. Mpilila Is., 13.i.1959, *Killick & Leistner* 3375 (K; PRE). **Botswana**. N: Nata to Kazungula Rd., 14.iv.1983, *Smith* 4311 (K; SRGH). SW: Deception Valley, 19.iii.1983, *Smith* 4209 (K; SRGH). SE: Serowe Distr., Selebi Pikwe, 9.iv.1959, *de Beer* 880a (K; SRGH). **Zambia**. C: 27 km. SW. of Lusaka, 15.ii.1970, *Drummond & Williamson* 9602 (K; SRGH). S: Kalomo, 14.i.1964, *Astle* 2848 (K). **Zimbabwe**. N: Hurungwe (Urungwe) Distr., Chewore River mouth, iii.1972, *Guy* 1970 (K; SRGH). W: Hwange Distr., Kazuma Range, 19.v.1972, *Simon* 2186 (K; SRGH). C: Harare, 5.iii.1965, *Crook* 702 (K; SRGH). E: Chipinge Distr., 5 km. S. of Rusongo, 1.ii.1975, *Gibbs-Russell* 2744 (K; PRE; SRGH). S: Chiredzi Distr., Hippo Valley Estate, 27.x.1972, *Taylor* 220 (K; SRGH). **Malawi**. C: Salima, Citala Rd. junction, 27.iv.1951, *Jackson* 484 (K).
Also from S. Africa to the Mediterranean, eastward to India. Disturbed or overgrazed places in damp grassland, particularly on black clays; 300–1500 m.

B. schoenfelderi Hubbard & Schweick. is a very similar species from Angola and Namibia, distinguished by its densely villous spikelets.

19. **Brachiaria malacodes** (Mez & K. Schum.) Scholz in Willdenowia **8**: 384 (1978). Type from Angola.
    Panicum malacodes Mez & K. Schum. in Not. Bot. Gart. Berlin **7**: 70 (1917). Type as above.
    Brachiaria poaeoides Stapf in F.T.A. **9**: 554 (1919). —Chippindall in Meredith, Grasses & Pastures of S. Afr.: 377 (1955). —Launert in Merxm., Prodr. Fl. SW. Afr. **160**: 44 (1970). Type from Angola.

Annual. Culms 25–100 cm. high. Inflorescence paniculate, the primary branches bearing pedunculate racemes, these 0.5–1 cm. long. Otherwise like *B. eruciformis*.

Zimbabwe. N: Binga Distr., L. Kariba, Sibilohilo Is., 17.ii.1966, *Jarman* 356 (K; SRGH).
Also from Angola and Namibia. Damp and shady places.
B. eruciformis occasionally has a sessile secondary racemelet at the base of the raceme. In *B. malacodes* this tendency is exaggerated to the point that most of the spikelets are borne on pedunculate secondary racemes. There is some intergradation particularly in Zimbabwe, and the cited specimen is closer to the intermediates than to typical members of the species.

20. **Brachiaria glomerata** (Stapf) A. Camus in Bull. Soc. Bot. Fr. **77**: 640 (1930). —Chippindall in Meredith, Grasses & Pastures of S. Afr.: 379 (1955). —Launert in Merxm., Prodr. Fl. SW. Afr. **160**: 43 (1970). Type from Namibia.
    Panicum glomeratum Hackel in Verh. Bot. Ver. Prov. Brand. **30**: 141 (1888) non Moench (1794) nec Buckley (1866). Type as above.
    Leucophrys glomerata Stapf in F.T.A. **9**: 504 (1919). Based on *Panicum glomeratum* Hack.

Annual. Culms 15–60 cm. high. Inflorescence of 4–12 racemes, these 1–2.5 cm. long, globose to oblong, bearing the spikelets in an irregular mass of tightly packed clusters on a triquetrous rhachis. Spikelets 2.5–3.5 mm. long, pilose, acute. Inferior glume c. $\frac{1}{2}$ length of spikelet. Superior lemma finely granulose and dully glossy, obtuse, subtending a subulate rhachilla extension.

Botswana. SW: 9 km. NE. of Twee Rivieren, 17.iii.1969, *Rains & Yalala* 37 (K; SRGH).   SE: 6.5 km. NW. of Lephepe, iii.1969, *Kelaole* 540 (K; SRGH).
Also in Namibia and S. Africa. Among sand dunes.
Related to *B. psammophila* (Rendle) Dandy from Angola and Namibia, which has tomentose leaves and longer (3.5–5 mm.) spikelets. Note the unusual presence of a rhachilla extension.

21. **Brachiaria marlothii** (Hackel) Stent in Bothalia **1**: 263 (1924). —Chippindall in Meredith, Grasses & Pastures of S. Afr.: 376 (1955). —Launert in Merxm., Prodr. Fl. SW. Afr. **160**: 43 (1970). Type from S. Africa.
    Panicum marlothii Hackel in Bot. Jahrb. **11**: 398 (1889). Type as above.

Decumbent or stoloniferous mat-forming annual. Culms 5–40 cm. high. Inflorescence of 2–5 racemes, these 0.5–4 cm. long, usually distant but the smallest sometimes condensed into a subglobose head, bearing paired spikelets on a triquetrous rhachis. Spikelets 2–2.5 mm. long, with a few tubercle-based tufts of hair above, otherwise glabrous or hispidulous, acute. Inferior glume c. $\frac{1}{2}$ length of spikelet. Superior lemma striate, obtuse.

Botswana. N: Rakops to Toromoja, 16.ii.1980, *Smith* 3055 (K; SRGH). SW: Tshane, 19.iii.1980, *Skarpe* S-414 (K).
Also in S. Africa. Weedy and overgrazed places; 1000 m.

22. **Brachiaria umbellata** (Trin.) Clayton in Kew Bull. **34**: 559 (1980). —Clayton & Renvoize in F.T.E.A., Gramineae: 592 (1982). Type from Mauritius.
    Panicum umbellatum Trin., Gram. Pan.: 238 (1826). Type as above.

Sward-forming perennial. Culms procumbent, c. 10 cm. high. Inflorescence ovate, of 4–8 closely spaced racemes, these 0.5–3 cm. long, bare at the base, bearing pedicelled spikelets singly on a slender rhachis, the pedicels c. as long as the spikelet. Spikelets 1.5–2 mm. long, glabrous, acute. Inferior glume $\frac{1}{8}$–$\frac{1}{4}$ length of spikelet, obtuse to emarginate. Superior lemma granulose, acute.

Zimbabwe. C: Chipinge Distr., New Year's Gift Tea Estate, 17.v.1962, *Chase* 7780 (K; SRGH).
Malawi. S: Mulanje, foot of Great Ruo Gorge, 18.iii.1970, *Brummitt & Banda* 9199 (K; SRGH).
Mozambique. MS: Beira, 1.v.1968, *Crook* (K; SRGH).

Also in Madagascar and Indian Ocean islands; introduced to southern Africa. Planted as a lawn grass and as a cover plant in tea gardens, but also occurs as an escape in forest clearings; sea level–1600 m.

23. **Brachiaria scalaris** Pilger in Not. Bot. Gart. Berlin **10**: 269 (1928). —Jackson & Wiehe, Annot. Check List Nyasal. Grass.: 32 (1958). —Clayton & Renvoize in F.T.E.A., Gramineae : 594 (1982). Type from Tanzania.
  Panicum scalare Mez in Bot. Jahrb. **34**: 138 (1904) non Schweinf. (1894). Type as above.
  Panicum heterocraspedum Peter in Abh. Königl. Ges. Wiss. Göttingen, Math.-Phys, Kl., n.f. **13**, 2: 45 (1928). Type from Tanzania.
  Brachiaria heterocraspeda (Peter) Pilger in Not. Bot. Gart. Berlin **13**: 263 (1936). Type as above.
  Brachiaria pilgeriana Scholz in Willdenowia **8**: 385 (1978). Based on *Panicum scalare* Mez.

Annual. Culms 10–60 cm. high, decumbent or ascending. Inflorescence of 3–20 racemes, these 0.5–3 cm. long, subsecund, bearing single or paired spikelets in 1–2 irregular rows on a triquetrous rhachis, sometimes with a few short branchlets at base of longer racemes. Spikelets 1.5–2 mm. long, glabrous or pubescent, obtuse or acute. Inferior glume $\frac{1}{3}$ length of spikelet. Superior lemma granulose to striate, acute.

Zambia. N: L. Mweru, vi.1961, *Chongo* 14 (K). **Zimbabwe**. E: Chimanimani Distr., Haroni-Lusitu R. junction, 29.v.1969, *Müller & Kelly* 1166 (K; SRGH). **Malawi**. N: Nkhata Bay, Chikale Beach, 15.iv.1977, *Pawek* 12588 (K; MA; MO; SRGH; UC). S: Mulanje, Mimosa, 24.vii.1951, *Jackson* 588 (K). Northwards to Zaire and Ethiopia. Roadsides and weedy places; 500–1500 m.

24. **Brachiaria reptans** (L.) Gardner & Hubbard in Ic. Pl. (Hook.) **34**: t.3363 (1938). —Bor, Grasses of B.C.I. & P.: 285 (1960). —Clayton & Renvoize in F.T.E.A., Gramineae: 591 (1982). Type from Jamaica.
  Panicum reptans L., Syst. Nat., ed. 10, **2**: 870 (1759). Type as above.
  Urochloa reptans (L.) Stapf in F.T.A. **9**: 601 (1920). Type as above.
  Panicum brachythyrsum Peter, Fl. Deutsch Ost-Afr. **1**, Anh.: 30 (1930). Type from Tanzania.
  Echinochloa reptans (L.) Roberty in Bull. Inst. Fond. Afr. Noire, sér. A, **17**: 66 (1955). Type as for *Panicum reptans*.

Annual. Culms 15–60 cm. high, usually decumbent and rooting at nodes. Inflorescence ovate, of 5–15 closely spaced racemes, these 1–4 cm. long, bearing paired spikelets crowded on a triquetrous rhachis with hirsute pedicels. Spikelets 1.5–2.2 mm. long, glabrous, acute. Inferior glume short, $\frac{1}{8}-\frac{1}{4}$ length of spikelet, truncate (rarely a little longer and broadly ovate). Superior lemma rugose, mucronate.

Mozambique. GI: Bona Vista to Santaca, 12.iv.1949, *Myre & Balsinhas* 567 (K).
Native to tropical Asia, introduced throughout the tropics. Roadsides and weedy places; 100 m.

25. **Brachiaria leersioides** (Hochst.) Stapf in F.T.A. **9**: 551 (1919). —Clayton & Renvoize in F.T.E.A., Gramineae: 591 (1982). Type from Ethiopia.
  Panicum leersioides Hochst. in Flora **38**: 196 (1855). Type as above.

Annual. Culms 10–100 cm. high, ascending. Inflorescence of 3–14 widely spaced racemes, these 1–7 cm. long, slender, secund, spreading horizontally or deflexing at maturity, bearing mostly paired spikelets on a triquetrous rhachis. Spikelets 2–3.5 mm. long, glabrous, subacute. Inferior glume $\frac{1}{3}-\frac{1}{2}$ length of spikelet, separated from the superior by a distinct internode 0.2–0.5 mm. long. Superior lemma coarsely rugose, subacute.

Mozambique. Without locality, 13.ix.1887, *Scott* (K).
Northward to Chad and Arabia. Weedy places.

26. **Brachiaria chusqueoides** (Hackel) Clayton in Kew Bull. **34**: 558 (1980). —Clayton & Renvoize in F.T.E.A., Gramineae: 590 (1982). Type from S. Africa.
  Panicum chusqueoides Hackel in Bull. Herb. Boiss. **3**: 377 (1895). —Chippindall in Meredith, Grasses & Pastures of S. Afr.: 326 (1955). Type as above.
  Panicum obumbratum Stapf in Fl. Cap. **7**: 401 (1899). Type from S. Africa.

Scandent or creeping annual. Culms 30–75 cm. high, wiry, rooting at the nodes. Leaf laminae 3–12 cm. long, narrowly lanceolate. Inflorescence of 2–7 distant racemes, these 1.5–7 cm. long, bearing single or paired shortly pedicelled spikelets loosely and irregularly arranged on a triquetrous rhachis. Spikelets 3–5 mm. long, glabrous, acute, with a short stipe 0.2–0.5 mm. long. Inferior glume $\frac{1}{3}-\frac{1}{2}$ length of spikelet. Superior lemma rugose, acute.

Mozambique. MS: Beira, 3.v.1969, *Crook* 884 (K; SRGH). M: Maputo, 2.iv.1948, *Schweickerdt* 1899 (K; PRE).

Also from Kenya to S. Africa. Coastal bushland and forest in shade, often on sand dunes; 0–30 m.

Usually described as perennial, but none of the specimens seen has dormant buds at the base.

27. **Brachiaria grossa** Stapf in F.T.A. **9**: 547 (1919). —Clayton & Renvoize in F.T.E.A., Gramineae: 596 (1982). TAB. **18**, fig. A. Types from Angola.

    Panicum nudiglume var. *major* Rendle, Cat. Afr. Pl. Welw. **2**: 170 (1899). Type from Angola.

Annual. Culms 30–100 cm. high. Leaf laminae cordate or not, the margins scaberulous. Inflorescence of 3–14 racemes, these 3–10 cm. long, bearing paired loosely continguous spikelets on a stiff triquetrous rhachis. Spikelets 3–4 mm. long, obovate-elliptic, glabrous, obtuse to subacute, without a stipe. Inferior glume $\frac{1}{4}$–$\frac{1}{2}$ length of spikelet. Superior lemma coarsely rugose, obtuse.

    Caprivi Strip. Bagoni, 19.i.1956, *de Winter* 4327 (K; PRE). **Botswana**. N: 21°S, 22°24.75'E, 13.iii.1980, *Smith* 3154 (K; SRGH). SW: Ghanzi, Old Winkel, 10.i.1970, *Brown* (K). **Zambia**. N: Mbala Distr., Crocodile Is., 9.ii.1964, *Richards* 18998 (K). C: Mumbwa Distr., Sala Res., 19.iii.1963, *Vesey-FitzGerald* 3970 (K). S: Mapanza Mission, 29.i.1953, *Robinson* 66 (K). **Zimbabwe**. N: Gokwe Distr., Sengwa Res. Sta., 4.ii.1976, *Guy* 2383 (K; SRGH). W: Shangani Distr., Gwampa Forest Reserve, ii.1956, *Goldsmith* 8/56 (K; SRGH). S: Bikita Distr., Birchenough Bridge, 13.ii.1974, *Davidse, Simon & Pope* 6630 (K; MO; SRGH). **Malawi**. N: Rumphi Gorge, 12.ii.1968, *Simon & Williamson* 1780 (K; SRGH).

Also in Kenya to S. Africa. Open places in savanna woodland; 500–1500 m.

Uncommon to the north of our area, where it is replaced by the closely related *B. serrifolia*.

28. **Brachiaria serrifolia** (Hochst.) Stapf in F.T.A. **9**: 548 (1919). —Clayton in F.W.T.A. **3**: 444 (1972). —Clayton & Renvoize in F.T.E.A., Gramineae: 596 (1982). TAB. **18**, fig. B. Type from Ethiopia.

    Panicum serrifolium Hochst. in Flora **38**: 196 (1855). Type as above.

Like *B. grossa* but leaf laminae cordate to amplexicaul, the margins cartilaginous and serrately crinkled; spikelets 4–5 mm. long.

    Zimbabwe. E: Mutambara Tribal Trust Land, Umvumvumvu, 30.xii.1968, *Crook* 830 (K; SRGH). S: Lower Sabi, Devure R., 2.ii.1948, *Rattray* 1307 (K; SRGH).

Northwards to Ethiopia and Niger Republic. Disturbed places in savanna woodland; 700 m.

29. **Brachiaria xantholeuca** (Schinz) Stapf in F.T.A. **9**: 541 (1919). —Launert in Merxm., Prodr. Fl. SW. Afr. **160**: 44 (1970). —Clayton in F.W.T.A. **3**: 444 (1972). —Clayton & Renvoize in F.T.E.A., Gramineae: 597 (1982). Type from Namibia.

    Panicum xantholeucum Schinz in Verh. Bot. Ver. Prov. Brand. **30**: 141 (1888). Type as above.

Tufted annual. Culms 10–60 cm. high. Leaf laminae velvety pubescent. Inflorescence of 2–8 racemes, these 1–7 cm. long, bearing single (rarely paired at base of longer racemes) spikelets in 2 neat rows on a triquetrous rhachis. Spikelets 2.7–4 mm. long, hispidulous (very rarely glabrous) acute to pungently cuspidate, borne on a stipe c. 0.5 mm. long. Inferior glume $\frac{1}{3}$–$\frac{1}{2}$ length of spikelet. Superior lemma granulose to rugulose, subacute.

    Botswana. N: Shamatoka, 15.ii.1983, *Smith* 4086 (K; SRGH). **Zambia**. B: Mongu airport, 22.iii.1964, *Verboom* 1311 (K). C: Chisamba, Kalangwa Farm, 23.iii.1953, *Hinds* 25 (K). S: Kalomo, Senkolo, 14.i.1964, *Astle* 2825 (K). **Zimbabwe**. N: Hurungwe (Urungwe) Distr., 17 km. ESE. of Chirundu, Mensa Pan, 29.i.1958, *Drummond* 5516 (K; SRGH). W: Matobo Distr., Figtree, 25.i.1974, *Mangena & Nyathi* 74/25 (K; SRGH). C: Gweru Distr., Mlezu Govt. Agric. School, 14.ii.1966, *Biegel* 912 (K; SRGH). E: Nyanga (Inyanga) Distr., Cheshire, 15.i.1931, *Norlindh & Weimarck* (K). S: Masvingo Distr., Glyntor, 27.xii.1947, *Robinson* 25 (K; SRGH). **Mozambique**. MS: Gorongosa Nat. Park, 13.iii.1969, *Tinley* 1855 (K; SRGH). M: Goba Fronteira, 11.i.1980, *de Koning* 7996 (K).

Tropical Africa. Savanna woodland (often mopane woodland on black clay), usually in disturbed or overgrazed places; 300–1300 m.

    B. xantholeuca is easily confused with *B. ramosa*, in which the spikelets are mostly borne in pairs and are less pointed at the apex.

30. **Brachiaria leucacrantha** (K. Schum.) Stapf in F.T.A. **9**: 540 (1919). —Clayton & Renvoize in F.T.E.A., Gramineae: 597 (1982). Type from Tanzania.

    Panicum leucacranthum K. Schum. in Pflanzenw. Ost-Afr. **C**: 102 (1895). Type as above.

Like *B. xantholeuca* but spikelets 4–5 mm. long, pubescent, becoming pilose above with the hairs forming a loose tuft on either side of the midnerve, caudate-acuminate.

Tab. 18. A.—BRACHIARIA GROSSA. A1, habit (×½); A2, spikelet showing inferior glume (× 6); A3, superior lemma (× 6), all from *Robinson* 500. B.—BRACHIARIA SERRIFOLIA. B1, portion of leaf lamina (× 1), from *Crook* 830.

Mozambique. Without locality, 5.iv.1894, *Kuntze* (K).
Northward to Kenya, Uganda and Zaire. Coastal bushland.
Perhaps no more than a local variant of *B. xantholeuca*.

31. **Brachiaria ramosa** (L.) Stapf in F.T.A. **9**: 542 (1919). —Jackson & Wiehe, Annot. Check List
Nyasal. Grass.: 31 (1958). —Bor, Grasses of B.C.I. & P.: 284 (1960). —Clayton in F.W.T.A. **3**: 444
(1972). —Clayton & Renvoize in F.T.E.A., Gramineae: 599 (1982). Type from India, cult. at
Uppsala.
　　Panicum ramosum L., Mant. Pl. **1**: 29 (1767). Type as above.
　　Panicum pallidum Peter in Abh. Königl. Ges. Wiss. Göttingen, Math.-Phys. Kl., n.f., **13**, 2: 45
(1928). Type from Tanzania.
　　Echinochloa ramosa (L.) Roberty in Bull. Inst. Fond. Afr. Noire **17**: 64 (1955). Type as for
Panicum ramosum.
　　Urochloa ramosa (L.) Nguyen in Nov. Sist. Pl. Vasc. Acad. Sci. U.R.S.S. **1966**: 13 (1966). Type as
above.

Loosely tufted annual. Culms 10–70 cm. high. Inflorescence of 3–15 racemes, these 1–3
cm. long, simple or the longest with branchlets at the base, bearing mostly paired loosely
contiguous spikelets appressed to a triquetrous rhachis; pedicels shorter than spikelets,
1–2 mm. long. Spikelets 2.5–3.5 mm. long, glabrous to pubescent, acute, with or without a
short stipe up to 0.5 mm. long. Inferior glume $\frac{1}{3}$–$\frac{1}{3}$ length of spikelet. Superior lemma
rugose, subacute to acute.

Zimbabwe. N: Gokwe, 3.ii.1984, *Mahlangu* 857 (K; SRGH). W: Matobo Distr., Figtree, Vreigevicht
Farm, 22.i.1974, *Mangena & Nyathi* 74/34. C: Kadoma, 20.xii.1927, *Eyles* 5080 (K; SRGH). E: 26 km.
S. of Mutare on Rd. to Masvingo, 13.ii.1974, *Davidse, Simon & Pope* 6610 (K; MO; SRGH).
S: Masvingo, Glyntor, 28.xii.1947, *Robinson* 56 (K; SRGH). **Malawi**. S: Blantyre Distr., Matope
Mission, 12.ii.1970, *Brummitt & Banda* 8536 (K). **Mozambique**. T: Tete, ii.1859, *Kirk* (K). GI:
Massingir Distr., Zulo, 4.xii.1981, *White* 41 (K).
　　Tropical Asia; Arabia to Senegal with scattered records elsewhere in Africa. Roadsides and old
farmland in savanna woodland; 500–1200 m.
　　B. ramosa is an Asiatic species extending through Arabia into sub-saharan Africa. It is not clear why
it should also be well represented in Zimbabwe.

32. **Brachiaria deflexa** (Schumach.) Robyns in Bull. Jard. Bot. Brux. **9**: 181 (1932). —Sturgeon in
Rhod. Agric. Journ. **50**: 424 (1953). —Chippindall in Meredith, Grasses & Pastures of S. Afr.: 378
(1955). —Jackson & Wiehe, Annot. Check List Nyasal. Grass.: 31 (1958). —Bor. Grasses of B.C.I.
& P.: 281 (1960). —Clayton in F.W.T.A. **3**: 444 (1972). —Clayton & Renvoize in F.T.E.A.,
Gramineae: 598 (1982). Type from Ghana.
　　Panicum deflexum Schumach., Beskr. Guin. Pl.: 63 (1827). Type as above.
　　Panicum regulare Nees, Fl. Afr. Austr.: 41 (1841). Type from Ghana.
　　Brachiaria regularis (Nees) Stapf in F.T.A. **9**: 544 (1919). Type as above.
　　Panicum ramosum var. *deflexum* (Schumach.) Peter, Fl. Deutsch Ost-Afr. **1**: 177 (1930). Type as
for *Panicum deflexum*.
　　Pseudobrachiaria deflexa (Schumach.) Launert in Mit. Bot. Staatss. München **8**: 158 (1970); in
Merxm., Prodr. Fl. SW. Afr. **160**: 156 (1970). Type as above.

Loosely tufted annual. Culms 15–70 cm. high. Inflorescence of 7–15 racemes, these
2–10 cm. long, often compound, bearing mostly paired distant spikelets spreading from a
triquetrous rhachis and imitating a panicle; pedicels, or some of them, longer than
spikelet, up to 15 mm. long. Spikelets like *B. ramosa*.

Caprivi Strip. Katima Mulilo, 24.xii.1958, *Killick & Leistner* 3064 (K; PRE). **Botswana**. N: Kasane,
27.iii.1966, *Blair-Rains* 69 (K). SW: Ghanze, Old Winkel Farm, 25.i.1970, *Brown* 8240 (K).
Zambia. N: Mpulungu, 29.xii.1951, *Richards* 208 (K). C: Mt. Makulu Res. St., 25.i.1964, *Angus* 3836
(FHO; K). E: Lugomo, 3.ii.1958, *Stewart* 102 (K). S: Livingstone, 12.ii.1959, *McCallum-Webster* Z14
(K; SRGH). **Zimbabwe**. N: Makonde (Lomagundi) Distr., Silverside, 3.ii.1965, *Wild* 6800 (K; SRGH). W: Hwange
Distr., Victoria Falls village, 9.i.1974, *Gonde* 34/73 (K; SRGH). C: Kwekwe Distr., Sable Park, 5.ii.1977,
Chipunga G27 (K; SRGH). E: Mutare, 12.iii.1968, *Crook* P64 (K; SRGH). S: Masvingo Distr., Great
Zimbabwe Nat. Park, 21.i.1971, *Chiparawasha* 293 (K; SRGH). **Malawi**. N: Rumphi Distr., St. Patrick's
Mission, 2.iv.1977, *Pawek* 12563 (K; MO; SRGH; UC). C: Blantyre Distr., Chileka, 22.i.1957, *Banda*
342 (K). S: Murchison Falls, 8.iii.1951, *Jackson* 449 (K). **Mozambique**. N: Erati, Est. Exp. Namapa,
2.iii.1960, *Lemos & Macúacua* 3 (K). Z: Namagoa, *Faulkner* 27 (K; PRE). T: Estima to Caho, 27.i.1972,
Macêdo 4725 (K). MS: Chemba, 12.iv.1960, *Lemos & Macuácua* 64 (K). M: Costa do Sol, 3.iii.1949,
Barbosa, Myre & Carvalho 395 (K).
　　Tropical Africa; a few records in India. Weedy places, preferring light shade; sea level to 1100 m.
　　In their typical form there is little similarity between the narrow racemes of *B. ramosa* and the
paniculate inflorescence of *B. deflexa*, which is often mistaken for a *Panicum*. In fact the two forms
intergrade so completely that they might be better treated as geographical subspecies.

13. ECCOPTOCARPHA Launert

Eccoptocarpha Launert in Senck. Biol. **46**: 124 (1965).

Inflorescence of racemes along a central axis, the spikelets single and adaxial. Spikelets obovate. Inferior glume $\frac{3}{4}$ length of spikelet. Superior glume and inferior lemma as long as spikelet, with prominent cross-nerves and resembling woven fabric. Superior lemma shorter than spikelet, crustaceous, obtuse, supported on an S-shaped internode which eventually straightens to extrude the floret.

A monotypic genus. Tanzania and Zambia.

Eccoptocarpha obconiciventris Launert in Senck. Biol. **46**: 126 (1965). —Clayton & Renvoize in F.T.E.A., Gramineae: 575 (1982). TAB. **19**. Type: Zambia, Kasama, *Phipps & Vesey-FitzGerald* 2980 (SRGH, holotype; K, isotype).

Annual. Culms 40–75 cm. high. Inflorescence of 2–7 racemes, these 1–5 cm. long on a triquetrous rhachis. Spikelets 3–4 mm. long. Stipe of superior floret c. 1 mm. long.

Zambia. N: Mporokoso, 13.iv.1961, *Phipps & Vesey-FitzGerald* 3137 (K; SRGH). Also in Tanzania. Open places; 1000–1500 m.

14. UROCHLOA Beauv.

Urochloa Beauv., Ess. Agrost.: 52 (1812)

Inflorescence of racemes along a short central axis, the spikelets single and abaxial, or paired. Spikelets lanceolate to ovate, plano-convex, cuspidate (*U. platyrrhachis* acute). Inferior glume mostly shorter than spikelet. Inferior lemma awnless. Superior lemma shorter than spikelet, coriaceous, broadly obtuse, mucronate (except *U. platyrrhachis*), the mucro sometimes puberulous; superior palea obtuse.

A genus of 11 species. Old World tropics, mainly African.

Urochloa is distinguished from *Brachiaria* by the orientation and shape of the spikelets, though the boundary is blurred by intermediates. Most of the species are closely related and difficult to separate. In particular the acceptance of species pairs differing only in their annual or perennial habit, and indistinguishable unless the base is well collected, is open to question, but no experimental evidence is available to clarify their status.

1. Raceme rhachis foliaceous - - - - - - - - - - 1. *platyrrhachis*
 – Raceme rhachis triquetrous or narrowly winged - - - - - - - 2
2. Inferior glume up to $\frac{1}{2}$ length of spikelet - - - - - - - - 3
 – Inferior glume over $\frac{2}{3}$ length of spikelet - - - - - - - - 4
3. Raceme rhachis fringed with stiff brown bristles 3–7 mm. long, the spikelets also
 bristly - - - - - - - - - - - - - 2. *echinolaenoides*
 – Raceme rhachis glabrous or with a few pallid bristles - - - - 3. *panicoides*
4. Inferior glume 3-nerved, oblong to narrowly oblong, slightly cartilaginous, often with 1–5 stiff
 hairs from middle of back, usually obtuse - - - - - - - - - 5
 – Inferior glume 5-nerved, lanceolate, membranous, without a central tuft of hairs, narrowly
 obtuse to acute - - - - - - - - - - - - - - 6
5. Plant perennial - - - - - - - - - - - 4. *mosambicensis*
 – Plant annual - - - - - - - - - - - - 5. *trichopus*
6. Plant perennial - - - - - - - - - - - 6. *oligotricha*
 – Plant annual - - - - - - - - - - - - 7. *brachyura*

1. **Urochloa platyrrhachis** C.E. Hubbard in Kew Bull. **1934**: 112 (1934). Type from Zaire.

Annual. Culms 20–60 cm. high, ascending. Inflorescence of 1–3 racemes, these 1–5 cm. long, bearing the spikelets singly on a foliaceous rhachis 2–5 mm. wide and terminating in a spikelet. Spikelets 3.5–4 mm. long, broadly elliptic, glabrous. Inferior glume less than 0.5 mm long, broadly obtuse to truncate. Inferior lemma with a hyaline groove down the midline. Superior lemma granulose, mucronulate, shorter than spikelet.

Zambia. N: Mporokoso Distr., 15 km. W. of Muzombwe, 16.iv.1961, *Phipps & Vesey-FitzGerald* 3240 (K; SRGH). W: 16 km. W. of Mufulira, 5.iii.1960, *Robinson* 3376 (K).

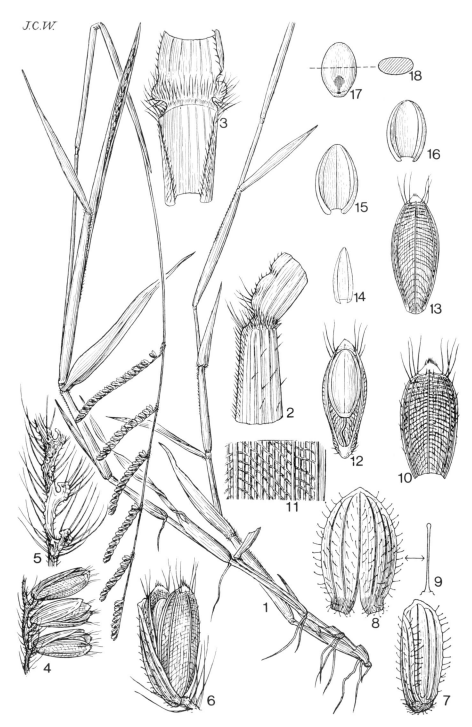

Tab. 19. ECCOPTOCARPHA OBCONICIVENTRIS. 1, habit (×⅔); 2, collar of leaf sheath; 3, ligule;
4, portion of raceme; 5, portion of raceme rhachis; 6, spikelet; 7, 8 & 9, inferior glume, detail of
clavate hair; 10 & 11, inferior glume (surface detail); 12, superior floret within inferior glume;
13, inferior lemma; 14, inferior palea; 15 superior lemma; 16, superior palea; 17–18, caryopsis,
all from *Boaler* 246, 2–4 (× 4), 5–18 (× 8).

Zaire (Katanga). Weedy places; 1000–1200 m.
An odd species, somewhat distant from the rest of the genus. It is not unlike *Paspalidium* but this genus has the rhachis ending in a point.

2. **Urochloa echinolaenoides** Stapf in Fl. Trop. Afr. **9**: 595 (1920). —Clayton & Renvoize in F.T.E.A., Gramineae: 601 (1982). Type from Zaire.
    Panicum fuscoviolaceum Peter, Fl. Deutsch Ost-Afr. **1**, Anh.: 25 (1930). Type from Tanzania.

Annual. Culms 30–100 cm. high. Inflorescence of 2–7 racemes, these 2–6 cm. long bearing single or paired spikelets on a subtriquetrous rhachis fringed by copious stiff brown tubercle-based bristles 3–7 mm. long. Spikelets 3.5–5.5 mm. long, lanceolate, bristly on the nerves of the inferior lemma. Inferior glume $\frac{1}{4}$ to almost $\frac{1}{2}$ length of spikelet, ovate. Superior lemma granulose, with a mucro 0.5 mm. long.

Zambia. N: Mporokoso Distr., 15 km. N. of Muzombwe, W. side of Mweru-Wantipa, 16.iv.1961, *Phipps & Vesey-FitzGerald* 3241 (K; SRGH). W: Ndola Distr., L. Kashila near St. Anthony's Mission, 14.ii.1975, *Williamson & Gassner* 2395 (K; SRGH). **Malawi**. N: 22 km. N. of Rumphi on Chilungero Rd., 26.ii.1978, *Pawek* 1390 (K; MO; SRGH; UC).

Tanzania and Zaire. Savanna woodland and thicket; 800–1100 m.

3. **Urochloa panicoides** Beauv., Ess. Agrost.: 53 (1812). —Sturgeon in Rhod. Agric. Journ. **50**: 429 (1953). —Chippindall in Meredith, Grasses & Pastures of S. Afr.: 385 (1955). —Jackson & Wiehe, Annot. Check List Nyasal. Grass.: 65 (1958). —Bor, Grasses of B.C.I. & P.: 372 (1960). —Launert in Merxm., Prodr. Fl. SW. Afr. **160**: 217 (1970). —Clayton & Renvoize in F.T.E.A., Gramineae: 602 (1982). Type from Mauritius.
    Panicum helopus Trin. in Spreng., Neue Entd. **2**: 84 (1821). Type from India.
    Panicum panicoides (Beauv.) Hitchc. in Journ. Wash. Acad. Sci. **9**: 551 (1919). Type as for *Urochloa panicoides*.
    Urochloa helopus (Trin.) Stapf in F.T.A. **9**: 595 (1920). Type as for *Panicum helopus*.
    Panicum oxycephalum Peter, Fl. Deutsch Ost-Afr. **1**, Anh.: 34 (1930). Type from Tanzania.
    Urochloa ruschii Pilger in Not. Bot. Gart. Berlin **15**: 449 (1941). Types from Namibia.

Annual. Culms 10–100 cm. high, often ascending from a prostrate rooting base. Leaf laminae linear to narrowly lanceolate, subamplexicaul, coarse, glabrous or pubescent. Inflorescence of 2–7 (rarely more) racemes, these 1–7 cm. long bearing single or sometimes paired spikelets on a narrowly winged rhachis, the pedicels (rarely also the rhachis) with long white hairs. Spikelets (2.5) 3.5–4.5 (5.5) mm. long, elliptic, glabrous, pubescent or setosely fringed. Inferior glume $\frac{1}{4}$–$\frac{1}{3}$($\frac{1}{2}$) length of spikelet, ovate, 3–5-nerved. Superior glume and inferior lemma usually with cross-nerves. Superior lemma rugulose, with a mucro 0.3–1.3 mm. long.

Botswana. SE: 43 km. W. of Lutlhe, 17.ii.1960, *de Winter* 7326 (K; PRE). **Zambia**. C: Mt. Makulu Res. St., 30.xii.1970, *Anton-Smith* in GHS 211785 (K; SRGH). **Zimbabwe**. N: Makonde (Lomagundi) Distr., Raffingora, xii.1972, *Davies* 3198 (K; SRGH). W: Matobo Distr., Matopos Res. Sta., 8.ii.1954, *Rattray* 1582 (K; SRGH). C: Harare, Newmarket farm, 10.i.1966, *Simon* 633 (K; SRGH). E: 26 km. S. of Mutare on Rd. to Masvingo, 13.ii.1974, *Davidse, Simon & Pope* 6615 (K; MO; SRGH). S: Gwanda Distr., Insindi Ranch, 28.iii.1976, *Nyathi & Mangena* RRI 76/106 (K; SRGH). **Malawi**. C: Salima-Citala Rd. junction, 27.iv.1951, *Jackson* 483 (K). S: Zomba, 23.i.1950, *Wiehe* N/418 (K). **Mozambique**. M: Quinta da Pedra, 9.i.1948, *Gomes e Sousa* 3650 (K).

Also in Africa and India. Weedy and overgrazed places; 0–1600 m.
The spikelet indumentum is variable but appears to be of no taxonomic significance. Similar variability occurs in the four following species.
Occasional specimens have the inferior glume half as long as the spikelet or even a little more. They seem to represent the extreme range of normal variation, and their removal to a separate species (*U. ruschii*) is not justified. They can be distinguished from *U. trichopus* by the cross-nerves and membranous (as opposed to slightly cartilaginous) texture of the superior glume.

4. **Urochloa mosambicensis** (Hackel) Dandy in Journ. Bot., Lond. **69**: 54 (1931). —Sturgeon in Rhod. Agric. Journ. **50**: 429 (1953). —Chippindall in Meredith, Grasses & Pastures of S. Afr.: 382 (1955). —Jackson & Wiehe, Annot. Check List Nyasal. Grass.: 65 (1958). —Clayton & Renvoize in F.T.E.A., Gramineae: 603 (1982). Type: Mozambique, *de Carvalho* (K, isotype).
    Panicum mosambicense Hackel in Bol. Soc. Brot. **6**: 140 (1888). Type as above.
    Urochloa pullulans Stapf in F.T.A. **9**: 590 (1920). —Sturgeon in Rhod. Agric. Journ. **50**: 428 (1953). —Jackson & Wiehe, Annot. Check List Nyasal. Grass.: 65 (1958). Type as above.
    Urochloa rhodesiensis Stent in Proc. Trans. Rhod. Sci. Ass. **32**: 26 (1933). Types: Zimbabwe, Nyamandhlovu, *Rattray* 500 (SRGH, syntype; K, isosyntype); Harare, *Stent* 3669 (SRGH, syntype; K, isosyntype), *Stent* 5416 (SRGH, syntype; K,

isosyntype), *Stent* 5547 (SRGH, syntype; K, isosyntype).

    Brachiaria stolonifera Goossens in Kew Bull. **1934**: 195 (1934). Type from S. Africa.

    Urochloa stolonifera (Goossens) Chippindall in Meredith, Grasses & Pastures of S. Afr.: 381 (1955). Type as above.

Tufted or stoloniferous perennial, the basal sheaths silky pubescent. Culms 20–170 cm. high, ascending. Leaf laminae broadly linear, ± hispid. Inflorescence of 3–20 racemes, these 1–14 cm. long, bearing usually single spikelets on a narrowly winged rhachis. Spikelets 2.5–5.5 mm. long, ovate, glabrous, pubescent or setosely fringed, acuminate sometimes to a subulate tip. Inferior glume $\frac{2}{3}$–$\frac{5}{6}$ length of spikelet, elliptic-oblong, 3-nerved, slightly cartilaginous and shiny, often with a tuft of hairs from middle of back. Superior glume and inferior lemma usually without cross-nerves. Superior lemma granulose to rugulose, with a mucro 0.5–1.2 mm. long.

    Botswana. N: 11 km. N. of Tsau, 11.iii.1965, *Wild & Drummond* 6854 (K; SRGH). SW: Central Kalahari Game Reserve, Deception Valley, 19.iii.1983, *Smith* 4204 (K; SRGH). SE: Mahalapye, xii.1963, *Yalala* 380 (K; SRGH). **Zambia**. C: Chilanga, Mt. Makulu Res. St., 15.i.1958, *Angus* 1817 (K). E: Chipata, 31.xii.1957, *Stewart* 100 (K). S: Mapanza Mission, 23.xii.1952, *Robinson* 21 (K). **Zimbabwe**. N: Hurungwe (Urungwe) Distr., Mana Pools, Nat. Park, 15.ii.1983, *Dunham* 256 (K; SRGH). W: Hwange Distr., Victoria Falls Nat. Park, Chamabondo, *Matika* 30/75 (K; SRGH). C: Harare, Agric. Exp. St., 1.ii.1954, *Sturgeon* in GHS 45321 (K; SRGH). E: 78 km. S. of Mutare, 13.ii.1974, *Davidse, Simon & Pope* 6623 (K; MO; SRGH). S: Chiredzi Distr., Magudu Ranch, 21.ii.1967, *Cleghorn* 1376 (K; SRGH). **Malawi**. N: Rumphi, 12.ii.1968, *Simon & Williamson* 1777 (K; SRGH). C: Lilongwe Agric. Res. St., 8.ii.1951, *Jackson* 395 (K). S: Zomba, 16.i.1950, *Wiehe* N/410 (K). **Mozambique**. N: Marrupa, 5.ii.1981, *Nuvunga* 462 (K; PRE). Z: Mocuba, Namagoa, *Faulkner* 26 (K; PRE). T: Estima to Inhacapirire, 26.i.1972, *Macêdo* 4692 (K). MS: Chemba, 12.iv.1960, *Lemos & Macuácua* 69 (K). GI: Massingir, 15 km. from Lagoa Nova, 10.iv.1972, *Myre, Lousã & Rosa* 5738 (K). M: Maputo, 9.i.1941, *Torre* 2476 (COI; K).

    Also in Uganda and Kenya to S. Africa; introduced to many other tropical countries as a forage plant. Savanna woodland and open grassland, often on seasonally flooded clays or disturbed sites; 0–1600 m.

    U. mosambicensis is variable, but Burt et al. (Austral. Journ. Bot. **28**: 343–356, 1980) comment that its morphological and agronomic characters intergrade so much that infraspecific taxa cannot be separated.

5. **Urochloa trichopus** (Hochst.) Stapf in F.T.A. **9**: 589 (1920). —Sturgeon in Rhod. Agric. Journ. **50**: 428 (1953). —Chippindall in Meredith, Grasses & Pastures of S. Afr.: 384 (1955). — Launert in Merxm., Prodr. Fl. SW. Afr. **160**: 217 (1970). —Clayton in F.W.T.A. **3**: 440 (1972). —Clayton & Renvoize in F.T.E.A., Gramineae: 604 (1982). TAB. **20**. Type from Sudan Republic.

    Panicum trichopus Hochst. in Flora **27**: 254 (1844). Type as above.

    Urochloa engleri Pilger in Not. Bot. Gart. Berlin **15**: 450 (1941). Type from Namibia.

Coarse annual, the basal sheaths glabrous to loosely pubescent. Otherwise like *U. mosambicensis*.

    Caprivi Strip. Mpilila Is., 13.i.1959, *Killick & Leistner* 3343 (K; PRE). **Botswana**. N: Maun, 13.iii.1975, *Smith* 1273 (K; PRE). SE: Gaborone, 20.xi.1973, *Mott* 21 (K; UBLS). **Zambia**. W: Kitwe, 17.ii.1960, *Fanshawe* 5366 (K). E: Chipata Distr., Chizombo, 17.ii.1969, *Astle* 5473 (K). S: Mapanza Mission, 31.i.1954, *Robinson* 504 (K). **Zimbabwe**. N: Darwin Distr., Mkumbura, 29.i.1970, *Simon* 2098 (K; SRGH). W: Nyamandhlovu Pasture Res. Sta., 15.i.1954, *Plowes* 1674 (K; SRGH). E: Chimanimani Distr., 8 km. S. of Hot Springs, 23.iv.1969, *Plowes* 3188 (K; SRGH). S: Mwenezi (Nuanetsi) Expt. St., 18.i.1987, *Kelly* 557 (K; SRGH). **Malawi**. N: 32 km. N. of Chilumba, 40 km. S. of Karonga, 14.iv.1976, *Pawek* 11019 (K; MO). **Mozambique**. T: Mutarara Velha to Sinjal, 18.vi.1949, *Barbosa & Carvalho* 3128 (K). GI: Chibuto, 11.ii.1959, *Barbosa & Lemos* 8374 (K).

    Africa, mainly in the east; Yemen. Savanna woodland and grassland, usually on sandy soils and following cultivation; 400–1200 m.

6. **Urochloa oligotricha** (Fig. & De Not.) Henr. in Blumea **4**: 502 (1941). —Clayton & Renvoize in F.T.E.A., Gramineae: 606 (1982). Type from Sudan.

    Panicum oligotrichum Fig. & De Not. in Mem. Accad. Sci. Torino, ser. 2, **14**: 333 (1853). Type as above.

    Helopus bolbodes Steud., Syn. Pl. Glum. **1**: 100 (1854). Type from Ethiopia.

    Panicum bolbodes (Steud.) Schweinf., Beitr. Fl. Aeth.: 300 (1867). Type as above.

    Eriochloa bolbodes (Steud.) Schweinf. in Bull. Herb. Boiss. **2**, App. 2: 17 (1894). Type as above.

    Panicum bulawayense Hackel in Proc. Trans. Rhod. Sci. Ass. **7**, 2: 69 (1908). Type: Zimbabwe, Bulawayo, *Jeffreys* 28 (W, holotype).

    Urochloa bolbodes (Steud.) Stapf in F.T.A. **9**: 593 (1920). —Sturgeon in Rhod. Agric. Journ. **50**: 429 (1953). —Chippindall in Meredith, Grasses & Pastures of S. Afr.: 384 (1955). —Launert in Merxm., Prodr. Fl. SW. Afr. **160**: 216 (1970). Type as for *Helopus bolbodes*.

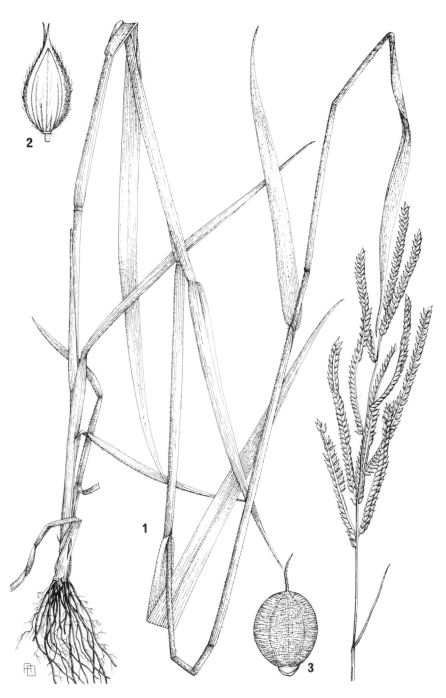

Tab. 20. UROCHLOA TRICHOPUS. 1, habit (× ½); 2, spikelet showing inferior glume (× 6); 3, superior glume (× 10), all from *Procter* 2381. From F.T.E.A.

Brachiaria bulawayensis (Hackel) Henr. in Blumea **3**: 436 (1940). Type as for *Panicum bulawayense.*

Tufted perennial, the base silky tomentose. Culms 20–120 cm. high, ascending. Leaf laminae broadly linear. Inflorescence of 2–20 racemes, these bearing mostly paired spikelets on a subtriquetrous rhachis, the latter with or without sparse bristles. Spikelets 3.5–6 mm. long, lanceolate to narrowly ovate, glabrous, pubescent or setosely fringed, acuminate. Inferior glume $\frac{2}{3}-\frac{3}{4}$ length of spikelet, lanceolate, 5-nerved, membranous to herbaceous, without a central tuft of bristles, narrowly obtuse to acute. Superior lemma granulose to rugulose, with a mucro 0.3–1 mm. long.

Botswana. N: Moremi Wildlife Reserve, Rd. to Maxara Pan, 5.iii.1976, *Smith* 1635 (K; SRGH). SE: 29 km. N. of Mahalapye, 15.iv.1931, *Pole-Evans* 3192(42) (K; PRE). **Zambia**. C: Lusaka, 9.i.1963, *van Rensburg* 1166 (K). E: Jumbe, 30.xii.1965, *Astle* 4224 (K). S: Choma Distr., Zeze, Sinazongwe, 29.xii.1958, *Robson* 1020 (K). **Zimbabwe**. N: Gokwe Distr., Sengwa Res. St., 5.ii.1976, *Guy* 2384 (K; SRGH). W: Waterford, 6.ii.1974, *Norrgrann* 502 (K; SRGH). C: 20 km. S. of Kadoma, 23.i.1967, *Biegel* 1802 (K; SRGH). S: Masvingo Distr., Mushandike Nat. Park, 28.i.1971, *Chiparawasha* 316 (K; SRGH). **Malawi**. N: Ngonga, c. 16 km. from Rumphi on Livingstonia Rd., 13.i.1976, *Phillips* 959 (K; MO).

Also in Ethiopia to S. Africa. Savanna woodland, roadsides and old farmland; 500–1500 m.

Very close to *U. mosambicensis,* being distinguished by the thinner 5-nerved inferior glume and narrower spikelets.

Occasional Zambian specimens have a superior glume as long as the spikelet and prominently nerved, the spikelets borne singly on few racemes. Similar plants occur sporadically in Zaire, Tanzania and Angola, and seem to represent the extreme range of normal variation rather than a separate taxon.

7. **Urochloa brachyura** (Hackel) Stapf in F.T.A. **9**: 592 (1920). —Sturgeon in Rhod. Agric. Journ. **50**: 429 (1954). —Chippindall in Meredith, Grasses & Pastures of S. Afr.: 384 (1955). —Launert in Merxm., Prodr. Fl. SW. Afr. **160**: 217 (1970) —Clayton & Renvoize in F.T.E.A., Gramineae: 606 (1982). Type from Namibia.
 Panicum brachyurum Hackel in Verh. Bot. Ver. Prov. Brand. **30**: 142 (1888). Type as above.
 Urochloa geniculata C.E. Hubbard in Kew Bull. **1933**: 499 (1933). Type from Tanzania.
 Urochloa novemnervia C.E. Hubbard in Kew Bull. **1933**: 500 (1933). Type from Tanzania.

Coarse annual, the basal sheaths loosely pubescent. Otherwise like *U. oligotricha.*

Botswana. N: 15 km. SW. of Maun, 24.i.1972, *Biegel & Gibbs-Russell* 3768 (K; SRGH). SW: 4 km. N. of Dondong bore hole, 23.i.1976, *Skarpe* 19 (K). SE: 9 km. N. of Muramosh (Morwamosu) on Rd. to Kan (Kang), 17.ii.1960, *de Winter* 7339 (K; PRE). **Zimbabwe**. W: Nyamandhlovu Distr., Edwaleni Farm, ii.1972, *Keogh* 16 (K; SRGH). C: Marondera, 6.ii.1954, *Corby* 7920 (K; SRGH).

Also in Ethiopia to S. Africa. Savanna woodland and bushland on sandy or sodic soils; 900–1100 m.

15. ERIOCHLOA Kunth

Eriochloa Kunth in Humb. & Bonpl., Nov. Gen. Sp. **1**: 94 (1816).

Inflorescence a panicle condensed about the primary branches or of racemes along a central axis, the spikelets single and abaxial, paired or on short side branchlets. Spikelets lanceolate to narrowly ovate, thinly biconvex, acute to aristate, with a little globose bead at the base. Inferior glume vestigial (except *E. meyeriana, E. rovumensis*), adnate to swollen basal internode of rhachilla. Superior glume as long as spikelet, often with an awn-point. Superior lemma coriaceous, granulose, usually mucronate.

A genus of 30 species. Tropics.

1. Inferior glume present - - - - - - - - - - - - 2
 – Inferior glume apparently absent: - - - - - - - - - 3
2. Inflorescence of 3–7 racemes; inferior glume $\frac{4}{5}$ length of spikelet - - 1. *rovumensis*
 – Inflorescence a panicle; inferior glume a little truncate cuff up to 0.5 mm. long, rarely longer - - - - - - - - - - - - 2. *meyeriana*
3. Inferior floret ♂ (occasionally barren), with a palea - - - - - 4
 – Inferior floret barren, without a palea; annual - - - - - - 5
4. Perennial; raceme rhachis seldom hairy - - - - - 3. *stapfiana*
 – Annual; raceme rhachis often hairy - - - - - - 4. *macclounii*
5. Spikelet tipped by an awn-point or awn 0.5–4 mm. long - - - - 5. *fatmensis*
 – Spikelet acute to acuminate - - - - - - - - - 6. *procera*

1. **Eriochloa rovumensis** (Pilger) Clayton in Kew Bull. **34**: 560 (1980). —Clayton & Renvoize in
   F.T.E.A., Gramineae: 568 (1982). Type from Tanzania.
   Panicum rovumense Pilger in Bot. Jahrb. **33**: 47 (1904). Type as above.
   Brachiaria rovumensis (Pilger) Pilger in Not. Bot. Gart. Berlin **13**: 263 (1936). Type as above.
   Eriochloa biglumis Clayton in Kew Bull. **30**: 107 (1975). Type from Tanzania.

Annual. Culms 45–60 cm. high, geniculately ascending. Inflorescence of few (3–7)
racemes, these 1.5–3 cm. long, bearing single or paired spikelets on a shortly pilose
triquetrous rhachis. Spikelets 3.5–5.5 mm. long, lanceolate, thinly pilose. Inferior glume $\frac{4}{5}$
length of spikelet, lanceolate, acuminate; superior glume acuminate or with an awn-point
up to 1 mm. long. Inferior floret with a palea. Superior lemma obscurely mucronulate.

Mozambique. N: Nampula, 21 km. towards Nametil, 1.iv.1964, *Torre & Paiva* 11554 (COI;
K). Z: Ile Mts., 2.iv.1943, *Torre* 5047 (COI; K).
Also in Tanzania. Shallow soil over rocks; 320 m.
The large inferior glume could lead to confusion with *Brachiaria*, but otherwise the spikelet
structure is typical of *Eriochloa*.

2. **Eriochloa meyeriana** (Nees) Pilger in Engl. & Prantl, Pflanzenfam., ed. 2, **14e**: 56 (1940). —Clayton
   in F.W.T.A. **3**: 437 (1972). —Clayton & Renvoize in F.T.E.A., Gramineae: 569 (1982). Type from
   S. Africa.
   Panicum meyerianum Nees, Fl. Afr. Austr.: 32 (1841). —Stapf in F.T.A. **9**: 650 (1920). —
   Sturgeon in Rhod. Agric. Journ. **50**: 433 (1953). —Chippindall in Meredith, Grasses & Pastures
   of S. Afr.: 331 (1955). —Jackson & Wiehe, Annot. Check List Nyasal. Grass.: 51 (1958). Type as
   above.
   Eriochloa borumensis Hackel in Bull. Herb. Boiss. sér. 2, **1**: 765 (1901). Type: Mozambique,
   Boroma, *Menyharth* 1114 (W, holotype).
   Panicum meyerianum var. *grandeglume* Stent & Rattray in Proc. Trans. Rhod. Sci. Ass. **32**: 28
   (1933). —Chippindall in Meredith, Grasses & Pastures of S. Afr.: 331 (1955). Type: Zimbabwe,
   Hunyani, *Eyles* 3138 (SRGH, holotype; K, isotype).

Perennial. Culms 30–200 cm. high, geniculately ascending and rooting at the nodes,
often rambling, sometimes almost woody. Inflorescence a panicle 8–18 cm. long, bearing
spikelets on branchlets appressed to the primary branches, these puberulous. Spikelets
2.3–3.5 mm. long, elliptic, glabrous or occasionally thinly pubescent. Inferior glume a
little truncate or cuspidate cuff up to 0.5 mm. long, but occasionally larger and at the
extreme becoming a broadly ovate scale up to 2 mm. long; superior glume acute. Inferior
floret with a palea. Superior lemma obscurely mucronate.

Botswana. N: Nata R., 9.iv.1983, *Smith* 4219 (K; SRGH). **Zambia**. E: Chipata, 11.v.1963, *van
Rensburg* 2136 (K). S: 47 km. N. of Choma, 20.v.1954, *Robinson* 773 (K). **Zimbabwe**. N: Gokwe Distr.,
Sengwa Res. St., 24.iii.1977, *Guy* 2490 (K; SRGH). W: Hwange Distr., near Sebungwe-Zambezi R.
confluence, 13.v.1955, *Plowes* 1841 (K; SRGH). E: Chipinge Distr., Chibuwe, 28.vi.1966, *West* 7378 (K;
SRGH). S: Gwanda Distr., Doddieburn Ranch, Umzingwane R., 7.v.1972, *Pope* 659 (K; SRGH).
Malawi. S: Tanga, Phalombe (Palombe) R., 6.v.1952, *Jackson* 800 (K). **Mozambique**. Z: Massingire,
21.v.1943, *Torre* 5352 (COI; K). GI: Chibuto, Baixa Changane, 24.viii.1963, *Macêdo & Macuácua* 1143
(K). M: Between Catuane, Umbeluzi & Goba, 20.iv.1944, *Torre* 6485 (COI; K).
Tropical and S. Africa; SE. Asia. Marshy places and by water; 0–1200 m.
The overall facies of the spikelets places this species in *Eriochloa*, but the presence of an inferior
glume allies it to *Brachiaria mutica*, and the paniculate infloresence can easily be confused with
Panicum.

3. **Eriochloa stapfiana** Clayton in Kew Bull. **30**: 109 (1975). —Clayton & Renvoize in F.T.E.A.,
   Gramineae: 569 (1982). Type: Mozambique, Messalo R., *Allen* 122 (K holotype).
   Eriochloa borumensis auct. non Hackel., —Stapf in F.T.A. **9**: 500 (1919). —Chippindall in
   Meredith, Grasses & Pastures of S. Afr.: 369 (1955). —Jackson & Wiehe, Annot. Check List
   Nyasal. Grass.: 41 (1958).

Tufted perennial. Culms 60–120 cm. high, robust. Inflorescence of 6–many racemes,
these 2–7 cm. long, bearing spikelets in pairs or on short appressed side branchlets, the
rhachis triquetrous and puberulous or pubescent, rarely pilose. Spikelets 3–4 mm. long,
narrowly ovate, appressedly pubescent. Inferior glume absent; superior glume acuminate
or with an awn-point up to 1 mm. long. Inferior floret with a palea. Superior lemma
usually with a mucro up to 0.5 mm. long.

Malawi. S: Mulanje, Tuchila, 23.i.1951, *Wiehe* N739 (K). **Mozambique**. N: L. Chilwa, Mecanhelas,
11.v.1971, *Bowbrick* L5 (K; SRGH). Z: Mocuba, Namagoa, *Faulkner* 23 (K; PRE). MS: Nova

Mambone to Buzi, 28.v.1941, *Torre* 2832 (COI; K). GI: Chibuto to Chaimite, 11.ii.1942, *Torre* 3962 (COI; K). M: Goba, 2.xi.1960, *Balsinhas* 167 (K).

Kenya, Tanzania & S. Africa. Marshy places and by water; 0–100 m.

Rarely the raceme branchlets may be so prolific as to approach a panicle, and a rudimentary inferior glume may occur as an obscure membranous rim. The species then intergrades with *E. meyeriana*; note that the latter seldom has hairy spikelets and never has an obviously mucronate superior lemma.

4. **Eriochloa macclounii** Stapf in F.T.A. **9**: 501 (1919). —Sturgeon in Rhod. Agric. Journ. **50**: 420 (1953). —Jackson & Wiehe, Annot. Check List Nyasal. Grass.: 41 (1958). —Clayton & Renvoize in F.T.E.A., Gramineae: 570 (1982). Type: Malawi, Mwanemba, *McClounie* 8 (K, holotype).

Annual. Culms 60–180 cm. high, erect. Inflorescence of 5–many racemes, these 3–6 cm. long bearing paired spikelets on a pubescent to triquetrous rhachis. Spikelets 3.5–5 mm. long, lanceolate, pubescent to pilose. Inferior glume absent; superior glume attenuate to an awn-point 1–3 mm. long. Inferior floret with a palea. Superior lemma with a mucro 1–2 mm. long.

Botswana. N: Sibuyu R. tributary, 13.iv.1983, *Smith* 4299 (K; SRGH). **Zambia**. C: Luangwa Valley Game Res., Mushilashi R., 31.iii.1967, *Prince* 423 (K). E: Lupande Munkanya, 1.iii.1968, *Phiri* 59 (K; SRGH). S: Mazabuka Distr., Namirongwe Hills, 13.ii.1960, *White* 7007 (FHO; K). **Zimbabwe**. N: Binga Distr., Chizarira Game Res., Lumdum Sidaga, 7.v.1972, *Cleghorn* 2552 (K; SRGH). **Mozambique**. Z: Posto Chire, 4.v.1972, *Bowbrick* J209 (K; SRGH).

Also from Tanzania. Mopane woodland, drainage courses and flood plains; 60–1100 m.

5. **Eriochloa fatmensis** (Hochst. & Steud.) Clayton in Kew Bull. **30**: 108 (1975). —Clayton & Renvoize in F.T.E.A., Gramineae: 571 (1982). TAB. **21**. Type from Saudi Arabia.

Panicum fatmense Hochst. & Steud. in sched., Schimper It. Un.: 806 (1837). Type as above.

Helopus nubicus Steud., Syn. Pl. Glum. **1**: 100 (1854). Type from Sudan.

Eriochloa nubica (Steud.) Thell. in Viert. Nat. Ges. Zürich **52**: 435 (1907). —Sturgeon in Rhod. Agric. Journ. **50**: 420 (1953). —Chippindall in Meredith, Grasses & Pastures of S. Afr.: 370 (1955). —Jackson & Wiehe, Annot. Check List Nyasal. Grass.: 41 (1958). —Bor, Grasses of B.C.I. & P.: 312 (1960). —Launert in Merxm., Prodr. Fl. SW. Afr. **160**: 115 (1970). —Clayton in F.W.T.A. **3**: 437 (1972). Type as above.

Eriochloa fouchei Stent in Bothalia **1**: 260 (1924). Type from S. Africa.

Eriochloa procera sensu Jackson & Wiehe, Annot. Check List Nyasal. Grass.: 41 (1958) non (Retz.) C.E. Hubbard.

Annual. Culms 10–120 cm. high, erect or geniculately ascending. Inflorescence of 3–20 racemes, these 1–5 cm. long, bearing paired spikelets on a puberulous (occasionally pilose) triquetrous or narrowly winged rhachis; pedicels commonly setose, those of a pair usually free. Spikelets (2.5)3–5 mm. long, lanceolate, thinly pubescent. Inferior glume absent; superior glume attenuate to an awnlet 0.5–4 mm. long. Inferior floret represented only by a lemma. Superior lemma with a mucro 0.3–1 mm. long.

Botswana. SE: 5 km. N. of Gaborone, 7.i.1978, *Hansen* 3329 (C; GAB; K; PRE; SRGH; WAG). **Zambia**. N: Mporokoso Distr., Mweru-Wantipa, 15.iv.1961, *Phipps & Vesey-FitzGerald* 3191 (K; SRGH). C: Kafue Bridge, 6.ii.1954, *Hinds* 90 (K). S: Mazabuka Distr., (Zambezi R.) 3 km. upstream from Chirundu Bridge, 1.ii.1958, *Drummond* 5422 (K; SRGH). **Zimbabwe**. W: Matopos Res. St., 12.ii.1954, *Rattray* 1604 (K; SRGH). S: Mwenezi Expt. St., 7.ii.1976, *Takaindisa* 12 (K; SRGH). **Malawi**. S: Blantyre Distr., Chileka, 23.i.1957, *Jackson* 2109 (K). **Mozambique**. Z: Mopeia ix.1974, *Bond* W563 (K; SRGH). MS: Gorongosa Nat. Park, Urema floodplain, 4.ii.1971, *Tinley* 2023 (K; SRGH).

Also in Tropical and S. Africa, Arabia, a few records in India. Damp depressions and streamsides, favouring clayey soils; 0–1500 m.

Intergrades with *E. procera*.

6. **Eriochloa procera** (Retz.) C.E. Hubbard in Kew Bull. **1930**: 256 (1930). —Bor, Grasses of B.C.I. & P.: 312 (1960). —Clayton & Renvoize in F.T.E.A., Gramineae: 571 (1982). Type from India.

Agrostis procera Retz., Obs. Bot. **4**: 19 (1786). Type as above.

Annual. Like *E. fatmensis* except: pedicels commonly glabrous, those of a pair often fused below; superior glume acute to acuminate.

Mozambique. Z: Luabo R., 25.v.1858, *Kirk* (K).

Also in Tropical Asia, introduced to eastern Africa. Damp places.

Tab. 21. ERIOCHLOA FATMENSIS. 1, habit (× ½); 2, raceme (× 5); 3, spikelet (× 10), all from *McCallum-Webster* K29. From F.T.E.A.

16. PASPALUM L.

Paspalum L., Syst. Nat., ed. 10: 855 (1759).

Inflorescence of single, paired, digitate or scattered racemes, the spikelets single and abaxial or paired, on a flat rhachis. Spikelets plano-convex, orbicular to ovate. Inferior glume absent, or rarely represented by a tiny scale. Inferior lemma awnless. Superior lemma coriaceous to crustaceous, smooth or finely striate, usually obtuse.

A genus of c. 330 species. Tropics, predominantly New World.

1. Spikelets with a ciliate fringe from the margins of the superior glume - - - 2
 - Spikelets glabrous or minutely pubescent, without a ciliate fringe - - - - 4
2. Racemes paired; plants stoloniferous - - - - - - - - - 1. conjugatum
 - Racemes 3–20; plants tufted - - - - - - - - - - - - 3
3. Racemes mostly 3–7; spikelets 2.8–4 mm. long - - - - - - 2. dilatatum
 - Racemes mostly 10–20; spikelets 2–2.8 mm. long - - - - - - - 3. urvillei
4. Superior floret pallid; racemes paired (rarely 3–5) - - - - - - - 5
 - Superior floret brown at maturity; racemes 1–20, usually borne on a short common axis: 7
5. Plant rhizomatous; superior glume cartilaginous, glabrous - - - - 4. notatum
 - Plant stoloniferous - - - - - - - - - - - - - - - 6
6. Superior glume papery, glabrous - - - - - - - - - 5. vaginatum
 - Superior glume thinly coriaceous, obscurely pubescent - - - - 6. distichum
7. Inferior glume present on at least some of the spikelets; spikelets 2.8–3.5 mm. long, sometimes paired - - - - - - - - - - - - - - 9. glumaceum
 - Inferior glume absent; spikelets 1.4–3 mm. long, single - - - - - - 8
8. Leaf laminae linear, acuminate to a filiform apex; culm nodes commonly exposed - - - - - - - - - - - - 7. scrobiculatum
 - Leaf laminae narrowly lanceolate, acute, often broadly rounded at base; culms leafy, the nodes concealed - - - - - - - - - - - 8. lamprocaryon

1. **Paspalum conjugatum** Berg. in Acta Helv. Phys.-Math. **7**: 129 (1762). —Stapf in F.T.A. **9**: 569 (1919). —Bor, Grasses of B.C.I. & P.: 336 (1960). —Clayton in F.W.T.A. **3**: 445 (1972). —Clayton & Renvoize in F.T.E.A., Gramineae: 607 (1982). Type from Surinam.

Stoloniferous perennial, often forming swards. Culms 30–60 cm. high. Racemes paired, each 5–17 cm. long, the spikelets borne singly in 2 rows on a narrow rhachis. Spikelets 1.5–1.7 mm. long, orbicular, pale yellow. Superior glume ciliate on the margins.

Malawi. S: Mulanje Distr., Great Ruo Gorge, 19.vi.1962, *Robinson* 5393 (K).
Also in tropical America, widely introduced throughout the tropics. Forest clearing; 800 m.

2. **Paspalum dilatatum** Poir., Encycl. Méth. Bot. **5**: 35 (1804). —Sturgeon in Rhod. Agric. Journ. **50**: 426 (1953). —Chippindall in Meredith, Grasses & Pastures of S. Afr.: 387 (1955). —Bor, Grasses of B.C.I. & P.: 338 (1960). —Clayton in F.W.T.A. **3**: 445 (1972). —Clayton & Renvoize in F.T.E.A., Gramineae: 608 (1982). Type from Argentina.

Tufted perennial. Culms 40–120 cm. high. Racemes usually 3–7, borne along an axis 4–20 cm. long, each raceme 3–11 cm. long, the spikelets paired in 2–4 rows on a rhachis c. 1.2 mm wide. Spikelets 2.8–4 mm. long, 1.8–2.5 mm. wide, ovate, yellowish green. Superior glume sparsely pilose on the surface, ciliate on the margins.

Zambia. C: Lusaka, Carruthers Farm on Leopards Hill Rd., v.1982, *Vernon & Carruthers* 1201 (K). S: Mazabuka Distr., Kafue Pilot Polder, 23.i.1963, *Angus* 3545 (FHO; K). **Zimbabwe**. C: Newmarsh's Farm, 20 km. NE. of Harare, 10.i.1966, *Simon* 634 (K; SRGH). E: Rd. up to Mt. Peni near Glencoe, 29.xi.1967, *Simon & Ngoni* 1377 (K; SRGH). S: Masvingo Distr., 9 km. NW. of Ndanga, 2.iii.1985, *Keller-Grein* Z-110-3 (K; SRGH).
Also in tropical America; widely introduced as a forage grass elsewhere in the tropics, but often becoming a serious weed. Roadsides and farmland; 900–1600 m.

3. **Paspalum urvillei** Steud., Syn. Pl. Glum. **1**: 24 (1854). —Sturgeon in Rhod. Agric. Journ. **50**: 426 (1953). —Chippindall in Meredith, Grasses & Pastures of S. Afr.: 387 (1955). —Bor, Grasses of B.C.I. & P.: 341 (1960). —Clayton in F.W.T.A. **3**: 445 (1972). Type from Brazil.

Tufted perennial. Culms 80–250 cm. high. Racemes usually 10–20, borne along an axis 10–20 cm. long, each raceme 4–13 cm. long, the spikelets paired in 2–4 rows on a rhachis c. 1 mm. wide. Spikelets 2–2.8 mm. long, 1.2–1.5 mm. wide, ovate, yellowish green. Superior glume sparsely pilose on the surface, ciliate on the margins.

Zimbabwe. C: Shurugwi Distr., Rd. to Ferny Creek, 1.iv.1967, *Biegel* 2036 (K; SRGH). E: Umvumvumvu Irrigation Project, *Robinson* 388 (K; SRGH). **Malawi.** C: Mwera Hill Res. St., 31.x.1950, *Jackson* 225 (K). S: Mulanje Distr., Ndala, 28.i.1978, *Seyani* 777 (K; SRGH).

Also in tropical America; widely introduced elsewhere. Roadsides and farmland, favouring damp soils; 1200 m.

P. urvillei closely approaches *P. dilatatum*, being distinguished by its robust habit, numerous racemes and smaller spikelets.

4. **Paspalum notatum** Fluegge, Monogr. Pasp.: 106 (1810). —Chippindall in Meredith, Grasses & Pastures of S. Afr.: 389 (1955). —Bor, Grasses of B.C.I. & P.: 339 (1960). —Clayton & Renvoize in F.T.E.A., Gramineae: 608 (1982). Type from West Indies.

Mat-forming perennial arising from stout rhizomes. Culms 15–50 cm. high. Racemes paired (rarely 3), each 2.5–12 cm. long, the spikelets borne singly in 2 rows on a narrow rhachis. Spikelets 2.5–3.8 mm. long, broadly elliptic, plumply plano-convex, green. Superior glume and inferior lemma cartilaginous, glabrous. Superior lemma pallid at maturity.

Zambia. C: Mt. Makulu, 13.ii.1965, *Lawton* 1202 (K). **Zimbabwe.** C: Harare Distr., Chibvuti Farm, 19.i.1982, *Nicholls* in GHS 278354 (K; SRGH). S: Masvingo Distr., Makaholi Expt. St., 13.iii.1978, *Senderayi* 193 (K; SRGH). **Malawi.** S: Zomba, 18.i.1950, *Wiehe* N/411 (K).

Also in South America. Introduced for grazing and erosion control; occasionally found as an escape.

5. **Paspalum vaginatum** Swartz, Prod. Veg. Ind. Occ.: 21 (1788). —Stapf in F.T.A. **9**: 570 (1919). —Chippindall in Meredith, Grasses & Pastures of S. Afr.: 389 (1955). —Bor, Grasses of B.C.I. & P.: 341 (1960). —Clayton in F.W.T.A. **3**: 445 (1972). —Clayton & Renvoize in F.T.E.A., Gramineae: 609 (1982). Type from Jamaica.

Creeping stoloniferous perennial. Culms 8–60 cm. high. Racemes paired (rarely up to 5), each 1.5–7.5 cm. long, the spikelets borne singly in 2 rows on a rhachis 1–2 mm. wide. Spikelets 3–4.5 mm. long, narrowly ovate-elliptic, markedly flattened, usually pale brownish green. Inferior glume very rarely present as a tiny scale. Superior glume and inferior lemma thinly papery, glabrous. Superior lemma pallid at maturity.

Mozambique. MS: Gorongosa Nat. Park, Cheringoma Plateau, i.1972, *Tinley* 2345 (K; SRGH). M: Maputo, Costa do Sol, 19.ii.1981, *de Koning & Boane* 8654 (K).

Throughout the tropics, extending into subtropical regions. Coastal salt marshes.

6. **Paspalum distichum** L., Syst. Nat. ed. 10, **2**: 855 (1759). —Chippindall in Meredith, Grasses & Pastures of S. Afr.: 389 (1955). —Bor, Grasses of B.C.I. & P.: 338 (1960). —Launert in Merxm., Prodr. Fl. SW. Afr. **160**: 146 (1970). Type from Jamaica.

Digitaria paspalodes Michaux, Fl. Bor. Am. **1**: 46 (1803). Type from U.S.A.

Paspalum paspalodes (Michaux) Scribner in Mem. Torr. Bot. Cl. **5**: 29 (1894). Type as above.

Creeping stoloniferous perennial. Culms 8–60 cm. high. Racemes paired (rarely 3), each 2–8 cm. long, the spikelets borne singly in 2 rows on a rhachis 1–2 mm. wide. Spikelets 3–4 mm. long, ovate, plump, green. Inferior glume present on at least some spikelets as a tiny scale. Superior glume and inferior lemma somewhat coriaceous, the glume nearly always obscurely pubescent. Superior lemma pallid at maturity.

Zimbabwe. W: Railway between Hwange and Victoria Falls, iii.1960, *West* in GHS 107816 (K; SRGH).

Tropics, but shunning the equatorial belt and extending to warm temperate regions. Damp depressions and marshy, but not saline, soils. It occurs only as an escape in the Flora Zambesiaca area, but has become a serious weed in S. Africa.

It closely resembles *P. vaginatum*, which has led to some nomenclatural confusion as the type sheet bears specimens of both species. This has now been resolved by lectotypification (Brummitt in Taxon **32**: 281 1983).

7. **Paspalum scrobiculatum** L., Mant. Pl. **1**: 29 (1767). —Bor, Grasses of B.C.I. & P.: 340 (1962). —Clayton in Kew Bull. **30**: 101–105 (1975). —Clayton & Renvoize in F.T.E.A., Gramineae: 610 (1982). —de Koning & Sosef in Blumea **30**: 279–318 (1985). Type from India.

Paspalum orbiculare Forst., Fl. Ins. Austr. Prodr.: 7 (1786). —Clayton in F.W.T.A. **3**: 446 (1972). Type from Society Is.

Paspalum commersonii Lam., Tab. Encycl. Méth. Bot. **1**: 175 (1791). —Sturgeon in Rhod. Agric. Journ. **50**: 425 (1953). —Chippindall in Meredith, Grasses & Pastures of S. Afr.: 389 (1955).

—Jackson & Wiehe, Annot. Check List Nyasal. Grass.: 53 (1958). —Launert in Merxm., Prodr. Fl. SW. Afr. **160**: 146 (1970). Type from Mauritius.

Paspalum polystachyum R.Br., Prodr. Fl. Nov. Holl.: 188 (1810). —Sturgeon in Rhod. Agric. Journ. **50**: 425 (1953). —Chippindall in Meredith, Grasses & Pastures of S. Afr.: 389 (1955). —Jackson & Wiehe, Annot. Check List Nyasal. Grass.: 53 (1958). —Bor, Grasses of B.C.I. & P.: 335 (1960). —Launert in Merxm., Prodr. Fl. SW. Afr. **160**: 145 (1970). —Clayton in F.W.T.A. **3**: 446 (1972). Type from Australia.

Paspalum auriculatum Presl, Rel. Haenk. **1**: 217 (1830). Type from Philippine Is.

Paspalum dissectum var. *grande* Nees, Fl. Afr. Austr. **1**: 15 (1841). Type from S. Africa.

Paspalum scrobiculatum var. *auriculatum* (Presl) Merr. in Philip. Journ. Sci. **1**: Suppl. 345 (1906). Type as for *P. auriculatum.*

Paspalum scrobiculatum var. *bispicatum* Hackel in Kneucker in Allg. Bot. Zeitschr. **20**: 146 (1914). Type from Philippine Is.

Paspalum scrobiculatum var. *commersonii* (Lam.) Stapf in F.T.A. **9**: 573 (1919). Type as for *P. commersonii.*

Paspalum scrobiculatum var. *polystachyum* (R.Br.) Stapf in F.T.A. **9**: 576 (1919). Type as for *P. polystachyum.*

Perennial. Culms 10–150 cm. high, 1–6 mm. in diam., the nodes mostly exposed, erect or ascending from a decumbent base and rooting at the nodes below. Leaf laminae 5–40 cm. long, 3–15 mm. wide, linear, tapering to a filiform apex. Racemes 1–20, digitate or borne on an axis up to 8 cm. long, the longest raceme 4–15 cm. long, the spikelets borne singly on a rhachis 1–2.5 mm. wide. Spikelets 1.4–3 mm. long, broadly elliptic to subrotund, green becoming brown. Inferior glume absent. Inferior lemma papery or rarely coriaceous, glabrous. Superior lemma brown at maturity.

Caprivi Strip. 11 km. S. of Katima Mulilo on Rd. to Nooma, 22.xii.1958, *Killick & Leistner* 3039 (K; PRE). **Botswana**. N: Maun, 16.i.1974, *Smith* 771 (K; SRGH). SE: Gaborone, 10.ii.1978, *Hansen* 3351 (C; GAB; K; PRE; SRGH; WAG). **Zambia**. B: Barotse Flood Plain, 26.iii.1964, *Verboom* 1156 (K). N: Isoka, 17.i.1962, *Astle* 1270 (K). W: Mwinilunga Distr., Kalenda Plain, 8.xii.1937, *Milne-Redhead* 3545 (K; SRGH). C: Mt. Makulu, 18.i.1965, *Lawton* 1188 (K). E: Chipata, i.1963, *Verboom* 503 (K). S: Mapanza Mission, 10.ii.1954, *Robinson* 523 (K). **Zimbabwe**. N: Gokwe Distr., Swiswi R., 24.i.1963, *Bingham* 446 (K; SRGH). W: Hwange Distr., Kazungula border, 8.iii.1981, *Gonde* 330 (K; SRGH). C: Harare, Mt. Hampden Rd., i.1955, *Davies* 938 (K; SRGH). E: Manyange's Kraal, 5 km. from Haroni-Lusitu confluence, 26.xi.1957, *Ngoni* 57 (K; SRGH). S: Lower Sabi, Devuli R., i.1948, *Rattray* 1304 (K; SRGH). **Malawi**. N: Mzuzu, Katoto, 22.i, *Pawek* 1648 (K). C: Nkhota Kota, Sani Rd., 1.x.1950, *Jackson* 194 (K). S: Zomba, 9.ii.1949, *Wiehe* N/7 (K). **Mozambique**. N: Messalo R., *Allen* 123 (K). Z: Posto Chiré, 4.iv.1972, *Bowbrick* J215 (K; SRGH). MS: Gorongosa Nat. Park, 28.iv.1973, *Tinley* 2793 (K; SRGH). GI: Chibuto, Baixo Changana, 24.viii.1963, *Macêdo & Macúacua* 1144 (K). M: Boane, 19.ii.1981, *de Koning & Boane* 8662 (K).

Old World tropics. Damp soils and by water; 0–1900 m. It is grown as a minor cereal in Asia (de Wet et al. in Econ. Bot. **37**: 154–163, 1983).

There is considerable variation in the size of vegetative and flowering parts and the species may comprise a swarm of apomicts, but variation in Africa is apparently continuous (Clayton l.c.) and the name is here taken in a wide sense. de Konig & Sosef (l.c.) have subdivided the Asian population, separating the cultivated races (var. *scrobiculatum* and var. *auriculatum*) from the common wild form (var. *bispicatum*).

Asian and Australian specimens sometimes have the nerves of the inferior lemma darker than the space between them; these have been segregated as *P. orbiculare*. Similar specimens are also found in Africa, where they intergrade with the rest of the species and do not seem to warrant separation.

8. **Paspalum lamprocaryon** K. Schum. in Pflanzenw. Ost-Afr. **C**: 100 (1895). Type from Tanzania.

Paspalum scrobiculatum var. *lanceolatum* de Koning & Sosef in Blumea **30**: 312 (1985). Type from Tanzania.

Paspalum auriculatum auct. non Presl., —Stapf in F.T.A. **9**: 572 (1919). —Jackson & Wiehe, Annot. Check List Nyasal. Grass.: 52 (1958). —Clayton in F.W.T.A. **3**: 445 (1972). —Clayton & Renvoize in F.T.E.A., Gramineae: 609 (1982).

Perennial. Culms 90–200 cm. high, 3–7 mm. in diam., stout, leafy with the sheaths usually overlapping and concealing the node, ascending, rooting at the nodes below. Leaf laminae 8–30 cm. long, 8–27 mm. wide, narrowly lanceolate, often broadly rounded at the base, acute at the apex. Racemes 2–11 on an axis up to 9 cm. long, each raceme 4–12 cm. long, the spikelets borne singly on a rhachis 2–3 mm. wide. Spikelets 2–3 mm. long, resembling *P. scrobiculatum.*

Zambia. N: Mbala Distr., Chilongolwelo, 28.ii.1959, *McCallum-Webster* A180 (K; SRGH). W: Mwinilunga Distr., tributary of Ysongailu R., 14.xi.1937, *Milne-Redhead* 3235 (K; SRGH). **Zimbabwe**. W: Mufulira, 28.v.1934, *Eyles* 8411 (K; SRGH). **Malawi**. N: Chisenga, foot of Mafinga Mts., 8.xi.1958, *Robson* 523 (K; SRGH). S: Nsanje, Malawe Hill, 23.iii.1960, *Phipps* 2640 (K; SRGH).

Tropical Africa. Streamsides and drainage hollows; 700–1600 m.

A segregate from *P. scrobiculatum*, distinct enough in its typical facies, but the weaker plants merging with robust forms of *P. scrobiculatum*. It has been confused with *P. auriculatum*, a broad but linear leaved Asiatic taxon now included in *P. scrobiculatum*.

9. **Paspalum glumaceum** Clayton in Kew Bull. **30**: 104 (1975). —Clayton & Renvoize in F.T.E.A., Gramineae: 612 (1982). TAB. **22**. Type: Zambia, Mfuwe, *Astle* 5458 (K, holotype).

Tufted perennial. Culms 30–100 cm. high, ascending. Leaf laminae 8–30 cm. long, 8–19 mm. wide, broadly linear to narrowly lanceolate. Racemes 2–3, borne on an axis 1–6 cm. long, each raceme 2–11 cm. long, the spikelets on a rhachis 1.5–2 mm. wide, usually a few of them paired though sometimes the second spikelet represented only by a tooth on the pedicel. Spikelets (2.5) 2.8–3.5 mm. long, broadly elliptic to rotund, light brown. Inferior glume present on at least some of the spikelets as a triangular scale up to 1.5 mm. long. Inferior lemma papery, glabrous to pubescent. Superior lemma brown at maturity.

Zambia. N: Mbesuma, Chambeshi R., 8.i.1962, *Astle* 1208 (K). W: Ndola, i.1934, *Trapnell* 1503 (K). C: Luangwa Valley, Mutinsase R., 10.xi.1965, *Astle* 4098 (K). **Zimbabwe**. N: Sebungwe Distr., Vulunduli camp, ix.1955, *Davies* 1530 (K: SRGH). C: Kwe Kwe, Naude's farm, 20.ii.1984, *Sheppard* in GHS 282179 (K; SRGH). **Malawi**. C: Nkhota Kota, Kanyenda, 24.ii.1953, *Jackson* 1101 (K). S: Zomba, i.1958, *Jackson* 2146 (K).

Northwards to Sudan and also Madagascar. Flood plains and forest margins; 600–1300 m.

A minor segregate from the *P. scrobiculatum* complex.

17. AXONOPUS Beauv.

Axonopus Beauv., Ess. Agrost.: 12 (1812). —Black in Adv. Front. Pl. Sci. **5**: 1–186 (1963).

Inflorescence of 2–many racemes, mostly subdigitate but sometimes along a central axis, the spikelets single and adaxial. Spikelets thinly biconvex, lanceolate to oblong-elliptic. Inferior glume absent. Superior glume as long as spikelet; inferior lemma similar, without a palea. Superior lemma crustaceous, granulose.

A genus of c. 110 species. Tropical and subtropical America; 1 species endemic in Africa.

1. Superior lemma almost as long as spikelet, this 1.6–2.1 mm. long;
   nodes glabrous - - - - - - - - - - - - - 3. *affinis*
   – Superior lemma distinctly shorter than spikelet - - - - - - - - 2
2. Spikelets 2.7–4 mm. long; nodes glabrous - - - - - - - 1. *flexuosus*
   – Spikelets 2–2.5 mm. long; nodes pubescent - - - - - - - 2. *compressus*

1. **Axonopus flexuosus** (Peter) Troupin, Fl. Garamba **1**: 18 (1956). —Clayton in F.W.T.A. **3**: 446 (1972). —Clayton & Renvoize in F.T.E.A., Gramineae: 613 (1982). TAB. **23**. Type from Tanzania.
   Digitaria flexuosa Peter, Fl. Deutsch Ost-Afr. **1**, Anh.: 60 (1930). Type as above.
   Axonopus compressus subsp. *congoensis* Henr. in Blumea **5**: 529 (1945). Type from Zaire.
   Axonopus compressus var. *congoensis* (Henr.) Black. in Adv. Front. Pl. Sci. **5**: 81 (1963). Type as above.

Stoloniferous perennial. Culms 25–120 cm. high, the nodes glabrous. Leaf laminae 5–22 mm. wide, obtuse to bluntly acute. Racemes 2–5(8), each 4–15 cm. long. Spikelets 2.7–4 mm. long, lanceolate, glabrous to pubescent. Superior lemma ⅔ length of spikelet, glabrous.

Zambia. N: Mpika Distr., Chambeshi R. on Mpika to Kasama Rd., 29.xii.1967, *Simon & Williamson* 1600 (K; SRGH). W: Mwinilunga Distr., Chitunta R., 23.xii.1969, *Simon & Williamson* 1974 (K; SRGH).

Tropical Africa. Damp or swampy soils; 1200–1500 m.

Apparently a native African species, probably resulting from secondary speciation after the introduction of *A. compressus* from America (Gledhill in Journ. W. Afr. Sci. Ass. **11**: 20–27 1966).

2. **Axonopus compressus** (Sw.) Beauv., Ess. Agrost.: 154, 167 (1812). —Stapf in F.T.A. **9**: 566 (1919). —Chippindall in Meredith, Grasses & Pastures of S. Afr.: 391 (1955). —Clayton in F.W.T.A. **3**: 446 (1972). —Clayton & Renvoize in F.T.E.A., Gramineae: 613 (1982). Type from Jamaica.
   Milium compressum Swartz, Prodr. Veg. Ind. Occ.: 24 (1788). Type as above.
   Axonopus kisantuensis Vanderyst in Bull. Agric. Congo Belge **16**: 667 (1925). Type from Zaire.

Tab. 22. PASPALUM GLUMACEUM. 1, habit (×⅔); 2, portion of raceme (× 6); 3, spikelet, lateral view (× 6), all from *Astle* 1208.

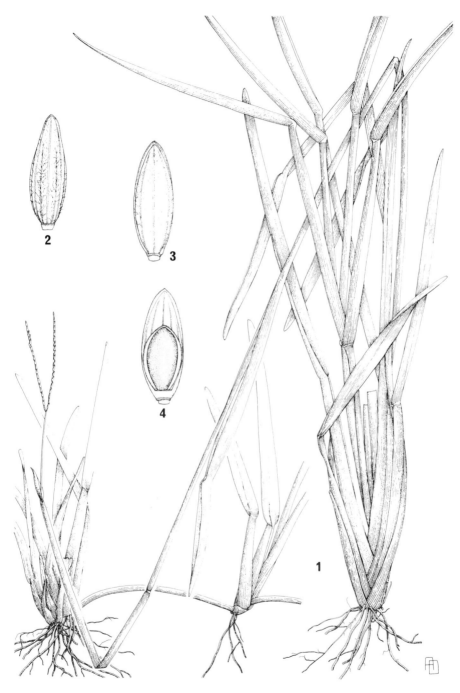

Tab. 23. AXONOPUS FLEXUOSUS. 1, habit (× ½); 2, spikelet showing superior glume (× 10); 3, spikelet showing inferior lemma (× 10); 4, spikelet with inferior lemma removed to show superior floret (× 10), all from *Bogdan* 3343. From F.T.E.A.

Stoloniferous sward-forming perennial. Culms 15–60 cm. high, the nodes pubescent. Leaf laminae 3–12 mm. wide, obtuse to bluntly acute. Racemes 2–5, each 3–10 cm. long. Spikelets 2–2.5 mm. long, oblong-elliptic, glabrous to pubescent. Superior lemma $\frac{4}{5}$ length of spikelet, with a short tuft of hairs at apex.

Zimbabwe. S: Chiredzi Distr., Triangle Estate, 21.iv.1976, *Cleghorn* 3085 (K; SRGH).
Tropical America; introduced to most tropical countries. Damp shady places, where it is commonly used as a lawn grass.

3. **Axonopus affinis** Chase in Journ. Wash. Acad. Sci. **28**: 180 (1938). —Sturgeon in Rhod. Agric. Journ. **50**: 425 (1953). —Clayton in F.W.T.A. **3**: 446 (1972). Type from U.S.A.

Like *A. compressus* except: nodes glabrous; leaf laminae narrower, 1.5–7 mm. wide; spikelets 1.6–2.1 mm. long; superior lemma almost filling spikelet, the apex at most obscurely hairy.

Zimbabwe. E: Mutare, Sheba Forest Estates, 17.v.1969, *Crook* 892 (K; SRGH).
Subtropical America; introduced as a lawn grass, preferring a slightly cooler climate than *A. compressus*. Damp shady places; 2000 m.

18. SETARIA Beauv.

Setaria Beauv., Ess. Agrost.: 51 (1812) nom. conserv.

Cymbosetaria Schweick. in Ic. Pl. (Hook.) **34**: t.3320 (1936).

Inflorescence a panicle, either spiciform or with spikelets ± contracted about primary branches, these occasionally reduced to racemes, all or most of the spikelets subtended by 1 or more scabrid (rarely hirsute) bristles which persist on axis. Spikelets oblong to ovate, ± gibbous. Glumes unequal, the inferior generally much smaller, ovate from a clasping base. Inferior lemma as long as spikelet, herbaceous. Superior lemma crustaceous, strongly convex on the back, often rugose.

A genus of c. 100 species. Tropics and subtropics. The characteristic subtending bristles are derived from sterile panicle branches (Butzin in Willdenowia **9**: 67–79, 1977).

1. Panicle spiciform, sometimes ± lobed; leaf laminae not sagittate: -   -   -   -   2
– Panicle branched   -   -   -   -   -   -   -   -   -   -   -   -   -   9
2. Spikelets not deciduous, the superior floret disarticulating tardily at maturity above the persistent inferior lemma; cultivated annual   -   -   -   -   -   -   1. *italica*
– Spikelets deciduous as a whole   -   -   -   -   -   -   -   3
3. Bristles retrorsely barbed, adhering tenaciously to clothing   -   -   -   2. *verticillata*
– Bristles antrorsely barbed   -   -   -   -   -   -   -   -   -   -   4
4. Bristles ciliate; inferior floret sterile, without palea   -   -   -   3. *atrata*
– Bristles glabrous; inferior floret male (sometimes sterile), with palea   -   -   -   5
5. Nodes nearly always pubescent; bristles pallid to purplish; superior glume mostly 7–9-nerved, covering much of the superior lemma, this usually punctate; spikelets gibbous, laterally compressed   -   -   -   -   -   -   -   -   -   6
– Nodes quite glabrous; bristles often fulvous; superior glume 3–5-nerved, much shorter than the superior lemma, this usually rugose; spikelets scarcely compressed   -   -   -   7
6. Basal sheaths straw-coloured to light brown; bristles scabrid and tapering to the apex   -   -   -   -   -   -   -   -   -   -   -   4. *incrassata*
– Basal sheaths dark chocolate brown; bristles slightly clavate and glutinous at the apex   -   -   -   -   -   -   -   -   -   -   5. *nigrirostris*
7. Plant perennial   -   -   -   -   -   -   -   -   -   -   6. *sphacelata*
– Plant annual   -   -   -   -   -   -   -   -   -   -   -   8
8. Panicle dense, cylindrical, the rhachis tomentellous; leaves not falsely petiolate   -   -   -   -   -   -   -   -   -   -   7. *pumila*
– Panicle loose, spiciform to lanceolate, the rhachis nearly always loosely hirsute; inferior leaves often falsely petiolate   -   -   -   -   -   -   9. *petiolata*
9. Leaf laminae, or at least the inferior, sagittate at the base; superior lemma laterally compressed, conspicuously gibbous, keeled   -   -   -   -   8. *sagittifolia*
– Leaf laminae not sagittate; superior lemma not keeled   -   -   -   10
10. Leaf lamina flat, not pleated   -   -   -   -   -   -   -   11
– Leaf lamina pleated fanwise (best seen at base of young leaves)   -   -   16
11. Bristles in clusters of 2–4   -   -   -   -   -   -   -   12
– Bristles occurring singly   -   -   -   -   -   -   -   13

12. Plant annual; inferior leaf laminae falsely petiolate   -   -   -   -   -   9. *petiolata*
  - Plant perennial; leaf laminae not petiolate   -   -   -   -   -   -   -   10. *grandis*
13. Superior lemma smooth or obscurely rugose below; spikelets 3–3.5 mm. long, the superior
  glume ⅔–¾ as long; panicle loose, flexuous, with well developed secondary
  branches   -   -   -   -   -   -   -   -   -   -   -   13. *pseudaristata*
  - Superior lemma distinctly rugose   -   -   -   -   -   -   -   -   -   -   - 14
14. Plant densely tufted, the base invested in tightly packed dead leaf-sheaths; spikelets 2–3 mm.
  long in a contracted (rarely open) panicle; superior glume often almost as long as superior
  lemma; inferior lemma not sulcate   -   -   -   -   -   -   -   17. *lindenbergiana*
  - Plant loosely tufted; spikelets 1.2–2.2 mm. long, clustered around the spreading primary
  branches; superior glume ⅔–¾ length of spikelet   -   -   -   -   -   -   -   - 15
15. Habit annual; spikelet and glumes generally obtuse; inferior lemma not or slightly grooved; axis
  of inflorescence hirsute   -   -   -   -   -   -   -   -   -   -   11. *orthosticha*
  - Habit perennial; spikelet and glumes generally acute; inferior lemma mostly with a conspicuous
  median groove; axis of inflorescence scaberulous or hirsute   -   -   -   12. *longiseta*
16. Plant annual   -   -   -   -   -   -   -   -   -   -   -   -   -   -   - 17
  - Plant perennial   -   -   -   -   -   -   -   -   -   -   -   -   -   - 19
17. Superior lemma smooth   -   -   -   -   -   -   -   -   -   -   - 16. *seriata*
  - Superior lemma rugose   -   -   -   -   -   -   -   -   -   -   -   -   - 18
18. Superior glume shorter than superior lemma; spikelets on pedicels or secondary branchlets
  along the primary branches   -   -   -   -   -   -   -   -   -   14. *barbata*
  - Superior glume as long as and concealing the superior lemma, or almost so; spikelets subsessile
  in 2 rows along the primary branches   -   -   -   -   -   -   -   15. *homonyma*
19. Plants compactly tufted, the base invested in tightly packed leaf-sheaths; superior lemma
  distinctly rugose; leaf lamina 2–10 mm. wide   -   -   -   -   -   17. *lindenbergiana*
  - Plants forming large clumps, but the base not noticeably clothed in dead leaf-sheaths; superior
  lemma smooth and often shining, or rugose below (rarely rugose throughout); leaf lamina
  5–110 mm. wide   -   -   -   -   -   -   -   -   -   -   -   18. *megaphylla*

1. **Setaria italica** (L.) Beauv., Ess. Agrost.: 178 (1812). —Stapf & Hubbard in F.T.A. **9**: 820 (1930).
  —Sturgeon in Rhod. Agric. Journ. **50**: 503 (1953). —Chippindall in Meredith, Grasses &
  Pastures of S. Afr.: 353 (1955). —Jackson & Wiehe, Annot. Check List Nyasal. Grass.: 58 (1958).
  —Bor, Grasses of B.C.I. & P.: 362 (1960). Type from India.
  Panicum italicum L., Sp. Pl.: 56 (1753). Type as above.

Stout annual. Culms up to 150 cm. high. Panicle 5–30 cm. long, spiciform, congested,
continuous or the lower branches forming spiciform lobes, rhachis villous; bristles in
groups of 2–5, up to c. 1 cm. long. Spikelets 2–3 mm. long, broadly elliptic. Superior
glume ⅔–¾ length of spikelet. Superior lemma smooth or rugulose.

Zimbabwe. N: Chinhoyi, 24.vi.1955, *Steele* in GHS. 53687 (K; SRGH). W: Bulawayo, *Rogers* 5878
(K; SRGH). S: Masvingo, 18.i.1948, *Robinson* 200 (K; SRGH). **Malawi**. S: Zomba, 20.ii.1950, *Wiehe*
N/426 (K). **Mozambique**. N: Nampula, 28.vii.1950, *Costa Rosa* 2 (COI). MS: Shupanga, i.1862, *Kirk*
(K). M: Umbeluzi, ii.1955, *Deveza* (COI).
  Cultivated as a major cereal crop in China, and grown as a minor cereal or for birdseed elsewhere.
It is occasionally cultivated in the Flora Zambesiaca area, and sometimes occurs as an escape near
habitations. It is derived from *S. viridis* (L.) Beauv., and was probably domesticated at several
independent European and Asian centres c. 7000 years ago (de Wet et al. in Journ. Agric. Trad. Bot.
Appl. **26**: 53–60, 1979).

2. **Setaria verticillata** (L.) Beauv., Ess. Agrost.: 178 (1812). —Stapf & Hubbard in F.T.A. **9**: 824 (1930).
  —Sturgeon in Rhod. Agric. Journ. **50**: 503 (1953). —Chippindall in Meredith, Grasses &
  Pastures of S. Afr.: 355 (1955). —Jackson & Wiehe, Annot. Check List Nyasal. Grass.: 60 (1958).
  —Launert in Merxm., Prodr. Fl. SW. Afr. **160**: 171 (1970). —Clayton in F.W.T.A. **3**: 421 (1972).
  —Clayton & Renvoize in F.T.E.A., Gramineae: 522 (1982). Type from southern Europe.
  Panicum verticillatum L., Sp. Pl. ed. 2: 82 (1762). Type as above.
  Panicum adhaerens Forssk., Fl. Aegypt.-Arab.: 20 (1775). Type from Yemen.
  Setaria adhaerens (Forssk.) Chiov. in Nuov. Giorn. Bot. Ital. n.s. **26**: 77 (1919). Type as above.

Loosely tufted annual. Culms 50–100 cm. high or more, geniculately ascending. Panicle
2–15 cm. long, spiciform, linear to untidily lobed, often entangled, the rhachis
hispidulous; bristles 3–8 mm. long, retrorsely barbed, tenaciously clinging. Spikelets
1.5–2.5 mm. long, elliptic. Superior glume as long as spikelet. Inferior floret sterile, the
palea minute. Superior lemma finely rugose.

Botswana. N: Ngamiland, Maun, iii.1958, *Robertson* 609 (K; SRGH). SW: Kalahari Game
Reserve, Kalk Pan, 26.iii.1975, *Owens* 7 (K; SRGH). **Zambia**. B: Mongu Distr., Lizulu R., 8.iv.1964,
Verboom 1377 (K; SRGH). N: Mweru-Wantipa, Kanjiri, 7.iv.1957, *Richards* 9076 (K). C: Lusaka, Mt.
Makulu Res. St., 9.vi.1956, *Angus* 1330 (K). E: Chipata, 23.iv.1968, *Phiri* 190 (SRGH). S: Mapanza

Mission, 7.iii.1954, *Robinson* 605 (K). **Zimbabwe**. N: Bindura, 27.iii.1948, *Rattray* 1484 (K; SRGH). W: Shangani, Gwampa Forest Reserve, i.1954, *Goldsmith* in GHS 48846 (K; SRGH). C: Harare, Nat. Botanic Garden, 12.ii.1965, *Simon* 133 (K; SRGH). E: Chipinge, Sabi Valley Expt. St., vii.1959, *Soane* 13 (SRGH). S: Great Zimbabwe Nat. Park, 18.i.1971, *Chiparawasha* 291 (K; SRGH). **Malawi**. N: Mzimba, Kasitu Valley, 24.i.1938, *Fenner* 201 (K). C: Lilongwe, 12.iii.1951, *Jackson* 400 (K). S: Zomba, *Cormack* 464 (K). **Mozambique**. N: Massangulo, *Gomes e Sousa* 1302 (K). Z: Alto Lugela, 1944, *Pimente* 52581 (COI). T: Kaimba Is., opposite Tete, 1860, *Kirk* (K). GI: Marracuene, 24.viii.1947, *Pedro & Pedrógão* 698 (COI). M: Matola, 11.ix.1940, *Torre* 1753 (K).

Tropical and warm temperate regions generally. Pathsides, old farmland and weedy places, with some preference for damp or shady sites; 850–1600 m.

Some authorities recognize two species: temperate *S. verticillata* with ciliate sheath-margins, glabrous laminae and spikelets over 2 mm. long; tropical *S. adhaerens* with glabrous sheath-margins, hairy laminae and spikelets under 2 mm. long (Henrard in Blumea **3**: 412, 1940; Belo-Correia & Costa in Rev. Biol. **13**: 117–143, 1987). However these are only two segregates from a number of intergrading populations, and it seems better to treat the whole complex as a single polymorphic species.

3. **Setaria atrata** Hackel in Abh. Preuss. Akad. Wiss. Berl. **2**: 122 (1891). —Stapf & Hubbard in F.T.A. **9**: 812 (1930). —Jackson & Wiehe, Annot. Check List Nyasal. Grass.: 58 (1958). —Clayton & Renvoize in F.T.E.A., Gramineae: 524 (1982). Type from Ethiopia.

Tussocky perennial. Culms 60–250 cm. high, the nodes glabrous. Leaf laminae narrow, often convolute. Panicle 6–40 cm. long, spiciform, cylindrical, the rhachis tomentellous to pilose; bristles 3–10 mm. long, ciliate but sometimes sparsely so, filiform and flexuous, rather scanty. Spikelets 2–3 mm long, lanceolate-elliptic, dorsally compressed. Inferior glume up to $\frac{1}{3}$, the superior to $\frac{1}{2}$ length of spikelet. Inferior floret sterile without a palea, its lemma membranous and light green to purple. Superior lemma punctate.

Malawi. N: Mzimba, *Jackson* 445 (K). S: Ntcheu, 17.iii.1955, *Jackson* 1527 (K). Northwards to Sudan and Ethiopia. Montane marshlands.

S. restioidea (Franch.) Stapf, ranging from Angola to the Sudan, has glabrous bristles, but otherwise resembles *S. atrata* and barely warrants separate recognition. *S. rigida* Stapf, a S. African species with slightly thicker and stiffer bristles, also intergrades with *S. atrata*.

4. **Setaria incrassata** (Hochst.) Hackel in Abh. Preuss. Akad. Wiss. Berl. **2**: 122 (1891). —Stapf & Hubbard in F.T.A. **9**: 790 (1930). —Clayton & Renvoize in F.T.E.A., Gramineae: 525 (1982). TAB. **24**. Type from Ethiopia.

Panicum incrassatum Hochst. in Flora **38**: 197 (1855). Type as above.
Setaria woodii Hackel in Bull. Herb. Boiss. **7**: 24 (Jan. 1899). —Chippindall in Meredith, Grasses & Pastures of S. Afr.: 346 (1955). Type from S. Africa.
Setaria gerrardii Stapf in Fl. Cap. **7**: 424 (Aug. 1899). Type from S. Africa.
Setaria merkeri Herrm. in Beitr. Biol. Pflanz. **10**: 44 (1910). —Stapf & Hubbard in F.T.A. **9**: 781 (1930). Type from Tanzania.
Setaria holstii Herrm. in Beitr. Biol. Pflanz. **10**: 45 (1910). —Stapf & Hubbard in F.T.A. **9**: 780 (1930). Type from Tanzania.
Setaria mombassana Herrm. in Beitr. Biol. Pflanz. **10**: 46 (1910). —Stapf & Hubbard in F.T.A. **9**: 783 (1930). —Jackson & Weihe, Annot. Check List Nyasal. Grass.: 59 (1958). Type from Kenya.
Setaria longissima Chiov. in Ann. Bot. Roma **13**: 46 (1914). Type from Zaire.
Setaria modesta Stapf in F.T.A. **9**: 781 (1930. Type from Tanzania.
Setaria phragmitoides Stapf in F.T.A. **9**: 782 (1930). —Jackson & Wiehe, Annot. Check List Nyasal. Grass: 59 (1958). Types: Malawi, Elephant Marsh, *Kirk* (K, syntype); Mozambique, Shupanga, *Kirk* (K, syntype); Zambia, Kafue Flats, *Collector unknown GHS* 3115 (K, syntype); Zimbabwe, cult. at Harare, *Collector unknown GHS* 3366 (K, syntype).
Setaria phanerococca Stapf in F.T.A. **9**: 784 (1930). —Sturgeon in Rhod. Agric. Journ. **50**: 501 (1953). Type: Zimbabwe, Makwiro, *Eyles* 2217 (K, holotype).
Setaria palustris Stapf in F.T.A. **9**: 785 (1930). —Sturgeon in Rhod. Agric. Journ. **50**: 501 (1953). —Jackson & Wiehe, Annot. Check List Nyasal. Grass.: 59 (1958). Type: Malawi, Chiromo, *Scott* (K, holotype).
Setaria rudifolia Stapf in F.T.A. **9**: 787 (1930). —Sturgeon in Rhod. Agric. Journ. **50**: 501 (1953). Types: Zimbabwe, Harare, *Eyles* 2890 (K, syntype) & Mazowe (Mazoe), *Eyles* 2246 (K, syntype).
Setaria porphyrantha Stapf in F.T.A. **9**: 788 (1930). —Sturgeon in Rhod. Agric. Journ. **50**: 501 (1953). Type from S. Africa.
Setaria pabularis Stapf in F.T.A. **9**: 789 (1930). —Sturgeon in Rhod. Agric. Journ. **50**: 501 (1953). Types: Zimbabwe, Harare, *Mundy* 36 (K, syntype), *Hobbs* 632 (K, syntype) & Mutare (Umtali), *Eyles* 3027 (K, syntype).
Setaria ciliolata Stapf & Hubbard in F.T.A. **9**: 807 (1930). —Sturgeon in Rhod. Agric. Journ. **50**: 502 (1953). —Clayton in F.W.T.A. **3**: 423 (1972). Type: Zimbabwe, Harare, *Eyles* 2889 (K, holotype).

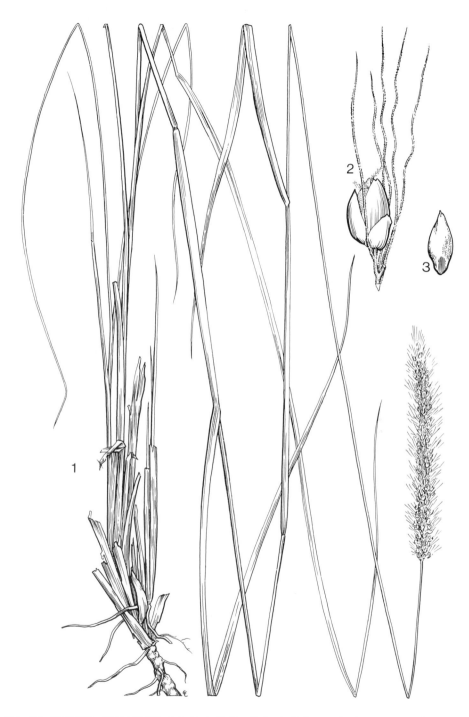

Tab. 24. SETARIA INCRASSATA. 1, habit (× $\frac{2}{3}$); 2, spikelet (× 6); 3, superior lemma (× 6), all from *Astle* 2838.

Setaria eylesii Stapf & Hubbard in F.T.A. **9**: 831 (1930). —Sturgeon in Rhod. Agric. Journ. **50**: 504 (1953). —Launert in Merxm., Prodr. Fl. SW. Afr. **160**: 172 (1970). Type: Zimbabwe, Mutare, *Eyles* 3035 (K, holotype).
Setaria ramulosa Peter, Fl. Deutsch Ost-Afr. **1**: Anh. 64 (1930). Type from Tanzania.
Setaria kersteniana Peter, Fl. Deutsch Ost-Afr. **1**: Anh. 67 (1930). Type from Tanzania.
Setaria breviseta Peter, Fl. Deutsch Ost-Afr. **1**: Anh. 68 (1930). Type from Tanzania.
Setaria lacunosa Peter, Fl. Deutsch Ost-Afr. **1**: Anh. 69 (1930). Type from Tanzania.
Setaria bequaertii Robyns in Bull. Jard. Bot. Brux. **9**: 190 (1932). Type from Zaire.
Setaria lindiensis Pilger in Notizbl. Bot. Gart. Berl. **14**: 94 (1938). Type from Tanzania.
Setaria perberbis de Wit in Bull. Bot. Gart. Buitenz., sér. 3, **17**: 31 (1941). —Chippindall in Meredith, Grasses & Pastures of S. Afr.: 347 (1953). Type from S. Africa.

Tufted perennial, arising from a short rhizome. Culms 20–200 cm. high, the nodes pubescent though the hairs sometimes fugacious, rarely glabrous (see note); basal sheaths pallid to light brown. Panicle 3–30 cm. long, spiciform, cylindrical or somewhat irregular, spikelets and bristles typically pallid green often with purple tips and sometimes wholly purple, the rhachis tomentellous to sparsely pilose; bristles 2–15 mm. long, in bundles of c. 4, tapering and scaberulous to the tip. Spikelets 2–3(4) mm. long, broadly ovate to gibbously suborbicular, laterally compressed. Inferior glume $\frac{1}{3}-\frac{2}{3}$, the superior $\frac{2}{3}$ to as long as spikelet and (5)7–9(11)-nerved. Inferior floret ♂, its lemma firmly membranous. Superior lemma punctate to obscurely rugose.

Botswana. N: Francistown to Maun, c. 97 km. from Francistown, 14.i.1959, *West* 3818 (SRGH). SW: Kgalogadi Distr., Lephepe, iv.1969, *Kelaole* 583 (K; SRGH). SE: Serowe, 9.iv.1959, *de Beer* 881a (K; SRGH). **Zambia**. N: Mbala Distr., Chipululu Farm, 6.i.1965, *Richards* 19418 (K). C: Lusaka, 15.viii.1972, *Kornaś* 2000 (K). E: Chipata, 25.vi.1962 *Verboom* 736 (K). S: Mapanza Mission, 14.ii.1954, *Robinson* 526 (K). **Zimbabwe**. N: Gokwe Distr., Chirisa, 12.ii.1984, *Mahlangu* 434 (K; SRGH). W: Nyamandhlovu Distr., Edwaleni Farm, ii.1972, *Keogh* 13 (K; SRGH). C: Chegutu, Poole Farm, 19.i.1944, *Hornby* 2338 (K; SRGH). E: Chipinge Distr., Giriwayo, 19.i.1957, *Phipps* 16 (K; SRGH). S: 16 km. from Chipinda Pools on Chiredzi Rd., 12.i.1971, *Kelly* 376 (K; SRGH). **Malawi**. C: Lilongwe, Agricultural Res. St., 8.ii.1966, *Salubeni* 407 (K). S: Chikwawa Distr., Lengwe Game Reserve, 1.ii.1970, *Hall-Martin* 542 (K; SRGH). **Mozambique**. N: Namapa, 23.ii.1960, *Lemos & Macuáca* 56 (K; COI). Z: Chinde, Luabo, 13.iii.1971, *Correia* 115 (COI). T: Estima-Inhacapirire, 26.i.1972, *Macêdo* 4690 (K; COI). MS: Gorongosa, Parque Nacional de Caça, 30.iv.1964, *Torre & Paiva* 12190 (K; COI). GI: Chibuto, Maniquenique, 13.ii.1959, *Barbosa & Lemos* 8394 (K; COI). M: Xai-xai (Vila João Belo), 12.vii.1944, *Torre* 6723 (K; COI).

Also in Nigeria and Ethiopia to S. Africa. Mainly black clay plains and marshy stream margins, but extending onto drier soils; sea level to 1500 m.

A polymorphic species varying considerably in overall size. It is best recognized by its hairy culm nodes but, especially in Mozambique, these are occasionally glabrous (*S. palustris* was based on this variant); the gibbous spikelets with long many-nerved superior glume then distinguish it from *S. sphacelata*. However, the two species are not always easy to separate, and may hybridise.

5. **Setaria nigrirostris** (Nees) Dur. & Schinz, Consp. Fl. Afr. **5**: 774 (1894). —Chippindall in Meredith, Grasses & Pastures of S. Afr.: 345 (1955). Type from S. Africa.
    Panicum nigrirostre Nees, Fl. Afr. Austr.: 55 (1841). Type as above.

Tufted perennial from a short rhizome. Culms 20–100 cm. high, the nodes pubescent and basal sheaths chocolate brown. Panicle 2–15 cm. long; bristles 5–10 mm. long, slightly clavate and glutinous at the apex. Spikelets 3–4 mm. long. Otherwise like *S. incrassata*.

Malawi. C: c. 48 km. N. of Kasungu, 21.i.1951, *Jackson* 378 (K).
Also in S. Africa. Savanna woodland.

6. **Setaria sphacelata** (Schumach.) Moss in Kew Bull. **1929**: 195 (1929). —Stapf & Hubbard in F.T.A. **9**: 795 (1930). —Sturgeon in Rhod. Agric. Journ. **50**: 501 (1953). —Chippindall in Meredith, Grasses & Pastures of S. Afr.: 351 (1955). —Jackson & Wiehe, Annot. Check List Nyasal. Grass.: 59 (1958). —Bor, Grasses of B.C.I. & P.: 364 (1960). —Launert in Merxm., Prodr. Fl. SW. Afr. **160**: 172 (1970). —Clayton in F.W.T.A. **3**: 423 (1972). —Clayton & Renvoize in F.T.E.A., Gramineae: 527 (1982). Type from Ghana.
    Panicum sphacelatum Schumach., Beskr. Guin. Pl.: 78 (1827). Type as above.
    Setaria aurea A. Br. in Flora **24**: 276 (1841). —Clayton in F.W.T.A. **3**: 423 (1972). Type cult. in Germany.
    Setaria perennis Hack in Bull. Herb. Boiss. **3**: 379 (1895) *non* Smyth (1892). —Chippindall in Meredith, Grasses & pastures of S. Afr.: 351 (1955). Types from S. Africa.
    Setaria flabellata Stapf in Fl. Cap. **7**: 425 (1899). —Chippindall in Meredith, Grasses & Pastures of S. Afr.: 348 (1955). Types from S. Africa.
    Setaria bussei Herrm. in Beitr. Biol. Pflanz. **10**: 46 (1910). —Stapf & Hubbard in F.T.A. **9**: 806

(1930). Type from Tanzania.

Setaria homblei De Wild. in Ann. Soc. Sci. Brux. **39**, Mém.: 134 (1920). —Stapf & Hubbard in F.T.A. **9**: 800 (1930). —Sturgeon in Rhod. Agric. Journ. **50**: 502 (1953). Type from Zaire.

Setaria anceps Stapf in F.T.A. **9**: 793 (1930). —Clayton in F.W.T.A. **3**: 423 (1972). Types from Ghana etc.

Setaria anceps var. *sericea* Stapf in F.T.A. **9**: 794 (1930). Type from Sudan.

Setaria angustifolia Stapf in F.T.A. **9**: 803 (1930). —Sturgeon in Rhod. Agric. Journ. **50**: 502 (1953). —Jackson & Wiehe, Annot. Check List Nyasal. Grass.: 58 (1958). Types many including: Zambia, Lukanda R., *Kässner* 2137 (K, syntype). Zimbabwe, Mrewa, *Appleton* 1 (K, syntype); Harare, *Eyles* 2492 (K, syntype); Bulawayo, *Appleton* 19 (K, syntype) & *Rogers* 13683 (K, syntype).

Setaria planifolia Stapf in F.T.A. **9**: 794 (1930). Type from Tanzania.

Setaria splendida Stapf in F.T.A. **9**: 799 (1930). —Sturgeon in Rhod. Agric. Journ. **50**: 502 (1953). —Jackson & Wiehe, Annot. Check List Nyasal. Grass.: 60 (1958). Types many including: Malawi, Shire Highlands, *Buchanan* 10 (K, syntype) & *Cameron* 18 (K, syntype); Mount Mulanje, *Buchanan* 408 (K, syntype).

Setaria stenantha Stapf in F.T.A. **9**: 804 (1930). —Jackson & Wiehe, Annot. Check List Nyasal. Grass.: 60 (1958). Type: Malawi, without precise locality, *Buchanan* 236 (BM, holotype).

Setaria stolzii Stapf in F.T.A. **9**: 805 (1930). Type from Tanzania.

Setaria torta Stapf in F.T.A. **9**: 801 (1930). —Sturgeon in Rhod. Agric. Journ. **50**: 502 (1953). Types: Zimbabwe, Harare, *Eyles* 2912 (K, syntype) & Charter, *Eyles* 4569 (K, syntype).

Setaria trinervia Stapf in F.T.A. **9**: 791 (1930). Types from Tanzania etc.

Setaria alpestris Peter, Fl. Deutsch Ost-Afr. **1**, Anh.: 66 (1936). Type from Tanzania.

Setaria myosuroides Peter, Fl. Deutsch Ost-Afr. **1**, Anh: 64 (1936). Type from Tanzania.

Setaria scalaris Peter, Fl. Deutsch Ost-Afr. **1**, Anh: 67 (1936). Type from Tanzania.

Setaria almaspicata de Wit in Bull. Bot. Gard. Buitenz., sér. 3, **17**: 47 (1941). Type from S. Africa.

Setaria cana de Wit in Bull. Bot. Gard. Buitenz., sér. 3, **17**: 30 (1941). Type from Namibia.

Setaria decipiens de Wit in Bull. Bot. Gard. Buitenz., sér. 3, **17**: 28 (1941). Type from S. Africa.

Setaria flabelliformis de Wit in Bull. Bot. Gard. Buitenz., sér.3, **17**: 40 (1941). Type from S. Africa.

Setaria neglecta de Wit in Bull. Bot. Gard. Buitenz., sér. 3, **17**: 49 (1941). —Chippindall in Meredith, Grasses & Pastures of S. Afr.: 349 (1955). Type from S. Africa.

Setaria sphacelata var. *aurea* (A.Br.) Clayton in Kew Bull. **33**: 505 (1979). —Clayton & Renvoize in F.T.E.A., Gramineae: 528 (1982). Type as for *Setaria aurea*.

Setaria sphacelata var. *sericea* (Stapf) Clayton in Kew Bull **33**: 506 (1979). —Clayton & Renvoize in F.T.E.A., Gramineae: 529 (1982). Type as for *Setaria anceps* var. *sericea* Stapf.

Setaria sphacelata var. *splendida* (Stapf) Clayton in Kew Bull. **33**: 506 (1979). —Clayton & Renvoize in F.T.E.A., Gramineae: 530 (1982). Type as for *Setaria splendida* Stapf.

Setaria sphacelata var. *torta* (Stapf) Clayton in Kew Bull. **33**: 506 (1982). —Clayton & Renvoize in F.T.E.A., Gramineae: 529 (1982). Type as for *Setaria torta* Stapf.

Tufted perennial, arising from a short rhizome. Culms 20–300 cm. high, the nodes quite glabrous. Panicle 3–50 cm. long, spiciform, cylindrical, spikelets mostly pallid to purple with fulvous bristles, the rhachis tomentellous; bristles 1.5–12 mm. long, 6–15 below each cluster of 1–4 spikelets. Spikelets 1.5–3.5 mm. long, elliptic, oblique but scarcely gibbous or laterally compressed. Inferior glume up to $\frac{1}{2}$, the superior $\frac{1}{3}$–$\frac{2}{3}$($\frac{3}{4}$) length of spikelet and (3)5-nerved. Inferior floret ♂, its lemma thinly papery. Superior floret rugose, usually strongly so, but occasionally almost smooth.

Caprivi Strip. Singalamwe, 31.xii.1958, *Killick & Leistner* 3223 (K; SRGH; PRE). **Botswana**. N: Shianikola Pan, 28.i.1978, *Smith* 2288 (K; SRGH). SE: Content Farm, 27.i.1978, *Hansen* 3342 (C; K; SRGH). **Zambia**. B: Senanga, 31.vii.1952, *Codd* 7300 (K; PRE). N: Mpulungu, 19.i.1952, *Richards* 479 (K). W: Mwinilunga, Dobeka Bridge, 17.ix.1937, *Milne-Redhead* 3291 (K). C: Mumbwa, 20.iii.1963, *van Rensburg* 1710 (K). E: Chipata, Senegallia Farm, 30.i.1953, *Grout* 114 (K; SRGH). S: Choma, 8.xii.1962, *Astle* 1782 (K). **Zimbabwe**. N: Hurungwe (Urungwe), Nyamnyetsi Estate, 5.ix.1978, *Nyariri* 324 (K; SRGH). W: Hwange, Victoria Falls Nat. Park, 29.i.1974, *Mukuya* 2/74 (K; SRGH). C: Chegutu, Poole Farm, 5.ii.1945, *Hornby* 2363 (K; SRGH). E: Chimanimani, Haroni-Lusitu confluence, 10.i.1969, *Biegel* 2790 (K; SRGH). S: Masvingo, 1.i.1948, *Robinson* 56 (K; SRGH). **Malawi**. N: Chitipa, c. 6.4 km. downstream from L. Kaulime, 11.ii.1968, *Simon, Williamson & Ball* 1751 (K; SRGH). C: Kasungu Nat. Park, 28.xii.1970, *Hall-Martin* 1597 (K; PRE). S: Mulanje, Litchenya Forestry Hut, 28.ii.1960, *Phipps* 2759 (K; SRGH). **Mozambique**. N: Namache, ii.1931, *Gomes e Sousa* 430 (K). Z: Mocuba, Namagoa, *Faulkner* 13 (K; PRE). T: Angónia, Ulongue, 17.xii.1980, *Macuáca* 1467 (K). MS: Between Matarara de Lucite and Dombe, 21.x.1953, *Gomes e Pedro* 4352 (K; COI). GI: Massingir, Banga, 3.xii.1981, *White* 20 (K). M: Namaacha, 20.i.1958, *Barbosa & Lemos* 8240 (K; COI).

Tropical and S. Africa. Favours dambos and other damp habitats, but extends onto drier savanna woodland soils; sea level to 2000 m.

A polymorphic species, varying greatly in overall size. It is customarily partitioned into a number of segregate taxa, either at species or, as here, at varietal level.

1. Basal sheaths fibrous; culms 2–7-noded, 1.5–6 mm. in diam.   -   -   -   - var. *aurea*
– Basal sheaths not fibrous   -   -   -   -   -   -   -   -   -   -   -   -   -   - 2
2. Culms 2–4-noded, 1–3 mm. in diam., up to 1 m. high; basal sheaths not flabellate; leaf laminae
   mostly 1–4 mm. wide   -   -   -   -   -   -   -   -   -   -   -   -   -   -   - 3
– Culms 4–16-noded; basal sheaths ± flabellate   -   -   -   -   -   -   -   - 4
3. Bristles fulvous   -   -   -   -   -   -   -   -   -   -   -   -   - var. *sphacelata*
– Bristles pallid to purplish, imitating *S. incrassata*   -   -   -   -   -   - var. *torta*
4. Culms 3–6 mm. in diam., up to 2 m. high; leaf laminae 3–10 mm. wide.   var. *sericea*
– Culms 6–12 mm. in diam., up to 3 m. high; leaf laminae 10–17 mm. wide   var. *splendida*

Unfortunately the characters are not as clear cut as the key implies, but vary continuously so that the 'varieties' represent arbitrary segments of a continuum rather than recognizable noda within it (Clayton in Kew Bull. **33**: 501–509 1979). In fact the species contains a polyploid series running from diploid to decaploid, in which the different ploidy levels cross freely but seemingly show little correlation with morphology (Hacker in Austr. Journ. Bot. **16**: 539–544, 551–554 1968). Present knowledge therefore provides no sound basis for sustaining the traditional segregates as formally circumscribed taxa, though they may have some use as an approximate indication of size class. For practical purposes the infra-specific classification of this important forage species is probably best conducted at cultivar level. Popular cultivars, such as 'Nandi' or 'Kazungula', lie in the *sericea-splendida* size range.

7. **Setaria pumila** (Poir.) Roem. & Schult., Syst. Veg. **2**: 891 (1817). —Rauschert in Fedde, Repert. **83**: 661 (1973). —Clayton in Kew Bull. **33**: 501–503 (1979). —Clayton & Renvoize in F.T.E.A., Gramineae: 530 (1982). Type from Mediterranean.

   Panicum pumilum Poir., Encycl. Méth. Bot., Suppl. **4**: 273 (1816). Type as above.
   Panicum pallide-fuscum Schumach., Beskr. Guin. Pl.: 58 (1827). Type from Ghana.
   Setaria pallide-fusca (Schumach.) Stapf & Hubbard in Kew Bull. **1930**: 259 (1930) & in F.T.A. **9**: 815 (1930). —Sturgeon in Rhod. Agric. Journ. **50**: 503 (1953). —Chippindall in Meredith, Grasses & Pastures of S. Afr.: 353 (1955). —Jackson & Wiehe, Annot. Check List Nyasal. Grass.: 59 (1958). —Bor, Grasses of B.C.I. & P.: 362 (1960). —Launert in Merxm., Prodr. Fl. SW. Afr. **160**: 172 (1970). —Clayton in F.W.T.A. **3**: 423 (1972). Type as above.
   Setaria ustilata de Wit in Bull. Bot. Gard. Buitenz., sér. 3, **17**: 59 (1941). —Chippindall in Meredith, Grasses & Pastures of S. Afr.: 355 (1955). Type from S. Africa.
   Setaria glauca var. *pallide-fusca* (Schumach.) Koyama in Journ. Jap. Bot. **37**: 237 (1962). Type as for *Panicum pallide-fuscum*.
   Setaria glauca subsp. *pallide-fusca* (Schumach.) B.K. Simon in Austrobaileya **2**: 22 (1984). Type as above.

Loosely tufted annual. Culms 5–130 cm. high, ascending, the nodes glabrous. Panicle 1–10(20) cm. long, spiciform, cylindrical, the rhachis tomentellous; bristles 3–12 mm. long, 6–8 per spikelet, commonly fulvous. Spikelets 1.5–3.5 mm. long (seldom over 2.5 mm. in the Flora Zambesiaca area), ovate. Glumes $\frac{1}{3}$–$\frac{2}{3}$ length of spikelet. Inferior floret ♂ or sterile, its palea almost as long as lemma. Superior lemma rugose to corrugate, rarely almost smooth.

Botswana. N: 6 km. SE. of Tsau, 18.iii.1965, *Wild & Drummond* 7131 (K; SRGH). SW: Ghanzi, Old Winkel, 25.i.1970, *Brown* 8237 (K). SE: Morale Res. St., 3.iii.1978, *Hansen* 3361 (C; K). **Zambia**. B: Mongu Airport, 22.iii.1964, *Verboom* 1305 (K). N: Mbala, Kalambo Falls, 15.vii.1964, *Richards* 19018 (K). W: Mwinilunga, Matonchi, 21.xii.1969, *Simon & Williamson* 1918 (K; SRGH). C: Kanona, Kundalila Falls, 17.iii.1974, *Davidse & Handlos* 7241 (K; MO). E: Nyika Plateau, v.1968, *Williamson* 951 (SRGH). S: Mazabuka, Siamambo stream c. 3.2 km. below confluence with Bunchele, 3.iii.1960, *White* 7593 (FHO; K). **Zimbabwe**. N: Hurungwe (Urungwe) Distr., Mensa pan, c. 18 km. ESE. of Chirundu bridge, 30.i.1958, *Drummond* 5369 (K; SRGH). W: Gwampa Forest Reserve, i.1954, *Goldsmith* in GHS 48830 (K; SRGH). C: Chegutu, Poole Farm, 14.i.1944, *Hornby* in GHS 12492 (K, SRGH). E: Mutare, Maranki Native Reserve Farm School, 10.ii.1953, *Chase* 4779. S: Makoholi Expt. St., 14.ii.1977, *Senderayi* 72 (K; SRGH). **Malawi**. N: Rumphi, 27.iii.1976, *Phillips* 1541 (K; MO). C: Lilongwe, Agric. Res. St., 1.ii.1951, *Jackson* 383 (K). S: Blantyre, Matope Mission, 12.ii.1970, *Brummitt* 8537 (K). **Mozambique**. N: Marrupa, 5.ii.1981, *Nuvunga* 466 (K). T: Planalto do Songo, 23.ii.1972, *Macêdo* 4892 (SRGH). GI: Alto Limpopo, Chifu R., 17.i.1965, *Macêdo* 170 (COI).

Tropical and warm temperate regions of the Old World; introduced to America. Weedy places; sea level to 1500 m.

S. pumila is a polymorphic weed segregating into regional populations distinguished by the statistical distribution of their spikelet lengths. The overlap between these populations is considerable and their assignation to separate taxa is impractical. Other characters, such as length of the inferior glume or rugosity of the superior lemma, have been proposed, but they seem to be uncorrelated with one another.

The nomenclature of the species is controversial (reviewed by Terrell in Taxon **25**: 297–304 1976), but it seems that the commonly used names *S. glauca* and *S. lutescens* should be referred to *Pennisetum*. *S. pumila* is an annual counterpart of the caespitose perennial *S. sphacelata*, and is indistinguishable from it if the base is missing. It is even more difficult to separate from the American species *S. gracilis* Kunth (= *S. geniculata* auct. non (Willd.) Beauv.); this is a short-lived perennial with short knotty rhizomes bearing dormant buds, though the perennial habit is not usually very obvious. *S. viridis*, an annual weed from southern Europe, has been found as an adventive in Harare. It is distinguished from *S. pumila* by the long superior glume completely covering the superior lemma.

8. **Setaria sagittiifolia** (A. Rich.) Walp., Ann. Bot. Syst. **3**: 721 (1852). —Stapf & Hubbard in F.T.A. **9**: 862 (1930). —Jackson & Wiehe, Annot. Check List Nyasal. Grass.: 59 (1958). —Clayton & Renvoize in F.T.E.A., Gramineae: 533 (1982). Types from Ethiopia.

    Pennisetum sagittifolium A. Rich., Tent. Fl. Abyss. **2**: 379 (1851). Type as above.
    Panicum barbigerum Bertol. in Mem. Acad. Sci. Bologna **3**: 250, fig. 19/1–2 (1851). Type: Mozambique, Inhambane, *Fornasini* (BOLO, holotype).
    Panicum sagittifolium (A. Rich.) Steud., Syn. Pl. Glum. **1**: 54 (1854). Type as for *S. sagittifolia*.
    Setaria barbigera (Bertol.) Stapf in F.T.A. **9**: 862 (1930). Type as for *P. barbigerum*.
    Cymbosetaria sagittifolia (A. Rich.) Schweick. in Ic. Pl. (Hook.) **34**: t. 3320 (1936). —Chippindall in Meredith, Grasses & Pastures of S. Afr.: 355 (1955). Type as for *S. sagittifolia*.

Weak-stemmed annual. Culms 20–80 cm. high. Leaf laminae broadly linear to narrowly lanceolate, at least the lower leaves falsely petiolate and sagittate with lobes 2–30 mm. long. Panicle 5–15 cm. long, narrowly oblong to ovate, the spikelets secund along scaberulous raceme-like primary branches; bristles 2–15 mm. long. Spikelets 2 mm. long, suborbicular, laterally compressed. Inferior glume ⅓, the superior ½ length of spikelet. Superior lemma conspicuously gibbous, keeled, rugose.

    Caprivi Strip. Andara to Bagani, 21.i.1956, *de Winter* 4372 (K; PRE). **Botswana**. N: 8 km. N. of Aha Hills, 13.iii.1965, *Wild & Drummond* 6970 (K; SRGH). SE: Palapye, Malede, 7.ii.1958, *de Beer* 594 (SRGH). **Zambia**. C: Katondwe, 27.ii.1966, *Fanshawe* 9589 (SRGH). S: Kazungulu, ii.1932, *Trapnell* 954 (K; SRGH). **Zimbabwe**. N: Gokwe, Sengwa/Lutope confluence road, 10.ii.1984, *Mahlangu* 867 (K; SRGH). W: Victoria Falls Game Res., Zambezi camp, 16.i.1974, *Gonde* 45/73 (SRGH). E: Chipinge, Mutema to Chisumbanje, 29.i.1975, *Gibbs-Russell* 2666 (K; PRE; SRGH). S: Gwanda, Tuli-Makwe dam, 8.iii.1967, *Cleghorn* 1521 (K; SRGH). **Malawi**. N: Karonga Distr., c. 32 km. N. of Chilumba, 14.iv.1976, *Pawek* 11016 (K; SRGH). S: Lengwe Game Reserve, 5.ii.1970, *Brummitt* 8879 (K; SRGH). **Mozambique**. T: Zambezi R., 1.iii.1961, *Vesey-FitzGerald* 3065 (SRGH). MS: Chemba, Chiou, 5.iv.1962, *Balsinhas & Macúaca* 555 (COI; K). GI: Massingir, 18.iv.1972, *Lousã & Rosa* 235 (K). M: Maputo, 8.ii.1920, *Borle* 5149b (K).
    Also in Sudan and Yemen to S. Africa. Shady places in bushland, woodland and riverine forest; sea level to 1200 m.
    S. appendiculata (Hackel) Stapf also has sagittate leaf laminae, but the superior lemma is not laterally compressed and keeled. Although present in East Africa and the Cape, it has not yet been recorded from the Flora Zambesiaca area.

9. **Setaria petiolata** Stapf & Hubbard in F.T.A. **9**: 813 (1930). —Clayton & Renvoize in F.T.E.A., Gramineae: 531 (1982). Type: Malawi, Lake Malawi, *Simons* (BM, holotype).

    Setaria interpilosa Stapf & Hubbard in F.T.A. **9**: 829 (1930). —Jackson & Wiehe, Annot. Check List Nyasal. Grass.: 58 (1958). Types from Kenya and Malawi, *Buchanan* 1273 (K, syntype).

Loosely tufted annual. Culms 15–100 cm. high, weak, the nodes glabrous. Leaf laminae broadly linear, flaccid, finely scabrid along the nerves below, the inferior narrowed at the base into a false petiole 0.5–5 cm. long. Panicle 3–17 cm. long, loosely spiciform to lanceolate and then shortly branched below, the rhachis loosely hirsute at least in the upper part (rarely puberulous only); bristles 5–10 mm. long in clusters of 2–5, slender, almost straight, scabrid throughout. Spikelets 1.5–2 mm. long, broadly elliptic. Inferior glume ½, the superior 1¾ length of spikelet. Superior lemma rugose.

    Zambia. N: Mbala Distr., L. Tanganyika, Crocodile Is., 9.ii.1964, *Richards* 18999 (K). E: Chipata Distr., Muchinja Hill, 24.iii.1963, *Verboom* 940 (K). **Zimbabwe**. N: Gokwe Distr., Chief Nemangine's, 22.i.1964, *Bingham* 1194 (K; SRGH). W: Matobo Distr., Toghwana Dam, 20.ii.1965, *Simon* 161 (K; SRGH). E: Mutambara Tribal Trust Land, main road E. of Umvumvumvu bridge, 26.ii.1969, *Crook* 852 (K; SRGH). S: Wanga Hill, c. 32 km. S. of Gwanda, 2.ii.1955, *Plowes* 1766 (K; SRGH). **Malawi**. N: Rumphi, 15.ii.1954, *Jackson* 1243 (K). C: Salima Distr., Namalenje Is., 21.iii.1977, *Grosvenor & Renz* 1295 (K; SRGH). S: Mangochi, Kapirinjoti, 26.v.1954, *Jackson* 1325 (K). **Mozambique**. T: Marueira to Songo, 24.iii.1972, *Macêdo* 5081 (SRGH).
    Also in Kenya and Tanzania. Wooded grassland and among rocks, in shade; 700–1500 m.
    Related to the Indian species *S. intermedia* Roem. & Schult. (= *S. tomentosa* (Roxb.) Kunth, which lacks the petiolate leaves and has puberulous inflorescence branches.

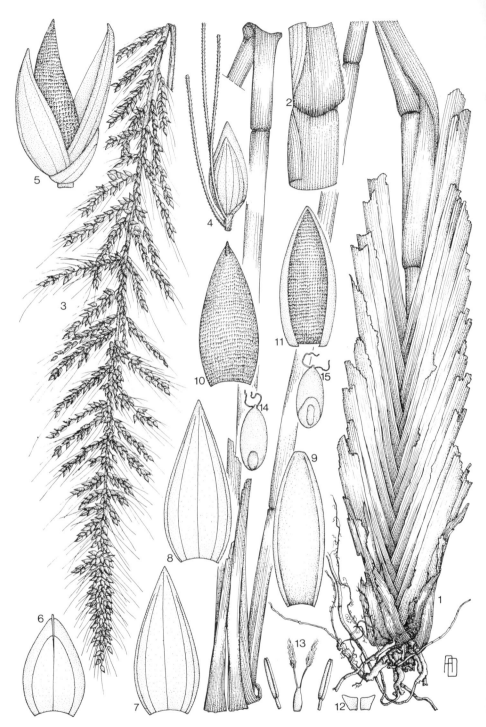

Tab. 25. SETARIA GRANDIS. 1, habit (× ½); 2, junction of leaf sheath and lamina showing ligule (× 2); 3, inflorescence (× ½); 4, spikelet (× 5); 5, spikelet, lateral view (× 10); 6, inferior glume (× 10); 7, superior glume (× 10); 8, inferior lemma (× 10); 9, inferior palea (× 10); 10, superior lemma (× 10); 11, superior palea (× 10); 12 lodicules (× 6); 13, ovary and stamens (× 5); 14 & 15, caryopsis (× 8), all from *Simon, Williamson & Ball* 1711, the inferior part of fig. 1 from *Tyler* 803.

10. **Setaria grandis** Stapf in F.T.A. **9**: 832 (1930). —Jackson & Wiehe, Annot. Check List Nyasal. Grass.: 58 (1958). TAB. **25**. Type: Malawi, Nyika Plateau, *Henderson* (BM; holotype).

Robust caespitose perennial with strongly flabellate basal sheaths. Culms 2–3 m. high. Leaf laminae up to 15 mm. wide. Panicle 20–30 cm. long, lanceolate, with stiffly spreading or ascending raceme-like primary branches up to 3 cm. long; bristles 10–25 mm. long, copious, rigid and almost pungent. Spikelets 3.5–4 mm. long, narrowly ovate. Inferior glume ½, the superior almost as long as spikelet. Superior lemma indistinctly rugulose.

Malawi. N: Nyika Plateau, L. Kaulime, 16.v.1970, *Brummitt* 10812 (K; SRGH).
Confined to the Nyika Plateau. Marshy places bordering lakes and streams; 2300-2500 m.
Loosely related to the Angolan endemic, *S. welwitschii* Rendle, which has spikelets only 2 mm. long.

11. **Setaria orthosticha** Herrm. in Beitr. Biol. Pflanz. **10**: 49 (1910). —Stapf & Hubbard in F.T.A. **9**: 839 (1930). —Sturgeon in Rhod. Agric. Journ. **50**: 504 (1953). —Jackson & Wiehe, Annot. Check List Nyasal. Grass.: 59 (1958). —Clayton & Renvoize in F.T.E.A., Gramineae: 533 (1982). TAB. **26**. Type: Malawi, Shire Highlands, *Buchanan* 3 (B, holotype; K, isotype).

Loosely tufted annual. Culms 10–150 cm. high, ascending. Leaf laminae not petiolate. Panicle 1–16 cm. long, lanceolate, with spikelets clustered around the raceme-like primary branches or on short secondary branchlets, the rhachis hirsute; bristles 2–10 mm. long, mostly single, stiffly bent, often hirsute in the lower half, smooth or weakly scaberulous above. Spikelets 1.2–1.7 mm. long, broadly elliptic, gibbous, generally obtuse. Glumes obtuse, the inferior ¼–⅓, the superior ⅔–¾ as long as spikelet. Inferior lemma not or indistinctly grooved. Superior lemma finely rugose, golden brown to black.

Zambia. N: Mbala, Simonwe Farm, 24.ii.1959, *McCallum-Webster* A227 (K; SRGH). W: Ndola, iii.1954, *ACRE* E34 (K). C: Kundalila Falls, 17.iii.1974, *Davidse & Handlos* 7268 (K; MO). S:Choma, 28.iii.1955, *Robinson* 1219 (K; SRGH). **Zimbabwe**. N: Mazowe (Mazoe), 8.iv.1928, *Eyles* 5847 (K; SRGH). W: Victoria Falls, 12.iii.1932, *Brain* 9231 (SRGH). C: Wedza Mt., 6.iv.1964, *Cleghorn* 910 (K; SRGH). E: Mutare, Vumba, 8.vii.1948, *Fisher & Schweickert* 205 (K; PRE; SRGH). S: Masvingo Distr., iv.1921, *Eyles* 2988 (K; SRGH). **Malawi**. N: Misubu, 28.vi.1951, *Jackson* 559 (K). C: Dedza Mt., 4.v.1968, *Salubeni* 1085 (K; SRGH). S: Zomba Mt., 2.iv.1934, *Lawrence* 154 (K). **Mozambique**. N: Massangulo, iv.1955, *Gomes e Sousa* 1372 (K). MS: Mt. Anuta, 5.ii.1954, *Chase* 5322 (K; SRGH).
Northward to Uganda. Weedy places and forest margins; 1000–2000m.
Related to *S. petiolata*, from which it is distinguished mainly by its raceme-like panicle branches and stiffly bent bristles. There is also some similarity to the perennial species *S. longiseta*.

12. **Setaria longiseta** Beauv., Fl. Owar. **2**: 81, fig.110/2 (1819). —Stapf & Hubbard in F.T.A. **9**: 836 (1930). —Sturgeon in Rhod. Agric. Journ. **50**: 504 (1953). —Jackson & Wiehe, Annot. Check List Nyasal. Grass.: 58 (1958). —Clayton in F.W.T.A. **3**: 423 (1972). —Clayton & Renvoize in F.T.E.A., Gramineae: 535 (1982). Type from Nigeria.
Setaria paniciformis Rendle, Cat. Afr. Pl. Welw. 2: 186 (1899). Types from Angola.

Loosely tufted perennial. Culms 50–150 cm. high, geniculately ascending. Leaf laminae 1–6(15) mm. wide, neither plicate nor petiolate. Panicle 7–25 cm. long, lanceolate to ovate with spikelets loosely clustered around the primary branches, often purple tinged, the rhachis scaberulous with occasional long hairs or rarely hirsute; bristles 2–12 mm. long. Spikelets 1.5–2.2 mm. long, elliptic, slightly gibbous, generally acute. Glumes often acute, the inferior ¼–⅓, the superior ⅔–¾ as long as spikelet. Inferior lemma firmly and smoothly membranous with a pronounced median groove. Superior lemma finely but distinctly rugose, pallid becoming golden.

Zambia. N: Mbala, Dulamidi, 29.i.1952, *Richards* 535 (K). W: Mufulira, 17.i.1959, *Fanshawe* 5081 (SRGH). C: Chakwenga Headwaters, 19.i.1964, *Robinson* 6220 (K; SRGH). E: Lugomo, 3.ii.1958, *Stewart* 105 (K; SRGH). S: Mazabuka, 29.iii.1963, *van Rensburg* 1870 (K; SRGH). **Zimbabwe**. N: Mpingi–Kildonan Rd., Mukwadi R., 2.iii.1965, *Simon* 181 (K; SRGH). W: Matobo, Efifi, 7.i.1973, *Simon* 2314 (SRGH). C: Harare, i.1921, *Eyles* 2949 (K; SRGH). E: Chimanimani, Nyahodi Valley, 25.i.1951, *Crook* 358 (K; SRGH). **Malawi**. N: Nyika Plateau, 26.viii.1970, *Hall-Martin* 1689 (K; PRE). C: Lilongwe–Chipata Rd., Namitete R., 5.ii.1959, *Robson* 1482 (K; SRGH). S: Thyolo, Tung St., 11.i.1950, *Wiehe* N/408 (K; SRGH). **Mozambique**. Z: Mocuba, Namagoa, *Faulkner* 3 (K; COI; SRGH). T: Angónia, Ulongue, 19.xii.1980, *Macuácua* 1479 (K). MS: Gorongosa, Mucoza to Vema, 5.xi.1968, *Macêdo* 1786 (COI).
Northwards to Sudan & Sierra Leone. Woodland, thicket and streamsides, often in shady places or on termite mounds; sea level to 2000 m.
Best recognized by the combination of grooved firm-textured inferior lemma and exposed rugose superior lemma. The acute spikelets provide a useful, but occasionally unreliable, distinction from *S. orthosticha*. The occurrence of long hairs on the panicle axis is very variable, and seems to be of no

104

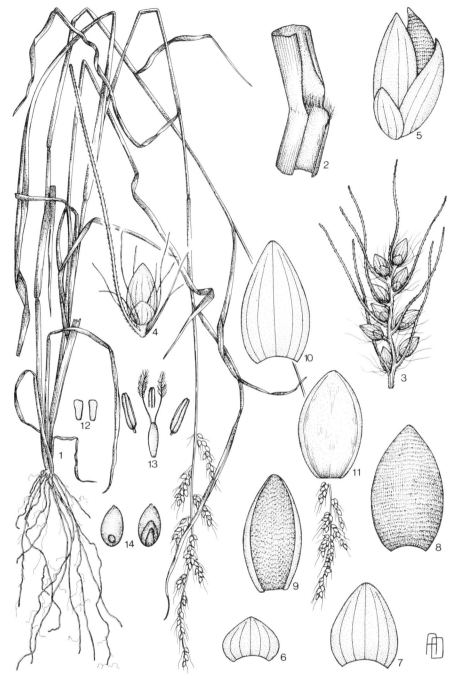

Tab. 26. SETARIA ORTHOSTICHA. 1, habit (× 1); 2, junction of leaf sheath and lamina showing ligule (× 8); 3, portion of inflorescence (× 5); 4, spikelet and subtending bristles (× 10); 5, spikelet, lateral view (× 20); 6, inferior glume (× 20); 7, superior glume (× 20); 8, inferior lemma (× 20); 9, inferior palea (× 20); 10, superior lemma (× 20); 11, superior palea (× 20); 12, lodicules (× 15); 13, ovary & stamens (× 10); 14, caryopsis in dorsal & ventral view (× 10), all from *Vesey-FitzGerald* 3313.

taxonomic significance.

It can be confused with narrow-leaved forms of *S. lindenbergiana*. These differ in their densely caespitose base, contracted panicle, usually longer superior glume, and thinner ungrooved inferior lemma.

13. **Setaria pseudaristata** (Peter) Pilger in Engl. & Prantl, Pflanzenfam. 14e: 72 (1940). —Clayton & Renvoize in F.T.E.A., Gramineae: 535 (1982). Type from Tanzania.

Acrochaete pseudaristata Peter, Fl. Deutsch Ost-Afr. 1, Anh.: 54 (1930). Type as above.

Setaria tenuiseta de Wit in Bull. Bot. Gard. Buitenz. sér.3, **17**: 15 (1941). —Chippindall in Meredith, Grasses & Pastures of S. Afr.: 343 (1955). Type: Botswana, Kabulabula, *Pole-Evans & van Rensburg* 24630 (PRE, holotype).

Loosely tufted perennial. Culms 50–110 cm. high. Leaf laminae 2–6 mm. wide, linear, not plicate. Panicle 15–30 cm. long, narrowly oblong to narrowly ovate, primary branches flexuous, the spikelets borne on secondary branches, these puberulous; bristles 3–20 mm. long. Spikelets 3–4 mm. long, narrowly ovate-elliptic, acute. Glumes acute, the inferior $\frac{1}{3}-\frac{1}{2}$, the superior $\frac{2}{3}-\frac{3}{4}$ as long as spikelet. Inferior lemma not or weakly sulcate. Superior lemma smooth above, indistinctly rugose below.

Botswana. N: Kabulabula, Chobi R., *Pole-Evans & van Rensburg* 24629 (K; PRE). **Zambia**. B: Masese, 13.iii.1961, *Fanshawe* 6414 (SRGH). S: Livingstone, Zambezi R., 24.i.1963, *Mitchell* 17/48 (K; SRGH). **Zimbabwe**. W: Victoria Falls, 7.iii.1977, *Elias* 35 (K; SRGH).

Also in Tanzania and Namibia. In shade of riverine forest; 800–900 m.

Similar to *S. longiseta* in habit, but with spikelets resembling those of *S. megaphylla*.

14. **Setaria barbata** (Lam.) Kunth, Rév. Gram. **1**: 47 (1829). —Stapf & Hubbard in F.T.A. **9**: 854 (1930). —Bor, Grasses of B.C.I. & P.: 360 (1960). —Clayton in F.W.T.A. **3**: 424 (1972). —Clayton & Renvoize in F.T.E.A., Gramineae: 536 (1982). Type from Mauritius.

Panicum barbatum Lam., Tab. Encycl. Méth. Bot. **1**: 171 (1791). Type as above.

Setaria flaccifolia Stapf in F.T.A. **9**: 850. Type: Mozambique, *Kirk* (K, holotype).

Loosely tufted annual. Culms 10–150(200) cm. high, often rooting at lower nodes. Leaf laminae 5–20(30) mm. wide, broadly linear to narrowly lanceolate, conspicuously plicate, firm, dark green. Panicle 3–25 cm. long, narrowly elliptic, the spikelets densely crowded on pedicels or secondary branchlets along the spreading primary branches, these pubescent to pilose; bristles 1–15 mm. long. Spikelets 2–3.2 mm. long, oblong elliptic, plump. Inferior glume obtuse, $\frac{1}{4}-\frac{1}{3}$, the superior $\frac{2}{3}-\frac{4}{5}$ as long as spikelet. Superior lemma rugose.

Mozambique. MS: Shupanga, *Kirk* (K).

Western Africa, with a few isolated records elsewhere in Africa; tropical Asia; introduced to the West Indies. A shade-loving weed of disturbed land.

S. flaccifolia is known from a single specimen. This appears to be part of a weakly growing and somewhat atypical plant of *S. barbata*.

15. **Setaria homonyma** (Steud.) Chiov. in Nuov. Giorn. Bot. Ital., n.s., **26**: 78 (1919). —Stapf & Hubbard in F.T.A. **9**: 857 (1930). —Sturgeon in Rhod. Agric. Journ. **50**: 505 (1953). — Chippindall in Meredith, Grasses & Pastures of S. Afr.: 343 (1955). —Jackson & Wiehe, Annot. Check List Nyasal. Grass.: 58 (1958). —Bor, Grasses of B.C.I. & P.: 361 (1960). —Launert in Merxm., Prodr. Fl. SW. Afr. **160**: 171 (1970). —Clayton & Renvoize in F.T.E.A., Gramineae: 536 (1982). TAB. **27**. Type from India.

Panicum homonymum Steud., Syn Pl. Glum. **1**: 48 (1854).

Panicum thollonii Franch. in Bull. Soc. Hist. Nat. Autun **8**: 351 (1893). Type from Congo (Brazzaville).

Panicum bongaense Pilger in Bot. Jahrb. **33**: 44 (1902). Type from Congo (Brazzaville).

Setaria kialaensis Vanderyst in Bull. Agric. Congo Belge **16**: 682 (1925). Type from Zaire.

Setaria bongaensis (Pilger) A. Camus in Bull. Soc. Bot. Fr. **74**: 633 (1927). Type as for *Panicum bongaense*.

Setaria aequalis Stapf in Kew Bull. **1927**: 267 (1927). —Stapf & Hubbard in F.T.A. **9**: 859 (1930). Type from Tanzania.

Setaria thollonii (Franch.) Stapf in Kew Bull. **1927**: 267 (1927). —Stapf & Hubbard in F.T.A. **9**: 860 (1930). Type as for *Panicum thollonii*.

Setaria microprolepis Stapf in F.T.A. **9**: 849 (1930). Type from Angola.

Loosely tufted annual. Culms 25–100 cm. high. Leaf laminae lanceolate, plicate, thin, flaccid, light green. Panicle 4–20 cm. long, narrowly pyramidal, the spikelets subsessile in 2 rows on raceme-like primary branches, these villous or sometimes merely pubescent; bristles 5–10 mm. long. Spikelets 2–2.5 mm. long, narrowly elliptic, slightly dorsally

Tab. 27. SETARIA HOMONYMA. 1, habit (× $\frac{2}{3}$), from *Osterkamp* 3; 2 & 3, spikelet (× 8); 4, superior lemma (× 8), all from *McCallum-Webster* A167a.

compressed. Inferior glume obtuse, $\frac{1}{8}-\frac{1}{4}$ length of spikelet, superior glume $\frac{4}{5}$ to as long as spikelet. Superior lemma rugose.

Caprivi Strip. Andara, 23.ii.1956, *de Winter & Marais* 4806 (K; PRE). **Botswana**. N: Zibadianja, 3.ii.1978, *Smith* 2350 (K; SRGH). **Zambia**. B: Mongu, 24.iii.1964, *Verboom* 1330 (K; SRGH). N: Mbala Distr., Kawimbe, 26.ii.1959, *McCallum-Webster* A167 (K; SRGH). W: Mwinilunga, Matonchi Farm, 19.ii.1938, *Milne-Redhead* 4633 (K; SRGH). C: Lusaka, 23.i.1954, *Best* 56 (K). S: Mazabuka Distr., Kanchomba to Munyona, 17.ii.1960, *White* 7139 (FHO; K). **Zimbabwe**. N: Lomagundi Distr., c. 34 km. past Karoi–Vuti Rd., ii.1972, *Davies* 3205 (K; SRGH). W: Shangani Distr., Gwampa Forest Reserve, iii.1955, *Goldsmith* 82/55 (K; SRGH). C: Shurugwi Distr., Ferny Creek, 2.iv.1967, *Biegel* 2043 (K; SRGH). E: Mutare, 18.iii.1973, *Crook* 1067 (K; SRGH). S: Chibi, Lundi R., 3.iii.1970, *Mavi* 1064 (K; SRGH). **Malawi**. N: Mzimba, Mzuzu, 2.iv.1971, *Pawek* 4565 (K). C: Ntchisi Mt., 6.v.1963, *Verboom* 963 (K). S: Zomba, 6.iv.1949, *Wiehe* N/46 (K). **Mozambique**. N: Mandimba, 29.iii.1942, *Hornby* 3344 (K). MS: Macequece, Serra de Vumba, 16.iii.1948, *Barbosa* 1191 (COI; K).

Also in Cameroon and Ethiopia to S. Africa; also in India. Shady places in forest, woodland and thicket, often as a weed of cultivation; 600–1500 m.

Plants with a sparse linear panicle of few appressed branches have been separated as *S. thollonii*, but this character seems indicative of impoverishment rather than taxonomic difference.

16. **Setaria seriata** Stapf in F.T.A. **9**: 853 (1930). Type from Zaire.
   Setaria gracilipes C.E. Hubbard in Kew Bull. **4**: 362 (1949). —Clayton in F.W.T.A. **3**: 423 (1972). Type from Nigeria.

Loosely tufted annual. Culms 50–200 cm. high. Leaf laminae broadly linear to narrowly lanceolate, plicate, thin. Panicle 10–20 cm. long, lanceolate, the spikelets briefly pedicelled and untidily appressed to the primary branches, these scaberulous; bristles 5–10 mm. long. Spikelets 2–2.5 mm. long, narrowly elliptic. Inferior glume $\frac{1}{8}-\frac{1}{4}$ length of spikelet, truncate to obtuse, the superior $\frac{3}{4}$ to as long as spikelet. Superior lemma smooth.

Zambia. N: Lake Mueru-Wantipa, on steep path from lake to Mpundu, 10.iv.1957, *Richards* 9116 (K; SRGH).

Extends northwest to Ivory Coast. Shady places in forest and thicket; 1000 m.

17. **Setaria lindenbergiana** (Nees) Stapf in Fl. Cap. **7**: 422 (1899). —Stapf & Hubbard in F.T.A. **9**: 848 (1930). —Sturgeon in Rhod. Agric. Journ. **50**: 505 (1953). —Chippindall in Meredith, Grasses & Pastures of S. Afr.: 343 (1955). —Clayton & Renvoize in F.T.E.A., Gramineae: 537 (1982). TAB. **28**. Types from S. Africa.
   Panicum lindbergianum Nees, Fl. Afr. Austr.: 47 (1841). Types as above.
   Setaria mauritiana var. *angustifolia* Rendle, Cat. Afr. Pl. Welw. **2**: 188 (1899). Types from Angola
   Setaria thermitaria Chiov. in Nuov. Giorn. Bot. Ital., n.s., **29**: 112 (1923). Type from Zaire.
   Setaria angustissima Stapf in F.T.A. **9**: 835 (1930). Type from Angola.
   Setaria subsetosa Stapf in F.T.A. **9**: 837 (1930). Type from Zaire.

Densely tufted perennial, the base invested in tightly packed dead sheaths. Culms 30–120 cm. high, the nodes glabrous or pubescent. Leaf laminae 1–10 mm. wide, linear, often narrow, harsh, plicate when young but soon flattening, often falsely petiolate. Panicle 5–20 cm. long, usually contracted and narrowly lanceolate but sometimes open and narrowly ovate, the branches pubescent or sometimes pilose; bristles 2–10 mm. long, often inconspicuous. Spikelets 2–3 mm. long, narrowly ovate to elliptic. Inferior glume $\frac{1}{4}-\frac{1}{2}$, the superior $\frac{1}{2}$ to as long as spikelet. Superior lemma rugose, pallid.

Botswana. SE: Gaborone, Mannyelanong Hill, 9.ii.1975, *Mott* 660 (SRGH). **Zambia**. N: Mbesuma Ranch, 9.xii.1961, *Astle* 1074 (K). W: Mwinilunga Distr., 5.xi.1937, *Milne-Redhead* 3101 (K; SRGH). **Zimbabwe**. N: Chiporiro Distr., Nyamnyetsi Estate, 3.i.1979, *Nyariri* 611 (K; SRGH). W: Matobo Distr., Efifi, 17.i.1973, *Simon* 2314 (K; SRGH). C: Shurugwi Distr., Good Hope Farm, *Davies* in GHS 22540 (K; SRGH). E: Nyanga (Inyanga), 13.i.1931, *Norlindh & Weimarck* 4255 (K). S: Masvingo, 17.iii.1972, *Wild* 7918 (K; SRGH).

Also in Zaire to S. Africa. Open woodlands, often on rocky slopes or termite mounds; 1000–1700 m.

A variable species characterized by its densely caespitose habit and rugose superior lemma. The main sources of variation are:
(a) The harsh narrow leaves tend to lose their plicate venation, which must be carefully sought at the base of young laminae. In the narrowest leaved forms (*S. angustissima*) it may be quite lacking, leading to confusion with *S. longiseta*.
(b) The superior glume commonly covers most of the superior lemma and is a useful aid to recognition, though rather unreliable as it can sometimes be much shorter.
(c) The panicle is usually contracted, but specimens from Zambia and Zaire mostly have the branches spreading (*S. thermitaria*); such plants do not seem to merit separation at species level.

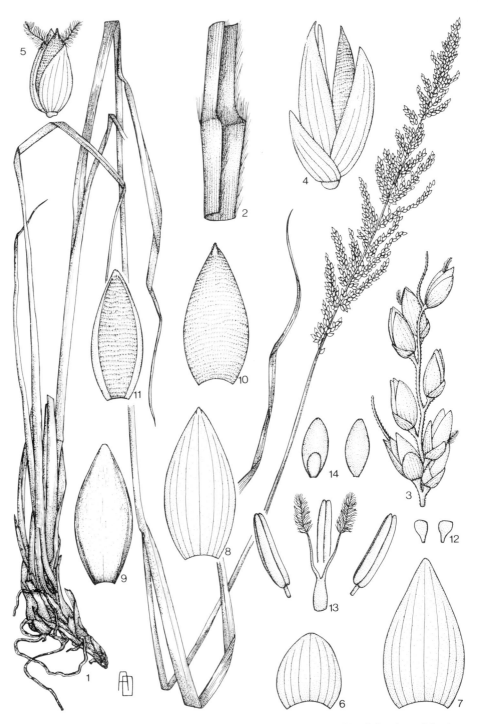

Tab. 28. SETARIA LINDENBERGIANA. 1, habit (× ½); 2, junction of leaf sheath and lamina showing ligule (× 2½); 3, portion of inflorescence (× 5); 4, spikelet (× 10); 5, spikelet, lateral view (× 15); 6, inferior glume (× 15); 7, superior glume (× 15); 8, inferior lemma (× 15); 9, inferior palea (× 15); 10 superior lemma (× 15); 11, superior palea (× 15); 12, lodicules (× 15); 13, ovary & stamens (× 10); 14, caryopsis in dorsal & ventral view (× 10), all from *Brain* 1248.

18. **Setaria megaphylla** (Steud.) Th. Dur. & Schinz, Consp. Fl. Afr. **5**: 773 (1894). —Stapf & Hubbard in F.T.A. **9**: 840 (1930). —Sturgeon in Rhod. Agric. Journ. **50**: 504 (1953). —Clayton in F.W.T.A. **3**: 424 (1972). —Clayton & Renvoize in F.T.E.A., Gramineae: 539 (1982). Type from Gabon.

   Panicum megaphyllum Steud., Syn. Pl. Glum. **1**: 53 (1854). Type as above.

   Panicum plicatile Hochst. in Flora **38**: 198 (1855). Type from Ethiopia.

   Setaria macrophylla Anderss. in Peters, Reise Mossamb., Bot. **2**: 550 (1864). Type: Mozambique, Boror, *Peters* (B, holotype).

   Setaria plicatilis (Hochst.) Engl., Hochgebirgsfl. Trop. Afr.: 121 (1891). —Stapf & Hubbard in F.T.A. **9**: 847 (1930). —Sturgeon in Rhod. Agric. Journ. **50**: 504 (1953). —Jackson & Wiehe, Annot. Check List Nyasal. Grass.: 59 (1958). —Clayton & Renvoize in F.T.E.A., Gramineae: 538 (1982). Type from Ethiopia.

   Setaria oligochaete K. Schum. in Pflanzenw. Ost-Afr. **C**: 105 (1895). Type from Tanzania.

   Panicum oligochaete (K. Schum.) Kneucker in Allg. Bot. Zeitschr. **21**: 28 (1915). Type as above.

   Setaria chevalieri Stapf in F.T.A. **9**: 842 (1930). —Sturgeon in Rhod. Agric. Journ. **50**: 504 (1953). —Chippindall in Meredith, Grasses & Pastures of S. Afr.: 341 (1955). —Jackson & Wiehe, Annot. Check List Nyasal. Grass.: 57 (1958). —Clayton in F.W.T.A. **3**: 424 (1972). Types from Nigeria.

   Setaria insignis de Wit in Bull. Bot. Gard. Buitenz., sér.3, **17**: 13 (1914). Type from S. Africa.

   Setaria natalensis de Wit in Bull. Bot. Gard. Buitenz., sér. 3, **17**: 19 (1941). Type from S. Africa.

   Setaria phillipsii de Wit in Bull. Bot. Gard. Buitenz., sér. 3, **17**: 21 (1941). Type from S. Africa.

   Setaria megaphylla var. *chevalieri* (Stapf) Berhaut, Fl. Sénégal, ed. **2**: 401 (1954). Types as for *Setaria chevalieri*.

Tufted perennial, often forming large clumps. Culms 0.5–3 m. high and 2–10 mm. in diam. at the base. Leaf laminae 5–110 mm. wide, broadly linear to narrowly lanceolate, conspicuously plicate, sometimes falsely petiolate. Panicle 10–60 cm. long, linear to lanceolate, the branches short (occasionally the lowest long and flexuous) and ascending close to the rhachis or projecting laterally like a pagoda; bristles 3–15 mm. long. Spikelets 2.2–3(3.5) mm. long, narrowly ovate to elliptic, acute. Inferior glume $\frac{1}{3}-\frac{1}{2}$, the superior $\frac{1}{2}-\frac{3}{4}$ length of spikelet. Inferior lemma firmly membranous, as long as the superior. Superior lemma usually smooth and often shining, sometimes rugose below, rarely rugose throughout, becoming light brown.

Zambia. N: Mbala, Kellett's Farm, 14.iv.1959, *McCallum-Webster* A317 (K; SRGH). W: Solwezi, 10.iv.1960, *Robinson* 3534 (K). C: Kabwe, Ralph's Farm, 25.iv.1953, *Hinds* 105 (K). S: Kalomo, Dundumwenzi, Shiadomo stream, 7.ii.1965, *Mitchell* 25/95 (K). **Zimbabwe**. C: Shurugwi Town, 1.iv.1967, *Biegel* 2035 (K; SRGH). E: Chimanimani, Rocklands, 16.x.1950, *Sturgeon* in GHS 30785 (K; SRGH). S: Masvingo, Glenlivet Hotel, 15.ii.1974, *Davidse, Simon & Pope* 6654 (K; MO; SRGH). **Malawi**. N: Nkhata Bay, on Rd. to Chinteche, 11.v.1970, *Brummitt* 10595 (K). C: Chipata Mt., 4.v.1963, *Verboom* 970 (K). S: Zomba, Rd junction to Zomba Plateau, 9.viii.1978, *Msiska* 84 (K; SRGH). **Mozambique**. N: Cabo Delgado, Macondes, 13.iv.1964, *Torre & Paiva* 11894 (COI; SRGH). Z: Gúruè, Licungo R., 6.iv.1943, *Torre* 5086 (COI; K). MS: Gorongosa, Nhandore Mt., 6.v.1964, *Torre & Paiva* 12271 (COI; K). GI: Inhambane, ii.1938, *Gomes e Sousa* 2084 (K). M: Maputo, Ponto do Ouro, 3.i.1980, *de Koning* 7897 (K).

From tropical and S. Africa, tropical America and a few records from India. Woodland shade, forest glades and streamsides; sea level to 1800 m.

The considerable variation in overall size between the smallest (*S. plicatilis*) and largest (*S. megaphylla*) plants appears to be quite continuous, affording no opportunity to partition the complex into several discrete species. In fact there is a case for treating this taxon as a subspecies of *S. palmifolia* (Koenig) Stapf, an Asiatic species distinguished only by its lax ovate inflorescence with long flexuous branches; a similar facies occurs sporadically in Africa, and it is clear that *S. megaphylla* and *S. palmifolia* intergrade. *S. megaphylla* also intergrades with *S. poiretiana* (Schult.) Kunth. The latter has a tenuous claim to separate recognition by virtue of spikelets 3–4.5 mm. long whose inferior lemma exceeds the superior, but is has not been recorded south of Tanzania. See Clayton in Kew Bull. **33**: 505–509 (1979).

The superior lemma is characteristically smooth, but may be ± rugose in the lower half under cover of the superior glume. Occasionally it is rugose throughout, blurring the boundaries with *S. lindenbergiana* (compact base, narrow leaves) and *S. longiseta* (non-plicate leaves).

19. PASPALIDIUM Stapf

Paspalidium Stapf in F.T.A. **9**: 582 (1920)

Inflorescence of short racemes along a central axis, these ending in an inconspicuous point, the spikelets borne singly and often in 2 neat rows, rarely the lower paired or clustered. Spikelets abaxial, ovate. Superior glume $\frac{1}{3}$ to as long as spikelet, the inferior shorter. Superior lemma crustaceous, as long as spikelet, acute; superior palea acute.

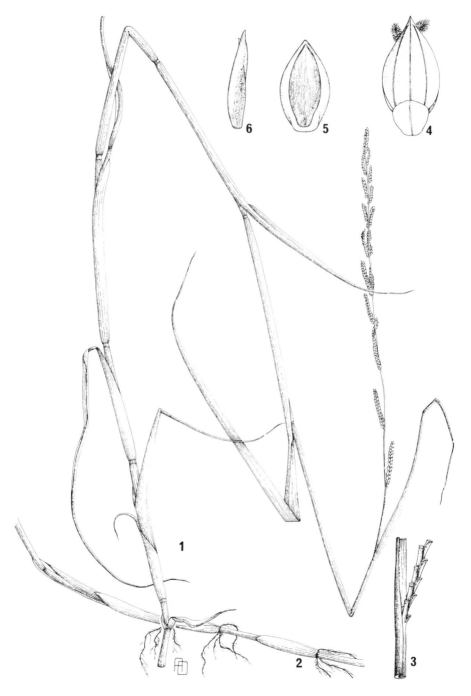

Tab. 29. PASPALIDIUM GEMINATUM. 1, habit (× ½), from *Harris* 3054; 2, stolon (× ½), from *Palmer* 21; 3, portion of main axis with rhachis of a raceme (× 2); 4, spikelet (× 10); 5, superior floret (× 10); 6, superior floret in lateral view (× 10), 3–6 from *Lind* 401a. From F.T.E.A.

A genus of c. 40 species. Tropics. The inconspicuous rhachis point of our species bears little resemblance to *Setaria,* but in fact there are many intermediate species and the two genera are quite difficult to separate.

1. Spikelets 1.6–2.6 mm. long; leaf laminae setaceously acuminate   -   -   -   1. *geminatum*
- Spikelets 3–4 mm. long; leaf laminae obtuse to bluntly acute   -   -   -   2. *obtusifolium*

1. **Paspalidium geminatum** (Forssk.) Stapf in F.T.A. **9**: 583 (1920). —Chippindall in Meredith, Grasses & Pastures of S. Afr.: 366 (1955). —Jackson & Wiehe, Annot. Check List Nyasal. Grass.: 52 (1958). —Bor, Grasses of B.C.I. & P.: 333 (1960). —Launert in Merxm., Prodr. Fl. SW. Afr. **160**: 144 (1970). —Clayton in F.W.T.A. **3**: 440 (1972). —Clayton & Renvoize in F.T.E.A., Gramineae: 552 (1982). TAB. **29**. Type from Egypt.
    Panicum geminatum Forssk., Fl. Aegypt.-Arab.: 18 (1775). Type as above.
    Echinochloa geminata (Forssk.) Roberty in Bull. I.F.A.N., sér. A, **17**: 64 (1955). Type as above.

Perennial with creeping or floating ± spongy rhizomes. Culms 10–100 cm. high, prostrate and rooting at the nodes below. Leaf laminae setaceously acuminate. Inflorescence 5–30 cm. long, bearing numerous short 2-rowed overlapping racemes, these 0.5–4 cm. long. Spikelets 1.6–2.6 mm. long. Inferior glume truncate, up to ⅓ length of spikelet; superior ⅔ to almost as long as spikelet.

Zambia. N: Mbala Distr., Kasaba Bay, 17.ii.1959, *McCallum-Webster* A 66 (K; SRGH). **Malawi**. N: Karonga, lakeside, 10.i.1959, *Robinson* 3135 (K). C: Nkhota Kota, 2.v.1963, *Verboom* 992 (K). S: Nsanje, Shire flood plain, 28.xi.1950, *Jackson* 341 (K). **Mozambique**. N: Mogincual, 30.iii.1964, *Torre & Paiva* 11482 (COI; K). Z: Massingire, 21.v.1943, *Torre* 5351 (COI; K). M: Maputo, 29.xii.1963 *Balsinhas* 651 (K).
    Old World tropics. Marshy soils, river sandbanks and in water; sea level to 1800 m.

2. **Paspalidium obtusifolium** (Del.) N.D. Simpson, Min. Agric. Egypt Bull. **93**: 10 (1930). —Chippindall in Meredith, Grasses & Pastures of S. Afr.: 366 (1955). —Clayton & Renvoize in F.T.E.A., Gramineae: 551 (1982). Type from Egypt.
    Panicum obtusifolium Del., Fl. Egypte: 150, fig.5/1 (1813). Type as above.
    Digitaria obtusifolia (Del.) Roem. & Schult., Syst. Veg. **2**: 889 (1817). Type as above.
    Paspalidium platyrrhachis C.E. Hubbard in Kew Bull. **1934**: 262 (1934). Type: Zambia, Mazabuka, *Trapnell* 1086 (K, holotype).

Like *P. geminatum* but leaf laminae broadly obtuse to acute, spikelets 3–4 mm. long and glumes shorter, the superior up to ½ length of spikelet.

Botswana. N: Lake Ngami floodplain, 20.iii.1976, *Ellis* 2731 (K; PRE). **Zambia**. B: Mongu Distr., Luena flats, 13.iv.1964, *Verboom* 1118 (K). S: Mazabuka, Lochinvar Ranch, 25.iv.1962, *Mitchell* 13/91 (K). **Zimbabwe**. S: Mwenezi, c. 4 km. S. of Chipinda pools, 31.iii.1961, *Phipps* 2903 (K; SRGH). **Mozambique**. MS: Gorongosa, Urema plains, iv.1970, *Tinley* 1907 (K; SRGH). GI: Massingir, 39 km. ESE. of Lagoa Nova, 5.v.1972, *Lousã & Rosa* 273 (K).
    Also in Kenya to S. Africa, Algeria and Egypt. Marshy places and in water; sea level to 1500 m.
    The leaf apex varies from subulate to rounded, but this seems to be of no taxonomic significance.

20. STENOTAPHRUM Trin.

Stenotaphrum Trin., Fund. Agrost.: 175 (1822). —Saur in Brittonia **24**: 202–222 (1972).

Inflorescence of very short racemes bearing a few single spikelets and sunk in pockets on one or both sides of a foliaceous or corky axis, falling entire or fracturing into irregular segments at maturity, the raceme rhachis ending in a point. Spikelets abaxial. Glumes both short, or the superior as long as the spikelet. Inferior lemma coriaceous; superior lemma chartaceous with flat margins.

A genus of 7 species. Tropics and subtropics.
    The inflorescence is spiciform and it requires careful examination to perceive its derivation from *Paspalidium* by enlargement of the central axis and extreme reduction of the racemes. Associated with this condensation is a transfer of protective function from the superior to the inferior lemma.

1. Axis of inflorescence herbaceous, toothed on the back; racemes with
    3–8 spikelets   -   -   -   -   -   -   -   -   -   -   -   1. *dimidiatum*
- Axis of inflorescence corky, entire; racemes commonly reduced to
    1 spikelet   -   -   -   -   -   -   -   -   -   -   -   2. *secundatum*

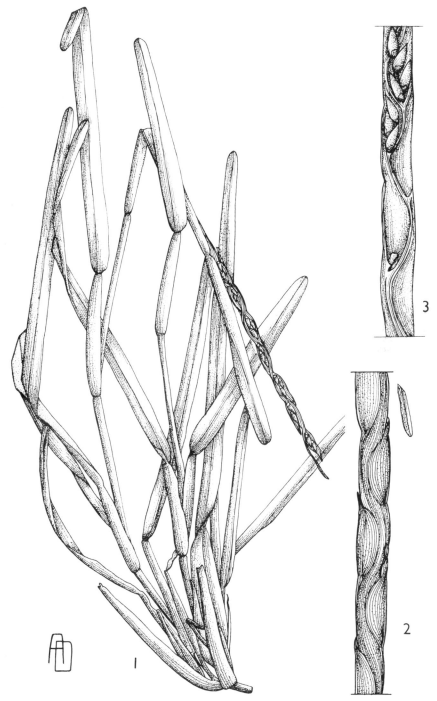

Tab. 30. STENOTAPHRUM DIMIDIATUM. 1, habit (× ⅔); 2, inflorescence, ventral view (× 2), 1–2 from *Drummond & Hemsley* 3996; 3, inflorescence showing lateral pockets, the two superior containing racemes (× 2), from *Lyne* 112. From F.T.E.A.

1. **Stenotaphrum dimidiatum** (L.) Brongn. in Duperrey, Bot. Voy. Coquille: 127 (1832). —Bor,
Grasses of B.C.I. & P.: 366 (1960). —Clayton in F.W.T.A. **3**: 435 (1972). —Clayton & Renvoize in
F.T.E.A., Gramineae: 549 (1982). TAB. **30**. Type from India.
Panicum dimidiatum L., Sp. Pl.: 57 (1753). Type as above.

Stoloniferous perennial forming a close sward. Culms 10–30 cm. high. Leaf laminae
broadly linear, folded when young, obtuse. Inflorescence 6–12 cm. long, the central axis
herbaceously winged, with racemes borne in anterior pockets on alternate sides of the
sinuous midrib, the wing of at least the upper pockets produced into a broad acute tooth;
racemes 7–15 mm. long, bearing 3–8 spikelets. Spikelets (3)4–5 mm. long, lanceolate to
narrowly ovate. Inferior glume up to $\frac{1}{4}$, the superior as long as spikelet.

    Zimbabwe. C: Harare Distr., Crowborough Sewage Farm, 6.xii.1973, *Biegel* 4447 (K; SRGH).
S: Chiredzi, xii.1969, *van Niekerk* in GHS 203962 (K; SRGH). **Mozambique**. Z: Mocuba, 6.vi.1949,
Barbosa & Carvalho 2981 (K). MS: Beira Distr., Savane, 29.iv.1968, *Crook* 825 (K; SRGH). GI: Inharrime,
Závora, 11.iv.1955, *Myre & Carvalho* 2267 (K).
    Also from the shores of the Indian Ocean. Coastal areas in partial shade, occurring inland as an
introduction.

2. **Stenotaphrum secundatum** (Walt.) Kuntze, Rev. Gen. Pl. **2**: 794 (1891). —Chippindall in Meredith,
Grasses & Pastures of S. Afr.: 367 (1955). —Clayton in F.W.T.A. **3**: 435 (1972). Type from USA.
Ischaemum secundatum Walt., Fl. Carol.: 294 (1788). Type as above.

Vegetative parts and spikelets like *S. dimidiatum*. Inflorescence 5–12 cm. long, the
central axis thickened and corky, with racemes embedded in its anterior face; racemes
4–10 mm. long, bearing 1–3 spikelets.

    Mozambique. M: Inhaca Is., Ponta Basa, 9.iii.1958, *Mogg* 31562 (K).
    Tropical shores of the Atlantic, rounding the Cape to Mozambique. Pioneer habitats on the coast,
and moist open sites further inland. Widely introduced as a lawn species where rainfall is adequate,
under the name St Augustine grass.
    The structure of the inflorescence is not immediately apparent, but a rudimentary raceme rhachis
can be found on the adaxial side of the spikelet when it is prised out of the corky central axis. The
latter is presumably an adaptation to dispersal by sea, but it only floats for about a week.

21. **TRICHOLAENA** Schrader ex Schultes

By G. Zizka

Tricholaena Schrader ex Schultes, Mant. **2**: 163 (1824). —Zizka in Bibl. Bot. **138**: 36–50
(1988).

Inflorescence a panicle. Spikelets slightly laterally compressed. Inferior glume small or
suppressed; superior glume and inferior lemma as long as spikelet, membranous, not
gibbous, emarginate to acute, awnless to mucronate. Superior floret dorsally compressed,
readily deciduous, its lemma cartilaginous.

    A genus of 4 species; Africa and the Mediterranean to India. One species endemic to Socotra.

Tricholaena monachne (Trin.) Stapf & Hubbard in F.T.A. **9**: 909 (1930). —Sturgeon in Rhod. Agric.
Journ. **50**: 105 (1953). —Chippindall in Meredith, Grasses & Pastures of S. Afr.: 434 (1955).
—Jackson & Wiehe, Annot. Check List Nyasal. Grass.: 63 (1958). —Bor, Grasses of B.C.I. & P.:
371 (1960). —Launert in Merxm., Prodr. Fl. SW. Afr. **160**: 208 (1970). —Clayton in F.W.T.A. **3**:
455 (1972). —Clayton & Renvoize in F.T.E.A., Gramineae: 502 (1982). —Müller, Gräser
SWA/Nam.: 264 (1985). TAB. **31**. Type from Réunion.
Panicum monachne Trin. in Sprengel in Neue Entd. **2**: 86 (1821). Type as above.
Panicum madagascariense var. *minus* Hackel in Bot. Jahrb. **11**: 400 (1889). Type from S. Africa.
Panicum gracillimum K. Schum. in Bot. Jahrb. **24**: 331 (1897). Type from Angola.
Panicum madagascariense var. *brevispiculum* Rendle in Cat. Afr. Pl. Welw. **2**: 183 (1899). Type
from Angola.
Tricholaena glabra Stapf in Fl. Cap. **7**: 446 (1899). Type from S. Africa.
Melinis glabra (Stapf) Hackel in Öst. Bot. Zeitschr. **51**: 464 (1901). Type as above.
Melinis monachne (Trin.) Pilger in Bot. Jahrb. **33**: 51 (1902). Type as for *Panicum monachne*.
Tricholaena arenaria var. *semiglabra* Hackel in Viert. Nat. Ges. Zürich **57**: 533 (1912). Type
from Angola.
Melinis trichotoma Mez in Bot. Jahrb. **57**: 200 (1921). Type from Namibia.

114

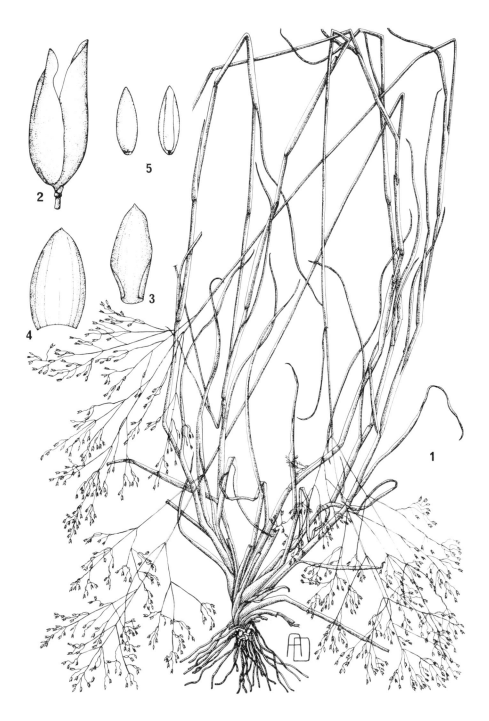

Tab. 31. TRICHOLAENA MONACHNE. 1, habit (×½); 2, spikelet (× 15); 3, superior glume (× 10); 4, inferior lemma (× 10); 5, superior floret (× 10), all from *Procter* 2383. From F.T.E.A.

Xyochlaena monachne (Trin.) Stapf in Ic. Pl. (Hook.) **31**: t. 3098 (1922). Type as for *Panicum monachne.*
Xyochlaena monachne var. *minus* Garab. in Ann. S. Afr. Mus. **16**: 395 (1925) nom. nud.
Eremochlamys arenaria Peter in Fl. Deutsch Ost-Afr. **1**, Anh.: 19, pl. 18, fig. 1 (1930). Type from Angola.
Eremochlamys littoralis Peter in Fl. Deutsch Ost-Afr. **1**, Anh.: 19, pl. 18, fig. 2 (1930). Type from Tanzania.
Tricholaena delicatula Stapf & Hubbard in F.T.A. **9**: 911 (1930). Based on *Panicum madagascariense* var. *brevispiculum.*
Tricholaena monachne var. *annua* J.G. Anderson in Kirkia **1**: 103 (1961). Type from Namibia.

Tufted annual to short-lived perennial. Culms (13)20–100(114) cm. high, geniculately ascending and often rooting at lower nodes. Panicle (4.7)6–14(18) cm. long, broadly ovate; pedicels glabrous. Spikelets (1.2)1.4–2.5(3) mm. long, oblong to narrowly ovate in side view, green to straw-coloured, often tinged with purple. Inferior glume a tiny nerveless scale 0.1–0.3(rarely 0.7) mm. long, inserted close to the superior; superior glume 5-nerved, broadly ovate, obtuse to emarginate, awnless, sometimes with a short mucro, glabrous or inconspicuously hairy on the margins. Inferior floret male, its lemma 5-nerved, resembling the superior glume but slightly wider; the palea ciliate on its keels.

Caprivi Strip. E. Caprivi, Singalamwe, 2.i.1959, *Killick & Leistner* 3261 (M). **Botswana**. N: Ngamiland, 64 km. E. of Maun, iv.1958, *Robertson* 636 (K; LISC). SW: 107 km. N. of Kang on Rd. to Ghanzi, 19.ii.1960, *de Winter* 7396 (K; PRE). SE: Kweneng Distr., Lephepe Village, iv.1969, *Kelaole* 579 (K; PRE). **Zambia**. B: Mongu, 20.xii.1965, *Robinson* 6737 (K). S: Kafue Nat. Park, Ngoma, 13.iii.1961, *Mitchell* 6/68 (BR; K). **Zimbabwe**. N: Gandavaruyi (Ganderowe) Falls, Sanyati R., 29.iv.1965, *Simon* 294 (K). W: Shangani Distr., Gwampa Forest Res., i.1955, *Goldsmith* 132/55 (BR). C: Marondera (Marandellas), near Red Leaf Gate, *West* 3226 (K). E: Mutare Distr., 58 km. S. of Mutare, 23.iv.1969, *Plowes* 3197 (K; LISC). S: Chiredzi Distr., Chipinda pools, edge of Lundi R., 28.ii.1971, *Taylor* 136 (BR; K). **Malawi**. C: Nkhota Kota, 24.ii.1953, *Jackson* 1109 (K). S: Mangochi (Fort Johnston), *Wiehe* N/61 (K). **Mozambique**. N: 16 km. from Meconta to Nacavala, 26.xi.1963, *Torre & Paiva* 9328 (LISC). Z: Zambezia Distr., between Quelimane and Mopeia, 11.x.1941, *Torre* 3621 (LISC). T: Msusa, Zambezi R. (Sambesi R.), *Chase* 683 (BM). MS: Beira, 2.xi.1919, *Shantz* 368 (K). GI: Bazaruto I., 24.ix.1958, *Mogg* 28619 (LISC). M: Maputo, Matola, 27.xii.1947, *Torre* 7003 (LISC).

Also from Tropical and S. Africa; also Madagascar and the Mascarene Is. Sandy places on the seashore, river banks and lake margins; in *Brachystegia* and *Julbernardia* woodland, often on anthropogenic sites; 0–1600 m.

The species shows considerable variability in habit and spikelet-length. Plants from western Africa, mostly from dunes near the coast, tend to have smaller and more hairy spikelets (described as *T. bicolor*). Plants from the Namib desert in southwest Africa are often delicate and small (described as *T. monachne* var. *annua*). There is no clear dividing line between these groups, and the variation seems to be due to environmental factors.

22. **MELINIS** Beauv.

By G. Zizka

Melinis Beauv., Ess. Agrost.: 54 (1812). —Zizka in Bibl Bot. **138**: 50–133 (1988).
Rhynchelytrum Nees in Lindley, Nat. Syst. ed. **2**: 446 (1836).
Mildbraediochloa Butzin in Willdenowia **6**: 288 (1971).

Inflorescence a panicle. Spikelets laterally compressed. Inferior glume small or suppressed; superior glume and inferior lemma as long as spikelet, membranous to coriaceous, sometimes gibbous, emarginate to bilobed, sometimes awned. Superior floret laterally compressed, readily deciduous, its lemma membranous to thinly cartilaginous.

A genus of 22 species; Africa (especially the southern tropical part), 1 species extending to SW Asia; 2 species introduced throughout the tropics.
The genera *Melinis* and *Rhynchelytrum* have usually been regarded as distinct, the nervation of the superior glume being the diagnostic criterion for separating them. This distinction coincides with a broad difference in facies, though troubled by some intergradation, but it is artificial and does not correctly reflect the disposition of species according to their overall similarity. Consequently *Rhynchelytrum* and *Melinis* (as well as *Mildbraediochloa* from Pagalu Island) have been united.

1. Palea of inferior floret present (rarely somewhat reduced)   -   -   -   -   -   -   2
- Palea of inferior floret absent   -   -   -   -   -   -   -   -   -   -   7
2. Palea of inferior floret ciliate on the keels   -   -   -   -   -   -   -   -   3
- Palea of inferior floret scaberulous on the keels   -   -   -   -   -   -   -   6
3. Plants densely caespitose; leaf laminae filiform; visible culm-nodes 2–4; perennial; inferior
   lemma equalling the superior glume   -   -   -   -   -   -   -   - 10. *nerviglumis*
- Plants sprawling or loosely tufted, rarely caespitose; leaf laminae flat; visible culm-nodes more
   than 5; annual to short-lived perennial or perennial; inferior lemma narrower and less gibbous
   than the superior glume   -   -   -   -   -   -   -   -   -   -   -   -   4
4. Plants perennial with a short knotty rootstock; spikelets mostly glabrous
   and mucronate   -   -   -   -   -   -   -   -   -   -   -   -   3. *subglabra*
- Plants annual to short-lived perennial, never with a knotty rootstock; spikelets glabrous or hairy,
   awned or mucronate   -   -   -   -   -   -   -   -   -   -   -   -   5
5. Pedicel glabrous, attached laterally to the base of the spikelet; spikelets 3.5–5 mm. long,
   conspicuously awned, glabrous to pubescent   -   -   -   -   -   - 2. *kallimorpha*
- Pedicel with long hairs towards the apex, attached centrally to the base of the spikelet; spikelets
   2–12 mm. long, mostly hairy   -   -   -   -   -   -   -   -   -   - 1. *repens*
6. Superior glume 5-nerved   -   -   -   -   -   -   -   -   -   -   - 4. *longiseta*
- Superior glume 7-nerved   -   -   -   -   -   -   -   -   -   -   - 5. *ambigua*
7. Pedicels smooth, except for long white hairs towards the apex; superior glume 5-nerved;
   spikelets 1–1.5 mm. long   -   -   -   -   -   -   -   -   - 8. *tenuissima*
- Pedicels scaberulous; superior glume 7-nerved; spikelets usually longer
   than 1.5 mm.   -   -   -   -   -   -   -   -   -   -   -   -   -   8
8. Superior glume with an awn 0.8–4.5 mm. long; spikelets 2.4–3.4 mm. long, mostly hairy; inferior
   glume 0.4–1 mm. long   -   -   -   -   -   -   -   -   -   - 5. *ambigua*
- Superior glume awnless or with an awn up to 0.5 mm. long (if awn longer then spikelet less than
   2.2 mm. long); spikelets 1.5–2.2 mm. long, glabrous to shortly pilose; inferior glume a tiny scale
   0.1–0.4 mm. long   -   -   -   -   -   -   -   -   -   -   -   -   9
9. Superior glume deeply grooved between the prominent nerves   -   -   - 6. *minutiflora*
- Superior glume not grooved, finely nerved   -   -   -   -   -   -   -   -   10
10. Pedicels glabrous; annual; inferior glume an inconspicuous scale up to 0.2 mm. long; inferior
    lemma 3–5-nerved   -   -   -   -   -   -   -   -   -   -   - 9. *macrochaeta*
-   Pedicels always with long hairs towards the apex; perennial; inferior glume a tiny scale 0.2–0.4
    mm. long; inferior lemma 5-nerved   -   -   -   -   -   -   -   -   - 7. *effusa*

1. **Melinis repens** (Willd.) Zizka in Bibl. Bot. **138**: 55 (1988). Type from Ghana.
 Saccharum repens Willd., Sp. Pl. ed. 4,1: 322 (1798). Type as above.
 Rhynchelytrum dregeanum Nees in Lindl., Nat. Syst., ed. 2: 447 (1836). Type from S. Africa.
 Tricholaena rosea Nees in Cat. Sem. Hort. Vratisl. **1835** (1836) and Linnaea **11**, Lit. Ber.: 129
(1837). Type from S. Africa.
 Tricholaena tonsa Nees in Cat. Sem. Hort. Vratisl. **1835** (1836) and Linnaea **11**, Lit. Ber.: 128
(1837). Type from S. Africa.
 Monachyron roseum (Nees) Parl. in Fl. Ital. **1**: 131 (1850). Type as for *Tricholaena rosea.*
 Monachyron tonsum (Nees) Parl. in Fl. Ital. **1**: 131 (1850). Type as for *Tricholaena tonsa.*
 Panicum roseum (Nees) Steudel in Syn. Pl. Glum. **1**: 92 (1854) non *P. roseum* Sprengel (1825).
Type as for *Tricholaena rosea.*
 Panicum tonsum (Nees) Steudel in Syn. Pl. Glum. **1**: 92 (1854). Type as for *Tricholaena tonsa.*
 Tricholaena dregeana (Nees) Dur. & Schinz, Consp. Fl. Afr. **5**: 769 (1894). Type as for
Rhynchelytrum dregeanum.
 Tricholaena grandiflora var. *collina* Rendle in Cat. Afr. Pl. Welw. **2**: 195 (1899). Type from
Angola.
 Melinis rosea (Nees) Hackel in Öst. Bot. Zeitschr. **51**: 464 (1901). Type as for *Tricholaena rosea.*
 Rhynchelytrum tonsum (Nees) Lanza & Mattei in Boll. Reale Orto Bot. Giardino Colon.
Palermo **9**: 49 (1910). Type as for *Tricholaena tonsa.*
 Melinis congesta Mez in Bot. Jahrb. **57**: 197 (1921). Type from Angola.
 Melinis nitens Mez in Bot. Jahrb. **57**: 198 (1921). Type from Tanzania.
 Melinis stolzii Mez in Bot. Jahrb. **57**: 199 (1921). Type from Tanzania.
 Rhynchelytrum gossweileri Stapf & Hubbard in F.T.A. **9**: 891 (1930). Type from Angola.
 Rhynchelytrum roseum (Nees) Stapf & Hubbard in F.T.A. **9**: 880 (1930). Type as for *Tricholaena
rosea.*
 Rhynchelytrum stolzii (Mez) Stapf & Hubbard in F.T.A. **9**: 885 (1930). Type as for *Melinis stolzii.*
 Rhynchelytrum repens (Willd.) Hubbard in Kew Bull. **1934**: 110 (1934). —Sturgeon in Rhod.
Agric. Journ. **50**: 108 (1953). —Chippindall in Meredith, Grasses & Pastures of S. Afr.: 430
(1955). —Jackson & Wiehe, Annot. Check List Nyasal. Grass.: 55 (1958). —Bor, Grasses of B.C.I.
& P.: 355 (1960). —Launert in Merxm., Prodr. Fl. SW. Afr. **160**: 163 (1970). —Clayton in F.W.T.A.
3: 454 (1972). —Clayton & Renvoize in F.T.E.A., Gramineae: 515 (1982). —Müller, Gräser
SWA/Nam.: 218 (1985). Type as for *Saccharum repens.*
 Tricholaena repens (Willd.) A.S. Hitchc. in U.S.D.A. Misc. Publ. **243**: 331 (1936). Type as above.
 Rhynchelytrum repens var. *roseum* (Nees) Chiov. in Miss. Biol. Borana Racc. Bot.: 275 (1939).
Type as for *Tricholaena rosea.*

Tufted annual to short-lived perennial. Culms 20–150 cm. high, geniculately ascending and often rooting at lower nodes. Leaf laminae 4–20(27) cm. long, 2–12(14) mm. wide, flat. Panicle (6)8–20 cm. long, broadly ovate; pedicels with a few long hairs. Spikelets 2–12 mm. long, ovate. Inferior glume (0.3)0.6–3(4.3) mm. long, narrowly ovate, 0–1-nerved, separated from the superior by an internode (0.1)0.2–1.7(2) mm. long; superior glume 5-nerved, membranous to subcoriaceous, usually gibbous on the back, hairy or glabrous, often tapering to a glabrous beak, emarginate, awned or mucronate. Inferior floret male, its lemma 5-nerved, resembling the superior glume but less gibbous, the palea ciliate on its keels.

Very variable and formerly divided into 2–4 species according to hairiness of the spikelet, growth-form and length of the internode between the glumes. Hairiness and colour of the spikelet, as well as growth-form, prove to be of no taxonomic value. Four weakly separated infraspecific groups (3 in the Flora Zambesiaca region) are recognized. Typical specimens can be identified easily, but frequent intergradation occurs between the infraspecific groups.

The length of the internode between the glumes varies quite continuously but, together with spikelet length, is the best character for separating the three groups. Single specimens approach *M. ambigua* subsp. *longicauda* and *M. nerviglumis* (introgression?). If the basal parts are missing, discrimination between *M. repens* and *M. subglabra* is hardly possible.

1. Culms 0.9–2.5(3.1) mm. in diam., (20)30–100(130) cm. high; leaf laminae up to 3–6(11) mm. wide; spikelets 2–12 mm. long, brown to straw-coloured, mostly hairy - - - - - - 2
   - Culms 2–3.2 mm. in diam., (60)100–150 cm. high; leaf laminae up to (6)7–12(14) mm. wide; spikelets 3–4.5(5) mm. long, dark olive green to straw-coloured, mostly glabrous or pilose - - - - - - - - - - - subsp. *nigricans*
2. Spikelets (4)5–12 mm. long, glabrous to hairy; internode between the glumes (0.5)0.7–1.7(2) mm. long; superior glume and inferior lemma often drawn out into a narrow beak - - - - - - - - - - subsp. *grandiflora*
   - Spikelets 2–4(4.5) mm. long, mostly densely hairy; internode between the glumes 0.1–0.5(0.6) mm. long; superior glume and inferior lemma not drawn out into a narrow beak - - - - - - - - - - - subsp. *repens*

Subsp. repens

Tufted or caespitose annual or short-lived perennial. Culms 25–120(150) cm. high, (0.9)1.1–2.5(3.1) mm. in diam. Spikelets 2–4(4.5) mm. long, slightly laterally compressed, usually densely hairy. Inferior glume (0.3)0.6–1.3(1.5) mm. long, separated from the superior by an internode 0.1–0.5(0.6) mm. long. Superior glume and inferior lemma membranous.

Botswana. SE: Kweneng Distr., Lephepe Pasture Res. St., i.1964, *Yalala* 494 (BR). **Zambia**. W: Solwezi, 6.vi.1930, *Milne-Redhead* 227 (BR). C: 100–129 km. E. of Lusaka, Chakwenga headwaters, 27.iii.1965, *Robinson* 6487 (B). E: Lundazi Distr., 62 km. on Rd. from Lundazi to Chama, 19.x.1958, *Robson* 168 (LISC). **Zimbabwe**. N: Hurungwe (Urungwe) Safari Area, 311 km. on Harare to Chirundu Rd., *Philcox, Leppard & Dini* 8580 (SRGH). W: Hwange Game Res., 22.ii.1956, *Wild* 4794 (BR). C: Gweru (Gwelo), 11.i.1967, *Biegel* 1809 (LISC). E: Nyanga (Inyanga) Distr., near Nyamingura R., 23.iv.1958, *Phipps* 1217 (BR). S: Bikita Distr., Matsai Res., *Cleghorn* 622 (SRGH). **Malawi**. N: Mzimba Distr., 4 km. S. of Chikangawa, 2.vii.1978, *Phillips* 3411 (BR). S: Mt. Mulanje, Tuchila plateau, 29.vii.1956, *Newman & Whitmore* 308 (BR; LISC). **Mozambique**. N: Mozambique Prov., Nacala Nova, 21.xi.1963, *Correia* 43 (LISC). T: Tete, Songo, 23.xi.1972, *Macêdo* 4904 (LISC). MS: Manica Distr., Mavita, 8.xi.1942, *Salbany* 48 (LISC). GI: Santa Carolina Is., x-xi.1958, *Mogg* 28800 (LISC). M: 21 km. on Rd. between Moamba and Boane, 15.ii.1952, *Myre & Carvalho* 1303 (LISC).

Probably native in Africa and western Asia, but now established throughout the tropics and subtropics. A weed on anthropogenic and otherwise disturbed sites, often forming extensive stands; 0–2500 m.

Subsp. **grandiflora** (Hochst.) Zizka in Bibl. Bot. **138**: 60 (1988). Type from Sudan Republic.
   Rhynchelytrum grandiflorum Hochst. in Flora **27**: 249 (1844). Type as above.
   Monachyron villosum Parl., in Hook., Niger Fl.: 191 (1849). Type from Cape Verde Is.
   Tricholaena grandiflora A. Rich., Tent. Fl. Abyss. **2**: 445 (1851). Type from Ethiopia.
   Saccharum grandiflorum (A. Rich.) Walp. in Ann. Bot. Syst. **3**: 792 (1852). Type as above.
   Panicum insigne Steudel, Syn. Pl. Glum. **1**: 92 (1854). Based on *Tricholaena grandiflora*.
   Tricholaena rosea var. *setosa* Peters, Reise Mossamb. Bot. **2**: 561 (1864). Type: Mozambique, Tette, *Peters* s.n. (K).
   Monachyron grandiflorum (A. Rich.) Martelli, Fl. Bogos.: **94** (1886). Type as for *Tricholaena grandiflora*.

Tricholaena brevipila Hackel in Verh. Bot. Ver. Prov. Brand. **30**: 143 (1888). Type from Namibia.

Tricholaena villosa (Parl.) Dur. & Schinz, Consp. Fl. Afr. **5**: 771 (1894). Type as for *Monachyron villosum*.

Tricholaena uniglumis Dur. & Schinz, Consp. Fl. Afr. **5**: 770 (1894). Based on *Rhynchelytrum grandiflorum*.

Tricholaena monachyron Oliver in Hook. Ic. Pl. **24**: t. 2374 (1895) nom. superfl. Based on *Monachyron villosum*.

Tricholaena grandiflora var. *glabrescens* Rendle, Cat. Afr. Pl. Welw. **2**: 196 (1899). Type from Angola.

Melinis brevipila (Hackel) Hackel in Öst. Bot. Zeitschr. **51**: 464 (1901). Type as for *Tricholaena brevipila*.

Melinis grandiflora (Hochst.) Hackel in Ost. Bot. Zeitschr. **51**: 464 (1901). Type as for *Rhynchelytrum grandiflorum*.

Melinis villosa (Parl.) Hackel in Ost. Bot. Zeitschr. **51**: 464 (1901). Type as for *Monachyron villosum*.

Panicum setinsigne Mez in Bot. Jahrb. **34**: 133 (1904). Type from S. Africa.

Rhynchelytrum villosum (Parl.) Chiov. in Ann. Ist. Bot. Roma **8**: 310 (1907). —Stapf & Hubbard in F.T.A. **9**: 875 (1930). —Sturgeon in Rhod. Agric. Journ. **50**: 108 (1953). —Chippindall in Meredith, Grasses & Pastures of S. Afr.: 431 (1955). —Bor, Grasses of B.C.I. & P.: 355 (1960). —Launert in Merxm., Prodr. Fl. SW. Afr. **160**: 162 (1970). —Clayton in F.W.T.A. **3**: 454 (1972). —Müller, Gräser SWA/Nam.: 220 (1985). Type as for *Monachyron villosum*.

Rhynchelytrum brevipilum (Hackel) Chiov. in Ann. Ist. Bot. Roma **8**: 310 (1907). —Sturgeon in Rhod. Agric. Journ. **50**: 108 (1953). Type as for *Tricholaena brevipila*.

Melinis affinis Mez in Bot. Jahrb. **57**: 196 (1921). Type from Namibia.

Melinis bertlingii Mez in Bot. Jahrb. **57**: 195 (1921). Type from Namibia.

Melinis ejubata Mez in Bot. Jahrb. **57**: 196 (1921). Type from Namibia.

Melinis mutica Mez in Bot. Jahrb. **57**: 197 (1921). Type from Namibia.

Melinis otaviensis Mez in Bot. Jahrb. **57**: 195 (1921). Type from Namibia.

Melinis pulchra Mez in Bot. Jahrb. **57**: 196 (1921). Type from Namibia.

Melinis rangei Mez in Bot. Jahrb. **57**: 195 (1921). Type from Namibia.

Melinis seineri Mez in Bot. Jahrb. **57**: 195 (1921). Type from Namibia.

Rhynchelytrum costatum Stapf & Hubbard in F.T.A. **9**: 874 (1930) nom. superfl. Based on *Melinis ejubata*.

Rhynchelytrum suberostratum Stapf & Hubbard in F.T.A. **9**: 879 (1930). Type from Namibia.

Tufted annual or rarely a short-lived perennial. Culms 25–90 cm. high, 0.9–1.6(2) mm. in diam. Spikelets (4)5–12 mm. long, glabrous or hairy, conspicuously laterally compressed. Inferior glume (0.6)1.5–3(4.3) mm. long, separated from the superior by an internode (0.5)0.7–1.7(2) mm. long. Superior glume and inferior lemma membranous to coriaceous, often drawn out into a narrow and glabrous beak.

Botswana. N: Ngamiland, Maun, iii.1958, *Robertson* 624 (LISC). SW: 85 km. N. of Kang on Rd. to Ghanzi, 19.ii.1960, *De Winter* 7387 (M). SE: Lobatse, 16.i.1977, *Giess* 11 (M). **Zambia**. N: Mbala Distr., Mpulungu, Crocodile Is., 12.iv.1959, *McCallum-Webster* A309 (K). E: W. of Sasare, 10.xii.1958, *Robson* 895 (LISC). S: Victoria Falls, xi.1933, *Meebold* 13808 (M). **Zimbabwe**. W: Hwange Nat. Park, 14.ii.1969, *Rushworth* 1538 (BR). C: Gweru Distr., 29 km. SSE. of Kwekwe, 14.ii.1966, *Biegel* 904 (BR). E: Chipinge Distr., Giriwayo, 19.i.1957, *Phipps* 28 (BR). S: Beitbridge, Pioneer Memorial, 21.iii.1959, *Drummond* 5893 (LISC; M; PRE). **Mozambique**. T: 6 km. from Tete, on Rd. to Changara, 19.iii.1966, *Torre & Correia* 15251 (LISC).

From tropical and S. Africa, extending through Arabia to India. Dry and sunny places, on sand or boulders, in grassland and savanna woodland, rarely on disturbed sites; 200–1400 m.

Subsp. **nigricans** (Mez) Zizka in Bibl. Bot. **138**: 64 (1988). Type from Angola.

Melinis nigricans Mez in Bot. Jahrb. **57**: 198 (1921). Type as above.

Rhynchelytrum nigricans (Mez) Stapf & Hubbard in F.T.A. **9**: 887 (1930). Type as above.

Vigorous tufted annual. Culms 60–160 cm. high, (1.7)2–3.2 mm. in diam. Spikelets 3–4.5(5) mm. long, dark olive green, rarely green to straw-coloured, shortly hairy or rarely glabrous. Inferior glume 0.5–1(1.2) mm. long, separated from the superior by an internode 0.1–0.5 mm. long. Superior glume and inferior lemma membranous and subcoriaceous, not drawn out into a narrow beak.

Zambia. N: Mporokoso, between Musesha and Musombwe, 14.iv.1961, *Phipps & Vesey-FitzGerald* 3182 (SRGH). **Zimbabwe**. N: Mazowe (Mazoe), Spelonken Farm, 12.iv.1981, *Burrows* 1718 (SRGH). **Malawi**. C: Dowa Distr., Mwera Hill St., 8.v.1951, *Jackson* 495 (K). S: Zomba, 30.vii.1949, *Wiehe* N/178 (K). **Mozambique**. T: Tete, Vila Vasco da Gama, 12.viii.1941, *Torre* 3266 (LISC).

Also from Zaire, Tanzania and Angola. On anthropogenic sites and embankments; 300–1700 m.

2. **Melinis kallimorpha** (Clayton) Zizka in Bibl. Bot. **138**: 84 (1988). Type: Zambia, Mbala, *McCallum-Webster* A289 (K).
 Rhynchelytrum kallimorphon Clayton in Kew Bull. **33**: 22 (1978). —Clayton & Renvoize in F.T.E.A., Gramineae: 513 (1982). Type as above.

Tufted annual to short-lived perennial. Culms 40–100 cm. high, geniculately ascending. Leaf laminae 4–16 cm. long, 2–4(5) mm. wide, flat. Panicle 5–12 cm. long, linear in outline. Pedicels glabrous, attached laterally to the base of the spikelet. Spikelets 3.5–5 mm. long, ovate. Inferior glume 0.7–1.5(1.7) mm. long, ovate, separated from the superior by an internode 0.4–0.7(0.9) mm. long; superior glume 5-nerved, gibbous on the back, coriaceous with the nerves slightly raised, without a distinct membranous beak, broadly ovate, emarginate, awned. Inferior floret male, its lemma 5-nerved, resembling the superior glume but less gibbous, the palea ciliate on its keels.

Caprivi Strip. E. Caprivi, Katima Mulilo, 5.i.1959, *Killick & Leistner* 3300 (M). **Botswana**. N: Shakawe, 24.iv.1975, *Beigel, Müller & Gibbs-Russell* 4996 (PRE). **Zambia**. B: Masese, 11.iii.1960, *Fanshawe* 5453 (BM). N: Shiwa Ngandu, 4.vi.1956, *Robinson* 1591 (K; SRGH). C: near Serenje on Great North road, 5.iv.1961, *Phipps & Vesey-FitzGerald* 2947 (BM). S: Namwala Distr., Kafue Nat. Park, Nkala R., 12.ii.1969, *Day* 101/69 (SRGH). **Zimbabwe**. N: Gokwe Distr., Charama Plateau, 27.iv.1965, *Simon* 271 (SRGH). W: Hwange Distr., Kazuma Forest Area, 22.i.1974, *Gonde* 6/74 (SRGH). C: Charter Distr., Rd. to Wiltshire Estate, 12.vi.1963, *Cleghorn* 789 (SRGH). E: Mutare, 14.xi.1948, *Chase* (K). S: Masvingo, ii–iii.1945, *Clarke* 54 (SRGH).
 Also from Kenya, Tanzania and Angola. On Kalahari Sand in *Brachystegia*, *Colophospermum* and *Baikiaea* wooded grassland; 1000–1700 m.
 Easily recognized by its spikelet morphology and habit, and clearly distinct from *M. subglabra* with which it is often confused.

3. **Melinis subglabra** Mez in Bot. Jahrb. **57**: 197 (1921). Type from Tanzania.
 Melinis denudata Mez in Bot. Jahrb. **57**: 200 (1921). Type from Angola.
 Melinis merkeri Mez in Bot. Jahrb. **57**: 199 (1921). Type from Tanzania.
 Rhynchelytrum denudatum (Mez) Stapf & Hubbard in F.T.A. **9**: 889 (1930). Type as for *Melinis denudata*.
 Rhynchelytrum eylesii Stapf & Hubbard in F.T.A. **9**: 889 (1930). —Sturgeon in Rhod. Agric. Journ. **50**: 109 (1953). —Type: Zimbabwe, Harare, *Eyles* 3430 (K, holotype).
 Rhynchelytrum merkeri (Mez) Stapf & Hubbard in F.T.A. **9**: 887 (1930). Type as for *Melinis merkeri*.
 Rhynchelytrum subglabrum (Mez) Stapf & Hubbard in F.T.A. **9**: 886 (1930). —Sturgeon in Rhod. Agric. Journ. **50**: 109 (1953). —Jackson & Wlehe, Annot. Check List Nyasal. Grass.: 56 (1958). —Clayton & Renvoize in F.T.E.A., Gramineae: 513 (1982). Type as for *Melinis subglabra*.

Perennial from a knotty rootstock. Culms 40–130 cm. high, erect or rarely geniculately ascending. Leaf laminae 3–17 cm. long, (2.3)3–8(10) mm. wide. Panicle 6–14(20) cm. long, broadly ovate. Pedicels usually with a few long hairs. Spikelets 3.2–5 mm. long, ovate. Inferior glume (0.3)0.6–1(1.3) mm. long, inconspicuously 1-nerved, narrowly ovate, inserted close to the superior; superior glume 5-nerved, usually gibbous on the back, subcoriaceous to coriaceous, glabrous, rarely hairy, emarginate, usually mucronate. Inferior floret male, its lemma 5-nerved, resembling the superior glume but less gibbous, the palea ciliate on its keels.

Zambia. N: Luangwa Valley, Lokwa stream, 31.iii.1966, *Astle* 4730 (SRGH). C: Serenje Distr., 14 km. SE. of Kanona, 5.iv.1961, *Phipps & Vesey-FitzGerald* 2962 (SRGH). **Zimbabwe**. N: Lomagundi Distr., Silverside Mine, 14.ii.1968, *Jacobsen* 3371 (PRE; SRGH). W: Matobo Distr., Besna Kobila Farm, x.1958, *Miller* 5462 (PRE; SRGH). C: Wedza Distr., Wedza Mt., 64 km. S. of Marondera, 22.v.1968, *Simon, Rushworth & Mavi* 1823 (K; SRGH). E: Mutare, Rd. to Mutare Heights, 18.v.1970, *Crook* 927 (B; BR; LISC; PRE). S: Masvingo Distr., Kyle Nat. Park, 23.v.1971, *Ngoni* 115 (K). **Malawi**. N: Mzimbo Viphya Mts., 15.ii.1968, *Simon, Williamson & Ball* 1811 (LISC; PRE; SRGH). C: Kasungu, 12.iii.1953, *Jackson* 1139, (B; LISC; SRGH).
 Also from Kenya, Tanzania, Angola and S. Africa. Often under shade and by streamsides, in grassland and *Brachystegia* woodland; 800–2100 m.
 The most reliable character is the knotty rootstock. Spikelets are usually glabrous and mucronate, but specimens with hairy or awned spikelets have also been collected. Plants without basal parts may be difficult to separate from *M. repens*.

4. **Melinis longiseta** (A. Rich.) Zizka in Bibl. Bot. **138**: 73 (1988). Type from Ethiopia.
 Tricholaena longiseta A. Rich., Tent. Fl. Aybss. **2**: 446 (1851). Type as above.
 Saccharum longisetum (A. Rich.) Walp. in Ann. Bot. Syst. **3**: 793 (1852). Type as above.
 Panicum macrotrichum Steudel, Syn. Pl. Glum. **1**: 92 (1854). Based on *Tricholaena longiseta*.
 Tricholaena minutiflora Rendle, Cat. Afr. Pl. Welw. **2**: 198 (1899). Type from Angola.

Tricholaena melinioides Stent in Kew Bull. **1925**: 364 (1925). Type: Zimbabwe, Harare, *Eyles* 2191 (K, holotype; BM, isotype; PRE, isotype).
Rhynchelytrum longisetum (A. Rich.) Stapf & Hubbard in F.T.A. **9**: 902 (1930). —Clayton & Renvoize in F.T.E.A., Gramineae: 512 (1982). Type as for *Tricholaena longiseta*.
Rhynchelytrum minutiflorum (Rendle) Stapf & Hubbard in F.T.A. **9**: 903 (1930). —Sturgeon in Rhod. Agric. Journ. **50**: 110 (1953). —Jackson & Wiehe, Annot. Check List Nyasal. Grass.: 55 (1958). Type as for *Tricholaena minutiflora*.
Rhynchelytrum minutiflorum var. *melinioides* (Stent) Stapf & Hubbard in F.T.A. **9**: 904 (1930). —Sturgeon in Rhod. Agric. Journ. **50**: 110 (1953). —Chippindall in Meredith, Grasses & Pastures of S. Afr.: 432 (1955). —Jackson & Wiehe, Annot. Check List Nyasal. Grass.: 55 (1958). Type as for *Tricholaena melinioides*.

Tufted to loosely caespitose perennial. Culms 20–100 cm. high, geniculately ascending or erect. Leaf laminae (3)6–20 cm. long, 1.5–9 mm. wide, flat. Panicle 4–17 cm. long, narrowly oblong to linear. Pedicels with a few long hairs. Spikelets 2.3–8.6 mm. long, laterally compressed, oblong to linear. Inferior glume 0.6–1.4 mm. long, inconspicuously 1-nerved, ovate, separated from the superior by an internode 0.3–0.5 mm. long; superior glume 5-nerved, subcoriaceous to coriaceous, emarginate to bilobed, awned, densely hairy on the keel, glabrous on the sides. Inferior floret male, its lemma 5-nerved, similar to the superior glume but glabrous on the keel and hairy on the sides, the palea scabrous on its keels.

Easily recognized by its 5-nerved superior glume and the scabrous palea keels of the inferior floret. The pattern of hairiness of superior glume and inferior lemma is also very characteristic, but in a few specimens from Zimbabwe their nervation and texture intergrade with *M. ambigua*.
Two subspecies, differing in spikelet morphology, distribution and ecological preferences, can be distinguished, but they are insufficiently distinct to justify the species rank commonly accorded them.

Spikelets 2.8–3.8(4.2) mm. long; leaf laminae up to 5–9 mm. wide, narrowly
　　lanceolate　-　-　-　-　-　-　-　-　-　-　-　-　-　-　subsp. *longiseta*
Spikelets (3.4)4–8.5 mm. long; leaf laminae up to 2.5–6 mm. wide,
　　linear　-　-　-　-　-　-　-　-　-　-　-　-　-　-　subsp. *bellespicata*

Subsp. longiseta

Leaf laminae 4–12 cm. long, 4–9 mm. wide. Spikelets 2.3–3.8(4.2) mm. long, usually ovate.

Caprivi Strip. Andara, 23.v.1939, *Volk* 2159 (M; PRE). **Zambia**. W: Misaka, 7.v.1969, *Mutimushi* 3064 (SRGH). C: Serenje, 5.iv.1961, *Vesey-FitzGerald* 2950 (SRGH). E: Lukusuzi, 12.iv.1971, *Sayer* 1200 (SRGH). S: Choma, Mochipapa, 16.iii.1964, *Astle* 2979 (K). **Zimbabwe**. N: Darwin Distr., Msengedzi, 9.v.1955, *Whellan* 560 (SRGH). W: Nkayi Distr., Gwampa Forest Res., iii.1956, *Goldsmith* 80/56 (M; PRE). C: Charter Distr., Wiltshire, 10.v.1963, *Cleghorn* 779 (SRGH). E: Mutare Distr., 58 km. S. of Mutare, 23.iv.1969, *Plowes* 3196 (SRGH). S: Masvingo Distr., Mushandike, 27.iii.1984, *Mahlangu* 1076 (SRGH). **Malawi**. C: Kasungu Game Res., 14.vi.1960, *Hall-Martin* 1732 (SRGH). S: Zomba, 1936, *Cormack* 205 (K). **Mozambique**. N: Niassa Prov., Mandimba, 16.iv.1942, *Hornby* 3363 (K).
Also from Sudan, Tanzania and Angola. Sandy ground in open *Brachystegia* woodland; (600)1000–2000 m.

Subsp. bellespicata

(Rendle) Zizka in Bibl. Bot. **138**: 78 (1988). Type from Angola.
Tricholaena bellespicata Rendle, Cat. Afr. Pl. Welw. **2**: 196 (1899). Type as above.
Melinis chaetophora Mez in Bot. Jahrb. **57**: 196 (1921). Types from Angola.
Melinis secunda Mez in Bot. Jahrb. **57**: 196 (1921). Types from Namibia.
Rhynchelytrum bellespicatum (Rendle) Stapf & Hubbard in F.T.A. **9**: 900 (1930). —Sturgeon in Rhod. Agric. Journ. **50**: 111 (1953). —Chippindall in Meredith, Grasses & Pastures of S. Afr.: 431 (1955). —Launert in Merxm., Prodr. Fl. SW. Afr. **160**: 160 (1970). —Clayton in F.W.T.A. **3**: 455 (1972). Type as for *Tricholaena bellespicata*.

Leaf laminae (3)6–20 cm. long, 1.5–5(6) mm. wide. Spikelets (3.5)4–8.5 mm. long, usually linear.

Botswana. N: 18°35.5'S, 24°4.7'E, 21.v.1977, *Smith* 2073 (PRE; SRGH). **Zambia**. N: Mpika Distr., Lubi, 19.iii.1969, *Astle* 5618 (SRGH). S: Mapanza, 19.iv.1954, *Robinson* 688 (M). **Zimbabwe**. N: Binga Distr., Chizarira Game Res., 16.i.1968, *Thomson* 22 (SRGH). W: Nkayi Distr., Gwampa Forest Res., iii.1956, *Goldsmith* 135/56 (LISC; PRE). C: Gweru Distr., 29 km. SSE. of Kwekwe, 4.iv.1966, *Biegel* 1080 (PRE; SRGH). E: Mutare, 23.iv.1969, *Crook* 878 (BR; M; PRE; SRGH). S: Masvingo Distr., Kyle Nat. Park, 21.v.1971, *Ngoni* 88 (BR; LISC; PRE; SRGH). **Malawi**. S: Blantyre Distr., Mpatamanga

Gorge, 28.ii.1961, *Vesey-FitzGerald* 3062 (SRGH). **Mozambique**. T: Tete Distr., Songo, 11.ii.1973, *Torre et al.* 19058 (LISC).

Also from Nigeria, Cameroon, Angola and S. Africa. Stony ground in open, sunny locations in Mopane woodland, often in crevices; 300–2100 m.

5. **Melinis ambigua** Hackel in Öst. Bot. Zeitschr. **51**: 462 (1901). —Stapf & Hubbard in F.T.A. **9**: 921 (1930). —Sturgeon in Rhod. Agric. Journ. **50**: 106 (1953). —Jackson & Wiehe, Annot. Check List Nyasal. Grass.: 48 (1958). —Clayton & Renvoize in F.T.E.A., Gramineae: 510 (1982). Type from Ethiopia.

    Melinis hirsuta Mez in Bot. Jahrb. **57**: 201 (1921). Type from Namibia.
    Melinis diminuta Mez in Bot. Jahrb. **57**: 200 (1921). Type from Tanzania.
    Melinis eylesii Stapf & Hubbard in Kew Bull. **1926**: 441 (1926). —Sturgeon in Rhod. Agric. Journ. **50**: 106 (1953). Type: Zimbabwe, Harare, *Eyles* 2276 (K, holotype).
    Melinis mollis Stapf & Hubbard in Kew Bull. **1926**: 440 (1926). Type from Angola.
    Melinis pallida Stapf & Hubbard in Kew Bull. **1926**: 442 (1926). —Clayton in F.W.T.A. **3**: 509 (1982). Type from Angola.
    Melinis intermedia Stapf & Hubbard in F.T.A. **9**: 922 (1930). Type from Tanzania.
    Melinis inamoena Pilger in Notizbl. Bot. Gart. Berl. **13**: 262 (1936). Type from Tanzania.

Caespitose or rarely tufted perennial. Culms (30)60–120(150) cm. high, erect or rarely geniculately ascending. Leaf laminae (3.5)5–30 cm. long, 2–8(11) mm. wide, flat or involute. Panicle 9–25 cm. long, narrowly ovate. Pedicels with a few long hairs. Spikelets (2)2.4–4.4(5.2) mm. long, narrowly ovate. Inferior glume 0.4–1.2 mm. long, ovate, 0–1-nerved, inserted close to the superior; superior glume 7-nerved, usually hairy, emarginate to bilobed with an awn (0.1)0.5–6.8(8.3) mm. long. Inferior floret male or barren, its lemma 5-nerved with an awn (0.3)2–13(14) mm. long, equalling the superior glume; palea absent or when present scabrous on its keels.

M. *ambigua*, M. *minutiflora*, M. *tenuissima*, M. *effusa* and M. *macrochaeta* form a group of similar species with rare intermediate forms occurring between neighbouring species. The weakest delimitation is between M. *minutiflora* on the one side and M. *ambigua*, M. *effusa* and M. *tenuissima* on the other.

M. *ambigua* is a very variable species. Subsp. *ambigua* and subsp. *longicauda* differ in spikelet morphology and habit with many intermediate forms occurring. Forms of subsp. *longicauda* may approach M. *longiseta*, but the nervation and hairiness of the superior glume are reliable characters for identification.

Rare specimens with the keels of the inferior palea ciliate, possibly introgression products from M. *repens*, have been described as *Rhynchelytrum bequaertii*.

Spikelets 2–3(3.4) mm. long; superior glume with an awn (0.1)0.8–4.5 mm. long; palea of inferior
    floret sometimes reduced - - - - - - - - - subsp. *ambigua*
Spikelets (2.7)3.3–4.4(5.2) mm. long; superior glume with an awn (1.8)3.5–8.3 mm.
    long - - - - - - - - - - - - subsp. *longicauda*

Subsp. **ambigua**. TAB. **32**.

Leaf laminae usually flat, (3.5)5–20(23) cm. long, 3–8(11) mm. wide. Spikelets (2)2.4–3(3.4) mm. long. Inferior glume 0.4–1 mm. long; superior glume with an awn (0.1)0.8–4.5 mm. long. Inferior floret barren, its lemma with an awn (0.3)2–12 mm. long.

Zambia. N: Mbala Distr., Kalambo Falls, 21.iv.1959, *McCallum-Webster* A347 (K). W: Mufulira, 29.iv.1964, *Fanshawe* 8609 (SRGH). C: Mkushi Distr., Mkushi R., 16.xii.1967, *Simon & Williamson* 1396 (K). **Zimbabwe**. N: Gokwe, 1.vii.1963, *Bingham* 739 (SRGH). W: Matobo Distr., Besna Kobila Farm, iii.1957, *Miller* 4235 (K; SRGH). C: Marondera, iv.1919, *Eyles* 1635 (BM). E: Chipinge, 21.v.1972, *Simon* 2234 (BR; LISC; SRGH). **Malawi**. N: Nkhata Bay Distr., 8 km. E. of Mzuzu, 13.vi.1960, *Pawek* 6832 (SRGH). C: Nkhota Kota Distr., Ntchisi Mts., 1.viii.1946, *Brass* 17092 (K). S: Blantyre Distr., 3 km. NE. of Blantyre, 10.vi.1970, *Staples* 594 (SRGH). **Mozambique**. T: Angonia, Monte Domue, 1.iv.1980, *Stefanesco & Nyongani* 478 (SRGH). MS: Manica, Zuira, Tsetserra Mt., 13.xi.1965, *Torre & Paiva* 12969 (LISC).

Also from Ethiopia, Zaire, Burundi, Tanzania, Angola and Namibia. Sandy or stony soils in open *Brachystegia* woodland or grassland at higher elevations, often at anthropogenic sites and streamsides; 1000–2500 m.

Subsp. **longicauda** (Mez) Zizka in Bibl. Bot. **138**: 90 (1988). Type from Tanzania.
    Panicum longicaudum Mez in Bot. Jahrb. **34**: 133 (1904). Type as above.
    Rhynchelytrum longicaudum (Mez) Chiov. in Nuovo Giorn. Bot. Ital. n.s. **26**: 78 (1919). Type as above.

Tab. 32. MELINIS AMBIGUA subsp. AMBIGUA. 1, habit (×½), from *Biegel* 1152; 2, junction of leaf
sheaths and lamina showing ligule (× 6); 3, part of inflorescence (× 3); 4, spikelet (× 10); 5,
inferior glume (× 15); 6, superior glume (× 15); 7, inferior lemma (× 15); 8, superior floret with
ovary (× 15); 9, caryopsis (× 15), 2–9 from *Schimper* 800.

Melinis goetzenii Mez in Bot. Jahrb. **57**: 198 (1921). Type from Ruanda.
Melinis longicauda (Mez) Stapf & Hubbard in F.T.A. **9**: 919 (1930). —Sturgeon in Rhod. Agric. Journ. **50**: 106 (1953). —Jackson & Wiehe, Annot. Check List Nyasal. Grass.: 48 (1955). Type as for *Panicum longicaudum*.
Rhynchelytrum shantzii Stapf & Hubbard in F.T.A. **9**: 892 (1930). Type from Zaire.
Rhynchelytrum bequaertii Robyns in Bull. Jard. Bot. Brux. **9**: 193, pl. 3 (1932). Type from Zaire.

Leaf laminae often involute, (11)14–30 cm. long, 2–4(7) mm. wide. Spikelets (2.7)3.3–4.4(5.2) mm. long. Inferior glume 0.5–1.2 mm. long; superior glume with an awn (1.8)2.5–7(8.3) mm. long. Inferior floret male, rarely barren, its lemma with an awn 6–13(14) mm. long; palea present, scabrous on its keels.

Zambia. N: Mbala Distr., Chilongowelo, 29.iii.1952, *Richards* 1308 (K). W: Kitwe, 25.iii.1966, *Mutimushi* 1343 (SRGH). C: Mt. Makulu Res. St., Chilanga, 17.iv.1953, *Hinds* 90 (K). E: Nyika Plateau, v.1968, *Williamson* 933 (SRGH). **Zimbabwe**. C: Harare, Domboshawa, 3.vii.1929, *Eyles* 4975 (BR; H). E: Chimanimani, Martin Forest Res., 20.viii.1955, *Barrett* 10/55 (SRGH). **Malawi**. N: Nyika Plateau, 11.iv.1969, *Pawek* 2098 (K). C: Dedza Mts., 10.ix.1929, *Davy* 1499 (K). S: Ntcheu, Kirk Range, *Young* 230 (SRGH). **Mozambique**. N: Massangulo, iv.1955, *Sousa* 1410 (K).

Zaire, Rwanda, Tanzania and Angola. Secondary grassland at higher elevations, rarely *Brachystegia* woodland or anthropogenic sites; 1000–2500 m.

6. **Melinis minutiflora** Beauv., Ess. Agrost.: 54, pl. 11, fig. 4 (1812). —Stapf & Hubbard in F.T.A. **9**: 931 (1930). —Sturgeon in Rhod. Agric. Journ. **50**: 107 (1953). —Chippindall in Meredith, Grasses & Pastures of S. Afr.: 427 (1955). —Jackson & Wiehe, Annot. Check List Nyasal. Grass.: 48 (1958). —Clayton in F.W.T.A. **3**: 455 (1972). —Clayton & Renvoize in F.T.E.A., Gramineae: 506 (1982). Type from Brazil.
Panicum minutiflorum (Beauv.) Raspail in Ann. Sci. Nat. Bot. **5**: 299 (1825). Type as above.
Panicum melinis var. *inerme* Doell in Mart., Fl. Bras. **2**: 2: 242 (1877). Type from Brazil.
Melinis minutiflora var. *inermis* (Doell) Rendle, Cat. Afr. Pl. Welw. **2**: 200 (1899). Type as above.
Melinis minutiflora var. *pilosa* Stapf in Fl. Cap. **7**: 447 (1899). Type from S. Africa.
Melinis minutiflora f. *mutica* Chiov. in Nuovo Giorn. Bot. Ital. n.s. **26**: 79 (1919). Type from Zaire.
Melinis tenuinervis Stapf in Kew Bull. **1922**: 310 (1922). —Stapf & Hubbard in F.T.A. **9**: 929 (1930). —Chippindall in Meredith, Grasses & Pastures of S. Afr.: 427 (1955). —Jackson & Wiehe, Annot. Check List Nyasal. Grass.: **48** (1958). Based on *M. minutiflora* var. *pilosa*.
Melinis purpurea Stapf & Hubbard in Kew Bull. **1926**: 444 (1926); in F.T.A. **9**: 928 (1930). Type from Tanzania.
Melinis minutiflora f. *inermis* (Doell) Stapf & Hubbard in F.T.A. **9**: 932 (1930). Type as for *M. minutiflora* var. *inermis*.
Melinis maitlandii f. *mutica* (Chiov.) Robyns in Bull. Jard. Bot. Brux. **9**: 197 (1932). Type as for *M. minutiflora* f. *mutica*.

Tufted perennial. Culms (50)80–150 cm. high, ascending. Leaf laminae (2)4–20 cm. long, (2.5)5–11(19) mm. wide, flat, these and the sheaths densely tomentose and usually sticky with a strong smell of linseed oil. Panicle (8)10–20(36) cm. long, narrowly ovate. Pedicels glabrous, rarely with a few hairs towards the apex, scabrous. Spikelets (1.5)1.7–2.2(2.4) mm. long, narrowly ovate to narrowly oblong. Inferior glume 0.1–0.4 mm. long, ovate, 0–1-nerved, inserted close to the superior; superior glume prominently 7-nerved, awnless or with a short mucro (rarely conspicuously awned), membranous, obtusely bilobed, glabrous, rarely hairy. Inferior floret barren without a palea, its lemma prominently 5-nerved, acutely bilobed, equalling the superior glume but narrower, awnless or with an awn up to 14 mm. long.

Zambia. N: 4.8 km. W. of Kasama, 15.vi.1950, *Jackson* 48 (LISU). E: Nyika Plateau, v.1968, *Williamson* 930 (SRGH). **Zimbabwe**. C: Makoni Distr., Mt. Dombo Pass, 13.vi.1972, *Biegel* 3953 (B; LISC). E: Nyanga (Inyanga), Nyamingura R., 23.iv.1958, *Phipps* 1231 (BR). **Malawi**. N: Nkhata Bay Distr., Chikla beach, 15.v.1977, *Pawek* 12762 (BR). S: Blantyre Distr., 2 km. N. of Limbe, 1.v.1970, *Brummitt* 10308 (SRGH). **Mozambique**. Z: Zambezia, Gúruè, 28.vi.1943, *Torre* 5598 (LISC). MS: Beira Distr., Gorongosa, 6.v.1964, *Torre & Paiva* 12253 (BR; LISC).

Grassland or *Brachystegia* woodland at higher elevations, and in montane forest, often at anthropogenic sites; (300)1000–2300 m.

The species can usually be recognised by the prominently nerved superior glume and inferior lemma and by the scabrous pedicels. Intergradation with *M. ambigua* subsp. *ambigua* (spikelets hairy, superior glume and inferior lemma awned) and *M. effusa* (pedicels with a few long hairs) may occur.

It is very variable as far as hairiness and length of awns is concerned, but variation is continuous. Investigation of 4 common variants revealed that cross-fertilization between them is rare and the species is probably apomictic (Bogdan in E. Afr. Agric. Journ. **26**: 49, 1960).

7. **Melinis effusa** (Rendle) Stapf. in Kew Bull **1926**: 444 (1926). —Stapf & Hubbard in F.T.A. **9**: 925 (1930). —Clayton in F.W.T.A. **3**: 457 (1972). —Clayton & Renvoize in F.T.E.A., Gramineae: 509 (1982). Type from Angola.
   Melinis minutiflora var. *effusa* Rendle, Cat. Afr. Pl. Welw. **2**: 200 (1899). Type as above.

Tufted perennial. Culms (59)70–130(160) cm. high, ascending, rarely erect. Leaf laminae (1.5)3–11(14) cm. long, (2)3–8(11) mm. wide, flat, often densely tomentose. Panicle (6)8–20(27) cm. long, ovate to narrowly ovate. Pedicels with a few long hairs towards the apex, scabrous. Spikelets 1.4–1.8 mm. long, narrowly ovate to oblong. Inferior glume (0.1)0.2–0.4(0.5) mm. long, ovate, inserted close to the superior; superior glume 7-nerved, awnless or mucronate, membranous, obtusely bilobed, glabrous to sparsely hairy. Inferior floret barren without a palea, its lemma 5-nerved, acutely bilobed, equalling the superior glume but narrower with an awn 2–10 mm. long, rarely awnless.

   Zambia. B: 112 km. S. of Mwinilunga on Kabompo Rd., 25.xii.1969, *Simon & Williamson* 2015 (K). N: Mbala Distr., Itimbwe gorge, 24.iv.1959, *McCallum-Webster* A361 (K; SRGH). **Zimbabwe**. E: Chipinge, Ngungunyana Forest Res., vi.1963, *Goldsmith* 25/63 (K; LISC). **Malawi**. N: Mzimba Distr., Marymount, 11.vi.1967, *Pawek* 1176 (SRGH). S: Zomba, 1936, *Cormack* 455 (K).
   Northwards to Ghana and Kenya; also in Angola. Grassland, rarely *Brachystegia* woodland, often at anthropogenic sites and streamsides; 1000–2200 m.
   M. effusa is intermediate between *M. minutiflora* and *M. tenuissima*, and its species rank is doubtful. Morphology and distribution support the suggestion that it may be of hybrid origin.

8. **Melinis tenuissima** Stapf in Ic. Pl. **27**: t.2660 (1900). —Stapf & Hubbard in F.T.A. **9**: 926 (1930). —Sturgeon in Rhod. Agric. Journ. **50**: 106 (1953). —Chippindall in Meredith, Grasses & Pastures of S. Afr.: 427 (1955). —Jackson & Wiehe, Annot. Check List Nyasal. Grass.: 48 (1958). —Clayton in F.W.T.A. **3**: 455 (1972). —Clayton & Renvoize in F.T.E.A., Gramineae: 508 (1982). Type: Malawi, Namasi, *Cameron* 33, (K, holotype).

Straggling perennial. Culms (28)50–110(140) cm. high, ascending. Leaf laminae (1.5)2–8(10) cm. long, 3–6 mm. wide, flat, glabrous or rarely pubescent. Panicle 10–20 cm. long, broadly ovate, delicately branched. Pedicels smooth, with a few long hairs towards the apex. Spikelets 1.1–1.5 mm. long, narrowly ovate to oblong. Inferior glume an obscure rim up to 0.1 mm. long, inserted close to the superior; superior glume 5-nerved, thinly membranous, irregularly dentate, glabrous to sparsely hairy. Inferior floret barren without a palea, its lemma 3–5-nerved, emarginate or irregularly dentate, equalling the superior glume but narrower and with an awn (1.5)4–10 mm. long.

   Zambia. N: Mbala, 16.vi.1955, *Siame* 665 (SRGH). W: Ndola, 3.vi.1962, *Fanshawe* 6852 (SRGH). E: Chipata, 23.vi.1963, *Verboom* 1109 (SRGH). **Zimbabwe**. N: Mazowe (Mazoe), *Eyles* 2362 (K). C: Wedza Mt., 6.iv.1964, *Cleghorn* 909 (SRGH). E: Chimanimani, *Longden* 1087 (SRGH). S: Great Zimbabwe, 9.iv.1973, *Chiparawasha* 695 (SRGH). **Malawi.** N: Mzimba Distr., Mzuzu, Marymount, 17.vi.1973, *Pawek* 6874 (SRGH). C: 5 km. NE. of Mchinji, 27.iv.1970, *Brummitt* 10218 (K; SRGH). S: Zomba, Naisi Rd., 15.v.1949, *Wiehe* N/108 (K). **Mozambique**. T: Planalto de Angonia, 12.v.1948, *Mendonça* 4201 (LISC).
   Also from Tropical Africa. Grassland and *Brachystegia* woodland, very often at anthropogenic sites and near river margins; (700)1000–2000 m.
   The most reliable characters are the smoothness of the pedicels and the 5-nerved superior glume. The broadly ovate, delicately branched panicle is an aid to recognition.

9. **Melinis macrochaeta** Stapf & Hubbard in Kew Bull. **1926**: 443 (1926); in F.T.A. **9**: 927 (1930). —Sturgeon in Rhod. Agric. Journ. **50**: 107 (1953). —Chippindall in Meredith, Grasses & Pastures of S. Afr.: 427 (1955). —Jackson & Wiehe, Annot. Check List Nyasal. Grass.: 48 (1958). —Clayton in F.W.T.A. **3**: 455 (1972). —Clayton & Renvoize in F.T.E.A., Gramineae: 508 (1982). TAB. **33**. Type from Nigeria.

Tufted annual to short-lived perennial. Culms (35)50–100(150) cm. high, geniculately ascending and often rooting at the lower nodes. Leaf laminae (3)5–15(20) cm. long, (3)5–10(13) mm. wide, flat, thin, softly pilose. Panicle 10–25 cm. long, narrowly ovate. Pedicels glabrous, rarely with a few long hairs towards the apex, scabrous. Spikelets 1.5–2 mm. long, narrowly oblong. Inferior glume an obscure rim up to 0.1 mm. long, inserted close to the superior; superior glume 7-nerved, delicately membranous, bilobed, the lobes irregularly dentate, awnless or mucronate. Inferior floret barren without a palea, its lemma 3–5-nerved, acutely bilobed, equalling the superior glume but narrower and with an awn (5)8–20 mm. long.

Tab. 33. MELINIS MACROCHAETA. 1, habit (×⅔); 2, spikelet (× 10); 3, superior glume (× 20); 4, inferior lemma (× 20), all from *Mitchell* 14/35.

Zambia. B: Kaoma Distr., 40 km. N. of Sitaka, 16.iv.1964, *Verboom* 1382 (SRGH). N: Chipili, 5.vi.1957, *Robinson* 2233 (BR; SRGH). W: Kitwe, 14.i.1958, *Fanshawe* 4197 (SRGH). S: Kafue Gorge, 14.iv.1956, *Robinson* 1475 (M). **Zimbabwe**. N: Trelawney, Tobacco Res. St., 24.iv.1943, *Moffett* G119 (SRGH). W: Matobo Distr., Besna Kobila farm, iv.1959, *Miller* 5905 (SRGH). C: Makoni Distr., Rusape, iv.1955, *Davies* 1073 (SRGH). E: Chipinge Distr., Gungunyama Forest Res., *Goldsmith* 27/63 (SRGH). **Malawi**. S: Zomba Distr., 6.vi.1946, *Brass* 16290 (BR). **Mozambique**. N: Lichinga (Vila Cabral), 16.v.1934, *Torre* 77 (LISC).

Also from Tropical Africa. In grassland or *Brachystegia* woodland, often in disturbed places, or near streams; 800–1800 m.

Typical members of *M. macrochaeta* are well characterised by the glabrous pedicels and the annual habit. A further aid to recognition is the number of culm-nodes, which in *M. macrochaeta* is less than 10, in *M. minutiflora* and *M. effusa* more than 10.

10. **Melinis nerviglumis** (Franchet) Zizka in Bibl. Bot. **138**: 111(1988). Type from Congo.
   Tricholaena nerviglumis Franchet in Bull. Soc. Hist. Nat. Autun. **8**: 357 (1895). Type as above.
   Tricholaena setifolia Stapf in Fl. Cap. **7**: 442 (1899). Type from S. Africa.
   Melinis setifolia (Stapf) Hackel in Öst. Bot. Zeitschr. **51**: 464 (1901). Type as above.
   Panicum busseanum Mez in Bot. Jahrb. **34**: 131 (1904). Type from Tanzania.
   Panicum elongatum Mez in Bot. Jahrb. **34**: 132 (1904). Type from Zaire.
   Rhynchelytrum setifolium (Stapf) Chiov. in Ann. Ist. Bot. Roma. **8**: 310 (1907). —Stapf & Hubbard in F.T.A. **9**: 898 (1930). —Sturgeon in Rhod. Agric. Journ. **50**: 110 (1953). — Chippindall in Meredith, Grasses & Pastures of S. Afr.: 432 (1955). Type as for *Tricholaena setifolia*.
   Tricholaena rhodesiana Rendle in Journ. Linn. Soc. Bot. **40**: 232 (1911). Types: Zimbabwe, Chirinda Forest, *Swynnerton* 1632 (K, isosyntype); Nyahodi R., *Swynnerton* 1663 (BM, syntype).
   Tricholaena rhodesiana var. *glabrescens* Rendle in Journ. Linn. Soc. Bot. **40**: 233 (1911). Type: Zimbabwe, Northern Chimanimani (Melsetter), *Swynnerton* 1685 (BM, holotype).
   Rhynchelytrum nerviglume (Franchet) Chiov. in Nuovo Giorn. Bot. Ital. n.s. **26**: 78 (1919). —Jackson & Wiehe, Annot. Check List Nyasal. Grass.: 55 (1958). —Clayton & Renvoize in F.T.E.A., Gramineae: 514 (1982). Type as above for *Tricholaena nerviglumis*.
   Melinis bachmannii Mez in Bot. Jahrb. **57**: 198 (1921). Type from S. Africa.
   Melinis munzneri Mez in Bot. Jahrb. **57**: 198 (1921). Type from Tanzania.
   Melinis nyassana Mez in Bot. Jahrb. **57**: 199 (1921). Type from Tanzania.
   Melinis villosipes Mez in Bot. Jahrb. **57**: 199 (1921). Types from Tanzania and Malawi, *Buchanan* 244 (B, syntype).
   Rhynchelytrum nyassanum (Mez) Stapf & Hubbard in F.T.A. **9**: 892 (1930). —Sturgeon in Rhod. Agric. Journ. **50**: 110 (1953). —Chippindall in Meredith, Grasses & Pastures of S. Afr.: 433 (1955). —Jackson & Wiehe, Annot. Check List Nyasal. Grass.: 55 (1958). —Launert in Merxm., Prodr. Fl. SW. Afr. **160**: 161 (1970). Type as for *Melinis nyassana*.
   Rhynchelytrum ramosum Stapf & Hubbard in F.T.A. **9**: 895 (1930). —Chippindall in Meredith, Grasses & Pastures of S. Afr.: 433 (1955). Type from Angola.
   Rhynchelytrum rhodesianum (Rendle) Stapf & Hubbard in F.T.A. **9**: 895 (1930). —Sturgeon in Rhod. Agric. Journ. **50**: 109 (1953). Type as for *Tricholaena rhodesiana*.
   Rhynchelytrum stuposum Stapf & Hubbard in F.T.A. **9**: 897 (1930). Type: Malawi, Mt. Malosa, *Whyte* (K, holotype).
   Tricholaena busseana (Mez) Peter in Fl. Deutsch Ost-Afr. **40**: 219 (1931). Type as for *Panicum busseanum*.

Densely caespitose perennial. Culms (25)40–120(150) cm. high, erect. Leaf laminae (3)10–30(44) cm. long, (1.3)2–3.5(4.5) mm. wide, often involute and setaceous. Panicle (5)8–25(31) cm. long, oblong or narrowly ovate. Spikelets (3.2)3.6–5(5.7) mm. long, narrowly ovate. Inferior glume (0.3)0.5–1(1.9) mm. long, 0–1-nerved, ovate, inserted close to the superior; superior glume 5-(rarely 7-)nerved, membranous to weakly coriaceous, straight on the back, usually hairy, emarginate, awned or mucronate. Inferior floret male, its lemma 5-nerved, equalling the superior glume, the palea ciliate on its keels.

Caprivi Strip. Andara, Kukakamo Rocks, 15.i.1956, *De Winter & Wiss* 4265 (M). **Botswana**. SE: Lobatsi, 1920, *Sandwith* 138 (K). **Zambia**. B: Mwinilunga Distr., Matonchi dambo, 26.x.1937, *Milne-Redhead* 2969 (B; BR). N: Luwingu, Chisinga Ranch, 4.xii.1961, *Astle* 1057 (B). W: Ndola, 1933, *Duff* 98 (K). C: 8 km. W. of Serenje, 16.xii.1967, *Simon & Williamson* 1405 (K). S: Namwala, 13.i.1962, *Mitchell* 12/61 (LISC). **Zimbabwe**. W: Matobo, xi.1957, *Miller* 4804 (BR; LISC). C: Rusape, 28.x.1972, *Crook* 1057 (B; BR; LISC; M). E: Gazaland, Chirinda, 7.xi.1906, *Swynnerton* 1632 (K). **Malawi**. N: Mzimba Distr., 5 km. W. of Mzuzu, 15.vii.1973, *Pawek* 7196 (K). C: Lilongwe, Dzalanyama, 4.xii.1951, *Jackson* 617 (BR). S: Mulanje Mts., Litchenya Plateau, 8.vi.1967, *Robinson* 5316 (BR; K). **Mozambique**. N: Lichinga, 15.xii.1934, *Torre* 233 (LISC). Z: Gúruè, 30.ix., *Torre* 3550 (LISC). MS: Baruè, 9.xii.1965, *Torre & Correia* 13481 (LISC). M: Namaacha, 9.xii.1947, *Torre* 7081 (LISC).

Also from Gabon to Kenya and southwards to S. Africa (Cape) and in Madagascar, Thailand and Vietnam. Grassland and *Brachystegia* woodland; (200)1000–3000 m.

Tab. 34. HYLEBATES CORDATUS. 1, habit (× ½); 2, spikelet showing inferior glume (× 10); 3, spikelet showing superior glume (× 10); 4, superior floret (× 10), all from *Hinds* 294.

The species has a remarkable disjunct occurrence in SE. Asia. It is very variable, the spikelets ranging from glabrous to villous at the same locality. It can be recognised by its densely caespitose habit and the narrow, often involute and setaceous, leaf-blades. Separation from *M. repens* may sometimes be difficult but the shape of the superior glume and the lemma of the inferior floret, as well as the narrowly ovate panicle, provide additional characters.

23. HYLEBATES Chippindall

Hylebates Chippindall in Journ. S. Afr. Bot. **11**: 127 (1945).

Inflorescence a panicle with some or all of the branches whorled. Spikelets elliptic. Inferior glume up to ⅓, the superior as long as spikelet. Inferior lemma tipped by an awnlet. Superior lemma thinly chartaceous, acute, its flat thin margins enfolding and concealing palea.

A genus of 2 species. Kenya to Zimbabwe.
Outwardly it is easily mistaken for *Panicum*, but the superior floret proclaims its affinity to *Digitaria*.

Hylebates cordatus Chippindall in Journ. S. Afr. Bot. **11**: 128 (1945). —Clayton & Renvoize in F.T.E.A., Gramineae: 660 (1982). TAB. **34**. Type: Mozambique, Mocuba to Milange, *Torre* 4945 (K, isotype).

Straggling annual, rooting at the lower nodes. Culms 40–150 cm. high. Leaf laminae narrowly lanceolate, cordate and amplexicaul at the base. Panicle 8–25 cm. long, ovate, effuse. Spikelets 2.5–3.5 mm. long. Inferior glume 0.5–1 mm. long, ovate, separated from the superior by a short columnar internode. Inferior lemma with an awnlet 0.5–1 mm. long.

Zambia. W: L. Kashiba near St Anthony's Mission, 14.ii.1975, *Williamson & Gassner* 2391 (K; SRGH). C: c. 13 km. E. of Lusaka, 22.ii.1954, *Best* 53 (K). E: Petauke Distr., Cholongozi, ii.1963, *Verboom* 912 (K). S: Mapanza Mission, 5.ii.1955, *Robinson* 1085 (K). **Zimbabwe**. N: Darwin Distr., Nyarandi R. c. 1.6 km. N. of Mavuradonha Mission, 29.i.1970, *Simon* 2095 (K; SRGH). E: Mutare, on main Rd. to Beira, 15.ii.1969, *Crook* 848 (K; SRGH). **Mozambique**. Z: Mocuba to Milange, 18.iii.1943, *Torre* 4945 (K). MS: Amatongas Forest, 1.iv.1952, *Chase* 4431 (K; SRGH).
Also in Tanzania. Shady places in thicket, woodland and riverine forest; 400–1300 m.

24. STEREOCHLAENA Hack.

Stereochlaena Hack. in Proc. Rhod. Sci. Ass. **7**, 2: 65 (1908). —Clayton in Kew Bull. **33**: 295–297 (1978).
Chloridion Stapf in Ic. Pl. (Hook.) **28**: t.2640 (1900) non Chloridium Link (1824).

Inflorescence of paired or digitate racemes with narrowly winged triquetrous rhachis, bearing paired spikelets. Spikelets narrowly lanceolate to oblong-elliptic. Inferior glume minute or suppressed, the superior from very short to almost as long as spikelet and sometimes awned. Inferior lemma scabrid, prominently nerved, entire to tridentate, terminating in a straight awn. Superior lemma cartilaginous with flat thin margins enfolding and concealing palea, acute to acuminate.

A genus of 5 species. Tanzania to S. Africa.

Superior glume up to ¼ length of spikelet; awn usually 5–20 mm. long   -   -   1. *cameronii*
Superior glume ⅓ length of spikelet or more; awn up to 3 mm. long   -   -   -   2. *caespitosa*

1. **Stereochlaena cameronii** (Stapf) Pilger in Engl. & Prantl, Pflanzenfam. ed. 2, **14e**: 45 (1940). —Chippindall in Meredith, Grasses & Pastures of S. Afr.: 425 (1955). —Clayton & Renvoize in F.T.E.A., Gramineae: 656 (1982). TAB. **35**. Type: Malawi, Namasi, *Cameron* 15 (K, holotype).
   Chloridion cameronii Stapf in Ic. Pl. (Hook.) **28**: t.2640 (1900); in F.T.A. **9**: 480 (1919). —Sturgeon in Rhod. Agric. Journ. **50**: 284 (1953). —Jackson & Wiehe, Annot. Check List Nyasal. Grass.: 32 (1958). Type as above.
   Stereochlaena jeffreysii Hackel in Proc. Rhod. Sci. Ass. **7**, **2**: 66 (1908). Types: Zimbabwe, Bulawayo, *Gardner* 46 & 83 (W, syntypes).

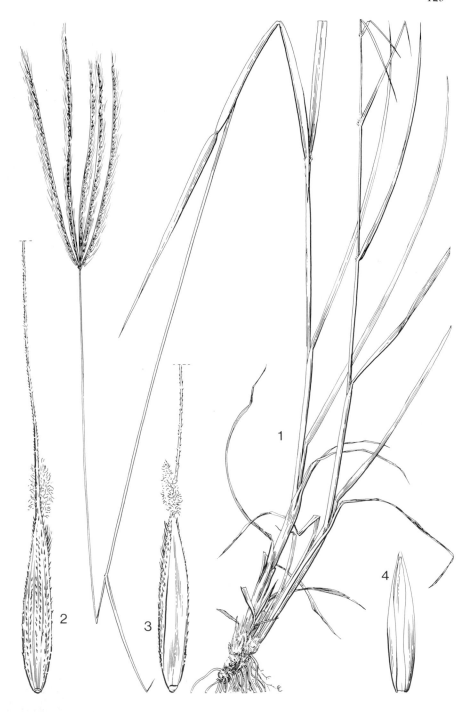

Tab. 35. STEREOCHLAENA CAMERONII. 1, habit (×$\frac{1}{2}$); 2, spikelet showing inferior lemma (× 15); 3, spikelet showing vestigial superior glume (× 15); 4, superior floret (× 15), all from *Davidse & Handlos* 7214.

Tufted perennial, the basal sheaths pubescent to tomentose. Culms 30–100 cm. high. Inflorescence of (2)3–8(10) racemes, each 3–16 cm. long. Spikelets 2.5–3.5 mm. long, narrowly lanceolate. Superior glume 0.1–0.5 mm. long, obtuse to acute. Inferior lemma entire, the awn 5–20 mm. long, rarely shorter on some of the spikelets. Superior floret dark brown.

Botswana. SE: Mahalapye, 15.ii.1958, *de Beer* 582 (K; SRGH). **Zambia**. N: Kasama, Misamfu Agric. Sta., 4.iv.1961, *Angus* 2641 (FHO; K). W: c. 24 km. S. of Luanshya, 12.iii.1960, *Robinson* 3397 (K). C: Kundalila Falls, Kanona, 17.iii.1974, *Davidse & Handlos* 7251 (K; MO). S: Mochipapa, 13.ii.1962, *Astle* 1504 (K). **Zimbabwe**. N: Gokwe Distr., Chirisa, Sengwa Res. Area, 12.ii.1981, *Mahlangu* 439 (K; SRGH). W: Hwange Distr., Matetsi Safari Area, 28.i.1980, *Gonde* 288 (K; SRGH). C: Harare, Twenty Dales Estate, 10.ii.1974, *Davidse & Simon* 6454 (K; MO; SRGH). E: Nyanga, Lawley's Concession, 21.ii.1984, *West* 3391 (K; SRGH). S: Masvingo Distr., Glyntor, 14.i.1948, *Robinson* 199 (K; SRGH). **Malawi**. N: Viphya, c. 34 km. SW. of Mzuzu, 10.iv.1971, *Pawek* 4610 (K). C: Nkhota Kota, Matichi, 27.ii.1953, *Jackson* 1128 (K). S: Zomba, *Wiehe* N/56 (K). **Mozambique**. N: Lichinga, road to Matama, 5.iv.1983, *Nuvunga* 1350 (K). Z: Mocuba to Milange, 18.iii.1943, *Torre* 4960 (K). MS: Garuso, iv.1935, *Gilliland* 1887 (K).
Also in Tanzania to Transvaal. Open places in savanna woodland, the habitats ranging from drainage hollows to stony slopes; 750–2100 m.

2. **Stereochlaena caespitosa** Clayton in Kew Bull. **33**: 296 (1978). —Clayton & Renvoize in F.T.E.A., Gramineae: 656 (1982). Type from Tanzania.

Densely tufted perennial, the basal sheaths pubescent. Culms 100–150 cm. high. Inflorescence of (2)3–8 racemes, each 8–13 cm. long. Spikelets 3–4.5 mm. long, narrowly elliptic. Superior glume ⅓ to almost as long as spikelet, finely acuminate to shortly awned. Inferior lemma entire, the awn 1–3 mm. long or some of the spikelets awnless.

Malawi. S: Chiuta Plain, Nampeya Dambo, 15.v.1952, *Jackson* 826 (K).
Also in Tanzania. Savanna woodland; 800 m.

25. BAPTORHACHIS Clayton & Renvoize

Baptorhachis Clayton & Renvoize in Gen. Gram.: 377 (1986).

Inflorescence a solitary raceme with broad colourful foliaceous rhachis, bearing paired spikelets. Spikelets elliptic-oblong. Inferior glume suppressed. Superior glume and inferior lemma as long as spikelet, 3-nerved, the laterals thickened and bearing a dense tuft of hair near the base, bilobed with an awn from the sinus. Superior lemma cartilaginous with flat thin margins enfolding and concealing palea, acute.

A monotypic genus. Mozambique.

Baptorhachis foliacea (Clayton) Clayton in Kew Bull. **42**: 401 (1987). TAB. **36**. Type: Mozambique, *Carvalho* 508 (K, holotype)
    Stereochlaena foliacea Clayton in Kew Bull. **33**: 296 (1978). Type as above.

Annual. Culms 30–60 cm. high. Raceme 2–4.5 cm. long, the rhachis membranously winged, 3–4 mm. wide, purple. Spikelets 2.5–3 mm. long. Superior glume and inferior lemma obscured by tufts of white hair c. 1 mm. long, terminating in an awn 1.5 mm. long.

Mozambique. N: Ribáuè, near Serra Nametere, 6.ii.1962, *Carvalho* 508 (K).

26. MEGALOPROTACHNE C.E. Hubbard

Megaloprotachne C.E. Hubbard in Kew Bull. **1929**: 320 (1929).

Inflorescence of digitate racemes bearing paired spikelets. Spikelets narrowly elliptic. Inferior glume as long as spikelet, (3)5-nerved, with obtuse hyaline apex, the superior shorter and 3-nerved. Superior lemma cartilaginous with flat thin margins enfolding and concealing palea, acuminate.

A monotypic genus. Zambia to S. Africa.
Often mistaken for *Digitaria*, but distinguished by its large inferior glume.

Tab. 36. BAPTORHACHIS FOLIACEA. 1, habit (× ⅔); 2, superior glume (× 12); 3, inferior lemma (× 12); 4, superior floret (× 12), all from *Carvalho* 508.

132

Tab. 37. MEGALOPROTACHNE ALBESCENS. 1, habit (× ⅔); 2, spikelet showing inferior glume (× 8); 3, spikelet showing superior glume (× 8); 4, superior floret (× 8), all from *de Winter* 7376.

Megaloprotachne albescens C.E. Hubbard in Kew Bull. **1929**: 321 (1929). —Chippindall in Meredith, Grasses & Pastures of S. Afr.: 422 (1955). —Launert in Merxm., Prodr. Fl. SW. Afr. **160**: 128 (1970). TAB. **37**. Type from S. Africa.

Megaloprotachne glabrescens Roiv. in Ann. Bot. Fenn. **11**: 40 (1974). Type from Namibia.

Annual. Culms 15–100 cm. high. Inflorescence of 2–9 racemes, each 4–16 cm. long. Spikelets 3–4.5 mm. long. Inferior glume glabrous. Inferior lemma villous on the margin and with glandular patches there.

Botswana. N: 77 km. N. of Aha Hills on Namibian border, 13.iii.1965, *Wild. & Drummond* 6991 (K; SRGH). SW: c. 85 km. N. of Kang on Rd. to Ghanzi, 19.ii.1960, *de Winter* 7376 (K; PRE). SE: Mahalapye Distr., Lephepe Pasture Res. St., *Yalala* 490 (K; SRGH). **Zambia**. S: Kazungula Quarantine Area, ii.1932, *Trapnell* 970 (K). **Zimbabwe**. W: Hwange Nat. Park, main Rd. to Shapi, 9.v.1967, *Cleghorn* 1672 (K; SRGH).

Also in S. Africa and Angola.

Sandy soils in savanna woodland and bushland; 1000 m.

27. DIGITARIA Hall.
by P. Goetghebeur & P. Van der Veken

Digitaria Hall., Stirp. Helv. **2**: 244 (1768) nom. conserv.

Spikelets pedicelled, 2–3(5)-nate, rarely solitary, 2-flowered, inferior floret barren, superior floret bisexual, slightly dorsiventrally compressed, more or less flattened on the front, convex to gibbous or basally spurred on the back, typically hairy between the nerves of the superior glume and inferior lemma, hairs unusually variable, rarely glabrous. Glumes 2, inferior glume small to absent, abaxial, superior glume from as long as (rarely much longer than) the inferior lemma to much shorter and exposing the superior lemma, membranous, often 3-nerved. Inferior lemma usually as long as the spikelet (rarely much reduced), membranous, mostly 5–7-nerved, palea vestigial or absent. Superior lemma cartilaginous to chartaceous, rounded on the back, the flat, thin, semi-hyaline margins enfolding and concealing most of the palea, acute to acuminate, rarely aristate, longitudinal ornamentation more or less pronounced, pale to dark brown in fruit, often purplish tinged. Palea nearly as long as superior lemma, not or faintly 2-nerved. Lodicules 2, cuneate, flat. Anthers 3. Styles 2, shortly connate at the base, stigmas plumose, often purplish. Caryopsis mostly planoconvex, embryo $\frac{1}{5}-\frac{1}{2}$ as long, hilum subbasal, small, punctiform to ellipsoid. Annuals or perennials with variable habit. Inflorescence of 1-many racemes, variably arranged, from solitary along a common axis to digitate or in superposed whorls, sometimes with secondary branchlets; rhachis triquetrous, but often winged on two sides, bearing the alternate spikelet groups along the non-winged side. Spikelet pedicels smooth to scabrous or variously hairy, sometimes with a coronula of long cilia, apex variously shaped.

A genus of c.230 species in tropical and warm temperate regions.

One of the more important key characters is the form of aggregation of the spikelets: solitary, 2-nate, or 3(5)-nate. These grouplets should be considered as reduced secondary racemes. This condition can be recognised in species where the racemes occasionally to frequently bear small branchlets (e.g. *D. nitens, D. pearsonii, D. phaeotricha*). The character state is to be observed near the middle of the raceme, unless explicitly stated otherwise. The longest pedicel of a 3-nate grouplet may be coalescent with the rhachis over a considerable part of its length.

Notwithstanding the existence of a recent monograph, no satisfactory infrageneric system can be given here. Henrard (1950) explicitly applied the typological approach, and in his overemphasis of type specimens he neglected to a high degree the variability presented by the non-types. He also heavily relied upon the presence or absence of different hair types. Sometimes, when he was dealing with highly specialized types (= synapomorhies), this resulted in the recognition of sections whose holophyletic nature is not contested. In other instances when working with symplesiomorphies or with non-homologous character states (e.g. absence of hairs), evidently unnatural sections were described. In particular his "glabrous" sections are highly suspect. Although we have up until now mainly studied African taxa, we hope to write an elaborate critique of Henrard's monograph, accompanied by a corrected classification.

1. Coronula hairs (long cilia on the apex of the pedicels) nearly as long as or overtopping the spikelet　-　-　-　-　-　-　-　-　-　-　-　-　-　-　-　-　2
- Coronula hairs smaller or absent　-　-　-　-　-　-　-　-　-　-　3
2. Perennial; spikelets 1.4 mm. long　-　-　-　-　-　-　-　1. *diagonalis*
- Annual; spikelets 1.3 mm. long　-　-　-　-　-　-　-　2. *pseudodiagonalis*
3. Spikelets 3-nate, at least partly　-　-　-　-　-　-　-　-　-　-　4
- Spikelets 2-nate or solitary　-　-　-　-　-　-　-　-　-　-　-　22
4. Inflorescence of 2–3 subdigitate racemes, conspicuously shining, often pinkish; spikelets less than 2.5 mm. long, conspicuously overtopped by stiff hairs; adaxial base gibbous; ripe superior lemma pale　-　-　-　-　-　-　-　-　-　-　-　-　-　-　-　5
- Features not as combined above　-　-　-　-　-　-　-　-　-　-　7
5. Densely caespitose perennial, surrounded at the base by densely woolly cataphylls　-　-　-　-　-　-　-　-　-　-　-　-　-　20. *brazzae*
- Features not combined as above　-　-　-　-　-　-　-　-　-　-　6
6. Annual; spikelets 2.5–2.7 mm. long; inferior lemma deeply sulcate; hairs overtopping the spikelet by 2–3 mm.　-　-　-　-　-　-　-　-　-　-　22. *gayana*
- Plant with creeping stem; spikelets 2.9–3.0 mm. long; inferior lemma nearly flat; hairs only slightly (0.5 mm.) longer than the spikelet　-　-　-　-　-　-　21. *procurrens*
7. Rhizomatous or densely caespitose perennials; rhachis scarcely winged　-　-　-　8
- Annuals (rarely short lived perennial, but then rhachis winged); rhachis often broadly winged　-　-　-　-　-　-　-　-　-　-　-　-　-　14
8. Rhizomatous perennial　-　-　-　-　-　-　-　-　-　-　-　9
- Caespitose perennial　-　-　-　-　-　-　-　-　-　-　-　-　11
9. Spikelets glabrous or very loosely hairy, hairs pale; racemes 1–2(3)　-　-　-　14. *eylesii*
- Spikelets densely hairy, hairs (partly) rufous; racemes (2)3–4　-　-　-　-　10
10. Racemes (5)10–25 cm. long; spikelets 3.0–4.0 mm. long　-　-　-　13. *hyalina*
- Racemes 4–8 cm. long; spikelets 2.3–2.5 mm. long　-　-　-　-　15. *fuscopilosa*
11. Spikelets (sub)glabrous, 1.8–2.2 mm. long; superior glume less than ⅓ of the spikelet in length　-　-　-　-　-　-　-　-　-　-　-　-　-　9. *maitlandii*
- Spikelets hairy, rarely glabrous, more than 2.6 mm. long; superior glume more than ⅓ of the spikelet in length　-　-　-　-　-　-　-　-　-　-　-　-　12
12. Inflorescence long and narrow, with up to 20 (very) small (1–3 cm.) appressed racemes, the terminal raceme often much longer　-　-　-　-　-　-　-　12. *phaeotricha*
- Inflorescence fan-shaped, with mostly fewer and longer racemes　-　-　-　13
13. Leaf lamina involute; racemes (2)3–4(5); spikelets 3–3.6 mm. long　-　-　11. *setifolia*
- Leaf lamina flat; racemes (5)9–15(20); spikelets 2.6–3.3 mm. long　-　-　10. *compressa*
14. Rhachis not or only slightly winged　-　-　-　-　-　-　-　-　15
- Rhachis conspicuously winged　-　-　-　-　-　-　-　-　-　-　17
15. Superior glume nearly as long as spikelet; spikelets whitish to purplish hairy; ripe superior lemma yellowish to pale brown, or purplish　-　-　-　-　-　42. *angolensis*
- Superior glume less than ½ the spikelet in length; spikelets often with brown hairs; ripe superior lemma dark brown　-　-　-　-　-　-　-　-　-　-　16
16. Racemes 7–11; spikelets 1.5–1.7 mm. long　-　-　-　-　-　-　7. *minoriflora*
- Racemes 2–4; spikelets 2.3–2.5 mm. long　-　-　-　-　-　-　6. *siderograpta*
17. Superior lemma pale or purplish when ripe; superior glume (nearly) as long as spikelet; spikelet hairs white, often purplish tinged　-　-　-　-　-　-　-　-　18
- Superior lemma dark brown when ripe; superior glume mostly less than ⅔ of the spikelet in length; spikelet hairs white never purplish, or absent　-　-　-　-　-　19
18. A cuff of fine hairs overtopping the spikelet by 0.5–1 mm.; inferior lemma with (1)3 prominent nerves; spikelet 2.0–3.0 mm. long; racemes (2)3 or 4　-　-　-　44. *argyrotricha*
- Hairs not forming a clear, overtopping cuff; inferior lemma with 5(7) less prominent nerves; spikelet (1.2)1.5–2.0(2.2) mm. long; racemes 2 (3 or 4)　-　-　-　43. *longiflora*
19. Top of pedicel smooth to very slightly scaberulous; inferior lemma conspicuously sulcate, often with swollen margins　-　-　-　-　-　-　-　-　-　-　-　5. *comifera*
- Top of pedicel conspicuously scabrous to ciliate, inferior lemma flat　-　-　-　20
20. Spikelets 1.2–1.8 mm. long, obovate to broadly elliptical　-　-　-　4. *thouaresiana*
- Spikelets 1.9–2.6 mm. long, oblong　-　-　-　-　-　-　-　21
21. Top of pedicel densely ciliate; inferior lemma with the 3 central nerves more closely together　-　-　-　-　-　-　-　-　-　-　-　-　-　3. *ternata*
- Top of pedicel only scabrous; inferior lemma with evenly spaced nerves　-　-　-　-　-　-　-　-　-　-　-　-　-　-　8. *atrofusca*
22. Perennial with 6–15 long racemes, solitary or partly whorled along a common axis; spikelets 3–3.5 mm. long; superior glume less than 0.7 mm. long, truncate　-　-　-　-　-　-　-　-　-　58 *gymnostachys*
- Features not as combined above　-　-　-　-　-　-　-　-　-　23
23. Superior glume much longer than inferior lemma, 7-nerved, long acuminate　-　-　-　-　-　-　-　-　-　-　-　-　-　-　24
- Superior glume equal to or shorter than inferior lemma, 3–5-nerved　-　-　-　-　25

24. Spikelets 2.7–3.6 mm. long - - - - - - - - - - - 29. *debilis*
 – Spikelets 8–11(15) mm. long - - - - - - - - - 30. *remotigluma*
25. Annuals; inflorescence with (20)30–100 short (3–8 cm. long) racemes, partly in 1 or more superposed whorls along an elongate central axis; spikelets small (1.3–2.2 mm. long), finely hairy, not spurred - - - - - - - - - - - - - 26
 – Features not as combined above - - - - - - - - - - 27
26. Racemes of the lowermost whorls conspicuously peduncled, barren at the base; spikelets 1.7–2.2 mm. long; inferior lemma 5-nerved - - - - - - - - 59. *perrottetii*
 – Racemes not peduncled, bearing spikelets to the base; spikelets 1.3–1.6(1.7) mm. long; inferior lemma 7-nerved - - - - - - - - - - 60. *floribunda*
27. Base of raceme peduncles swollen, smooth; racemes (6)10–100, short (1–7 cm. long), rather few-flowered, conspicuously peduncled; pedicels ciliate at least near the apex (rarely nearly smooth); spikelet hairs stiff, shining white, often purplish tinged - - - - 28
 – Features not as combined above - - - - - - - - - - 31
28. Caespitose perennials, thickened at the base, often surrounded by woolly cataphylls; pedicels ciliate only near the apex or glabrous - - - - - - - - - - 29
 – Slender to robust annuals; pedicels ciliate throughout - - - - - - 30
29. Pedicels scabrous, ciliate near the apex; rhachis pale to purplish, ripe superior lemma pale - - - - - - - - - - - - - - - - 17. *nitens*
 – Pedicels smooth, apex ciliate or nearly smooth; rhachis orange-yellow; ripe superior lemma dark brown - - - - - - - - - - - - - - - - 16. *pulchra*
30. Lowermost 4–10 racemes worled; half of the spikelets 2-nate - - - 18. *poggeana*
 – All racemes solitary along the axis; (nearly) all spikelets solitary - - - 19. *redheadii*
31. Densely caespitose perennial; inflorescence a single upright raceme 4–19 cm. long (rarely a few short racemes at its base) - - - - - - - - - 23. *monodactyla*
 – Features not as combined above - - - - - - - - - - - 32
32. Annual; inflorescence a single short raceme; leaf lamina involute, conspicuously papillose on both surfaces - - - - - - - - - - - 41. *tenuifolia*
 – Features not as combined above - - - - - - - - - - 33
33. Inflorescence of 1–2 (very rarely 3–4) racemes, less than 6(7) cm. long; leaves mostly 3 cm. long, inferior leaves often pinkish; annuals with hairy spikelets - - - - - 34
 – Features not as combined above (some depauperate forms may cause problems) - - - - - - - - - - - - - - 40
34. Inferior glume present, sometimes rim-like; superior glume as long as spikelet - - - - - - - - - - - - - - - 35
 – Inferior glume absent; superior glume less than $\frac{2}{3}$ of the spikelet in length - - 36
35. Inferior glume rim-like; spikelets 1.8–2.1 mm. long - - - - - 40. *parodii*
 – Inferior glume c. $\frac{1}{4}$ of the spikelet in length, truncate or slightly bilobed; spikelets 2.5–2.7 mm. long - - - - - - - - - - - - 39. *maniculata*
36. Raceme single; rhachis symmetrically winged, margin ciliate - - - 24. *ventriosa*
 – Racemes 1–2(3); rhachis asymmetrically winged, margin smooth to finely fimbriolate - - - - - - - - - - - - - - 37
37. Spikelets adaxially gibbous; racemes 1–2 - - - - - - - - 38
 – Spikelets adaxially spurred; racemes 2 - - - - - - - - - 39
38. Pedicels mostly solitary; hairs exceeding the spikelet by 1 mm. - - - 26. *bidactyla*
 – Pedicels 2-nate in the lower half of the raceme; hairs exceeding spikelet by 0.5 mm. - - - - - - - - - - - - - 25. *complanata*
39. Pedicels scaberulous only; spikelets 2.5–2.8 mm. long - - - - 27. *calcarata*
 – Pedicels ciliate in the upper half; spikelets 3.0–3.5 mm. long - - - 28. *sacculata*
40. Spikelets (sub)glabrous, nerves smooth; rhachis of racemes not or scarcely winged - - - - - - - - - - - - - - - 41
 – Spikelets appressed hairy, or if subglabrous then the nerves scaberulous or spikelets with large stiff hairs - - - - - - - - - - - - - - - 44
41. Inferior lemma with the 3 central nerves conspicuously approximate; spikelets lanceolate, slender; rhachis up to 0.2 mm. wide - - - - - - - - - 42
 – Inferior lemma with all nerves equidistant; spikelets oblong-ovate, inflated when ripe; rhachis up to 0.4 mm. wide - - - - - - - - - - - - 43
42. Annual with strictly erect stems; spikelets 1.7–2 mm. long - - - 33. *appropinquata*
 – Annual (or short lived perennial ?) with creeping stem; spikelets 2.1–2.4 mm. long - - - - - - - - - - - 34. *trinervis*
43. Inferior glume c. $\frac{1}{4}$ of the spikelet in length, (rounded) triangular; superior glume 3(5)-nerved; leaf lamina 4–8 × 0.5–1.2 cm. - - - - - - - - - 32. *abyssinica*
 – Inferior glume typically polymorphous, $\frac{1}{10}-\frac{1}{3}(\frac{1}{2})$ of the spikelet, ovate, often erose to truncate, with a conspicuous hyaline fringe; superior glume 5(7)-nerved; leaf lamina 5–10 × 0.2–0.4(0.6) cm. - - - - - - - - - 31. *scalarum*
44. Plants annual (stem often creeping) - - - - - - - - - 45
 – Plants perennial, rhizomatous or caespitose - - - - - - - - 51
45. Nerves of inferior lemma scaberulous (use strong lens!) - - - - - 46
 – Nerves of inferior lemma smooth or nearly so - - - - - - 47

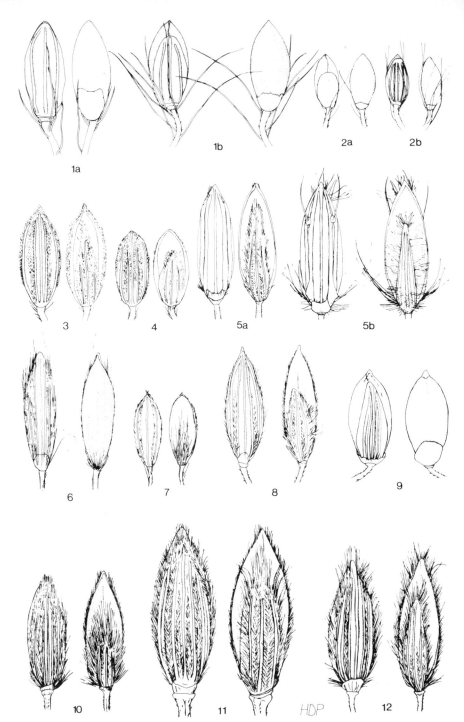

Tab. 38. DIGITARIA spikelets, species numbered as in text (a = abaxial, b = adaxial views) 1.—D. DIAGONALIS. 1a, *Nyariri* 675; 1b, *Simon* 2111. 2.—D. PSEUDODIAGONALIS. 2a, *Eyles* 8381; 2b, *Vesey-FitzGerald* 2738. 3.—D. TERNATA, *Banda* 189. 4.—D. THOUARESIANA, *Brain* 8861. 5.—D. COMIFERA. 5a, *Jackson* 2159; 5b, *Bingham* 502. 6.—D. SIDEROGRAPTA, *Fanshawe* 9946. 7.—D. MINORIFLORA, *Anton-Smith* 485. 8.—D. ATROFUSCA, *Astle* 3352; 9.—D. MAITLANDII, *West* 4609. 10.—D. COMPRESSA, *Macêdo* 2903. 11.—D. SETIFOLIA, *Crook* 287. 12.—D. PHAEOTRICHA, *Simon & Williamson* 994. (all × 6, except 10, 12: × 4½).

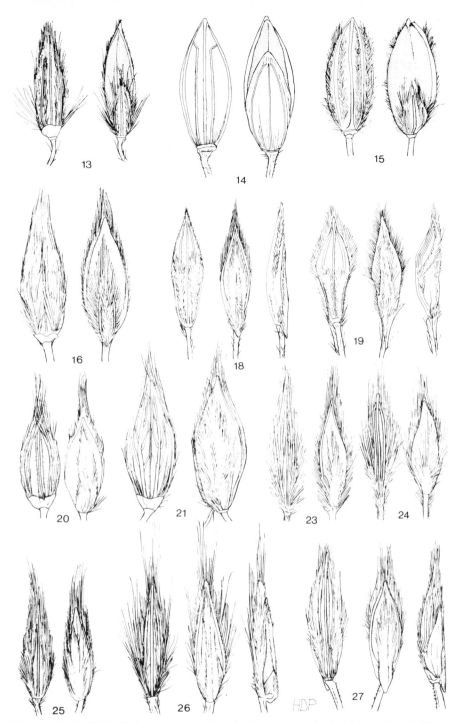

Tab. 39. DIGITARIA spikelets, species numbered as in text (abaxial, adaxial, lateral views). 13.—D. HYALINA, *Mutimushi* 1101. 14.—D. EYLESII, *White* 6807. 15.—D. FUSCOPILOSA, *Torre & Correia* 15642. 16.—-D. PULCHRA, *Astle* 346. 18.—D. POGGEANA, *Phipps & Vesey-FitzGerald* 3146. 19. D. REDHEADII, *Brummitt et al.* 14170. 20.—D. BRAZZAE, *Gilges* 867. 21.—D. PROCURRENS, *Phipps* 3212. 23.—D. MONODACTYLA, *Astle* 521. 24.—D. VENTRIOSA, *Simon & Williamson* 1995. 25.—D. COMPLANATA, *Drummond & Williamson* 9993. 26.—D. BIDACTYLA, *Astle* 312. 27.—D. CALCARATA, *Robinson* 4411. (all × 6, except 13, 20, 23: × 4½).

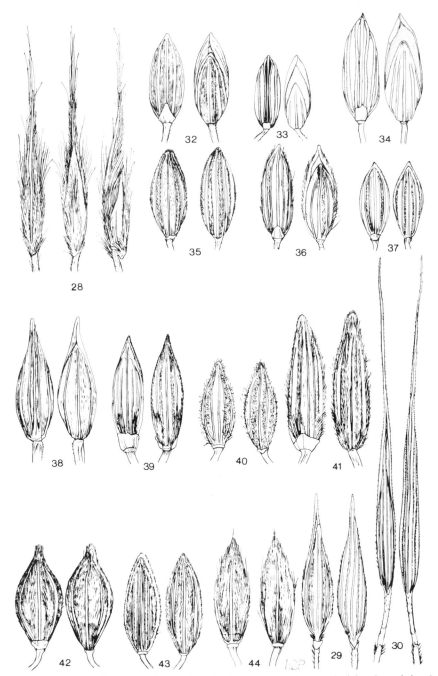

Tab. 40. DIGITARIA spikelets, species numbered as in text(a = abaxial,b = adaxial,c = lateral views). 28.—D. SACCULATA, *Robinson* 4352. 29.—D. DEBILIS, *Pawek* 8216. 30.—D. REMOTIGLUMA, *Drummond & Cookson* 6645. 32.—D. ABYSSINICA, *McCallum-Webster* A 321. 33.—D. APPROPINQUATA, *Torre* 5162. 34.—D. TRINERVIS, *Robinson* 5321. 35.—D. GAZENSIS, *Phipps* 844. 36.—D. RUKWAE, *Astle* 1193. 37.—D. LEPTORRHACHIS, *Phipps & Vesey-FitzGerald* 3184. 38.—D. DUNENSIS, *Verboom* 989. 39.—D. MANICULATA, *Robinson* 4338. 40.—D. PARODII, *Robinson* 4310. 41.—D. TENUIFOLIA, *Robinson* 4274. 42. —D. ANGOLENSIS, *Anton-Smith* 124/2. 43.—D. LONGIFLORA, *Siame* 303. 44.—D. ARGYROTRICHA, *Torre & Correia* 16999. (all × 6, except 30: × 4$\frac{1}{2}$).

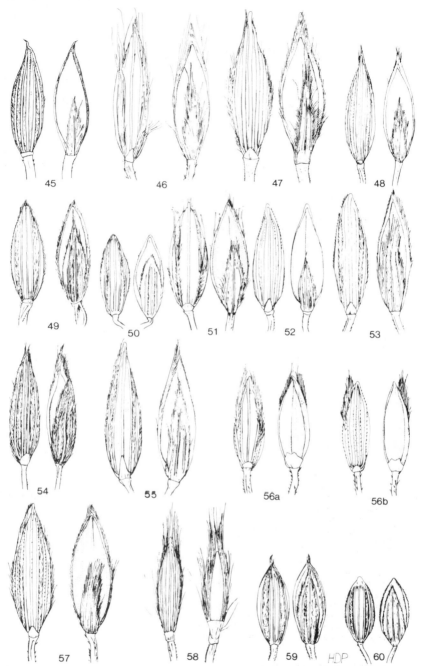

Tab. 41. DIGITARIA spikelets, species numbered as in text (abaxial, adaxial views). 45.—D. SANGUINALIS, *Balsinhas & Marrime* 446. 46.—D. ACUMINATISSIMA, *Torre & Paiva* 9512. 47.—D. CILIARIS, *Balsinhas & Marrime* 345. 48.—D. NUDA, *Drummond & Vernon* 11053. 49.—D. VELUTINA, *Davies* 3210. 50.—D. PEARSONII, *Brain* 10445. 51.—D. ERIANTHA, *Davies* 87. 52.—D. SWAZILANDENSIS, *Banda* 966. 53.—D. POLYPHYLLA, *Myre & Balsinhas* 767. 54.—D. ARGYROGRAPTA, *Myre & Carvalho* 2052. 55.—D. NATALENSIS, *Torre* 6625. 56a & b.—D. MEGASTHENES, *Torre & Correia* 14749. 57.—D. MILANJIANA, *Torre* 3977. 58.—D. GYMNOSTACHYS, *Mogg* 27746. 59.—D. PERROTTETII, *Verboom* 1314. 60. —D. FLORIBUNDA, *Siame* 609. (all × 6, except 54, 58: × 4½).

46. Inferior lemma longer than superior lemma, both acuminate, apex involute; mostly tall, robust plants with (very) long racemes; spikelets mostly more than
2.8 mm. long   -   -   -   -   -   -   -   -   -   -   46. *acuminatissima*
  – Inferior lemma as long as superior lemma, both acute; slender to robust plants, racemes short to long; spikelets mostly less than 2.7 mm. long   -   -   -   -   -   - 45. *sanguinalis*
47. Superior glume (nearly) as long as the spikelet, 3–5-nerved; rhachis not winged,
0.3 mm. wide   -   -   -   -   -   -   -   -   -   -   -   -   - 48
  – Superior glume $\frac{3}{4}$ of the spikelet in length (very rarely longer), 3-nerved; rhachis slightly to markedly winged, 0.5 mm. wide   -   -   -   -   -   -   -   -   -   - 49
48. Rhachis and pedicels (partly) scabrous; racemes mostly more than
5 cm. long   -   -   -   -   -   -   -   -   -   -   - 37. *leptorrhachis*
  – Rhachis and pedicels (nearly) smooth; inflorescence with (3)7–13 small (2–4 cm. long) racemes, solitary along a large central axis   -   -   -   -   -   -   - 38. *dunensis*
49. Racemes generally slender, often all solitary along a well developed central axis, rarely a few in a basal whorl   -   -   -   -   -   -   -   -   -   -   -   - 49. *velutina*
  – Racemes more robust, mostly digitate in 1-few superposed whorls, or only a few solitary along the (short) central axis   -   -   -   -   -   -   -   -   -   -   - 50
50. Inferior glume ovate to triangular, more than 0.3 mm. long; spikelets
2.7–3.4 mm. long   -   -   -   -   -   -   -   -   -   -   - 47. *ciliaris*
  – Inferior glume absent or rim-like, less than 0.2 mm. long; spikelets
2.0–2.5(2.8) mm. long   -   -   -   -   -   -   -   -   -   - 48. *nuda*
51. Most racemes solitary along a well developed central axis; rhachis not or
scarcely winged   -   -   -   -   -   -   -   -   -   -   -   -   - 52
  – Most racemes digitately in 1-few false whorls along a short central axis; rhachis slightly to markedly winged   -   -   -   -   -   -   -   -   -   -   -   - 54
52. Superior glume as long as the spikelet; stem at the base covered by velvety hairy cataphylls, often stoloniferous; nodes bearded   -   -   -   -   -   -   -   -   - 35. *gazensis*
  – Superior glume shorter than spikelet; cataphylls at stem base hairy or not; nodes
not bearded   -   -   -   -   -   -   -   -   -   -   -   -   - 53
53. Caespitose herb with creeping to ascending stems; leaf lamina 0.5–1.5 cm. wide, loosely hairy on both surfaces; spikelet pairs loosely scattered, often patent, or replaced by short
secondary racemes   -   -   -   -   -   -   -   -   -   - 50. *pearsonii*
  – Rhizomatous perennial with erect culms; leaf lamina 0.3–0.7 cm. wide, glabrous on lower surface; spikelet pairs imbricate to a varying degree, not patent   -   -   - 36. *rukwae*
54. Rhizome knotty, with rather slender, delicate straggling stems; leaf lamina narrow and often short, inferior leaves reduced to their sheath; racemes 2–3(4); spikelets
(2)2.2–2.7 mm. long   -   -   -   -   -   -   -   -   - 52. *swazilandensis*
  – Features not as combined above   -   -   -   -   -   -   -   -   -   - 55
55. Rhizome much branched, with rather short internodes; stems upright, often branched from the middle nodes; inferior leaves reduced to their sheath; middle and superior leaves crowded a few nodes above ground level   -   -   -   -   -   -   -   -   - 53. *polyphylla*
  – Features not as combined above   -   -   -   -   -   -   -   -   -   - 56
56. Racemes 2–3, very closely erect, stiff, often forming a silvery pencil; inferior lemma with the 3 central nerves approximate   -   -   -   -   -   -   -   - 54. *argyrograpta*
  – Inflorescence different   -   -   -   -   -   -   -   -   -   -   -   - 57
57. Densely caespitose to tussocky perennial; stem base with rusty-brown persistent leaf sheaths; leaf lamina 10–40 × 0.3–0.7 cm., ligule 3–12 mm. long; spikelets 2.8–4.2 mm. long; inferior lemma often densely beset with very fine spinules   -   -   -   -   -   - 55. *natalensis*
  – Features not as combined above   -   -   -   -   -   -   -   -   -   - 58
58. Superior glume $\frac{1}{5}$–$\frac{1}{3}$ (rarely in a few spikelets up to $\frac{1}{2}$) of the spikelet in length, nerveless or obscurely 3-nerved; inferior lemma scaberulous all over   -   -   -   - 56. *megasthenes*
  – Features not as combined above   -   -   -   -   -   -   -   -   -   - 59
59. Rhizomatous perennial, rhizomes branched, rarely with runners; nerves of inferior lemma mostly scaberulous, sometimes nearly smooth   -   -   -   -   - 57. *milanjiana*
  – Densely caespitose perennial, often with long-creeping runners; nerves of inferior lemma mostly smooth, sometimes slightly scaberulous   -   -   -   -   - 51. *eriantha*

1. **Digitaria diagonalis** (Nees) Stapf in F.C. **7**: 381 (1898). —Rendle, Cat. Afr. Pl. Welw. **2**: 163 (1899). —Eggeling, Annot. List Grass. Uganda: 14 (1947). —Henrard, Monogr. Digit.: 175 (1950). —Sturgeon in Rhod. Agric. Journ. **50**: 288 (1953). —Chippindall in Meredith, Grasses & Pastures of S. Afr.: 419, fig. 349 (1955). —Bogdan, Rev. List Kenya Grass.: 49 (1958). —Jackson & Wiehe, Annot. Check-list Nyasal. Grass.: 35 (1958). —Harker & Napper, Illustr. Guide Grass. Uganda: 26, t. 60 (1960). —Napper, Grass. Tangan.: 74 (1965). —Launert in Merxm., Prodr. Fl. SW. Afr. **160**: 63 (1970). —Simon in Kirkia **8**: 31, 65 (1971). —Clayton in F.W.T.A. ed. 2, **3**: 450 (1972); in Kew Bull. **29**: 527 (1974). —Rose Innes, Man. Ghana Grass.: 135, fig. 29 (1977). —Bennett in Kirkia **11**: 235 (1980). —Clayton in F.T.E.A., Gramineae: 624, fig. 145 (1982). TAB. **38**, fig. 1. Type from S. Africa.
  Panicum diagonale Nees, Fl. Afr. Austr.: 23 (1841). —Durand & Schinz, Consp. Fl. Afr. **5**: 746 (1894). Type as above.

Panicum uniglume Hochst. ex A. Rich., Tent. Fl. Abyss. **2**: 370 (1851). Type from Ethiopia.
Panicum diagonale var. *uniglume* (Hochst. ex A. Rich.) Hack. in Engl., Hochgebirgsfl. Trop. Afr.: 117 (1892). Type as above.
Panicum diagonale var. *glabrescens* K. Schum. in Engl., Pflanzenw. Ost-Afr. **C**: 100 (1895). Type from E. Africa.
Digitaria uniglumis (Hochst. ex A. Rich.) Stapf in F.T.A. **9**: 474 (1919). —Robyns, Esp. Cong. Digit.: 47 (1931). —Robyns, Fl. Agrost. Congo Belge **2**: 51, t. 21 (1934). —Henrard, Monogr. Digit.: 766 (1950). —Robyns & Tournay, Fl. Parc Nat. Alb. **3**: 74 (1955). —Andrews, Fl. Pl. Sudan **3**: 433 (1956). —Jacques-Felix, Gram. Afr. Trop. **1**: 234, fig. 159 (1962). Type as for *Panicum uniglume*.
Digitaria grantii C.E. Hubb. in Kew Bull. **1926**: 246, fig. 2 (1926). —Henrard, Monogr. Digit.: 300 (1950). —Harker & Napper, Illustr. Guide Grass. Uganda: 25, t. 59 (1960). —Napper, Grass. Tangan.: 74 (1965). Type from Tanzania.
Digitaria diagonalis var. *uniglumis* (Hochst. ex A. Rich.) Pilg. in Notizbl. Bot. Gard. Berl. **10**: 266 (1928). —Clayton in Kew Bull. **29**: 529 (1974); in F.T.E.A., Gramineae: 626 (1982). Type as for *Panicum uniglume*.
Digitaria trichopodia Stent in Bothalia **3**: 155 (1930). —Henrard, Monogr. Digit.: 756 (1950). Type from S. Africa.
Digitaria lasiostachya Peter in Repert. Spec. Nov. Beih. 40, 1, **2**: 207 (1930). —Henrard, Monogr. Digit.: 379 (1950). Type from E. Africa.
Digitaria diagonalis var. *glabrescens* (K. Schum.) Peter in Repert. Spec. Nov. Beih. 40, 1, **2**: 215 (1930). Type as for *Panicum diagonale* var. *glabrescens*.
Digitaria diagonalis var. *hirsuta* Troupin, Fl. Parc Nat. Garamba **1**: 29 (1956). —Napper, Grass. Tangan.: 75 (1965). —Clayton in F.W.T.A. ed. 2, **3**: 450, fig. 444 (1972); in Kew Bull. **29**: 528 (1974); in F.T.E.A., Gramineae: 626 (1982). Type from Zaire.

A tightly caespitose perennial, base surrounded by densely hairy cataphylls and fibrous leaf sheaths. Culms 50–120 cm., erect, glabrous, nodes dark and glabrous. Leaf sheaths scaberulous, loosely to densely hairy. Ligule 0.5 mm. long, rim-like, irregularly ciliate. Leaf laminae 10–15 × 0.2–0.4 cm., flat or involute, scabrous on both surfaces, scabrous along the margins. Inflorescence composed of 10–30 racemes, 8–14 cm. long, erect or sometimes arching, few together or solitary along a well developed common axis. Rhachis triquetrous, narrowly winged, up to 0.3 mm. broad, smooth to scaberulous with scabrous margins. Pedicels 3(5)-nate, 0.3–3.0 mm. long, triangular to compressed, scaberulous, much broadened at the apex, with a well developed coronula composed of (2)5–10 long bristles, these overtopping the spikelet. Spikelets 1.4–2.1 mm. long, ovate. Inferior glume absent or extremely short. Superior glume $\frac{1}{4}$–$\frac{1}{3}$ of the spikelet, ovate, nerveless, glabrous, hyaline. Inferior lemma as long as or somewhat shorter than the spikelet, ovate, with narrow recurved margins, 3-nerved, hyaline, glabrous. Superior lemma as long as the spikelet, ovate, apex acute, pale to dark brown.

Zambia. B: Bulosi Plain, 7.i.1960, *Gilges* 864 (P; SRGH). N: Mbala, 1800 m., 21.iii.1958, *Vesey-FitzGerald* 1559 (BR; K; SRGH). W: Solwezi, 7.i.1969, *Mutimushi* 2948 (SRGH). C: Mkushi Distr., 1430 m., 1.iv.1963, *Vesey-FitzGerald* 4078 (BM; COI). E: Nyika Plateau, v.1968, *Williamson* 928 (SRGH). S: Ngwezi Paddocks, 30.i.1964, *Astle* 2906 (K; SRGH). **Zimbabwe**. N: Nyamunyeche Estate, Karoi Vlei, 7.ii.1979, *Nyariri* 675 (SRGH). C: Harare, University Grounds, 1600 m., 1.i.1971, *Crook* 952 (BR; K; PRE; SRGH). E: Sibu, 14 km. W. of Chimanimani (Melsetter), 18.vi.1953, *Crook* 472 (K; SRGH). S: Mt. Buhwa, 1100 m., 1.v.1973, *Simon, Pope & Biegel* 2410 (K; SRGH). **Malawi**. N: Nyika Plateau, 2250 m., 13.iii.1961, *Robinson* 4475 (BM; K; PRE; SRGH). S: Thambani, 12.xii.1950, *Jackson* 328 (K; SRGH). **Mozambique**. N: 8 km. Marrupa-Lichinga, 800 m., 19.ii.1982, *Jansen & Boane* 7903 (SRGH). Z: 7 km. Mualama-Naburi, 100 m., 16.i.1968, *Torre & Correia* 17169 (LISC). T: Mt. Tsangano, 26.vi.1956, *Myre* 2534 (LMA). MS: Manica, Mavita, 9.xii.1965, *Pereira & Marques* 956 (LMU).
Tropical and S. Africa. Miombo woodland, (degraded) savanna, mountain grassland, vlei and wet rocky outcrops.

2. **Digitaria pseudodiagonalis** Chiov. in Nuov. Giorn. Bot. Ital. n.s. **26**: 63 (i.1919). —Robyns in Fl. Agrost. Congo Belge **2**: 55 (1934). —Clayton in F.T.E.A., Gramineae: 624 (1982). TAB. **38**, fig. 2. Type from Zaire.
Panicum minutiflorum Hochst. ex A. Rich., Tent. Fl. Abyss. **2**: 362 (1851) non (P. Beauv.) Rasp. (1825). —Steudel in Syn. Pl. Glum. **1**: 41 (1853). —Engl. in Hochgebirgsfl. Trop. Afr.: 117 (1892). —Durand & Schinz, Consp. Fl. Afr. **5**: 754 (1894). —Hack., Öst. Bot. Zeitschr. **1901**: 330 (1901). Type from Ethiopia.
Digitaria intecta Stapf in F.T.A. **9**: 478 (vi.1919). —Henrard, Monogr. Digit.: 351 (1950). —Andrews, Fl. Pl. Sudan **3**: 433 (1956). —Harker & Napper, Illustr. Guide Grass. Uganda: 25, t. 57 (1960). —Simon in Kirkia **8**: 65 (1971). —Clayton in F.T.E.A., Gramineae: 623 (1982). Type from Ethiopia.

Digitaria minutiflora Stapf in F.T.A. **9**: 476 (vi.1919). —Vanderyst in Bull. Agric. Congo Belge **16**: 661 (1925). —Robyns, Esp. Cong. Digit.: 51 (1931); in Fl. Agrost. Congo Belge **2**: 54 (1934). —Henrard, Monogr. Digit.: 461 (1950). —Robyns & Tournay in Fl. Parc Nat. Alb. **3**: 74 (1955). —Napper, Grass. Tangan.: 75 (1965). —Simon in Kirkia **8**: 66 (1971). —Clayton in F.W.T.A. ed. 2, **3**: 452 (1972). Type as for *Panicum minutiflorum*.

Digitaria eglumis Peter in Repert. Spec. Nov. Beih. 40, **1**: 62 (1930). —Henrard, Monogr. Digit.: 209 (1950). —Robyns & Van der Veken in Bull. Jard. Bot. Brux. **22**: 144 (1952). —Bennett in Kirkia **11**: 235 (1980). Type from Tanzania.

Digitaria eglumis var. *reducta* P. Goetghebeur in Bull. Nat. Plantentuin Belg. **45**: 397 (1975). Type: Zambia, Mazabuka, Kafue Gorge, 1060 m., 14.iv.1956, *Robinson* 1478 (K, isotype; SRGH, holotype).

A loosely caespitose or solitary growing, slender to rather robust annual. Culms 20–80 cm., erect or ascending, glabrous, nodes dark, hairy. Leaf sheaths glabrous to loosely hairy. Ligule 0.5–1.5 mm. long, truncate, ciliate. Leaf laminae 5–25 × 0.2–0.5 cm., linear, flat, scaberulous on the superior surface, glabrous on the inferior surface, scabrous along the margin. Inflorescence composed of (3)5–17 racemes, erect to very slightly arching, paired or solitary along a well developed common axis. Rhachis triquetrous, scarcely winged, up to 0.3 mm. wide, smooth or scabrous, margins scabrous. Pedicels 3–5-nate, triangular to compressed, scabrous, broadened at the apex, bearing a coronula of (1)2–6 long bristle hairs, these often overtopping the spikelet. Spikelets 0.6–1.3 mm. long, ovate. Inferior glume absent. Superior glume very short, truncate, nerveless, glabrous, hyaline. Inferior lemma as long as the spikelet or somewhat shorter, rarely only ⅓ of the spikelet, ovate, 3-nerved, glabrous, hyaline. Superior lemma as long as the spikelet, ovate, acute, slightly to prominently papillose, conspicuously shining, dark brown.

Zambia. N: Mbeshi Estate, 1600 m., 20.iv.1959, *McCallum-Webster* A332 (K; LISC; SRGH). W: Mufulira Distr., 11.v.1934, *Eyles* 8381 (K; P; SRGH). S: Kafue Gorge, 18.iii.1960, *Vesey-FitzGerald* 2738 (SRGH). **Zimbabwe**. N: Chizarira Escarpment, 5.v.1972, *Cleghorn* 2522 (SRGH). **Mozambique**. N: Maniamba, Malulo, 27.v.1948, *Pedro & Pedrogão* 3988 (LMA).

Scattered from Guinee to Ethiopia and south to Mozambique. On shallow soil over rocks in miombo woodland, in dambos, weed along roadsides and in plantations.

The amount of variability with respect to the length of the inferior lemma, and to the surface structure of the superior lemma does not seem correlated with other characters nor does it with distributional patterns. However more material of this rare species, which has a scattered distribution, is needed to assess the nature of these patterns.

3. **Digitaria ternata** (A. Rich.) Stapf in F.C. **7**: 376 (1898). —Rendle in Cat. Afr. Pl. Welw. **2**: 162 (1899). —Stapf in F.T.A. **9**: 452 (1919). —Stent in Bothalia **1**: 269 (1921). —Robyns, Esp. Cong. Digit.: 30 (1931); in Fl. Agrost. Congo Belge **2**: 33 (1934). —Eggeling, Annot. List Grass. Uganda: 15 (1947). —Henrard, Monogr. Digit.: 737 (1950). —Chiovenda in Webbia **8**: 70 (1951). —Sturgeon in Rhod. Agric. Journ. **50**: 285 (1953). —Chippindall in Meredith, Grasses & Pastures of S. Afr.: 418, fig. 348 (1955). —Bor in Webbia **11**: 337 (1956). —Andrews, Fl. Pl. Sudan **3**: 434 (1956). —Bogdan, Rev. List Kenya Grass.: 49 (1958). —Jackson & Wiehe, Annot. Check-list Nyasal. Grass.: 36 (1958). —Bor, Grass. B.C.I. & P.: 306 (1960). —Harker & Napper, Illustr. Guide Grass. Uganda: 25, t. 58 (1960). —Napper, Grass. Tangan.: 75 (1965). —Rugolo de Agrasar in Bol. Soc. Arg. Bot. **12**: 389, fig. 3 (1968). —Launert in Merxm., Prodr. Fl. SW. Afr. **160**: 66 (1970). —Simon in Kirkia **8**: 31, 65 (1971). —Clayton in F.W.T.A. ed. 2, **3**: 452 (1972). —Veldkamp in Blumea **21**: 59, fig. 14a (1973). —Rugolo de Agrasar in Darwiniana **19**: 122, fig. 14 (1974). —Rose Innes, Man. Ghana Grass.: 138 (1977). —Bennett in Kirkia **11**: 236 (1980). —Clayton in F.T.E.A., Gramineae: 630, fig. 146 (1982). TAB. **38**, fig. 3; TAB. **42**. Type from Ethiopia.

Panicum phaeocarpum var. *gracile* Nees, Fl. Afr. Austr. **1**: 23 (1841). Type from S. Africa.

Cynodon ternatus A. Rich., Tent. Fl. Abyss. **2**: 205 (1851). Type as for *Digitaria ternata*.

Panicum ternatum (Hochst. ex A. Rich.) Steud., Syn. Pl. Glum. **1**: 40 (1854). —Engl. in Hochgebirgsfl. Trop. Afr.: 118 (1892). —Durand & Schinz, Consp. Fl. Afr. **5**: 766 (1894). —Hackel in Öst. Bot. Zeitschr. **1901**: 331 (1901). Type as above.

An annual with loosely caespitose or solitary culms. Culms 20–60 cm., erect, glabrous or rarely loosely hairy, nodes dark and glabrous. Leaf sheaths glabrous. Ligule 1–1.5 mm. long, rounded. Leaf laminae 2–20 × 0.1–0.5 cm., linear, flat, glabrous to loosely hairy on both sides, scabrous along the often crisped margin. Inflorescence composed of 2–5 racemes, (5)8–12(15) cm. long, solitary along a short common axis. Rhachis triquetrous, broadly winged, 0.5–1 mm. broad, smooth with smooth or scabrous margins. Pedicels 3-nate, 0.5–2 mm. long, terete to compressed, scabrous, conspicuously broadened at the apex, bearing a well developed coronula of large, broad hairs up to 0.8 mm. long. Spikelets 1.9–2.6 mm. long, oblong. Inferior glume absent. Superior glume c. ⅔ of the spikelet, oblong, 3-nerved, hyaline, appressed hairy, hairs white, capitate-apiculate.

143

Tab. 42. DIGITARIA TERNATA. 1, habit (× ½); 2, portion of raceme (× 5); 3, ternate group of spikelets (× 6); 4, spikelet showing superior glume (× 10); 5, spikelet showing inferior lemma (× 10); 6, superior floret, dorsal and ventral views (× 10), all from *Bullock* 2542. From F.T.E.A.

Inferior lemma as long as the spikelet, oblong with recurved margins, 5-nerved, hyaline, appressed hairy. Superior lemma as long as the spikelet, oblong, acute, pale to dark brown.

Zambia. N: Mbala, 1900 m., 20.ii.1959, *McCallum-Webster* A105 (K; SRGH). W: Kitwe, 7.iv.1963, *Fanshawe* 7756 (SRGH). E: Nyika Plateau, iv.1972, *Williamson* 2170 (SRGH). S: Mazabuka-Kafue, 1.iii.1963, *van Rensburg* 1549 (K: PRE; SRGH). **Zimbabwe**. N: Chininga Farm, 1000 m., 18.ii.1971, *Davies* 3024 (SRGH). W: Victoria Falls, 900 m., 23.iv.1970, *Simon & Hill* 2139 (GENT; K; SRGH). C: Harare, Nat. Botanic Gardens, 1300 m., 21.ii.1957, *Phipps* 518 (BR; COI; LISC; SRGH). E: Troutbeck, 20.iii.1948, *Rattray* 1426 (K; SRGH). **Malawi**. N: Nyika Plateau, 1800 m., 11.iii.1978, *Pawek* 14056A (SRGH). C: Lilongwe, 1.iii.1966, *Salubeni* 419 (K; P). S: Zomba Mt., 22.ii.1956, *Banda* 189 (BM; LISC; SRGH). **Mozambique**. N: Lichinga (Vila Cabral), 1200 m., 28.ii.1964, *Torre & Paiva* 10853 (COI; LISC; SRGH). GI: W. of Sul do Save, *Hornby* 3038 (LMA; LISC). M: Namaacha, 28.ii.1948, *Torre* 7443B (K; LISC).

Also in tropical and S. Africa, extending eastwards to China and Indonesia. Weed of arable land, disturbed sites.

4. **Digitaria thouaresiana** (Fluegge) A. Camus in Bull. Soc. Bot. France **75**: 914 (1928). —Henrard, Monogr. Digit.: 739 (1950). —Simon in Kirkia **8**: 31 (1971). —Bennett in Kirkia **11**: 237 (1980). —Clayton in F.T.E.A., Gramineae: 631 (1982). TAB. **38**, fig. 4. Type from Madagascar.
    Paspalum thouaresianum Fluegge, Gram. Monogr. **1**: 148 (1810). Type as above.
    Panicum puberulum var. *tricostulatum* Hackel in Öst. Bot. Zeitschr. **51**: 332 (1901). Type from S. Africa.
    Digitaria melanochila Stapf in F.T.A. **9**: 453 (1919). —Eggeling, Annot. List Grass. Uganda: 14 (1947). —Henrard, Monogr. Digit.: 446 (1950). —Robyns & Van der Veken in Bull. Jard. Bot. Brux. **22**: 149 (1952). —Sturgeon in Rhod. Agric. Journ. **50**: 286 (1953). —Chippindall in Meredith, Grasses & Pastures of S. Afr.: 419 (1955). —Jackson & Wiehe, Annot. Check-list Nyasal. Grass.: 35 (1958). —Harker & Napper, Illustr. Guide Grass. Uganda: 25, t. 59 (1960). —Napper, Grass. Tangan.: 75 (1965). Type from Uganda.
    Digitaria scaettae Robyns, Esp. Cong. Digit.: 31, t. 5 (1931); in Fl. Agrost. Congo Belge **2**: 34 (1934). —Henrard, Monogr. Digit.: 666 (1950). —Robyns & Tournay in Fl. Parc Nat. Alb. **3**: 76 (1955). Type from Rwanda.
    Digitaria tricostulata (Hack.) Henrard, Blumea **1**: 101 (1934). —Henrard, Monogr. Digit.: 758 (1950). —Robyns & Van der Veken in Bull. Jard. Bot. Brux. **22**: 149 (1952). —Chippindall in Meredith, Grasses & Pastures of S. Afr.: 419 (1955). —Bogdan, Rev. List Kenya Grass.: 49 (1958). —Van der Veken in Bull. Jard. Bot. Brux. **32**: 126 (1962). Type from S. Africa.
    Digitaria scaettae var. *glabra* Robyns & Van der Veken in Bull. Jard. Bot. Brux. **22**: 148 (1952). —Robyns & Tournay in Fl. Parc Nat. Alb. **3**: 76 (1955). —Van der Veken in Bull. Jard. Bot. Brux. **32**: 126 (1962). Type from Zaire.

A loosely caespitose annual. Culms 20–40(80) cm., erect, glabrous, nodes dark and glabrous. Leaf sheaths glabrous. Ligule c. 2 mm. long, rounded triangular, entire. Leaf laminae 3–20 × 0.2–0.6 cm., linear, flat, glabrous, sometimes scaberulous on both surfaces, basally often with a few bulbous-based bristles, scaberulous along the margin. Inflorescence of 3–7 racemes, 7–15 cm. long, a few together or all solitary along a short common axis, often pseudoviviparous. Rhachis triquetrous, broadly winged, up to 1 mm. broad, smooth with scabrous margins. Pedicels 3-nate, 0.5–2.5 mm. long, terete to compressed, minutely scaberulous, broadened at the apex, bearing a poorly developed coronula with cilia up to 0.3 mm. long. Spikelets 1.2–1.8 mm. long, ovate. Inferior glume absent. Superior glume $\frac{1}{3}$–$\frac{4}{5}$ of the spikelet in length, ovate-oblong, 3-nerved, hyaline, glabrous or appressed hairy, hairs white, capitate-apiculate. Inferior lemma as long as the spikelet, ovate with recurved margins, 5-nerved, hyaline, glabrous or appressed hairy. Superior lemma as long as the spikelet, ovate, apiculate, dark brown.

Zimbabwe. N: Umboe Flats, 1200 m., 12.iii.1931, *Brain* 2861 (SRGH). C: Harare Distr., 18.v.1934, *Cleghorn* 156 (SRGH). S: Gutu, iv.1921, *Eyles* 3000 (SRGH). **Mozambique**. MS: Chimoio, 13.xi.1946, *Pedro & Pedrogão* 220 (LMA).

From Cameroon to Kenya, south to S. Africa. On disturbed soils.

In this species the inflorescence is often pseudoviviparous, the spikelets are replaced by plantlets, i.e. glumes and lemmas are growing abnormally into foliaceous organs instead of normally bracteate scales.

5. **Digitaria comifera** Pilg. in Notizbl. Bot. Gard. Berl. **15**: 708 (1942). —Henrard, Monogr. Digit.: 141 (1950). —Harker & Napper, Illustr. Guide Grass. Uganda: 25, pl. 58 (1960). —Napper, Grass. Tangan.: 76 (1965). —Simon in Kirkia **8**: 31, 65 (1971). —Bennett in Kirkia **11**: 234 (1980). —Clayton in F.T.E.A., Gramineae: 630 (1982). TAB. **38**, fig. 5. Type from Tanzania.
    Digitaria lunularis Henrard, Monogr. Digit.: 414 (1950). —Simon in Kirkia **8**: 31, 65 (1971). Type: Zambia, Kazungula, 1100 m., ii.1932, *Trapnell* 957 (K, holotype).

An annual with loosely caespitose or solitary culms. Culms 15–60 cm., erect or ascending, glabrous, nodes dark, glabrous. Leaf sheaths glabrous to loosely hairy. Ligule 0.5–1.5 mm. long, rim-like, conspicuously ciliate. Leaf laminae (2)5–10 × 0.1–0.6 cm., linear, flat or sometimes involute, scaberulous on both sides, densely hairy near the ligule, scabrous along the margin. Inflorescence composed of 2–4(6) racemes, 4–14(20) cm. long, solitary along a short common axis. Rhachis triquetrous, winged, up to 0.7 mm. broad, smooth, with scabrous margins. Pedicels 3–4-nate, 0.2–2.5 mm. long, terete to subtriangular, scaberulous, much broadened at the apex, without coronula. Spikelets 2.0–2.9 mm. long, oblong-lanceolate. Inferior glume very short, bilobed, nerveless and hyaline. Superior glume c. $\frac{2}{3}$ of the spikelet, oblong-triangular, 3(5)-nerved, hyaline, appressed hairy, hairs capitate apiculate. Inferior lemma as long as the spikelet, oblong-lanceolate with recurved margins, centrally sulcate, 5(7)-nerved, hyaline, basally and apically appressed hairy, sometimes with conspicuous lateral swellings and bristlen hairs. Superior lemma as long as the spikelet, oblong-lanceolate, apiculate, dark brown.

Zambia. B: Mongu Lealui Distr., 20.ii.1966, *Robinson* 6853 (BM; K; SRGH). N: Saisi R., 27 km. from Mbala, 1700 m., 14.iv.1959, *McCallum-Webster* A324 (K; SRGH). C: Kanona, 1600 m., 5.iv.1961, *Phipps & Vesey-FitzGerald* 2964 (SRGH). S: Livingstone, 1000 m., 6.iv.1956, *Robinson* 1439 (BR; K; SRGH). **Zimbabwe**. N: Gokwe village, 12.iii.1964, *Bingham* 1144 (SRGH). W: Hwange Distr., 7–10 km. E. of Halfway Hotel, iii.1956, *Rattray* 1782 (K; SRGH). C: Sable Park, 30.xii.1980, *Stephens* G22 (SRGH). E: near Inverness, N. of Dutch Settlement, 1200 m., 5.iv.1962, *Cleghorn* 812 (SRGH). **Malawi**. N: Mzimba, 1937, *Wilson* 7A (K). C: Lilongwe, ii.1958, *Jackson* 2159 (BM; K; PRE). **Mozambique**. N: Amaramba, 30 km. of Mecanhelas, 680 m., 17.ii.1964, *Torre & Paiva* 10639 (LISC). Z: 30 km. on Mocuba to Maganja da Costa Rd., 17.v.1949, *Barbosa* 2690 (LMA; P).
Also in Zaire, Uganda, Burundi, Tanzania and Angola. In miombo woodland, also on disturbed sandy soils along roadsides and river banks.

6. **Digitaria siderograpta** Chiov. in Ann. Bot. Roma **13**: 39 (1914). —Robyns, Esp. Cong. Digit.: 32 (1931); in Fl. Agrost. Congo Belge **2**: 35 (1934). —Henrard, Monogr. Digit.: 693 (1950). —Van der Veken in Bull. Jard. Bot. Brux. **25**: 326 (1955). —Clayton in F.T.E.A., Gramineae1: 630 (1982). TAB. **38**, fig. 6. Type from Zaire.

An annual with loosely caespitose or solitary culms. Culms 20–70 cm., erect, glabrous, nodes dark, with a few erect hairs. Leaf sheaths glabrous, sometimes scaberulous. Ligule 0.5–1 mm. long, rim-like, shortly ciliate. Leaf laminae 5–15 × 0.1–0.3 cm., linear, involute or flat, densely scabrous on both surfaces, scabrous along the margin. Inflorescence composed of 2–4 racemes, 2–10 cm. long, solitary along a well developed common axis. Rhachis triquetrous, very narrowly winged, up to 0.3 mm. broad, undulate, smooth with scabrous margins. Pedicels 3-nate, 0.2–1.5 mm. long, triangular, scabrous, broadened at the apex, bearing a coronula. Spikelets 2.3–2.5 mm. long, oblong-lanceolate. Inferior glume absent. Superior glume c. $\frac{1}{3}$ of the spikelet, 1–3-nerved, hyaline, appressed hairy, hairs thick, slightly undulating, capitate-apiculate, often red brown. Inferior lemma as long as the spikelet, oblong-lanceolate with narrow recurved margins, 5-nerved, appressed hairy. Superior lemma as long as the spikelet, oblong-lanceolate, acute, red brown.

Zambia. N: Mbala Distr., 11.iv.1961, *Phipps & Vesey-FitzGerald* 3087 (K). W: Kitwe, 7.iii.1967, *Fanshawe* 9946 (SRGH). **Malawi**. S: 50 km. on Zomba to Namwera Rd., 13.iii.1964, *Correia* 201 (LISC).
Also in Zaire, Burundi and Tanzania. Laterite outcrops in miombo woodland and roadsides.
A somewhat similar species *D. schmitzii* Van der Veken, known from Zaire and Uganda, could well occur in N. Zambia. This species is easily recognized by the smaller spikelets (1.9–2 mm. long), the pale and conspicuously capitate hairs and the winged rhachis.

7. **Digitaria minoriflora** P. Goetghebeur in Bull. Nat. Plantentuin Belg. **45**: 398, fig. 2 (1975). TAB. **38**, fig. 7. Type: Zambia, 30 km. from Mansa on Chembe Rd., 10.iii.1969, *Anton-Smith* 485 (SRGH, holotype).

A loosely caespitose annual. Culms 10–40 cm., erect, glabrous, nodes dark and bearded to subglabrous. Leaf sheaths loosely hairy with bulbous-based bristles. Ligule 1–2 mm. long, truncate, ciliolate. Leaf laminae 5–15 × 0.2–0.5 cm., linear, flat, conspicuously scabrous on both sides, often loosely hairy. Inflorescence composed of 7–11 racemes, 3–7 cm. long, erect, scattered along the well developed common axis. Rhachis triquetrous, narrowly winged, up to 0.3 mm. broad, smooth with scabrous margins. Pedicels 3-nate, 0.5–2.5 mm. long, triquetrous, scabrous, with long cilia near the broadened apex.

Spikelets 1.5–1.7 mm. long, oblong. Inferior glume absent. Superior glume $\frac{1}{3}$–$\frac{1}{2}$ of the spikelet, oblong, 1-nerved, hyaline, appressed hairy, hairs slightly undulating, rounded at the apex. Inferior lemma as long as the spikelet, oblong, 3–5-nerved, appressed hairy. Superior lemma as long as the spikelet, oblong, acuminate, reddish brown.

Zambia. N: 25 km. on the Mansa-Chembe Rd., 10.iii.1969, *Anton-Smith* 485 (SRGH). Known from the type collection only. Weed in a garden on sandy soil.

8. **Digitaria atrofusca** (Hack.) Camus in Bull. Mus. Nat. Hist. Nat. **30**: 106 (1924). —Henrard, Monogr. Digit.: 53 (1950). —Robyns & Van der Veken in Bull. Jard. Bot. Brux. **22**: 149 (1952). —Clayton in Kew Bull. **29**: 519 (1974). —Rose Innes, Mann. Ghana Grass.: 134 (1977). —Clayton in F.T.E.A., Gramineae: 628 (1982). TAB. **38**, fig. 8. Type from Madagascar.
   Panicum atrofuscum Hack. in Scott Elliot, Journ. Linn. Soc. Bot. **29**: 63 (1891).
   Digitaria seminuda Stapf in F.T.A. **9**: 446 (1919). —Henrard, Monogr. Digit.: 675 (1950). —Clayton in F.W.T.A. ed. 2, **3**: 452 (1972). Type from Sierra Leone.
   Digitaria masambaensis Vanderyst, Bull. Agric. Congo Belge **16**: 660 (1925). —Robyns, Esp. Cong. Digit.: 44, t. 5 (1931); in Fl. Agrost. Congo Belge **2**: 47 (1934). —Henrard, Monogr. Digit.: 437 (1950). —Robyns & Van der Veken in Bull. Jard. Bot. Brux. **22**: 150 (1952). —Compére in Bull. Jard. Bot. Brux. **33**: 389 (1963). Type from Zaire.

A loosely caespitose annual in appearance or a short-living (?) perennial on a short, weak rhizome. Culms 60–100 cm., erect to ascending, glabrous, nodes dark and glabrous. Leaf sheaths glabrous. Ligule up to 2 mm. long, truncate, ciliolate. Leaf laminae 15–25 × 0.3–0.5 cm., linear, flat, glabrous and smooth on both surfaces, sometimes with a few bulbous-based bristles near the base, scaberulous along the margin. Inflorescence composed of (3)4–8 racemes, 9–20 cm. long, erect, solitary on a well developed common axis. Rhachis conspicuously winged, up to 1.0 mm. broad, smooth to slightly scaberulous, with scabrous margins. Pedicels 3-nate, 0.2–1.75 mm. long, subterete to subtriquetrous, scabrous, broadened at the apex. Spikelets 2.0–2.5 mm. long, lanceolate. Inferior glume absent to obsolete, truncate, hyaline. Superior glume $\frac{1}{3}$–$\frac{2}{3}$ of the spikelet, oblong triangular, 3-nerved, appressed hairy, hairs fine, sometimes capitellate, white-hyaline. Inferior lemma as long as the spikelet, oblong-lanceolate, with recurved margins, 5–7-nerved, nerves equidistant, appressed hairy. Superior lemma as long as the spikelet, oblong-lanceolate, slightly apiculate, dark brown.

Zambia. N: Chishinga Ranch, Lufubu dambo, 1600 m., 12.ix.1965, *Astle* 3352 (SRGH). Also in W. Africa, Zaire, Tanzania and Madagascar. Swamps and wet places. The single specimen from the Flora Zambesica area is somewhat atypical in having a rhachis only 0.6 mm. wide as opposed to up to 1.1 mm.

9. **Digitaria maitlandii** Stapf & C.E. Hubbard in Kew Bull. **1927**: 266 (1927). —Robyns, Esp. Cong. Digit.: 26 (1931); in Fl. Agrost. Congo Belge **2**: 26 (1934). —Eggeling, Annot. List Grass. Uganda: 14 (1947). —Henrard, Monogr. Digit.: 420 (1950). —Sturgeon in Rhod. Agric. Journ. **50**: 285 (1953). —Bogdan, Rev. List Kenya Grass.: 49 (1958). —Harker & Napper, Illustr. Guide Grass. Uganda: 26, t. 60 (1960). —Van der Veken in Bull. Jard. Bot. Brux. **32**: 127 (1962). —Napper, Grass. Tangan.: 76 (1965). —Simon in Kirkia **8**: 31 (1971). —Bennett in Kirkia **11**: 235 (1980). —Clayton in F.T.E.A., Gramineae: 629 (1982). TAB. **38**, fig. 9. Type from Uganda.
   Digitaria keniensis Pilg. in Notizbl. Bot. Gard. Berl. **10**: 267 (1928). —Henrard, Monogr. Digit.: 367 (1950). Type from Kenya.
   Digitaria apiculata Stent in Bothalia **3**: 155 (1930). —Henrard, Monogr. Digit.: 42 (1950). —Chippindall in Meredith, Grasses & Pastures of S. Afr.: 418 (1955). —Van der Veken in Bull. Jard. Bot. Brux. **32**: 127 (1962). Type from S. Africa.
   Digitaria maitlandii var. *glabra* Van der Veken in Bull. Jard. Bot. Brux. **32**: 127 (1962). Type from Zaire.
   Digitaria apiculata var. *hirta* P. Goetghebeur in Bull. Nat. Plantentuin Belg. **45**: 401 (1975). Type: Zimbabwe, Nyanga–Mutare Rd., 15.xi.1958, *West* 3801 (SRGH, holotype).

A tightly caespitose perennial, base surrounded by cataphylls and leaf sheath remnants. Culms 20–60 cm., erect, glabrous, nodes dark and glabrous. Leaf sheaths scaberulous and sometimes loosely hairy. Ligule 0.5 mm. long, truncate, shortly ciliate. Leaf laminae 5–15 × 0.2–0.4 cm., linear, flat or involute, densely papillose and loosely hairy on the superior surface, scaberulous and loosely to densely hairy on the inferior, smooth to scaberulous along the margins. Inflorescence composed of 3–8 racemes, 4–9 cm. long, paired or solitary along a short common axis. Rhachis triquetrous, narrowly winged, up to 0.4 mm. broad, smooth with scabrous margins. Pedicels 3-nate, 0.2–3 mm. long, subterete to subtriangular, scabrous, broadened at the apex, which bears no coronula. Spikelets 1.8–2.2 mm. long, oblong. Inferior glume absent or obsolete, rim-like.

Superior glume c. $\frac{1}{4}-\frac{1}{3}$ of the spikelet, ovate, (0)1–3-nerved, glabrous or with a few short, thick hairs, broadened and rounded at the apex. Inferior lemma somewhat shorter than to as long as the spikelet, 3–5(7)-nerved, oblong, hyaline, glabrous or with a few scattered hairs. Superior lemma as long as the spikelet, oblong, apiculate, dark red brown.

Zimbabwe. E: Glencoe For. Res., 1000 m., 27.xi.1967, *Simon & Ngoni* 1329 (BM; LISC; PRE; SRGH). **Malawi**. N: Lake Kaulime, 4.i.1959, *Robinson* 3049 (BM; PRE; SRGH). S: Mt. Mulanje For. Res., 1950 m., 27.xii.1981, *Chapman* 6080 (SRGH). **Mozambique**. Z: Gúruè, near sources of Malema R., 1700 m., 3.i.1968, *Torre & Correia* 16857 (LISC). MS: 7 km. Tsetserra-Chimoio, (Vila Pery) 1840 m., 12.xi.1965, *Torre & Pereira* 12923 (LISC).

Also in Kenya, Uganda, Rwanda, Burundi, Zaire, Tanzania and S. Africa. Mountain grassland.

10. **Digitaria compressa** Stapf in F.T.A. 9: 443 (1919). —Henrard, Monogr. Digit.: 144 (1950). —Sturgeon in Rhod. Agric. Journ. **50**: 288 (1953). —Simon in Kirkia **8**: 31, 65 (1971). —Bennett in Kirkia **11**: 234 (1980). —Clayton in F.T.E.A., Gramineae, 3: 628 (1982). TAB. **38**, fig. 10. Type: Zimbabwe, Insiza Distr., Matabeleland, 1912–1913, *Mundy* 2170 (K, holotype).

Digitaria capitipila Stapf in F.T.A. 9: 445 (1919). —Henrard, Monogr. Digit.: 113 (1950). —Jackson & Wiehe, Annot. Check-list Nyasal. Grass.: 35 (1958). Type: Malawi, 1891, *Buchanan* 1443 (K, holotype).

Digitaria buchananii Mez in Engl., Bot. Jahrb. Syst. 57: 192 (1921) nom. superfl. Type as for *Digitaria capitipila*.

Digitaria homblei Robyns, Esp. Cong. Digit.: 27, t. 4 (1931); in Fl. Agrost. Congo Belge **2**: 28 (1934). —Henrard, Monogr. Digit.: 328 (1950). —Simon in Kirkia **8**: 31, 65 (1971). —Clayton in F.W.T.A. ed. 2, **3**: 452 (1972). —Bennett in Kirkia **11**: 234 (1980). Type from Zaire.

Digitaria katangensis Robyns, Esp. Cong. Digit.: 33, t. 6 (1931); in Fl. Agrost. Congo Belge **2**: 36 (1934). —Henrard, Monogr. Digit.: 365 (1950). Type from Zaire.

Digitaria katangensis var. *hirta* P. Goetghebeur in Bull. Nat. Plantentuin Belg.: 401 (1975). Type: Zambia, SW. Mweru-Wantipa, Mukupa-Kipundu, 24.i.1938, *Brédo* 2331 (BR, holotype).

A tightly caespitose perennial, base surrounded by fibrous, old leaf sheath remnants. Culms 40–100 cm., erect, glabrous to sparsely hairy, nodes (pale or) dark, glabrous to upwards hairy. Leaf sheaths minutely scaberulous. Ligule c. 0.5 mm. long, rim-like, shortly ciliate. Leaf laminae 5–20 × 0.2-0.7 cm., linear, flat or somewhat involute, scaberulous on the superior surface and hairy near the ligule, scaberulous and loosely hairy on the inferior surface, scabrous on the margins. Inflorescence composed of (5)6–13 racemes, 5–18 cm. long, sometimes branched, erect, solitary along a short, robust common axis. Rhachis triquetrous, scarcely winged, up to 0.4 mm. broad, smooth with scabrous margins. Pedicels (2)3-nate, 0.5-4.5 mm. long, subtriangular, scabrous, broadened at the apex. Spikelets 2.6–3.5 mm. long, oblong-lanceolate. Inferior glume short, truncate, nerveless, hyaline. Superior glume $\frac{1}{3}-\frac{2}{3}$ ($\frac{4}{5}$) of the spikelet, ovate, (1)3-nerved, hyaline, glabrous or appressed hairy, hairs hyaline-white, slightly undulating, apex obpyriform or broadened and obtuse. Inferior lemma as long as the spikelet, oblong-lanceolate, with recurved margins, 7-nerved, nerves equidistant, hyaline, glabrous or appressed hairy. Superior lemma as long as the spikelet, oblong-lanceolate, acuminate, dark brown.

Zambia. N: Mweru-Wantipa, 24.i.1938, *Brédo* 2331 (BR). C: near Kapamba R., 800 m., 3.iii.1970, *Astle* 5796 (K; SRGH). E: Nsefu Game Reserve, ii.1958, *Stewart* 113 (BM; K; SRGH). S: Ngwezi Area, 21.i.1965, *Phiri* 131 (SRGH). **Zimbabwe**. N: 18 km. W. of Gokwe, Charama Rd. Fly Gate, 28.i.1964, *Bingham* 1110 (BM; K; LISC; P; PRE; SRGH). **Malawi**. N: Chombe Estate, Limpasa Dambo, 500 m., 21.ii.1961, *Vesey-FitzGerald* 3019 (SRGH). C: Kasungu Nat. Park, 1100 m., 23.xii.1970, *Hall-Martin* 1375 (SRGH). **Mozambique**. N: 55 km. on Imala to Mecuburi Rd., 4.xii.1967, Macêdo 2903 (LMA). MS: 27 km. Chimoio (Vila Pery) to Vila Gouveia, 750 m., 7.xii.1965, *Torre & Correia* 13395 (LISC).

Also in Nigeria, Ethiopia, Tanzania, Zaire. In degraded *Brachystegia* woodland, dambos and vlei verges.

Species 10–12 are closely related and are not always easy to identify. As with many perennial species of *Digitaria*, also here several geographically and morphologically ill-defined swarms can be distinguished, but the presence of many intermediates prevents us from a formal recognition. In these species (plus 13–14) more or less regularly a remarkable monstrosity can be observed: a browning and hardening of the inferior lemma, which comes to resemble the superior lemma.

11. **Digitaria setifolia** Stapf in F.C. **7**: 376 (1898). —Henrard, Monogr. Digit.: 682 (1950). — Chippindall in Meredith, Grasses & Pastures of S. Afr.: 416 (1955). —Simon in Kirkia **8**: 31 (1971). —Clayton in Kew Bull. **29**: 521 (1974). —Bennett in Kirkia **11**: 236 (1980). TAB. **38**, fig. 11. Type from S. Africa.

A densely caespitose perennial, base surrounded by glabrous, fibrous, often partly burnt, pale to dark brown, shining leaf sheath remnants. Culms 20–50 cm., erect, glabrous, nodes dark and subglabrous to bearded. Leaf sheaths glabrous to loosely hairy. Ligule c. 0.5 mm. long, truncate, ciliolate. Leaf laminae 5–15 × 0.2–0.3 cm., linear, involute, densely papillose to scaberulous on the superior surface, smooth on the inferior surface, scabrous along the margin. Inflorescence composed of 2–3(5) racemes, up to 20 cm. long, solitary along a short common axis, often branched at the base. Rhachis triquetrous, up to 0.4 mm. broad, smooth to slightly scaberulous, with scabrous margins. Pedicels 2–3-nate, 1–4 mm. long, triquetrous, scabrous, broadened at the apex. Spikelets (3.0)3.2–3.8 mm., oblong. Inferior glume very short to short, truncate, nerveless, glabrous, hyaline. Superior glume $\frac{2}{3}$–$\frac{4}{5}$ of the spikelet, oblong triangular, 3-nerved, appressed hairy, hairs fine, slightly undulating, often capitellate, reddish brown. Inferior lemma as long as the spikelet, oblong, 7-nerved, appressed hairy. Superior lemma as long as the spikelet, oblong acute to slightly acuminate, mostly very dark brown.

Zambia. W: Chingola, Kafue Catchment, 22.x.1962, *Vesey-FitzGerald* 3797 (BM; SRGH). **Zimbabwe**. E: Chimanimani (Melsetter), Pork Pie Hill, 18.xi.1950, *Crook* 287 (K; PRE; SRGH). **Mozambique**. MS: Chimanimani Mountains, Skeleton Pass, 1700 m., 27.ix.1966, *Simon* 877 (BM; K; P; SRGH).
Also in Angola, Zaire and S. Africa. In mountain grassland.
This species is closely related to *D. bovonei* Chiov. (= *D. nardifolia* Stapf), with smaller spikelets, and known only from Angola and adjacent Zaire. Contrary to the opinion of Clayton (F.T.E.A., Gramineae, **3**: 629 (1982)) all Flora Zambesiaca specimens proved to belong to *D. setifolia*.

12. **Digitaria phaeotricha** (Chiov.) Robyns, Esp. Cong. Digit.: 28, t. 4 (1931); in Fl. Agrost. Congo Belge **2**: 29 (1934). —Henrard, Monogr. Digit.: 554 (1950). —Simon in Kirkia **8**: 66 (1971). —Clayton in F.W.T.A. ed. 2, **3**: 452 (1972); in Kew Bull. **29**: 521 (1974). TAB. **38**, fig. 12. Type from Zaire.
    Digitaria parlatorei var. *phaeotricha* Chiov., Ann. Bot. Roma **13**: 41 (1914). —Chiovenda, Nuov. Giorn. Bot. Ital. **26**: 75 (1919). Type as above.
    Digitaria phaeotricha var. *paucipilosa* Ballard & C.E. Hubbard in Kew Bull. **1934**: 11 (1934). Type: Zambia, Mwinilunga Distr., Dobeka R. near Matonchi Farm, 1.ix.1930, *Milne-Redhead* 1003 (BR, isotype; K, holotype).

A tightly caespitose perennial, base surrounded by often densely hairy, pale greyish leaf sheath remnants. Culm 60–120 cm., erect, glabrous, nodes dark, (sub)glabrous. Leaf sheaths glabrous to loosely hairy, scaberulous. Ligule c. 1 mm. long, truncate, ciliate. Leaf laminae 10–30 × 0.3–0.7 cm., linear, flat or sometimes involute, papillose to scabrous on the superior surface, densely hairy on the inferior surface, scabrous and hairy along the margin. Inflorescence conspicuously linear, composed of (5)10–20 racemes, 1–3 cm. long (the terminal one often much longer, up to 10 cm., and well developed), appressed, solitary along a well developed common axis. Rhachis triaquetrous, not winged, c. 0.3 mm. broad, smooth with scabrous margins. Pedicels 2–3nate, 1–8 mm. long, terete to subtriangular, scabrous, broadened at the apex. Spikelets 3.5–4.3 mm. long, oblong. Inferior glume short, ovate, nerveless, hyaline. Superior glume c. $\frac{2}{3}$ of the spikelet, oblong triangular, 3(5)nerved, hyaline, appressed hairy, hairs long, slightly undulating, red brown, apex obpyriform. Inferior lemma as long as the spikelet, oblong with recurved margins, 7-nerved, hyaline, appressed hairy. Superior lemma as long as the spikelet, oblong, acute, dark brown.

Zambia. N: Muyumvya Dambo, 1800 m., 26.i.1961, *Vesey-FitzGerald* 2937 (SRGH). W: Mwinilunga Distr., 1.ix.1930, *Milne-Redhead* 1003 (BR). C: 7 km. W. of Serenje, 1500 m., 14.x.1967, *Simon & Williamson* 989 (BM; K; P; PRE; SRGH).
Also in Zaire. In dambos, seasonal bogs and vlei grassland.

13. **Digitaria hyalina** Robyns & Van der Veken in Bull. Jard. Bot. Brux. **22**: 150, fig. 13 (1952). —Goetghebeur & Van der Veken in Misc. Papers Landb. Wag. **19**: 147 (1980). TAB. **39**, fig. 13. Type from Zaire.

A robust, rhizomatous perennial, base surrounded by cataphylls and hairy leaf sheath remnants. Culms 40–100 cm., erect, glabrous or loosely appressed hairy, nodes dark, glabrous or loosely appressed hairy. Leaf sheaths loosely or densely hairy or glabrous. Ligule 1–1.5 mm. long, truncate, shortly ciliate. Leaf laminae 5–20 × 0.3–0.4 mm., linear, flat, densely scabrous and loosely hairy on the superior surface, loosely to densely hairy on the inferior surface, scabrous along the margins. Inflorescence composed of (2)3–4 racemes, (5)10–25 cm. long, erect, sometimes closely together, solitary on a short common axis. Rhachis triquetrous, narrowly winged, up to 0.5 mm. broad, undulating, smooth with scabrous margins. Pedicels 3–4-nate, 0.5–3 mm. long, triangular, somewhat broadened at the apex. Spikelets 3.0–4.0 mm. long, oblong-lanceolate. Inferior glume short, ovate, truncate, nerveless, hyaline. Superior glume c. $\frac{1}{2}\frac{2}{3}$ of the spikelet, oblong-triangular, 3(5)-nerved, hyaline, appressed hairy, hairs long, hyaline to rufous, capitate-apiculate at the apex. Inferior lemma as long as the spikelet, oblong-lanceolate with recurved margins, 7-nerved, the central and two pseudomarginal nerves well developed, appressed hairy, hyaline. Superior lemma as long as the spikelet, oblong-lanceolate, acute, dark brown.

Zambia. N: Ndundu, Small Dambo, 1900 m., 23.ii.1959, *McCallum-Webster* A139 (K; SRGH). W: Ndola, 27.xi.1947, *Brenan* s.n. (BR; K). C: Mungala embankment, x.1940, *Stohr* 378 (SRGH). Also in Zaire. In dambos, swamps and near waterfalls.

14. **Digitaria eylesii** C.E. Hubbard in Kew Bull. **1926**: 246, fig. 1 (1926). —Henrard, Monogr. Digit.: 241 (1950). —Sturgeon in Rhod. Agric. Journ. **50**: 285 (1953). —Chippindall in Meredith, Grasses & Pastures of S. Afr.: 419 (1955). —Launert in Merxm., Prodr. Fl. SW. Afr. **160**: 64 (1970). —Simon in Kirkia **8**: 31, 65 (1971). —Bennett in Kirkia **11**: 235 (1980). —Goetghebeur & Van der Veken in Misc. Papers Landb. Wag. **19**: 147 (1980). TAB. **39**, fig. 14. Type: Zimbabwe, Harare, 1530 m., 12.ii.1922, *Eyles* 3277 (K, holotype; SRGH, isotype).

   Digitaria eylesii var. *hirta* P. Goetghebeur in Bull. Nat. Plantentuin Belg. **45**: 401 (1975). Type: Zimbabwe, Lutope R., 32 km. from Gokwe, 18.iii.1963, *Bingham* 542 (SRGH, holotype; BM, isotype).

A rhizomatous perennial, base surrounded by cataphylls. Culms 40–60 cm., erect, glabrous, nodes dark, glabrous. Leaf sheaths glabrous. Ligule c. 0.5 mm. long, truncate, shortly ciliate. Leaf laminae 5–10 × 0.2–0.4 cm., linear, flat, superior surface glabrous or loosely hairy near the ligule, inferior surface glabrous, sometimes scaberulous, smooth on the margins. Inflorescence composed of 1–2(3) racemes, (8)10–20 cm. long, erect, solitary along a short common axis. Rhachis triquetrous, narrowly winged, up to 0.3 mm. broad, smooth with scabrous margins. Pedicels 3-nate, 0.5–2.5 mm. long, subtriangular, scaberulous near the broadened apex, bearing a poorly developed coronula. Spikelets 2.5–3.0 mm. long, oblong-lanceolate. Inferior glume very short, ovate, truncate, nerveless, hyaline. Superior glume c. $\frac{1}{2}$ of the spikelet, oblong-triangular, 3–5-nerved, glabrous or very loosely hairy, hairs white, capitate-apiculate. Inferior lemma as long as the spikelet, oblonglanceolate with recurved margins, 7-nerved, the central and two pseudomarginal nerves well developed, glabrous or very loosely hairy. Superior lemma as long as the spikelet, oblong-lanceolate, acute, dark brown.

Botswana. N: Xaudum drainage line, Okovango, 1100 m., 15.iii.1961, *Vesey-FitzGerald* 3244 (SRGH). **Zambia**. W: 7 km. S. of Lusongwa School, 1300 m., 26.xII.1969, *Simon & Williamson* 2053 (SRGH). S: Mazabuka, 4.xi.1960, *White* 6807 (BM; K). **Zimbabwe**. N: Sengwa R. floodplain, 24.i.1983, *Mahlangu* 659 (SRGH). C: Somabula Flats, 19.iii.1964, *West* 4794 (SRGH). S: Masvingo (Victoria) Distr., 30.iii.1973, *Chiparawasha* 668 (SRGH).
Also in Angola, Namibia and S. Africa. In wet vlei grassland, floodplains and dambos.

15. **Digitaria fuscopilosa** P. Goetghebeur in Misc. Papers Landb. Wag. **19**: 146, fig. 1 (1980). TAB. **39**, fig. 15. Type: Mozambique, Manica e Sofala, serra Zuira, Tsetserra, 1800 m., 3.iv.1966, *Torre & Correia* 15642 (LISC, holotype).

A very loosely tufted, rhizomatous perennial. Culms 40–60 cm., erect, glabrous, nodes dark and glabrous. Leaf sheaths loosely hairy. Ligule up to 1 mm. long, truncate. Leaf laminae 7–10 × 0.2–0.4 cm., linear, flat or mostly involute, scaberulous on both surfaces, scabrous along the margin. Inflorescence composed of 3–4 racemes, 4–8 cm. long, erect, solitary along a short common axis. Rhachis triquetrous, scarcely winged, c. 0.25 mm. broad, scaberulous with scabrous margins. Pedicels 3(5)-nate, 0.5–5 mm. long, subtriangular, subglabrous, broadened at the apex. Spikelets 2.3–2.5 mm. long, oblong. Inferior glume very small, truncate, nerveless, hyaline. Superior glume $\frac{1}{2}$ of the spikelet, triangular, 3(5)-nerved, appressed hairy, hairs slightly undulating, acute, dark purplish brown. Inferior lemma as long as the spikelet, oblong, 7-nerved, the central and submarginal nerves well developed, appressed hairy. Superior lemma as long as the spikelet, oblong, acuminate, dark purplish brown.

Mozambique. MS: 6 km. on Tsetserra to Chimoio (Vila Pery) Rd., 1800 m., 3.iv.1966, *Torre & Correia* 15642 (LISC).
Not known from elsewhere. Savanna grassland.

16. **Digitaria pulchra** Van der Veken in Bull. Nat. Plantentuin Belg. **45**: 400, fig. 3 (1975). TAB. **39**, fig. 16. Type from Zaire.

A loosely caespitose perennial on a short rhizome, base surrounded by hairy cataphylls and old leaf sheath remnants. Culms 30–70 cm., erect, slender, glabrous, nodes dark and hairy. Leaf sheaths glabrous or loosely hairy with bulbous based bristles. Ligule small, up to 0.7 mm. long, shortly ciliate. Leaf laminae 2–6 × 0.1–0.3 cm., linear, flat to involute, glabrous or with scattered bulbous based bristles along the minutely scaberulous margin. composed of (3)6–16 racemes, 4–8 cm. long, sometimes branched at their base, racemes and branches conspicuously peduncled, erect to appressed, solitary along a well developed common axis. Rhachis at the base subterete, subtriquetrous in the distal half, up to 0.3 mm. broad, glabrous and smooth, with an undulating, smooth margin. Pedicels 2-nate, 0.5–3.5 mm. long, terete to subtriangular, broadened at the apex, sometimes bearing a poorly developed coronula of adaxial cilia. Spikelets 2.2–2.6 mm. long, ovate oblong. Inferior glume very short, truncate to bifid. Superior glume $\frac{2}{3}$–$\frac{3}{4}$ of the spikelet, oblong-triangular, 3-nerved, appressed hairy, hairs fine, sometimes slightly undulating, acute, white to reddish brown. Inferior lemma as long as the spikelet, oblong, 5(7)-nerved, the two marginal nerves well developed, appressed hairy, hairs of two types: broad, stiff, white or reddish brown hairs concealing smaller, finer, white, strongly undulating (pseudoverrucose) hairs. Superior lemma as long as the spikelet, oblong, acute, dark brown.

Zambia. N: Chishinga Ranch, 30.i.1961, *Astle* 346 (K; SRGH).
Also in Zaire and Burundi. Dambo.
Although described as a member of section *Calvulae*, we prefer now to include this species in section *Flaccidulae*, notwithstanding the dark brown fruit and spurless spikelet. *Flaccidulae*-characters are the peculiar inflorescence structure, the long, subterete, dark yellow raceme peduncles, the swollen and glabrous base of the racemes, the asymmetrically widened pedicel apex. *D. pulchra* may actually be closing the gap between the two sections.

17. **Digitaria nitens** Rendle, Cat. Afr. Pl. Welw. **2**: 165 (1899). —Stapf in F.T.A. **9**: 473 (1919). —Henrard, Monogr. Digit.: 495 (1950). —Sturgeon in Rhod. Agric. Journ. **50**: 287 (1953). —Jackson & Wiehe, Annot. Check-list Nyasal. Grass.: 36 (1958). —Simon in Kirkia **8**: 31, 66 (1971). —Goetghebeur & Van der Veken in Misc. Papers Landb. Wag. **19**: 149 (1980). TAB **43**. Type from Angola.
    Digitaria elegans Stapf in F.T.A **9**: 474 (1919). —Chiovenda, Nuov. Giorn. Bot. Ital. **29**: 110 (1922). —Vanderyst in Bull. Agric. Congo Belge **16**: 660 (1925). —Robyns, Esp. Cong. Digit.: 46 (1931); in Fl. Agrost. Congo Belge **2**: 49 (1934). —Henrard, Monogr. Digit.: 214 (1950). —Napper, Grass. Tangan.: 75 (1965). Type from Zaire.
    Digitaria melinioides Mez in Engl., Bot. Jahrb. Syst. **57**: 191 (1921). Type: Malawi, *Buchanan* 244 (B, holotype).
    Digitaria stolzii Mez in Engl., Bot. Jahrb. Syst. **57**: 191 (1921). —Robyns in Fl. Agrost. Congo Belge **2**: 50 (1934). —Henrard, Monogr. Digit.: 711 (1950). Type: Malawi, *Stolz* 1824 (B, hololectotype; L, isolectotype).

A tightly caespitose perennial, bulbously thickened base surrounded by woolly cataphylls and old leaf sheath remnants. Culms 40–70 cm., erect, glabrous, nodes dark, glabrous. Leaf sheaths glabrous. Ligule c. 0.5 mm. long, truncate, shortly ciliate. Leaf laminae 5–10 × 0.3–0.4 cm., linear, flat, scaberulous on the superior surface, glabrous on the inferior surface, scaberulous on the margin. Inflorescence composed of (6)10–17 racemes, 2–7 cm. long (incl. pedicel), few-flowered, long pedicelled, sometimes branched, a few together or all solitary along a long, well developed common axis. Rhachis triquetrous, scarcely winged, up to 0.4 mm. broad, smooth with smooth to scabrous margins. Pedicels 2-nate, 1–3 mm. long, triangular to compressed, basally smooth, asymmetrically broadened at the apex, bearing a setulose coronula. Spikelets (2.5)2.8–3.2(4.1) mm. long, oblong. Inferior glume absent or obsolete. Superior glume c. $\frac{2}{3}$ of the spikelet, oblong-triangular, 3-nerved, the central nerve poorly developed, basally conspicuously spurred, appressed hairy, hairs white to purple, stiff, acute. Inferior lemma as long as the spikelet, oblong with recurved margins, 5-nerved, appressed hairy, densely on the inferior half, in 4 rows on the superior half. Superior lemma as long as the spikelet, oblong, acuminate, basally spurred, but not gibbous, pale yellowish green.

Zambia. N: Mungwi, 16.xii.1960, *Robinson* 4196 (K; SRGH). W: Kitwe, 3.ii.1958, *Fanshawe* 4241 (SRGH). C: Chakwenga Headwaters, 10.i.1964, *Robinson* 6163 (K; SRGH). S: 34 km. SE. of Choma, 1400 m., 17.xii.1956, *Robinson* 1777 (K; SRGH). **Zimbabwe**. N: Nyamayanetsi (Nyamnetsi) Estate, 3.i.1979, *Nyariri* 612 (GENT; PRE; SRGH). C: Chinamora Reserve, near Ngomakurira, 1700 m.,

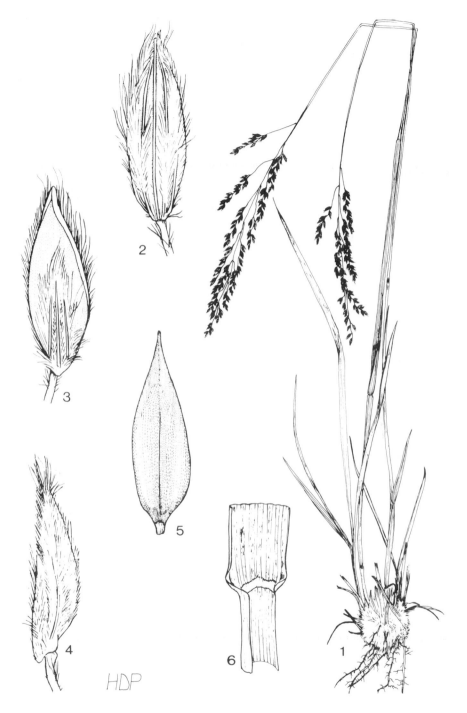

HDP

Tab. 43. DIGITARIA NITENS. 1, habit (×⅗), from *Torre & Correia* 13317; 2, spikelet in abaxial view (× 16); 3, spikelet in adaxial view (× 16); 4, spikelet in lateral view (× 16), 2–4 from *Torre & Correia* 16605; 5, fertile lemma (× 16); 6, ligule (× 8), 5–6 from *Torre & Correia* 13317.

5.i.1967, *Simon* 952 (BM; K; LISC; PRE; SRGH). E: Lookout Point, 1600 m., 13.xi.1975, *Crook* 2093 (GENT; SRGH). S: 2 km. from Masvingo (Fort Victoria) to Gienlivet, 30.i.1972, *Ellis* 829 (SRGH). **Malawi**. N: Nyika Plateau, 1600 m., 23.xii.1977, *Pawek* 13347 (SRGH). C: Mt. Dowa, 18.xii.1951, *Jackson* 712 (BR; K). S: Mulanje, 15.xi.1949, *Wiehe* 311 (K). **Mozambique**. Z: 67 km. Alto Molocue Gile, 400 m., 19.xii.1967, *Torre & Correia* 16605 (LISC). MS: Rotanda, 1850 m., 26.xi.1965, *Torre & Correia* 13317 (LISC).

Also in Zaire, Tanzania and Angola. Miombo woodland, often degraded, or on rocky outcrops and dambos.

Contrary to the opinion of Clayton (F.T.E.A., Gramineae: 638–639 (1982)) *D. nitens* is here kept separate from the S. African *D. flaccida* Stapf, with strictly not woolly hairy basal cataphylls.

18. **Digitaria poggeana** Mez in Engl., Bot. Jahrb. **57**: 191 (1921). —Robyns, Esp. Cong. Digit.: 52 (1931); in Fl. Agrost. Congo Belge **2**: 56 (1934). —Henrard, Monogr. Digit.: 578 (1950). —Scholz in Willdenowia **8**: 481 (1979). —Goetghebeur & Van der Veken in Misc. Papers Landb. Wag. **19**: 149, fig. 3 (1980). —Clayton in F.T.E.A., Gramineae: 39 (1982). TAB. **39**, fig. 18. Type from Zaire.

   Digitariopsis major Van der Veken in Bull. Jard. Bot. Brux. **25**: 330 (1955). —Van der Veken in Bull. Jard. Bot. Brux. **27**: 731 (1957). Type from Zaire.
   Digitaria major (Van der Veken) Clayton in Kew Bull. **29**: 523 (1974). Type as above.

A loosely caespitose or solitary growing, slender annual. Culms 40–120 cm., erect, glabrous, nodes dark and glabrous. Leaf sheaths glabrous. Ligule up to 0.5 mm. long, rim-like, minutely ciliolate. Leaf laminae 5–20 × 0.3–0.8 cm., linear, flat, glabrous and smooth on both surfaces, scabrous along the margin. Inflorescence composed of 20–100 racemes, 1–4 cm. long, erect to patent, 4–10 lowermost ones in a pseudowhorl, the other racemes solitary along a tall common axis, all racemes conspicuously peduncled. Rhachis at the base subterete and with long white hairs, in the distal half triquetrous, narrowly winged, up to 0.3 mm. broad, glabrous, smooth with conspicuously scabrous margins. Pedicels solitary or more rarely 2-nate, 0.5–1 mm. long, subtriangular, long ciliate, asymmetrically broadened at the apex. Spikelets 2.5–2.9 mm. long (without the spur), narrowly rhomboidal. Inferior glume absent. Superior glume $\frac{2}{3}$–$\frac{3}{4}$ of the spikelet, oblong triangular, 3-nerved, appressed hairy, hairs long, stiff, acute, dark purplish, basal spur acute, 0.2–0.4 mm. long. Inferior lemma as long as the spikelet, narrowly rhomboidal, with recurved margins, 5-nerved, appressed hairy, hairs slightly overtopping the spikelet. Superior lemma somewhat shorter than the spikelet, narrowly rhomboidal, acute, pale yellow to pale brown.

Zambia. N: 4 km. W. of Mukunsa, c. 50 km. W. of Mporokoso, 1400 m., 13.iv.1961, *Phipps & Vesey-FitzGerald* 3146 (BM; K; PRE; SRGH).

Also in Zaire and Tanzania. Miombo woodland.

19. **Digitaria redheadii** (C.E. Hubbard) Clayton in Kew Bull. **29**: 523 (1974). —Goetghebeur & Van der Veken in Misc. Papers Landb. Wag. **19**: 149 (1980). TAB. **39**, fig. 19. Type: Zambia, Mwinilunga Distr., near source of Matonchi R., 16.ii.1938, *Milne-Redhead* 4591 (K, holotype; BM, isotype).

   Digitariopsis redheadii C.E. Hubbard, Ic. Pl. 35, t. 3420 (1940). —Chippindall, Journ. S. Afr. Bot. **11**: 132 (1945). —Van der Veken in Bull. Jard. Bot. Brux. **25**: 328 (1955); loc. cit. **27**: 731 (1957). —Jacques-Felix, Gram. Afr. Trop. **1**: 234, fig. 160 (1962). Type as above.

A loosely caespitose or solitary growing, slender annual. Culms 10–60 cm., erect, glabrous, nodes dark, glabrous. Leaf sheaths glabrous. Ligule 0.5–3 mm. long, truncate, erose. Leaf laminae 1–7 × 0.1–0.4 cm., linear, flat to involute, glabrous on both surfaces, nearly smooth along the margin, often dark reddish coloured. Inflorescence composed of up to 30 racemes, 1–3 cm. long, erect to patent, solitary along the tall common axis, distinctly peduncled. Rhachis subterete near the base, triquetrous and winged, up to 0.4 mm. broad in the upper part, smooth with smooth or slightly scaberulous margins. Pedicels solitary, 1–1.5 mm. long, subtriangular, with long cilia in the upper part, asymmetrically broadened at the apex. Spikelets 2.3–2.6 mm. long (excl. basal spur), rhomboidal-oblong. Inferior glume absent. Superior glume c. $\frac{3}{4}$ of the spikelet, oblong triangular, 3-nerved, appressed hairy, hairs stiff, acute, often purplish tinged, the basal, acute spur 0.6–0.8 mm. long. Inferior lemma as long as the spikelet, rhomboidal-oblong, with recurved margins, 5-nerved, the central and 2 marginal nerves well developed, appressed hairy, hairs slightly overtopping the spikelet. Superior lemma somewhat smaller than the spikelet, rhomboidal-oblong, slightly apiculate, pale yellow.

Zambia. B: Mongu, 20.ii.1966, *Robinson* 6855 (K; SRGH). N: Chishinga Ranch, 21.ii.1964, *Phiri* 82 (SRGH). W: 7 km. E. of Kabompo R., 1400 m., 19.xii.1969, *Simon & Williamson* 1869 (K; SRGH). Also in Zaire. Roadsides, miombo woodland.

20. **Digitaria brazzae** (Franch.) Stapf in F.T.A. **9**: 447 (1919). —Stent in Bothalia **1**: 269 (1924). —Robyns, Esp. Cong. Digit.: 28 (1931); in Fl. Agrost. Congo Belge **2**: 30 (1934). —Henrard, Monogr. Digit.: 87 (1950). —Sturgeon in Rhod. Agric. Journ. **50**: 285 (1953). —Chippindall in Meredith, Grasses & Pastures of S. Afr.: 403 (1955). —Jackson & Wiehe, Annot. Check-list Nyasal. Grass.: 35 (1958). —Napper, Grass. Tangan.: 75 (1965). —Launert in Merxm., Prodr. Fl. SW. Afr. **160**: 63 (1970). —Simon in Kirkia **8**: 30, 65 (1971). —Bennett in Kirkia **11**: 234 (1980). —Clayton in F.T.E.A., Gramineae: 627 (1982). TAB. **39**, fig. 20. Type from Congo.
 Panicum brazzae Franch. in Bull. Soc. Hist. Nat. Autun **8**: 355 (1895). Type as above.
 Digitaria moninensis Rendle, Cat. Afr. Pl. Welw. **2**: 164 (1899). Type from Angola.
 Digitaria sulcigluma Chiov. in Ann. Bot. Roma **13**: 41 (1914). —Chiovenda, Nuov. Giorn. Bot. Ital. **26**: 75 (1919). Type from Zaire.
 Digitaria lomanensis Mez in Engl., Bot. Jahrb. **57**: 192 (1921). Type from Zaire.

A tightly caespitose perennial, base surrounded by more or less woolly cataphylls and dark fibrous leaf sheath remnants. Culms 60–120 cm., erect, glabrous, nodes dark and hairy. Leaf sheaths minutely scaberulous, loosely to densely hairy, sometimes glabrous. Ligule 0.5–1 mm. long, rim-like, shortly ciliate. Leaf laminae 6–9 × 0.2–0.4 cm., linear, flat, scaberulous and loosely hairy on the superior surface, mostly densely hairy on the inferior surface, scabrous on the margins. Inflorescence composed of (2) 3 racemes, erect, dispersed along a short common axis. Rhachis triquetrous, broadly winged and up to 1 mm. wide, scaberulous, with scabrous margins. Pedicels 3–4-nate, 0.5–3.5 mm. long, subterete to subtriangular, minutely scaberulous, asymmetrically broadened at the apex. Spikelets 2.9–3.2 mm., oblong, basally adaxially gibbous. Inferior glume up to 0.8 mm. long, subcircular to truncate or slightly bilobed, hyaline. Superior glume c. $\frac{2}{3}$–$\frac{4}{5}$ of the spikelet, narrowly triangular, 3-nerved, appressed hairy, hairs white to purplish, smooth and acute. Inferior lemma as long as the spikelet, oblong with recurved margins, centrally deeply sulcate, margins inflated, 5-nerved, the central and 2 marginal nerves well developed, hairs in two basal bundles, basally swollen, smooth, acute, hyaline to purplish, also part of the margins with this type of hairs, although somewhat more slender, overtopping the spikelet for c. 0.5–1 mm. Superior lemma somewhat shorter than the spikelet, oblong, acuminate, basally gibbous and rather sharply keeled, pale yellow to very pale brown.

 Botswana. N: Kwando R. Plain, 23.i.1978, *Smith* 2228 (GENT; SRGH). **Zambia**. B: Mongu, 29.iii.1964, *Verboom* 1348 (K; SRGH). N: Luapula Escarpment, xii.1968, *Williamson* 1159A (K; SRGH). W: Luakela R., 21 km. N. of Mwinilunga, 1500 m., 23.xii.1969, *Simon & Williamson* 1964 (K; P; SRGH). C: Fiwila Mission, 1300 m., 11.i.1958, *Robinson* 2737 (P; PRE; SRGH). S: Kalomo Distr., 17.xii.1963, *Mitchell* 24–69 (BM; LISC; SRGH). **Zimbabwe**. N: Nyamayanetsi (Nyamnyetsi) Estate, 2.i.1979, *Nyariri* 609 (PRE; SRGH). W: Matopos Distr., iv.1949, *Plowes* 1126 (BR; K; SRGH). C: Marondera (Marandellas) Distr., Digglefold, 24.xii.1948, *Corby* 309 (K; SRGH). E: Nyanga (Inyanga), 12.i.1931, *Norlindh & Weimarck* 4207 (BM; K; PRE). S: Kyle Game Reserve, 18.i.1965, *Blizard* 89 (SRGH). **Malawi**. C: Matundu Estate, 5.i.1950, *Wiehe* 398 (K; PRE; SRGH). **Mozambique**. T: Angonia, Mpasadzi R., 1300 m., 10.xii.1980, *Stefanesco & Nyongani* 592 (SRGH).
 From Congo, Zaire and Tanzania, south to Namibia and S. Africa. Disturbed sites in miombo woodland, floodplain grassland and dambo edges.
 Two closely related species are known from limitrophic regions, *D. pellita* Stapf from Angola, and *D. tricholaenoides* Stapf from S. Africa. Both may occur within the Flora Zambesiaca area. *D. pellita* has long persistent, leathery leaves, conspicuously scabrid to ciliate pedicel tips, spikelets 3.1–3.4 mm. long, and a flat (not sulcate) inferior lemma. In *D. tricholaenoides* the spikelets are longer than 3.5 mm., the inferior glume is up to 1.5 mm. long, the inferior lemma is beset with thick, stiff hairs between the central and subcentral nerves.

21. **Digitaria procurrens** P. Goetghebeur in Bull. Nat. Plantentuin Belg. **45**: 409, fig. 11 (1975). TAB. **39**, fig. 21. Type: Zambia, Mporokoso, near Muzombwe, western side of Mweru-Wantipa, 15.iv.1961, *Phipps & Vesey-FitzGerald* 3212 (BR, holotype; BM; K; SRGH, isotypes).

A creeping perennial(?), stoloniferous, mat-forming. Culms 20–30 cm., erect-ascending, from a creeping base, glabrous, nodes dark and glabrous. Leaf sheaths glabrous. Ligule 0.5–1 mm. long, rim-like, entire. Leaf laminae 2–3 × 0.4–0.7 cm., oblong-lanceolate, flat, glabrous on both surfaces, scabrous along the crisped margin. Inflorescence composed of 2, 7–11 cm. long, erect to patent, one shortly, the other long pedicellate. Rhachis triquetrous, winged, up to 0.5 mm. broad, smooth with scabrous margins. Pedicels 3-nate, 0.5–3.5 mm. long, subtriangular, smooth or minutely scaberulous, asymmetrically broadened at the apex. Spikelets 2.9–3.0 mm. long, oblong, basally adaxially gibbous. Inferior glume short, truncate, hyaline. Superior glume $\frac{2-4}{3\ 5}$ of the spikelet, oblong triangular, 3-nerved, appressed hairy, hairs fine, stiff, acute, white to purplish. Inferior lemma as long as the spikelet, oblong with recurved margins, centrally not sulcate, 5-nerved, the central nerve poorly developed, appressed hairy, hairs along the subcentral nerves long, stiff, acute, swollen at the base, often purplish, marginal hairs more slender, hairs overtopping the spikelet by less than 0.5 mm. Superior lemma as long as the spikelet, oblong, acuminate, basally gibbous, sharply keeled, pale yellow.

Zambia. N: Muzombwe, W. side of Mweru-Wantipa, 1200 m., 15.iv.1961, *Phipps & Vesey-FitzGerald* 3212 (BM; BR; K; PRE; SRGH).
Known from the type collection only. Roadside in mateshi thicket.

22. **Digitaria gayana** (Kunth) Stapf ex A. Chev., Sudania: 163 (1911). —Stapf in F.T.A. **9**: 449 (1919). —Vanderyst in Bull. Agric. Congo Belge **16**: 657 (1925). —Robyns, Esp. Cong. Digit.: 29 (1931); in Fl. Agrost. Congo Belge **2**: 32 (1934). —Eggeling, Annot. List Grass. Uganda: 14 (1947). —Henrard, Monogr. Digit.: 278 (1950). —Sturgeon in Rhod. Agric. Journ. **50**: 285 (1953). —Andrews, Fl. Pl. Sudan **3**: 435 (1956). —Bogdan, Rev. List Kenya Grass.: 49 (1958). —Harker & Napper, Illustr. Guide Grass. Uganda: 25, t. 56 (1960). —Napper, Grass. Tangan.: 75 (1965). —Launert in Merxm., Prodr. Fl. SW. Afr. **160**: 64 (1970). —Simon in Kirkia **8**: 31, 65 (1971). —Clayton in F.W.T.A. ed. 2, **3**: 453 (1972). —Rose Innes, Man. Ghana Grass.: 137 (1977). —Bennett in Kirkia **11**: 235 (1980). —Clayton in F.T.E.A., Gramineae: 627 (1982). TAB. **44**. Type from Senegal.
Panicum gayanum Kunth, Rev. Gram. **1**: 239, t. 31 (1830). —Durand & Schinz in Consp. Fl. Afr. **5**: 750 (1894). Type as above.
Panicum didymostachyum Steud., Syn. Pl. Glum. **1**: 97 (1854). Type from Senegal.
Digitaria elegantula Mez in Engl., Bot. Jahrb. **57**: 192 (1921). Type from E. Africa.

A loosely caespitose or solitary growing annual. Culms 15–40 cm., erect, glabrous, nodes dark and glabrous. Leaf sheaths scaberulous, loosely to densely hairy. Ligule 0.5–1 mm. long, rim-like, ciliate. Leaf laminae 3–7 × 0.4–0.8 cm., linear, flat, scaberulous on both surfaces, margins scaberulous, sometimes with scattered bulbous based bristles. Inflorescence composed of 2(3) racemes, 6–13 cm. long, erect, one subsessile, the other(s) pedicellate. Rhachis triquetrous, broadly winged, up to 1.4 mm. broad, smooth to minutely scaberulous, with scabrous margins. Pedicels 3–4-nate, 0.5–4.0 mm. long, subterete, smooth or scaberulous, asymmetrically broadened at the apex. Spikelets 2.5–2.7 mm. long, oblong, basally adaxially gibbous. Inferior glume short, hyaline, soon withering. Superior glume somewhat shorter than the spikelet, narrowly triangular, 3-nerved, appressed hairy, hairs white to purplish, smooth, acute. Inferior lemma as long as the spikelet, oblong with recurved margins, centrally deeply sulcate, margins inflated, 5-nerved, the central nerve poorly developed, hairs in two basal bundles and along the inflated margins with a loosely papillose wall and swollen base, the other, marginal hairs somewhat slenderer, overtopping the spikelet for 2–3 mm. Superior lemma somewhat shorter than the spikelet, oblong, acuminate, basally gibbous, keeled, pale yellow to very pale brown.

Caprivi Strip. 35 km. W. of Katima Mulilo, 300–500 m., 17.ii.1969, *de Winter* 9203 (SRGH). **Botswana**. N: Movombe Village, 15.ii.1983, *Smith* 4060 (GENT; SRGH). **Zambia**. B: Mongu, 6.ii.1966, *Robinson* 6833 (K; SRGH). N: Chambeshi (old) Pontoon, 27.ii.1960, *Vesey-FitzGerald* 2684 (SRGH). W: Zambezi Rapids, 6 km. N. of Kalene Hill Mission, 20.ii.1975, *Williamson & Gassner* 2451 (SRGH). C: Serenje, 1700 m., 5.iv.1961, *Phipps & Vesey-FitzGerald* 2951 (BM; SRGH). E: Mfuwe Air Strip, 700 m., 1.iii.1966, *Astle* 4574 (K; SRGH). S: Kalomo, 1.iii.1963, *Astle* 2220 (K; SRGH). **Zimbabwe**. N: Gokwe, 20.ii.1984, *Mahlangu* 908 (SRGH). W: Gwampa For. Res., ii.1956, *Goldsmith* in GHS 48897 (K; PRE; SRGH). C: Mlezu School Farm, 24 km. SSE. of Kwe Kwe (Que Que), 1400 m., 24.iii.1966, *Biegel* 1031 (BM; K; P; PRE; SRGH). S: Zimutu Reserve, 27.i.1948, *Robinson* 217 (SRGH).
From Senegal to Sudan, south to Namibia and Zimbabwe. In disturbed miombo woodland, weed of fallow land and roadsides, on sandy soil.

23. **Digitaria monodactyla** (Nees) Stapf in F.C. **7**: 373 (1898). —Stent in Bothalia **1**: 266 (1921). —Henrard, Monogr. Digit.: 468 (1950). —Chippindall in Meredith, Grasses & Pastures of S. Afr.: 415, fig. 347 (1955). —Van der Veken in Bull. Jard. Bot. Brux. **32**: 125 (1962). —Launert in Merxm., Prodr. Fl. SW. Afr. **160**: 65 (1970). —Simon in Kirkia **8**: 31, 65 (1971). —Clayton in Kew Bull. **29**: 523 (1974). —Bennett in Kirkia **11**: 235 (1980). —Goetghebeur & Van der Veken in Misc. Papers Landb. Wag. **19**: 150 (1980). —Clayton in F.T.E.A., Gramineae: 640 (1982). TAB. **39**, fig. 23. Type from S. Africa.

Tab. 44. DIGITARIA GAYANA. 1, habit (×⅖), from *Robinson* 6833; 2, spikelet in abaxial view (× 12);
3, spikelet in adaxial view (× 12); 4, spikelet in lateral view (× 12), 2–4 from *McCallum-Webster* A69;
5, fertile lemma in adaxial view (× 16); 6, fertile lemma in lateral view (× 16), 5–6 from *Robinson*
6833.

Panicum monodactylum Nees, Fl. Afr. Austr. **1**: 21 (1841). —Steudel, Syn. Pl. Glum. **1**: 56 (1853).
—Durand & Schinz in Consp. Fl. Afr. **5**: 755 (1894). Type as above.
Digitaria monodactyla var. *explicata* Stapf in F.T.A. **9**: 442 (1919). —Stent in Bothalia **1**: 266 (1924). —Sturgeon in Rhod. Agric. Journ. **50**: 284 (1953). Type from Angola.

A tightly caespitose perennial on a very short rhizome, base surrounded by cataphylls and old leaf sheath remnants. Culm 20–70 cm., erect, glabrous, nodes dark, glabrous. Leaf sheaths glabrous to loosely hairy. Ligule up to 0.5 mm. long, truncate, ciliate. Leaf laminae 2–5 × 0.2–0.3 cm., linear, flat, scaberulous and loosely hairy on the superior surface, loosely hairy on the inferior surface, smooth to scaberulous along the margin. Inflorescence of a single raceme, 4–19 cm. long, erect, terminal. Rhachis triquetrous, broadly winged, up to 0.8 mm. broad, conspicuously undulating and wings recurved, smooth to scaberulous, margin thickened, ciliate. Pedicels 2-nate, 0.5–2.5 mm. long, triangular, densely scabrous on the superior part, broadened at the apex, bearing a coronula. Spikelets 2.7–3 mm. long, oblong-lanceolate, basally adaxially gibbous. Inferior glume absent or obsolete. Superior glume c. ½ of the spikelet, oblong-triangular, 3-nerved, the central nerve poorly developed, appressed hairy, hairs white or rufous, stiff, acute, overtopping the superior glume. Inferior lemma as long as the spikelet, oblong-lanceolate, with recurved margins, 5-nerved, appressed hairy, hairs slightly overtopping the spikelet. Superior lemma as long as the spikelet, oblong-lanceolate, acute, basally gibbous, sharply keeled, pale brown.

Zambia. B: Mongu, 20.i.1966, *Robinson* 6818 (K; SRGH). N: 40 km. SE. of Mporokoso, 1600 m., 20.xii.1967, *Simon & Williamson* 1471 (BM; K; SRGH). W: 33 km. E. of Mwinilunga, 1500 m., 20.xii.1969, *Simon & Williamson* 1888 (BM; K; P; PRE; SRGH). C: Chakwenga Headwaters, 10.i.1964, *Robinson* 6203 (K; SRGH). S: Ngwezi Paddocks, 30.i.1964, *Astle* 2909 (K; SRGH). **Zimbabwe**. N: Mangwende Reserve, 14.i.1958, *Cleghorn* 359 (K; SRGH). W: Matopos Res. Stat., 1500 m., 4.iii.1954, *Rattray* 1674 (K; SRGH). C: Macheke, 1300 m., xii.1919, *Eyles* 2001 (BM; PRE; SRGH). E: "Sawerombi", 10 km. N. of Chimanimani (Melsetter), 16.i.1955, *Crook* 560 (K; P; PRE; SRGH). S: Ndanga Distr., 19.i.1960, *Goodier* 845 (P; SRGH).
Also in Tanzania, Zaire, Angola, Namibia and S. Africa. In dambos, wet grassland, vlei areas in open grassland on sandveld.

24. **Digitaria ventriosa** Van der Veken in Bull. Jard. Bot. Brux. **32**: 123 (1962). —Clayton in Kew Bull. **29**: 524 (1974). —Goetghebeur & Van der Veken in Misc. Papers Landb. Wag. **19**: 151 (1980). TAB. **39**, fig. 24. Type from Zaire.

A loosely caespitose or solitary growing annual. Culms 10–30 cm., glabrous, nodes dark, glabrous or loosely hairy. Leaf sheaths glabrous to loosely hairy. Ligule c. 0.5 mm. long, truncate, shortly ciliate. Leaf laminae 3–4 × 0.2–0.3 cm., linear, flat to somewhat involute, glabrous to loosely hairy on both surfaces, scaberulous along the margin. Inflorescence of a single raceme, 2–5 cm. long, erect, terminal. Rhachis triquetrous, broadly winged, up to 1 mm. broad, conspicuously undulating and wings recurved, smooth, margin thickened, ciliate. Pedicels 2-nate, 0.5–2.5 mm. long, triangular, densely scabrous on the superior part, broadened at the apex, bearing a coronula. Spikelets 2.0–2.2 mm. long, oblong-lanceolate, basally adaxially gibbous. Inferior glume absent or obsolete. Superior glume c. ½ of the spikelet, oblong-triangular, 3-nerved, the central nerve poorly developed, appressed hairy, hairs white, stiff, acute, overtopping the superior glume. Inferior lemma as long as the spikelet, oblong-lanceolate, with recurved margins, 5-nerved, appressed hairy, hairs slightly overtopping the spikelet. Superior lemma as long as the spikelet, oblong-lanceolate, acute, basally gibbous, sharply keeled, pale brown.

Zambia. N: Mbala Distr., St. Paul Rd., 22.ii.1968, *Sanane* 55 (K). W: Mwinilunga Distr., c. 6.5 km. SW. of Matonchi Farm, between Kamakonde and Kamulende Rivers, 17.ii.1938, *Milne-Redhead* 4621 (K).
Also in Zaire. On shallow soil overlying laterite, wet open sand and roadsides.

25. **Digitaria complanata** P. Goetghebeur in Bull. Nat. Plantentuin Belg. **45**: 411, fig. 12 (1975). —Goetghebeur & Van der Veken in Misc. Papers Landb. Wag. **19**: 152 (1980). TAB. **39**, fig. 25. Type: Zambia, Kasama, Malole, 2.ii.1961, *Robinson* 4332 (M, holotype; K; SRGH, isotypes).

A loosely caespitose or solitary growing annual. Culms 10–20 cm., erect, nodes dark and glabrous. Leaf sheaths glabrous. Ligule c. 1 mm. long, rimlike, shortly ciliate. Leaf laminae 2–5 × 0.2–0.4 mm., linear, flat, glabrous on both surfaces, with smooth margins. Inflorescence composed of 1–2 racemes, 2–4 cm. long, often patent, both shortly peduncled. Rhachis triquetrous, asymmetrically winged, undulating, up to 0.8 mm. broad, smooth with smooth margins. Pedicels 2-nate, solitary in the superior part, 0.5–1.5 mm. long, triangular, the superior part long-ciliate, broadened at the apex, which bears a coronula. Spikelets 1.9–2.1 mm. long, oblong-lanceolate, basally adaxially gibbous. Inferior glume absent. Superior glume c. $\frac{1}{2}$–$\frac{2}{3}$ of the spikelet, oblong-triangular, obsoletely 3-nerved, appressed hairy, hairs white, stiff, acute, overtopping the superior glume. Inferior lemma as long as the spikelet, oblong-lanceolate with recurved margins, 5-nerved, subcentral nerves poorly developed, appressed hairy, hairs scarcely overtopping the spikelet. Superior lemma somewhat shorter than the spikelet, oblong-lanceolate, acuminate, basally gibbous, sharply keeled, pale yellow or sometimes purplish tinged.

Zambia. N: Malole, 2.ii.1961, *Robinson* 4332 (K; M; PRE; SRGH). Also in Zaire. Dambos.

26. **Digitaria bidactyla** Van der Veken in Bull. Nat. Plantentuin Belg. **45**: 413, fig. 13 (1975). —Goetghebeur & Van der Veken in Misc. Papers Landb. Wag. **19**: 152 (1980). TAB. **39**, fig. 26. Type: Zambia, Luwingu, Chishinga Ranch, 1700 m., 29.i.1961, *Astle* 312 (K, isotype; SRGH, holotype).

A loosely caespitose or solitary growing, slender annual. Culms 10–20 cm., erect, glabrous, nodes dark and glabrous. Leaf sheaths glabrous. Ligule c. 1 mm. long, rim-like, ciliate. Leaf laminae 3–5 × 0.2–0.4 cm., linear, flat, glabrous on both surfaces, with smooth margins. Inlflorescence composed of 1–2 racemes, 2–4 cm. long, often patent, both shortly peduncled. Rhachis triquetrous, asymmetrically winged, undulating, up to 0.5 mm. wide, smooth with smooth and not thickened margins. Pedicels solitary (sometimes a reduced subsessile spikelet present), 0.5–1 mm. long, triangular, smooth, broadened at the apex, which bears a coronula. Spikelets 2.3–2.5 mm. long, lanceolate, basally adaxially gibbous. Inferior glume absent. Superior glume c. $\frac{2}{3}$ of the spikelet, oblong-triangular, 3-nerved, appressed hairy, hairs white, stiff, acute, overtopping the superior glume. Inferior lemma as long as the spikelet, lanceolate with recurved margins, 5-nerved, appressed hairy, hairs overtopping the spikelet for c. 1 mm. Superior lemma conspicuously shorter than the spikelet, lanceolate, slightly acuminate, basally gibbous, somewhat keeled, pale yellow or sometimes purplish tinged.

Zambia. N: Chishinga Ranch, 1600 m., 17.i.1963, *Astle* 1966 (BM; K; SRGH). Not known from elsewhere. In dambos and on wet sand.

27. **Digitaria calcarata** Clayton in Kew Bull. **29**: 524 (1974). —Goetghebeur & Van der Veken in Misc. Papers Landb. Wag. **19**: 152, fig. 3 (1980). TAB. **39**, fig. 27. Type: Zambia, Kasama, 19.ii.1961, *Robinson* 4411 (K, holotype).

A fasciculate or solitary growing, slender annual. Culm 10–25 cm., erect, glabrous, nodes dark and hairy. Leaf sheaths scaberulous. Ligule up to 2 mm. long, obtuse, erose and ciliolate. Leaf laminae 4–8 × 0.2–0.4 cm., linear, flat, glabrous and scaberulous on both surfaces, scaberulous along the margins. Inflorescence composed of 2 racemes, 3–6 cm. long, one lateral and subsessile, the other terminal, erect and pedunculate. Rhachis triquetrous, broadly winged, up to 0.75 mm. broad, smooth with ciliolate margins. Pedicels 2-nate, rarely a few near the top solitary, 0.5–3 mm. long, triangular, scabrous, scarcely broadened at the apex. Spikelets 2.5–2.8 mm. long, narrowly rhomboidal. Inferior glume absent. Superior glume c. $\frac{2}{3}$ of the spikelet, oblong triangular with a basal spur, 3-nerved, nerves poorly developed, appressed hairy, hairs low, white to pale rufous, stiff, acute. Inferior lemma as long as the spikelet, oblong triangular to narrowly rhomboidal, with recurved margins, 5(7)-nerved, the central and subcentral nerves prominent and well developed. Superior lemma slightly smaller than the spikelet, lanceolate, slightly acuminate, pale brown (unripe?).

Zambia. N: Zungu, iv.1961, *Verboom* G338 (PRE; SRGH). Not known from elsewhere. On shallow soil overlying rocks.

28. **Digitaria sacculata** Clayton in Kew Bull. **29**: 525 (1974). —Goetghebeur & Van der Veken in Misc. Papers Landb. Wag. **19**: 152, fig. 4 (1980). TAB. **40**, fig. 28. Type: Zambia, Kasama Distr., 4.ii.1961, *Robinson* 4352 (K, holotype; M; SRGH, isotypes).

A loosely caespitose or solitary growing, slender annual. Culms 10–25 cm., erect, glabrous, nodes dark and glabrous. Leaf sheaths glabrous. Ligule c. 1 mm. long, truncate. Leaf laminae 2–5 × 0.2–0.4 cm., lanceolate to linear, flat, glabrous and smooth to scaberulous on both surfaces, scaberulous along the crisped margin, basal leaves bright red when fresh, dark reddish purple when dry. Inflorescence composed of 2 racemes, 3–5 cm. long, both racemes equally long peduncled. Rhachis triquetrous, scarcely winged, c. 0.2 mm. broad, smooth with smooth margins. Pedicels 2-nate at the base, solitary in the superior half, 0.5–2.5 mm. long, triangular to compressed, with long cilia in the superior half, asymmetrically broadened at the apex, bearing a well developed coronula. Spikelets 3–3.5 mm. long (excl. overtopping hairs and basal spur), lanceolate. Inferior glume absent. Superior glume c.½ of the spikelet, oblong triangular, with a basal "spur", 3-nerved, appressed hairy, hairs long, white to purplish, stiff, acute, overtopping the glume. Inferior lemma as long as the spikelet, lanceolate, 5-nerved, midnerve prominent and well developed, densely appressed hairy, hairs overtopping the spikelet by 0.5-1 mm. Superior lemma c. ¾ of the spikelet, lanceolate, slightly acuminate, with a basal spur, pale brown (unripe?).

Zambia. N: 10 km. E. of Kasama, 4.ii.1961, *Robinson* 4352 (K; M; PRE; SRGH).
Known from the type collection only. On damp sand.

29. **Digitaria debilis** (Desf.) Willd., Enum. Pl. Hort. Reg. Bot. Berol.: 91 (1809). —Stapf in F.C. **7**: 377 (1898). —Rendle, Cat. Afr. Pl. Welw. **2**: 163 (1899). —Stapf in F.T.A. **9**: 464 (1919). —Stent in Bothalia **1**: 269 (1924). —Robyns, Esp. Cong. Digit.: 41 (1931); in Fl. Agrost. Congo Belge **2**: 43 (1934). —Henrard, Monogr. Digit.: 165 (1950). —Maire in Fl. Afr. Nord **1**: 302, fig. 169 (1952). —Sturgeon in Rhod. Agric. Journ. **50**: 286 (1953). —Chippindall in Meredith, Grasses & Pastures of S. Afr.: 396 (1955). —Andrews, Fl. Pl. Sudan **3**: 433 (1956). —Jackson & Wiehe, Annot. Check-list Nyasal. Grass.: 35 (1958). —Van der Veken in Bull. Jard. Bot. Brux. **32**: 125 (1962). —Napper, Grass. Tangan.: 77 (1965). —Launert in Merxm., Prodr. Fl. SW. Afr. **160**: 63 (1970). —Simon in Kirkia **8**: 31, 65 (1971). —Clayton in F.W.T.A. ed. 2, **3**: 454 (1972); in Kew Bull. **29**: 520 (1974). —Rose Innes, Mann. Ghana Grass.: 134 (1977). —Bennett in Kirkia **11**: 234 (1980). —Clayton in F.T.E.A., Gramineae: 637 (1982). TAB. **40**, fig. 29. Type from S. Europe.
  Panicum debile Desf., Fl. Atlant. **1**: 59 (1798). —Steudel, Syn. Pl. Glum. **1**: 41 (1853). —Durand & Schinz in Consp. Fl. Afr. **5**: 746 (1894). Type as above.
  Digitaria decipiens Fig. & De Not., Agrost. Aegypt. Fragm. **2**: 43, t. 24 (1853). Type from N. Africa.
  Digitaria variabilis Fig. & De Not., Agrost. Aegypt. Fragn **2**: 41, t. 23 (1853). Type from N. Africa.
  Panicum reimarioides Andersson in Peters, Naturwiss. Reise Mossamb., Bot. **2**: 547 (1864). Type: Mozambique, "Zambese", xi-xii, *Peters* s.n. (B, holotype).
  Panicum debile var. *reimarioides* (Andersson) Durand & Schinz in Consp. Fl. Afr. **5**: 746 (1894). Type as above.
  Digitaria debilis var. *gigantea* Rendle in Cat. Afr. Pl. Welw. **2**: 163 (1899). —Henrard, Monogr. Digit.: 168 (1950). Type from Angola.
  Digitaria bangweolensis Pilg. in Fries, Wiss. Ergebn. Schwed. Rhod.-Kongo-Exped. **1**: 199 (1915). Type: Zambia, Bangweulu near Kamindas, 1150 m., 11.xi.1911, *Fries* 944 (B, holotype).
  Digitaria debilis var. *reimarioides* (Anderss.) Henr., Monogr. Digit.: 168 (1950). Type as for *Panicum reimarioides*.

A loosely caespitose or solitary growing perennial, sometimes stoloniferous. Culms 10–40 cm., erect or creeping at the base, glabrous, nodes dark and glabrous. Leaf sheaths hairy. Ligule up to 2 mm. long, blunt or truncate. Leaf laminae 3–7 × 0.2–0.4 mm., linear, flat, hairy on both surfaces, margins scabrous, crisped. Inflorescence composed of 3–14 racemes, 5–15 cm. long, few together or solitary along a short common axis. Rhachis triquetrous, unwinged, up to 0.3 mm. broad, smooth with scabrous margins. Pedicels 2-nate, 0.3–2.5 mm. long, triangular to compressed, scabrous, somewhat broadened at the apex. Spikelets 2.7–3.5 mm. long, lanceolate, acuminate. Inferior glume at the base of the callus, remote from the superior glume, short, truncate, nerveless, hyaline. Superior glume as long as the spikelet, much longer than the superior lemma, lanceolate, acuminate, 7-nerved, pale green, appressed hairy, hairs fine, slightly undulating with an irregularly recurved or circinate apex. Inferior lemma somewhat shorter than the spikelet, lanceolate, acuminate, 7-nerved, pale green, appressed hairy. Superior lemma much shorter than the spikelet, lanceolate, acuminate, yellow green to bluish green.

Botswana. N: Lake Ngami, Toteng (Toten), 18.iii.1965, *Wild & Drummond* 7144 (BM; SRGH).
Zambia. B: Mongu, 10.vii.1962, *Robinson* 5418 (K; PRE; SRGH). N: Ndundu, 1900 m., 28.ii.1959, *McCallum-Webster* A183c (K; SRGH). W: 24 km. from Mwinilunga on Matonchi Rd., 1500 m., 21.xii.1969, *Simon & Williamson* 1937 (BM; GENT; K; P; PRE; SRGH). E: Nsefu Game Camp, 750 m., 15.x.1958, *Robson* 134 (BM; BR; K; LISC; PRE; SRGH). S: Livingstone, 12.ii.1959, *McCallum-Webster* Z13 (K). **Zimbabwe**. N: Mensa Pan, 15 km. ESE. of Chirundu Bridge, 460 m., 30.i.1958, *Drummond* 5383 (BR; COI; K; LISC; P; PRE; SRGH). W: Irisvale Ranch, Bala-Bala, 1100 m., xii.1961, *Poultney* 61

(SRGH). C: Gweru (Gwelo) Teachers College, 15.ii.1968, *Biegel* 2539 (BM; K; PRE; SRGH). E: Sibu, 13 km. W. of Chimanimani (Melsetter), 11.xii.1952, *Crook* 420 (K; SRGH). S: Muzero Farm, near Nemamwa School, 2.ii.1971, *Chiparawasha* 331 (GENT; K; PRE; SRGH; WAG). **Malawi**. N: Katoto, 5 km. W. of Mzuzu, 1500 m., 13.iii.1974, *Pawek* 8216 (K; PRE; SRGH). C: Kasungu Nat. Park, near Angombe Hill, 1080 m., 30.ix.1970, *Hall-Martin* 1720 (K; PRE; SRGH). S: Litchenya Forestry Hut, 1900 m., 29.iii.1960, *Phipps* 2785 (COI; K; PRE; SRGH). **Mozambique**. N: 4 km. from Montepuez on Namuno Rd., 430 m., 25.xii.1963, *Torre & Paiva* 9669 (COI; LISC; PRE). Z: Quelimane, 14.x.1965, *Mogg* 32275 (SRGH). T: Zambezi R., 16.x.1965, *Neves Rosa* 25 (LMA). MS: Gorongoza, 60 m., 15.xi.1963, *Torre & Paiva* 9236 (COI; LISC; SRGH). M: Incomati R., 30.v.1951, *Myre* 1088 (LISC; SRGH).

Also in Africa, the western Mediterranean and Madagascar. River banks, dambos, roadsides and damp places in general.

30. **Digitaria remotigluma** (de Winter) Clayton in Kew Bull. **29**: 529 (1974). —Bennett in Kirkia **11**: 236 (1980). —Clayton in F.T.E.A., Gramineae: 638 (1982). TAB. **40**, fig. 30. Type from Namibia.

Digitariella remotigluma de Winter in Bothalia **7**: 467 (1961). —Launert in Merxm., Prodr. Fl. SW. Afr. **160**: 66 (1970). Type as above.

A loosely caespitose or creeping annual. Culms (5)10–40(50) cm., decumbent or creeping at the base, glabrous, nodes dark and loosely to densely hairy. Leaf sheaths subglabrous to partly densely hairy, scaberulous. Ligule 1–2 mm. long, subtruncate, entire. Leaf laminae 3–8 × 0.2–0.4 cm., linear, flat, subglabrous and scaberulous on both surfaces, scaberulous along the crisped margin. Inflorescence composed of (1)2–5 racemes, 5–15 cm. long, mostly in one whorl, sometimes in two whorls, separated by a short common axis. Rhachis triquetrous, narrowly winged, up to 0.5 mm. broad, smooth with scabrous margins. Pedicels 2-nate, 0.5–2.5 mm. long, subterete to triangular, ciliate, not broadened at the apex, which can bear a poorly developed coronula. Spikelets 8–11(15) mm. long, linear. Inferior glume 0.25–0.75 mm. long, truncate, rarely oblong triangular, nerveless, hyaline, separated from the other scales by a narrow stipe, 1–3 mm. long. Superior glume 7–8(12) mm. long, lanceolate-linear, 7-nerved, the central and submarginal nerves well developed, pale green, appressed hairy, hairs fine, with an irregularly recurved apex. Inferior lemma c. 4 mm. long, lanceolate, 5-nerved, appressed hairy, pale green. Superior lemma c. 3 mm. long, lanceolate, acuminate, pale yellow to greenish or bluish purple.

Botswana. N: Mohembo, 15.ii.1979, *Smith* 2659 (SRGH). **Zambia**. B: Kaoma (Mankoya), 20.xi.1959, *Drummond & Cookson* 6645 (BM; SRGH). N: Mbesuma Ranch, 12.i.1961, *Astle* 211 (K; SRGH). W: 11 km. N. of Chingola, 3.i.1961, *Robinson* 4222B (SRGH). E: Lukusuzi Nat. Park near Kalindi, 1250 m., 4.vii.1971, *Sayer* 1232 (SRGH). S: Mochipapa Dam, 1400 m., 5.xii.1962, *Astle* 1764 (K; SRGH). **Zimbabwe**. N: near Binga, 1100 m., 6.xi.1958, *Phipps* 1355 (BM; BR; SRGH). C: McIlwaine Nat. Park, near Bushman Point, 1500 m., 14.xi.1965, *Crook* 756 (K; SRGH). **Malawi**. C: Kasungu, 24.xi.1951, *Jackson* 686 (BR; K).

Also from Tanzania, Zaire and Namibia. On sandy river banks and in dambos.

31. **Digitaria scalarum** (Schweinf.) Chiov., Res. Sci. Miss. Stef.-Paoli **1**: 225 (1916). —Eggeling, Annot. List Grass. Uganda: 14 (1947). —Chippindall in Meredith, Grasses & Pastures of S. Afr.: 397 (1955). —Robyns & Tournay in Fl. Parc Nat. Alb. **3**: 76, t. 10 (1955). —Harker, E. Afr. Agric. Journ. **23**: 109 (1957). —Bogdan, Rev. List Kenya Grass.: 49 (1958). —Jackson & Wiehe, Annot. Check-list Nyasal. Grass.: 36 (1958). —Harker & Napper, Illustr. Guide Grass. Uganda: 26, t. 57 (1960). —Van der Veken in Bull. Jard. Bot. Brux. **32**: 128 (1962). —Napper, Grass. Tangan.: 77 (1965). —Bennett in Kirkia **11**: 236 (1980). TAB. **45**. Type from Ethiopia.

Panicum muticum Hochst. ex A. Rich., Tent. Fl. Abyss. **2**: 362 (1851) non Forsk. (1775). Type as above.

Panicum scalarum Schweinf. in Bull. Herb. Boiss. 2, Appendix **2**: 20 (1894). —Durand & Schinz in Consp. Fl. Afr. **5**: 764 (1894). Type as above.

Panicum hackelii Pilg., Bot. Jahrb. Syst. **30**: 118 (1901). Type from Gabon.

Digitaria abyssinica var. *scalarum* (Schweinf.) Stapf in Kew Bull. **1907**: 213 (1907). Type as for *Panicum scalarum*.

Digitaria mutica Rendle in Journ. Linn. Soc. Bot. **40**: 229 (1911) nom. superfl. Type as for *Panicum muticum* Hochst. ex A. Rich.

Digitaria somalensis Chiov. in Res. Sci. Miss. Stef.-Paoli **1**: 224 (1916). —Henrard, Monogr. Digit.: 701 (1950). Type from Somalia.

Digitaria hackelii (Pilg.) Stapf in F.T.A. **9**: 459 (1919). —Robyns, Esp. Cong. Digit.: 38 (1931); in Fl. Agrost. Congo Belge **2**: 41 (1934). —Henrard, Monogr. Digit.: 307 (1950). —Koechlin in Fl. Gabon **5**: 48 (1962). —Van der Veken in Bull. Jard. Bot. Brux. **32**: 128 (1962). Type as for *Panicum hackelii*.

Tab. 45. DIGITARIA SCALARUM. 1, habit (× ⅓); 2, spikelet in abaxial view (× 25); 3, spikelet in adaxial view (× 25); 4, inferior glume (× 25); 5, superior glume (× 25); 6, sterile lemma (× 25); 7, sterile palea (× 25); 8, fertile lemma (× 25); 9, fertile palea (× 25); 10, sexual organs before anthesis (× 25), all from *Mogg* 30054.

Digitaria ciliaris Vanderyst in Bull. Agric. Congo Belge **16**: 659 (1925) non (Retz.) Koel. (1802). Type from Zaire.

Digitaria tangaensis Henrard in Blumea **1**: 103 (1934); Monogr. Digit.: 727 (1950). Type from Tanzania.

Digitaria scalarum var. *elgonensis* C.E. Hubbard & Snowden in Kew Bull. **1934**: 111 (1934). —Eggeling, Annot. List Grass. Uganda: 15 (1947). Type from Uganda.

Digitaria vestita var. *scalarum* (Schweinf.) Henrard, Monogr. Digit.: 785 (1950). —Sturgeon in Rhod. Agric. Journ. **50**: 286 (1953). Type as for *Panicum scalarum*.

Digitaria vestita subvar. *elgonensis* (C.E. Hubbard & Snowden) Henrard, Monogr. Digit.: 670 (1950). Type as for *Digitaria scalarum* var. *elgonensis*.

A rhizomatous perennial, rhizome conspicuously well developed, growing deeply, straight with vertically growing offshoots, much branched at groundlevel, sometimes stoloniferous. Culms (10)20–40 cm., ascending from a creeping base, glabrous, nodes dark, glabrous or hairy. Leaf sheaths often with many bulbous based bristles, sometimes scaberulous. Ligule 1–3 mm. long, truncate, erose or subentire. Leaf laminae 5–10 × 0.3–0.6 cm., linear, flat or involute, glabrous or sometimes loosely hairy on both surfaces, scabrous along the crisped margin. Inflorescence composed of (2)4–10(12) racemes, (1)3–10 cm. long, erect to patent, sometimes branched at the base, solitary along a well developed common axis. Rhachis triquetrous, narrowly winged, up to 0.4 mm. broad, smooth to scaberulous, with scaberulous margins. Pedicels 2nate, 0.5–2.5 mm. long, subterete to triangular, scaberulous, scarcely broadened at the apex. Spikelets (1.6)1.7–2.2(2.4) mm. long, oblong-ovate, swollen. Inferior glume $\frac{1}{10}$–$\frac{1}{4}$ of the spikelet, conspicuously heteromorphic, often erose or truncate, nerveless, glabrous, hyaline, sometimes purplish. Superior glume c. $\frac{4}{5}$ of to as long as the spikelet, ovate to oblong, 5(7)-nerved, glabrous or with some short, fine hairs in a marginal row, pale green, often purplish tinged. Inferior lemma as long as the spikelet, ovate-oblong with recurved margins, 7-nerved, glabrous or with a marginal row of short, fine hairs, pale green, often purplish tinged. Superior lemma as long as the spikelet, ovate-oblong, apiculate, swollen, bluish green to dark brown.

Botswana. N: Nokaneng, Taokhe R., 21.iii.1961, *Vesey-FitzGerald* 3306 (SRGH). **Zambia**. B: Mongu, 2.i.1966, *Robinson* 6763 (K; M; SRGH). N: Kawimbe, 1900 m., 25.ii.1959, *McCallum-Webster* A153 (K; SRGH). C: Mt. Makulu, 20.xii.1965, *van Rensburg* 3091 (BM; K; P; SRGH). **Zimbabwe**. C: Harare Distr., Agriculture Experiment Stat. (cult.?), 1.ii.1954, *Sturgeon* in GHS 45327 (K; PRE; SRGH). E: Mutare Distr., Vumba, i.1973, *Allison* in GHS 223435 (SRGH). **Malawi**, C; Dedza Mt., 1300 m., 27.iii.1950, *Wiehe* 468 (K; PRE; SRGH). **Mozambique**. M: Inhaca Is., 3.xi.1962, *Mogg* 30054 (K; PRE; SRGH).

Also from Cameroon to Ethiopia, southwards to S. Africa. Sandy flood plains, wet spots in disturbed sandy soils, weed in gardens.

32. **Digitaria abyssinica** (Hochst. ex A. Rich.) Stapf in Kew Bull. **1907**: 213 (1907). —Pilger in Fries, Wiss. Ergebn. Schwed. Rhod.-Kongo-Exped. **1**: 199 (1916). —Chiovenda, Nuov. Giorn. Bot. Ital. **26**: 75 (1919). —Stapf in F.T.A. **9**: 460 (1919) pro parte. —Robyns, Esp. Cong. Digit.: 39 (1931); in Fl. Agrost. Congo Belge **2**: 42 (1934). —Eggeling, Annot. List Grass. Uganda: 13 (1947). —Henrard, Monogr. Digit.: 1 (1950). —Harker & Napper, Illustr. Guide Grass. Uganda: 26, t. 57 (1960). —Clayton in F.W.T.A. ed. 2, **3**: 452 (1972). —Clayton in F.T.E.A., Gramineae: 641, fig. 147 (1982). TAB. **40**, fig. 32. Type from Ethiopia.

Panicum abyssinicum Hochst. ex A. Rich., Tent. Fl. Abyss. **2**: 360 (1851). —Durand & Schinz in Consp. Fl. Afr. **5**: 740 (1894). Type as above.

Digitaria vestita Fig. et De Not., Agrost. Aegypt. Fragm. **2**: 40, fig. 22 (1853). —Henrard, Monogr. Digit.: 784 (1950). —Andrews, Fl. Pl. Sudan **3**: 435 (1956). Type from N. Africa.

Digitaria eichingeri Mez in Engl., Bot. Jahrb. Syst. **57**: 193 (1921). —Henrard, Monogr. Digit.: 210 (1950). Type from Tanzania.

A loosely caespitose perennial on a short, slender rhizome. Culms 20–40 cm., erect from a creeping or decumbent base, glabrous, nodes dark, glabrous. Leaf sheaths loosely to densely hairy from minute tubercles. Ligule 1–2 mm. long, obtuse to truncate. Leaf laminae 4–8 × 0.5–1.2 cm., linear-lanceolate, flat, sometimes scaberulous on both surfaces, with scattered bulbous based bristles, scabrous along the crisped margin. Inflorescence composed of 6–12 racemes, 4–9 cm. long, ascendent-patent, solitary along a well developed common axis. Rhachis triquetrous, scarcely winged, up to 0.4 mm. broad, smooth, with scabrous margins. Pedicels 2-nate, 1–3 mm. long, triangular to compressed, scabrous, not or scarcely broadened at the apex. Spikelets 1.8–2.1 mm. long, oblong-ovate, inflated. Inferior glume c. $\frac{1}{4}$ of the spikelet, regularly triangular to rounded triangular, nerveless, glabrous, hyaline or purplish. Superior glume c. $\frac{4}{5}$ of the spikelet, oblong-ovate,

3(5)-nerved, glabrous or rarely minutely pubescent, pale green, sometimes purplish tinged. Inferior lemma as long as the spikelet, oblong-ovate with recurved margins, equidistantly 7-nerved, glabrous or rarely minutely pubescent, pale green or purplish tinged. Superior lemma as long as the spikelet, oblong-ovate, apiculate, pale to bluish green, inflated.

Zambia. N: Shiwa Ngandu, 1300 m., 1.vi.1956, *Robinson* 1526 (K; SRGH).
Also from Nigeria to Ethiopia, south to Zaire and Zambia. Forest edges, in shaded conditions.
The much more common species *D. scalarum* is very often included under this name. In our opinion however, these two species can readily be distinguished, in particular shape and texture of the inferior glume seem to yield a reliable diagnostic feature.

33. **Digitaria appropinquata** P. Goetghebeur, in Bull. Nat. Plantentuin Belg. **45**: 402, fig. 4 (1975). TAB. **40**, fig. 33. Type: Mozambique, Zambezia, Gúruè, near Pico Namuli, 1500 m., 9.iv.1943, *Torre* 5162 (K, isotype; LISC, holotype).

A loosely caespitose or solitary growing, slender annual. Culms 15–20 cm., strictly erect, glabrous, nodes dark, glabrous. Leaf sheaths sometimes scaberulous, loosely hairy from minute tubercles. Ligule 1.5–3 mm. long, truncate. Leaf laminae 3–6 × 0.1–0.2 cm., linear, flat, sometimes scaberulous on both surfaces, sparsely hairy from minute tubercles, scaberulous along the margin. Inflorescence composed of 2–4 racemes, 1–4 cm. long, erect, solitary along a short, slender common axis. Rhachis triquetrous, scarcely winged, up to 0.2 mm. broad, smooth, with scabrous margins and basally long white hairs. Pedicels 2-nate, 1–3 mm. long, triangular to compressed, scabrous, scarcely broadened at the apex. Spikelets 1.7–2 mm. long, lanceolate. Inferior glume c. $\frac{1}{10}$ of the spikelet, rounded, hyaline, yellowish green to bluish purple. Superior glume c. $\frac{2}{3}$ of the spikelet, oblong-triangular, 3-nerved, glabrous, yellowish green to bluish purple. Inferior lemma as long as the spikelet, lanceolate with recurved margins, 5(7)nerved, the central and 2 subcentrals closely together, glabrous, yellowish green to bluish purple. Superior lemma as long as the spikelet, lanceolate, apiculate, bluish purple.

Mozambique. Z: Gúruè, near Pico Namuli, 1500 m., 9.iv.1943, *Torre* 5162 (K; LISC; PRE).
Only known from the type collection. Shallow soil over rocks.

34. **Digitaria trinervis** Van der Veken in Bull. Nat. Plantentuin Belg. **45**: 402, fig. 5 (1975). TAB **40**, fig. 34. Type: Malawi, Mulanje Mt., Litchenya Plateau, 1950 m., 10.vi.1962, *Robinson* 5321 (K, isotype; SRGH, holotype).

A creeping, mat-forming (annual of short lived?) perennial. Culms 10–20 cm., glabrous, nodes dark and hairy. Leaf sheaths sometimes scaberulous, glabrous to loosely hairy. Ligule c. 2 mm. long, truncate, erose. Leaf laminae 3–5 × 0.1–0.3 cm., linear, flat, scaberulous on both surfaces, scaberulous along the margin. Inflorescence composed of 3–8 racemes, 1–4 cm. long, often branched, dispersed along a short common axis. Rhachis triquetrous, scarcely winged, up to 0.2 mm. broad, smooth with scaberulous margins. Pedicels 2-nate, 0.5–3 mm. long, triquetrous, scaberulous, broadened at the apex. Spikelets 2.1–2.4 mm. long, lanceolate. Inferior glume $\frac{1}{10}$–$\frac{1}{5}$ of the spikelet, often rounded triangular, sometimes truncate or slightly bifid, nerveless, yellowish green to purplish tinged. Superior glume c. $\frac{3}{4}$ of the spikelet, oblong triangular, 3(5)-nerved, glabrous, yellowish green or purplish tinged. Inferior lemma as long as the spikelet, oblong with recurved margins, 7-nerved, the central and 2 subcentral nerves very closely together, glabrous, yellowish green to purplish tinged. Superior lemma as long as the spikelet, acute to slightly acuminate, purplish tinged.

Malawi. S: W. slopes of Great Ruo Gorge, 1200 m., 19.vi.1962, *Robinson* 5378 (K; SRGH).
Not known from elsewhere. Mountain grassland.

35. **Digitaria gazensis** Rendle, Journ. Linn. Soc. Bot. **40**: 228, t. 6, fig. 1–5 (1911). —Stapf in F.T.A. **9**: 457 (1919). —Robyns, Esp. Congol. Digit.: 34 (1931); in Fl. Agrost. Congo Belge **2**: 37 (1934). —Eggeling, Annot. List Grass. Uganda: 14 (1947). —Henrard, Monogr. Digit.: 280 (1950). —Sturgeon in Rhod. Agric. Journ. **50**: 286 (1953). —Bogdan, Rev. List Kenya Grass.: 49 (1958). —Jackson & Wiehe, Annot. Check-list Nyasal. Grass.: 35 (1958). —Harker & Napper, Illustr. Guide Grass. Uganda: 26, t. 60 (1960). —Napper, Grass. Tangan.: 77 (1965). —Launert in Merxm., Prodr. Fl. SW. Afr. **160**: 64 (1970). —Simon in Kirkia **8**: 31, 65 (1971). —Bennett in Kirkia **11**: 235 (1980). —Clayton in F.T.E.A., Gramineae: 643 (1982). TAB. **40**, fig. 35. Type from Angola.

Digitaria herpoclados Pilg. in Fries, Wiss. Ergebn. Schwed. Rhod.-Kongo-Exped. **1**: 200 (1915). —Eggeling, Annot. List Grass. Uganda: 14 (1947). —Henrard, Monogr. Digit.: 313 (1950). —Harker & Napper, Illustr. Guide Grass. Uganda: 26, t. 57 (1960). —Van der Veken in Bull. Jard. Bot. Brux. **32**: 128 (1962). Type from Uganda.

Digitaria villosissima Chiov. in Nuov. Giorn. Bot. Ital. n.s. **26**: 61 (1919). Type from Zaire.

Digitaria antunesii Mez in Engl., Bot. Jahrb. **57**: 193 (1921). Type from Angola.

Digitaria nyassana Mez in Engl., Bot. Jahrb. **57**: 192 (1921) ("nyssana"). Type: Malawi, 1891, *Buchanan* 1449 (B, hololectotype; K, isolectotype).

Digitaria usambarica Mez in Engl., Bot. Jahrb. **57**: 193 (1921). Type from Kenya.

Digitaria proxima Henrard in Blumea **1**: 104 (1934); Monogr. Digit.: 593 (1950). Type from Tanzania.

A densely caespitose perennial on a very short rhizome, sometimes stoloniferous, base surrounded by velvety hairy cataphylls and withered leaf sheaths. Culms 40–120 cm., erect or creeping, glabrous, nodes dark and typically hairy. Leaf sheaths loosely to densely hairy. Ligule up to 3 mm. long, triangular, rounded. Leaf laminae 5–15 × 0.3–0.5 cm., linear, flat, loosely to densely hairy on both surfaces, scabrous along the margin. Inflorescence composed of 5–17 racemes, 8–20 cm. long, erect to patent, few together or solitary along a well developed common axis. Rhachis triquetrous, scarcely winged, up to 0.3 mm. broad, smooth, with scabrous margins. Pedicels 2-nate, 0.5–3.5 mm. long, triangular, scabrous, broadened at the apex. Spikelets 1.9–2.6 mm. long, oblong. Inferior glume up to 0.5 mm. long, ovate, truncate, nerveless, hyaline. Superior glume as long as the spikelet, oblong, membranous near the apex, 5-nerved, purplish, appressed hairy, hairs very slender, slightly undulating, often with incurved apex, often purplish. Inferior lemma as long as the spikelet, oblong, membranous near the apex, 7-nerved, purplish, appressed hairy. Superior lemma as long as the spikelet, oblong, acute, bluish green to bluish purple.

Zambia. B: 3 km. N. of Kabompo, 1000 m., 31.iii.1982, *Drummond & Vernon* 11105 (GENT; SRGH). N: Lake Chila For. Res., 1800 m., 6.ii.1958, *Vesey-FitzGerald* 1462 (SRGH). W: Chingola, 9.xii.1965, *Fanshawe* 9450 (SRGH). C: Mount Makulu, 28.ii.1973, *Rees* 1103 (SRGH). E: Kamkulo Dambo, 31.xii.1963, *Wilson* 41A (SRGH). S: Namwala, Ngoma burning experimental plots, 13.i.1962, *Mitchell* 12–60 (SRGH). **Zimbabwe**. N: Nyamayanetsi (Nyamnetsi) Estate, 2.i.1979, *Nyariri* 608 (PRE; SRGH). W: Gwampa For. Res., 1000 m., ii.1955, *Goldsmith* 73–55 (SRGH). C: Somabula Distr., 1600 m., 12.xii.1931, *Brain* 8521 (SRGH). E: 13 km. W. of Chimanimani (Melsetter), Sibu, 20.vi.1953, *Crook* 485 (BR; K; PRE; SRGH). S: Mazabuka Distr., iii.1932, *Trapnell* 863 (BM; K; SRGH). **Malawi**. N: Viphya Lwafwa area, 2000 m., 6.ii.1962, *Chapman* 1575 (K; LISC; PRE; SRGH). C: near Dedza, 6.i.1950, *Wiehe* 400 (K; LISC; SRGH). S: Zomba Plateau, 28.xi.1967, *Banda* 961 (SRGH). **Mozambique**. N: S. Antonio de Unango, i.1934, *Gomes e Sousa* 1648 (COI). Z: Nhamarroi, 12.vi.1943, *Torre* 5471 (K; LISC). T: Angonia, Dzenza R., 1400 m., 10.xii.1980, *Stefanesco & Nyongani* 583 (SRGH). MS: Beira-Dondo, 31.xii.1943, *Torre* 6333 (K; LISC).

Also from Zaire and Sudan southwards to Namibia and S. Africa, Madagascar. In degraded woodland, miombo woodland, also in dambos, vlei verges and salt marshes.

36. **Digitaria rukwae** Clayton in Kew Bull. **29**: 520 (1974); in F.T.E.A., Gramineae: 646 (1982). TAB. **40**, fig. 36. Type from Tanzania.

A caespitose perennial, with a well grown rhizome, covered at the base by cataphylls. Culms 40–110 cm., erect, glabrous, nodes dark and glabrous. Leaf sheaths glabrous, smooth to scaberulous. Ligule up to 3 mm. long, triangular to truncate. Leaf laminae 7–20 × 0.3–0.7 cm., linear, flat, glabrous and smooth to scaberulous on both surfaces, scabrous along the margins. Inflorescence composed of (9)11–21 racemes, 2–11 cm. long, sometimes with a inferior whorl of 3–4 racemes, or all racemes solitary along a well grown common axis, often as long as the largest raceme. Rhachis triquetrous, scarcely winged, up to 0.3 mm. broad, smooth to scabrous with a scaberulous margin. Pedicels 2-nate, 0.5–2.5 mm. long, subtriangular to compressed, scaberulous, broadened at the apex. Spikelets 2.3–2.9 mm. long, oblong. Inferior glume up to 0.5 mm. long, ovate to triangular, often with a hyaline margin. Superior glume c. $\frac{3}{4}$ of to as long as the spikelet, 3-nerved, appressed hairy, hairs very fine, white, smooth, acute. Inferior lemma as long as the spikelet, oblong, 7-nerved, central zone somewhat broadened, nerves smooth, appressed hairy. Superior lemma somewhat smaller than the spikelet, oblong, slightly acuminate, reddish to dark brown.

Zambia. N: Mbesuma Ranch, 12.i.1961, *Astle* 207 (BM; K; SRGH). S: Muckle Neuk, 16 km. N. of Choma, 1400 m., 27.xi.1954, *Robinson* 973 (BM; K; M). **Mozambique**. N: Angoche (Antonio Enes), San Luis de Malatane, 20 m., 24.i.1968, *Torre & Correia* 17340 (LISC). Z: 7 km. from Namacurra to

Vila Maganja da Costa, 40 m., 25.i.1966, *Torre & Correia* 14094 (LISC). M: Bela Vista—Santaca, 11.iv.1949, *Myre & Balsinhas* 550 (BM; K; LMA).
    Also in Tanzania. In flood plains, on seasonally wet sandy soils.

37. **Digitaria leptorrhachis** (Pilg.) Stapf in F.T.A. **9**: 462 (1919). —Henrard, Monogr. Digit.: 393 (1950). —Clayton in F.W.T.A. ed. 2, **3**: 454 (1972). —Rose Innes, Man. Ghana Grass.: 137 (1977). —Clayton in F.T.E.A., Gramineae: 644 (1982). TAB. **40**, fig. 37. Type from Senegal.
    Panicum leptorrhachis Pilg. in Engl., Bot. Jahrb. Syst. **30**: 119 (iii.1901). Type as above.
    Panicum nigritianum Hack., Bot. Zeitschr. **51**: 293 (viii.1901). Type from Nigeria.
    Digitaria chevalieri Stapf in F.T.A. **9**: 458 (1919). —Henrard, Monogr. Digit.: 119 (1950). Type from Ivory Coast.
    Digitaria nigritiana (Hack.) Stapf in F.T.A. **9**: 463 (1919). —Henrard, Monogr. Digit.: 494 (1950). —Robyns & Van der Veken in Bull. Jard. Bot. Brux. **22**: 153 (1952). Type as for *Panicum nigritianum.*
    Digitaria polybotrya Stapf in F.T.A. **9**: 462 (1919). —Robyns, Esp. Cong. Digit.: 34, t. 2 (1931); in Fl. Agrost. Congo Belge **2**: 38, t. 20 (1934). —Henrard, Monogr. Digit.: 581 (1950). —Koechlin in Fl. Gabon **5**: 48, t. 7 (1962). Type from Zaire.
    Digitaria richardsonii Mez in Engl., Bot. Jahrb. **57**: 193 (1921). Type from Nigeria.
    Digitaria bredoensis Robyns & Van der Veken in Bull. Jard. Bot. Brux. **22**: 146, fig. 12 (1952). —Simon in Kirkia **8**: 65 (1971). Type: Zambia, Mukuja, *Brêdo* 3860 (BR, holotype).
    Digitaria annua Van der Veken in Bull. Nat. Plantentuin Belg. **45**: 409, fig. 10 (1975). Type from Zaire.

A decumbent or creeping annual (or short lived perennial?). Culms (10)20–60(120) cm., ascending, glabrous, nodes dark and (sub)glabrous to bearded. Leaf sheaths loosely to densely covered by bulbous based bristles. Ligule 1–2 mm. long, ovate to truncate, entire. Leaf laminae (2)4–10(20) × (0.1)0.2–0.4(0.6) cm., linear, flat, loosely to densely covered by bulbous based bristles and scaberulous on both surfaces, scabrous along the crisped margin. Inflorescence composed of (1)5–10(20) racemes, (3)5–12(15) cm. long, solitary along a well developed common axis. Rhachis triquetrous, scarcely winged, up to 0.3 mm. broad, smooth with scabrous margins. Pedicels 2-nate, 0.3–2 mm. long, triquetrous to compressed, scabrous, slightly broadened at the apex. Spikelets 1.3–1.9 mm. long, oblong. Inferior glume very small, nerveless, hyaline. Superior glume as long as the spikelet, oblong, 3–5-nerved, appressed hairy, hairs fine, semiverrucose, with incurved apex. Inferior lemma as long as the spikelet, oblong, 7-nerved, appressed hairy. Superior lemma as long as the spikelet, oblong, acute, pale yellow (unripe) to purplish or bluish grey.

Zambia. N: Mweru-Wantipa, Muzambwe, 1100 m., 15.iv.1961, *Phipps & Vesey-FitzGerald* 3184 (BM; BR; K; PRE; SRGH). W: Matonchi Farm, 24.i.1938, *Milne-Redhead* 4308 (BR; K; PRE; SRGH).
    Also from Senegal to Sudan, south to Zambia. Dambos, on shallow soil overlying rocks and edaphic grasslands.
    The extraordinary amount of variability present in this complex of annual taxa was already mentioned by Clayton (loc. cit.). The recently described *D. annua* is here added to the synonymy. Recent collections and field studies in the salines of Mwashya (Shaba, Zaire) have shown that *D. annua* should be considered as an edaphic hunger-form. This is one extreme on the line of clinal variation from *D. polybotrya* to *D. annua*, with species as rather arbitrarily delimited units.

38. **Digitaria dunensis** P. Goetghebeur in Bull. Nat. Plantentuin Belg. **45**: 416, fig. 16 (1975). TAB. **40**, fig. 38. Type: Malawi, Nkhota Kota, Lake Malawi shore, 2.V.1963, *Verboom* 989 (BM; BR, holotype; K, isotype).

A mat-forming perennial (?). Culms 20–40 cm., with creeping base and erect flowering culms, glabrous, nodes dark and glabrous. Leaf sheaths glabrous. Ligule up to 0.5 mm. long, rim-like, erose. Leaf laminae 3–7 × 0.2–0.5 cm., linear, flat, glabrous and smooth on both surfaces, minutely scaberulous to almost smooth along the often crisped margin. Inflorescence composed of (3)7–13 racemes, 2–4 cm. long, erect, solitary along a common axis, mostly twice as long as the largest raceme. Rhachis triquetrous, not winged, smooth with smooth or scaberulous margins. Pedicels 2-nate, 0.5–2.5 mm. long, triangular, smooth to scaberulous, somewhat widened near the apex. Spikelets 2.3–2.6 mm. long, oblong-lanceolate. Inferior glume absent or obsolete. Superior glume somewhat shorter than the spikelet, oblong triangular, 3-nerved, appressed hairy, hairs very fine, smooth, acute. Inferior lemma as long as the spikelet, oblong-lanceolate, slightly acuminate, with recurved margins, 5-?nerved, nerves smooth, the central nerve and 2 marginal nerves well developed, appressed hairy. Superior lemma somewhat shorter than the spikelet, oblong-lanceolate, acute to acuminate, pale to dark brown.

Malawi. C: Salima Distr., iv.1972, *Williamson* 2175A (SRGH).
Not known elsewhere. Sand dune coloniser.

39. **Digitaria maniculata** Stapf in F.T.A. **9**: 466 (1919). —Vanderyst in Bull. Agric. Congo Belge **16**:
    660 (1925). —Robyns, Esp. Cong. Digit.: 41 (1931); in Fl. Agrost. Congo Belge **2**: 45 (1934).
    —Henrard, Monogr. Digit.: 425 (1950). —Launert in Merxm., Prodr. Fl. SW. Afr. **160**: 64 (1970).
    —Simon in Kirkia **8**: 31, 65 (1971). —Bennett in Kirkia **11**: 235 (1980). TAB. **40**, fig. 39. Type
    from Zaire.

A slender creeping annual, often stoloniferous. Culms 10–30 cm., ascending or erect,
glabrous, nodes dark and glabrous. Leaf sheaths glabrous. Ligule c. 0.5 mm. long,
truncate, shortly ciliate. Leaf laminae 1–2.5 × 0.2–0.4 mm., lanceolate to linear, flat,
glabrous on both surfaces, scaberulous along the margin, with a few bristle hairs near the
base. Inflorescence composed of 1–2 racemes, 2–6 cm. long, erect, one subsessile, the
other pedicellate. Rhachis asymmetrically winged, undulating, up to 0.75 mm. broad,
smooth with smooth margins. Pedicels 2-nate, 0.5–2.5 mm. long, triangular, smooth to
scaberulous, broadened at the apex. Spikelets 2.5–2.7 mm. long, oblong-lanceolate.
Inferior glume c. ¼ or less than ¼ of the spikelet, truncate or slightly bilobed, nerveless,
hyaline. Superior glume as long as the spikelet, oblong with recurved margins, 5-nerved,
pale green to purplish, appressed hairy, hairs long, fine, with incurved apex, hyaline to
bluish purple. Inferior lemma as long as the spikelet, oblong with recurved margins,
7-nerved, pale green to purplish, appressed hairy and with 2 basal bundles of long, white,
finely papillose hairs. Superior lemma shorter than the spikelet, oblong-lanceolate,
acuminate, pale green to bluish green.

Caprivi Strip. Katima Mulilo, 1000–1500 m., 11.ii.1969, *de Winter* 9116 (SRGH).
Botswana. N: western edge of Nxabega Isl., 13.iii.1982, *Smith* 3797 (GENT; SRGH).
Zambia. B: Machili, 2.iii.1961, *Fanshawe* 6362 (SRGH). N: 95 km. E. of Kasama, 2.ii.1961, *Robinson*
4338 (BR; K; M; PRE; SRGH). W: 50 km. W. of Kabompo, 1200 m., 26.xii.1969, *Simon & Williamson*
2037 (SRGH). S: Kabulamwanda, 90 km. N. of Choma, 1100 m., 13.ii.1955, *Robinson* 1107 (BR; K;
SRGH). **Zimbabwe.** W: Kazuma Depression, 1200 m., 25.i.1974, *Gonde* 25–74 (GENT; SRGH).
Also in Zaire and Namibia. In short, open grassland on seasonally wet sandy soil, sand banks and
dambo margins.

40. **Digitaria parodii** Jacques-Félix, Bol. Soc. Arg. Bot. **12**: 230, fig. 1–2 (1968). TAB. **40**, fig. 40. Type
    from Central African Republic.
    Digitaria spirifera P. Goetghebeur in Bull. Nat. Plantentuin Belg. **45**: 404, fig. 6 (1975). Type:
    Zambia, 8 km. S. of Mufulira, 5.iii.1960, *Robinson* 3377 (BR, holotype; K; M, isotypes).
    Digitaria kasamaensis Van der Veken in Bull. Nat. Plantentuin Belg. **45**: 406, fig. 8 (1975).
    Type: Zambia, Kasama Distr., 80 km. S. of Kasama, 28.i.1961, *Robinson* 4310 (BM, holotype; K; M,
    isotypes).

A slender creeping annual, mat-forming. Culms 10–20 cm., erect from a creeping base,
glabrous, nodes dark, glabrous. Leaf sheaths loosely to densely hairy. Ligule 0.3 mm.
long, rim-like, shortly ciliate. Leaf laminae 1–3 × 0.2–0.4 cm., lanceolate to linear, flat,
glabrous on both surfaces, scabrous along the margins, often with a few bristle hairs near
the base. Inflorescence composed of 2-(3–4) racemes, 3–7 cm. long, one shortly, the other
long pedicelled. Rhachis triquetrous, asymmetrically winged, undulating, up to 0.8 mm.
broad, smooth with smooth margins. Pedicels 2-nate, 0.5–3 mm. long, triangular, smooth,
asymmetrically broadened at the apex. Spikelets 1.8–2.1 mm. long, oblong. Inferior glume
very short, truncate, nerveless, hyaline. Superior glume as long as the spikelet, oblong,
5-nerved, appressed hairy, hairs white to purplish, fine, papillose, with a circinately
involute apex. Inferior lemma somewhat shorter than the spikelet, oblong with recurved
margins, 7-nerved, appressed hairy. Superior lemma somewhat shorter than the spikelet,
oblong, apiculate, pale yellowish green to bluish green, adaxially slightly gibbous.

Zambia. N: 80 km. S. of Kasama, 28.i.1961, *Robinson* 4310 (BM; K; M; PRE). W: Kitwe, 7.ii.1970,
Fanshawe 10747 (SRGH).
Also in Chad, the Central African Republic, and Zaire. Gravelly to sandy river banks.

41. **Digitaria tenuifolia** P. Goetghebeur in Bull. Nat. Plantentuin Belg. **45**: 407, fig. 9 (1975). TAB. **40**,
    fig. 41. Type: Zambia, Kasama Distr., 18 km. E. of Mungwi, 19.i.1961, *Robinson* 4274 (SRGH,
    holotype; K; M, isotypes).

A loosely caespitose, slender annual. Culm up to 20 cm., erect, glabrous, nodes dark,
slightly hairy. Leaf sheaths glabrous to densely papillose. Ligule 2–4 mm. long, bifid. Leaf

laminae 5–15 × 0.05–0.1 cm., linear, involute, densely papillose on both surfaces, papillose along the margin. Inflorescence of a single raceme, 3–5 cm. long, terminal, erect. Rhachis triquetrous, narrowly winged, up to 0.3 mm. broad, slightly undulate, smooth with smooth margins. Pedicels 2-nate, 1–3.5 mm. long, triangular, minutely scaberulous, scarcely broadened at the apex. Spikelets 2.8–3.4(4.0) mm. long, oblong-lanceolate. Inferior glume 0.3–0.5 mm. long, truncate, nerveless, hyaline. Superior glume as long as the spikelet, oblong-lanceolate with narrow recurved margins, 5(7)-nerved, pale green, appressed hairy, hairs long, fine, irregularly undulate, apex sometimes incurved. Inferior lemma as long as the spikelet, oblong-lanceolate with narrow recurved margins, 5(7)-nerved, pale green, appressed hairy, hairs as on the superior glume, but also with two basal bundles of much longer hairs. Superior lemma somewhat shorter than the spikelet, oblong-lanceolate, acute, pale green to bluish green.

Zambia. N: Malde Rocks, Kasama, 1400 m., 1.iii.1960, *Richards* 12679 (BR; K; P; SRGH).
Not known from elsewhere. Shallow depression in flat rocks.

42. **Digitaria angolensis** Rendle, Cat. Afr. Pl. Welw. **2**: 165 (1899). —Stapf in F.T.A. **9**: 467 (1919).
—Henrard, Monogr. Digit.: 39 (1950). —Sturgeon in Rhod. Agric. Journ. **50**: 287 (1953).
—Bennett in Kirkia **11**: 234 (1980). —Clayton in F.T.E.A., Gramineae: 633 (1982). TAB. **40**, fig.
42. Type from Angola.
    Digitaria yokoensis Vanderyst in Bull. Agric. Congo Belge **16**: 659 (1925). —Robyns, Esp.
Cong. Digit.: 42 (1931); in Fl. Agrost. Congo Belge **2**: 45 (1934). —Henrard, Monogr. Digit.: 803
(1950). —Compere in Bull. Jard. Bot. Brux. **33**: 389 (1963). —Napper, Grass. Tangan.: 75 (1965).
—Simon in Kirkia **8**: 31, 66 (1971). Type from Zaire.
    Digitaria verrucosa C.E. Hubbard in Kew Bull. **1928**: 40 (1928). —Sturgeon in Rhod. Agric.
Journ. **50**: 287 (1953). Type: Zimbabwe, Harare, 1440 m., iv.1927, *Eyles* 4859 (K, holotype).

A loosely caespitose or solitary growing annual. Culms 15–60 cm., erect or ascending, rarely from a creeping base, glabrous, nodes dark, glabrous. Leaf sheaths glabrous. Ligule 0.5–1 mm. long, rim-like. Leaf laminae 5–10 × 0.4–0.7 cm., linear, flat, glabrous on both surfaces, minutely scaberulous along the prominently crisped margin. Inflorescence composed of 2–5 racemes, (4)7–18 cm. long, erect, solitary along a short common axis. Rhachis triquetrous, scarcely winged, up to 0.5 mm. broad, smooth, with scabrous margins. Pedicels 3-nate, 0.4–3 mm. long, triangular, scabrous, slightly broadened at the apex. Spikelets 2.0–2.6 mm. long, ovate, inflated. Inferior glume 0.1–0.2 mm. long, blunt, hyaline. Superior glume somewhat shorter than the spikelet, oblong-triangular, 3(5)-nerved, appressed hairy, hairs white to purplish, undulating, with rough, papillose wall. Inferior lemma as long as the spikelet, ovate with recurved margins, 5-nerved, the central and 2 marginal nerves well developed, 2 subcentral nerves but poorly developed, appressed hairy, hairs scarcely overtopping the spikelet. Superior lemma somewhat shorter than the spikelet, ovate, apiculate, pale yellow to pale brown, sometimes purplish tinged.

Zambia. B: 7 km. from Kaoma on Kasempa Rd., Katembe, 1150 m., 30.iii.1982, *Drummond & Vernon* 11034 (GENT; SRGH). N: Mpika Boma, 1500 m., 17.iv.1959, *Vesey-FitzGerald* 2464 (SRGH). C: Mwembeshi Catchment, 1200 m., 17.iii.1963, *Vesey-FitzGerald* 3963 (BM; COI). S: Sibanyati, 22 km. ESE. of Choma, 1350 m., 25.iii.1982, *Drummond & Vernon* 11003 (GENT; SRGH). **Zimbabwe**. N: Gokwe, 16.v.1962, *Bingham* 260 (BM; SRGH). C: West Chiota (Shiota) Reserve, 21.i.1950, *Cleghorn* 17 (K; SRGH). E: Tanganda Tea Estate, Chipinge, 1000 m., iv.1934, *Brain* 10611 (SRGH). **Malawi**. C: 1 km. S. of Bua R., 1300 m., 15.ii.1968, *Simon, Williamson & Ball* 1812 (BM; K; PRE; SRGH).
From Zaire and Tanzania south to Angola and S. Africa. Weed of arable land and sandy roadsides.

43. **Digitaria longiflora** (Retz.) Pers., Syn. Pl. **1**: 85 (1805). —Rendle, Cat. Afr. Pl. Welw. **2**: 162 (1899).
—Chiovenda, Nuov. Giorn. Bot. Ital. **26**: 75 (1919). —Stapf in F.T.A. **9**: 469 (1919). —Stent in
Bothalia **1**: 269 (1924). —Hughes in Kew Bull. **1923**: 310 (1923). —Vanderyst in Bull. Agric.
Congo Belge **16**: 660 (1925). —Robyns, Esp. Cong. Digit.: 42 (1931); in Fl. Agrost. Congo Belge **2**:
46 (1934). Eggeling, Annot. List Grass. Uganda: 14 (1947). —Henrard, Monogr. Digit.: 408
(1950). —Chase in Hitchcock, Man. Grass. U.S.: 578, 855, fig. 833 (1951). —Sturgeon in Rhod.
Agric. Journ. **50**: 287 (1953). —Chippindall in Meredith, Grasses & Pastures of S. Afr.: 396, fig.
333 (1955). —Robyns & Tournay in Fl. Parc Nat. Alb. **3**: 75 (1955). —Andrews, Fl. Pl. Sudan **3**:
435 (1956). —Bor in Webbia **11**: 359 (1956). —Bogdan, Rev. List Kenya Grass.: 49 (1958).
—Jackson & Wiehe, Annot. Check-list Nyasal. Grass.: 35 (1958). —Bor, Grass. B.C.I. & P.: 302
(1960). —Harker & Napper, Illustr. Guide Grass. Uganda: 25, t. 56 (1960). —Napper, Grass.
Tangan.: 75 (1965). —Blake in Proc. Roy. Soc. Queensl. **81**: 17 (1969). —Simon in Kirkia **8**: 31,
65 (1971). —Clayton in F.W.T.A. ed. 2, **3**: 453 (1972). —Veldkamp in Blumea **21**: 66, fig. 14e

(1973). —Rose Innes, Man. Ghana Grass.: 137 (1977). —Bennett in Kirkia **11**: 235 (1980). —Clayton in F.T.E.A., Gramineae: 634 (1982). TAB. **40**, fig. 43. Type from Asia.

Paspalum longiflorum Retz., Obs. Bot. **4**: 15 (1786). Type as above.

Panicum propinquum R.Br., Prodr.: 193 (1810). Type from Australia.

Digitaria propinqua (R.Br.) Beauv., Ess. Agrost.: 51 (1812). —Blake in Proc. Roy. Soc. Queensl. **81**: 14 (1969). Type as for *Panicum propinquum*.

Milium filiforme Roxb., Fl. Ind. **1**: 317 (1820) non Lagasca (1816). Type from Asia.

Digitaria roxburghii Spreng., Syst. Veg. **1**: 270 (1825). Type as above.

Paspalum pubescens J.S. Presl in C.B. Presl, Reliq. Haenk. **1**: 214 (1830) non Muhlenberg (1809). Type from ?America.

Paspalum preslii Kunth, Enum. Pl. **1**: 47 (1833). Type as above.

Panicum pseudodurva Nees, Fl. Afr. Austr.: 21 (1841). —Steud. in Syn. Pl. Glum. **1**: 41 (1853). Type from S. Africa.

Digitaria pseudodurva (Nees) Schlechtend. in Linnaea **26**: 458 (1853). Type as above.

Digitaria friesii Pilg. in Fries, Wiss. Ergebn. Schwed. Rhod.-Kongo-Exped. **1**: 200 (1915). —Henrard, Monogr. Digit.: 267 (1950). —Simon in Kirkia **8**: 65 (1971). Type: Zimbabwe, *Fries* 952 (B, holotype).

Digitaria flexilis Henrard, Monogr. Digit.: 258 (1950). Type from Uganda.

Digitaria preslii (Kunth) Henrard, Monogr. Digit.: 589 (1950). —Bor in Webbia **11**: 360 (1956); Grass. B.C.I. & P.: 304 (1960). Type as for *Paspalum preslii*.

Digitaria corradii Chiov. in Chiarugi, Webbia **8**: 68, fig. 15B (1951). Type from Ethiopia.

A loosely caespitose annual or short lived perennial, sometimes with a short rhizome or matforming with creeping runners, rooting at the nodes. Culms 20–50(70) cm., erect, ascending or creeping, glabrous, nodes dark and glabrous. Leaf sheaths glabrous to loosely hairy. Ligule 0.5–1.5 mm. long, truncate, entire. Leaf laminae (1)5–10(15) × 0.2–0.6 cm., linear, flat, smooth or minutely scaberulous, subglabrous or with scattered bulbous based bristles on both surfaces, scabrous along the margin. Inflorescence composed of (1)2(4) racemes, (1.5)2.5–10(17) cm. long, one subsessile, the other pedicellate on a very short common axis, erect to patent. Rhachis broadly winged, up to 1.2 mm. broad, smooth with scabrous margins. Pedicels 3-nate, 0.5–2.5 mm. long, subterete, smooth, broadened at the apex. Spikelets (1.2)1.5–2.0(2.2) mm. long, ovate-oblong, broadest in or above the middle. Inferior glume obsolete, truncate, nerveless, hyaline. Superior glume as long as the spikelet, ovateoblong, 3–5-nerved, appressed hairy, hairs fine, slightly undulating, whitish or purplish tinged, with rough, papillose walls. Inferior lemma as long as the spikelet, ovate-oblong with narrow, recurved margins, (5)7-nerved, appressed hairy. Superior lemma as long as the spikelet, oblong,acute, pale yellow to bluish green.

Botswana. N: Okovango R., 15.ii.1979, *Smith* 2651 (SRGH). **Zambia**. N: Bwela Flats, 16.ii.1958, *Savory* 260 (SRGH). W: Kitwe, 7.iv.1963, *Fanshawe* 7757 (BM; SRGH). C: 34 km. S. of Kabwe, 1300 m., 4.iv.1961, *Phipps & Vesey-FitzGerald* 2933 (BM; K; SRGH). **Zimbabwe**. E: Nyamaropa, 1000 m., 16.i.1967, *Biegel* 1764 (K; SRGH). S: Matsai Reserve, 18.viii.1960, *Cleghorn* 625 (SRGH). **Malawi**. N: Mzuzu, Marymount, 1500 m., 17.iii.1974, *Pawek* 8225 (GENT; P; SRGH). S: Mulanje, 27.ii.1951, *Wiehe* N/713 (K; PRE). **Mozambique**. Z: Mopeia, 20.x.1974, *Barnes* J430 (SRGH). MS: Chimoio, 30.i.1948, *Barbosa* 918 (K; LISC). GI: Gaza, Macia, 21.ix.1948, *Myre & Carvalho* 270 (LMA; PRE). M: Inhaca Isl., 15.vii.1957, *Mogg* 27337 (LMA; PRE).

Widespread in the Palaeotropics. Disturbed or open sites in miombo woodland, fallow land and roadsides.

44. **Digitaria argyrotricha** (Andersson) Chiov., Res. Sci. Miss. Stefanini-Paoli. **1**: 183 (1916). —Stapf in F.T.A. **9**: 468 (1919). —Henrard, Monogr. Digit.: 49 (1950). —Andrews, Fl. Pl. Sudan **3**: 434, fig. 104 (1956). —Bogdan, Rev. List Kenya Grass.: 49 (1958). —Napper, Grass. Tangan.: 75 (1965). —Clayton in F.W.T.A. ed. 2, **3**: 453 (1972). —Rose Innes, Man. Ghana Grass.: 134 (1977). —Clayton in F.T.E.A., Gramineae: 634 (1982). TAB. **40**, fig. 44. Type: Mozambique, Querimba Isl., *Peters* s.n. (B, holotype).

Panicum argyrotrichum Andersson in Peters, Naturwiss. Reise Mossamb. Bot. **2**: 548 (1864). Type as above.

Panicum argyrotrichum var. *tenue* Andersson in Peters, Naturwiss. Reise Mossamb. Bot. **2**: 548 (1864). Type: Mozambique, Cabaceira, *Peters* s.n. (B, holotype).

A loosely caespitose annual (or short-living perennial?). Culms 40–70 cm., erect or ascending, glabrous, nodes dark, glabrous. Leaf sheaths glabrous. Ligule 0.5–1 mm. long, rim-like. Leaf laminae 4–9 × 0.4–1 cm., linear, flat, glabrous on both surfaces, sometimes minutely scaberulous, margins scaberulous, crisped. Inflorescence composed of 2–4 racemes, (5)9–14 cm. long, erect, solitary along a short common axis. Rhachis triquetrous, broadly winged, up to 1 mm. broad, smooth with scaberulous margins. Pedicel 3-nate, 0.5–3 mm. long, subterete, smooth, slightly broadened at the apex. Spikelets 2.0–3.0 mm. long, oblong-lanceolate. Inferior glume rim-like or absent. Superior glume somewhat shorter than the spikelet, oblong triangular, with 3 prominent nerves, appressed hairy, hairs white to purplish, undulating, with a rough, papillose wall. Inferior lemma as long as the spikelet, oblong-lanceolate with recurved margins, with 3 prominent nerves and sometimes 2 less developed marginal nerves, appressed hairy, hairs overtopping the spikelet. Superior lemma somewhat shorter than the spikelet, oblong-lanceolate, apiculate, pale yellow to pale brown.

Mozambique. N: Msatu R., 20.iii.1912, *Allen* 137 (K). Z: Vila de Maganja, Mocuba, km. 59, 100 m., 9.i.1968, *Torre & Correia* 16999 (LISC). GI: Panda Inharrime, 1.viii.1944, *Torre* 6814 (LISC). M: Inhaca Is., 20.vii.1980, *de Koning & Nuvunga* 8302 (SRGH).

Also in Kenya, Tanzania, and Ghana (introduced?). Coastal sands, also weed on fallow land.

45. **Digitaria sanguinalis** (L.) Scop., Fl. Carniol. ed. 2, **1**: 52 (1772). —Robyns, Esp. Cong. Digit.: 25 (1931). —Henrard, Monogr. Digit.: 647 (1950). —Chase in Hitchcock, Man. Grass. U.S.: 575, 855, fig. 827 (1951). —Maire, Fl. Afr. Nord **1**: 2983, fig. 167 (1952). —Chippindall in Meredith, Grasses & Pastures of S. Afr.: 399 (1955). —Port Res, Journ. Agric. Trop. Bot. Appl. **2**: 354 (1955). —Bor in Webbia **11**: 355 (1956); Grass. B.C.I. & P.: 304 (1960). —Ebinger in Brittonia **14**: 248 (1962). —Gould in Brittonia **15**: 241 (1963).—Bor in Fl. Iran. **70**: 491 (1970). —Rugolo de Agrasar in Darwiniana **19**: 149, fig. 24 (1974); in Hickenia **1**: 27 (1976). —Bennett in Kirkia **11**: 236 (1980). —Clayton in F.T.E.A., Gramineae: 650 (1982). TAB. **41**, fig. 45. Type from Europe.
Panicum sanguinale L., Sp. Pl. **1**: 57 (1753).—Kunth, Enum. Pl. **1**: 82 (1833). —Nees, Fl. Afr. Austr.: 28 (1841). —Steudel, Syn. Pl. Glum. **1**: 39 (1853). —Durand & Schinz in Consp. Fl. Afr. **5**: 761 (1894). Type as above.

A loosely caespitose or solitary growing annual. Culms 20–60 cm., decumbent or creeping at the base, glabrous, nodes dark and glabrous. Leaf sheaths smooth to scaberulous. Ligule 1–2 mm. long, truncate, erose. Leaf laminae 3–12 × 0.1–0.5 cm., linear, flat, scaberulous on both surfaces, with a few bulbous based bristles on the superior surface, near the base, scaberulous along the margin. Inflorescence composed of (2)4–8 racemes, (3)6–12 cm. long, erect, (2)3–4(5) in a inferior whorl, sometimes a second, superior whorl of 2–3 racemes. Rhachis triquetrous, broadly winged, up to 0.7 mm. wide, smooth to scaberulous with scabrous margins. Pedicels 2-nate, 0.3–3 mm. long, triangular, scabrous, scarcely widened at the apex. Spikelets (1.8)2.1–2.8 mm. long, ovate-oblong. Inferior glume short, triangular, sometimes truncate or even slightly bifid. Superior glume $1\frac{2}{3}$ of the spikelet, ovate to oblong triangular, 3-nerved, appressed hairy, hairs fine, smooth, acute. Inferior lemma as long as the spikelet, ovate-oblong, 7-nerved, nerves scabrous, especially the subcentral and submarginal nerves, central zone often broad, appressed hairy. Superior lemma as long as to somewhat shorter than the spikelet, oblong-lanceolate, slightly acuminate, pale yellow to pale brown, sometimes purplish tinged.

Botswana. N: Qangwa R., spring, 20.iv.1981, *Smith* 3612 (GENT; SRGH). SE: Gaborone Dam, 30.v.1982, *Smith* 3922 (SRGH). **Zambia**. B: Machili, 19.iv.1960, *Fanshawe* 5655 (SRGH). N: Mpulungu, 900 m., 18.ii.1959, *McCallum-Webster* A77 (K; SRGH). C: Lusaka, vii.1968, *Anton-Smith* in GHS 203922 (SRGH). E: Mkania (Mkhania), 600 m., 25.ii.1969, *Astle* 5534 (K; SRGH). S: Kafue Nat. Park, 7.iii.1969, *Hanks* 28–5 (SRGH). **Zimbabwe**. N: Mana Pools Game Reserve, 3.iv.1984, *Dunham* 301 (SRGH). W: Hwange Distr., E. of Halfway Hotel, iii.1956, *Rattray* 1778 (K; PRE; SRGH). C: Harare, 1500 m., 15.ii.1922, *Eyles* 3288 (SRGH). **Malawi**. N: 2 km. E. of Rumphi, 1200 m., 2.iv.1977, *Pawek* 12555 (SRGH). S: Mulanje Distr., 29.x.1958, *Jackson* 2245 (BM; K; SRGH). **Mozambique**. N: Mutuali, 25.iv.1961, *Balsinhas & Marrime* 446 (COI; K; LISC; LMA; PRE; SRGH). Z: 43 km. from Namacurra on Quelimane Rd., 30 m., 31.i.1966, *Torre & Correia* 14256 (LISC). T: 35 km. from Chioco on Chicoa Rd., 250 m., 15.ii.1968, *Torre & Correia* 17629 (LISC). MS: Gorongosa Nat. Park, iii.1969, *Tinley* 1739 (SRGH). M: Guija-Bilene, 13.v.1944, *Torre* s.n. (LISC).

Widespread in warm temperate and tropical regions, introduced in many parts. Weed.

Species 45–50 are closely related and they intergrade to an important degree. It seems to us that these species are defined rather arbitrarily. Again herbarium taxonomy alone cannot resolve this problem, help is needed from tropical experimental plots.

D. asthenes Clayton was described from Tanzania and Zambia. The only known Zambian specimen, *Astle* 4613 (K; SRGH) has scabrid nerves and no stiff bristles along the lemma margins, contrary to the description. We have included this specimen in *D. sanguinalis*, as a delicate, edaphic hunger form, quite comparable to the case of *D. annua* (see *D. leptorrhachis*, note).

46. **Digitaria acuminatissima** Stapf in F.T.A. **9**: 441 (1919). —Robyns, Esp. Cong. Digit.: 26 (1931); in Fl. Agrost. Congo Belge **2**: 25 (1934). —Henrard, Monogr. Digit.: 4 (1950). —Simon in Kirkia **8**: 65 (1971). —Clayton in F.W.T.A. ed. 2, **3**: 452 (1972); in Kew Bull. **29**: 517 (1974). —Rose Innes, Man. Ghana Grass.: 131 (1977). —Bennett in Kirkia **11**: 234 (1980). —Clayton in F.T.E.A.,

Gramineae: 650 (1982). TAB. **41**, fig. 46. Type from Nigeria.
Digitaria acuminatissima var. *conformis* Henrard, Monogr. Digit.: 7 (1950). Type from Burkina Faso.
Digitaria acuminatissima subvar. *grandiflora* Henrard, Monogr. Digit.: 8 (1950). Type from Chad.
Digitaria acuminatissima var. *typica* Henrard, Monogr. Digit.: 981 (1950) nom. inval. Type as for *Digitaria acuminatissima*.
Digitaria acuminatissima subsp. *inermis* P. Goetghebeur in Bull. Nat. Plantentuin Belg. **45**: 418 (1975). Type: Botswana, Lake Ngami near Sehitwa, 24.iii.1961, *Vesey-FitzGerald* 3329 (SRGH, holotype; BM, K, isotypes).

A creeping annual. Culms 30–80(100) cm., ascending from a creeping base, glabrous, nodes dark, glabrous. Leaf sheaths minutely scaberulous, sometimes loosely hairy. Ligule up to 2.5 mm. long, truncate to erose. Leaf laminae 5–12 × 0.3–0.7 cm., linear, flat, scabrous on both surfaces, with bulbous based bristles near the base, scabrous along the margin. Inflorescence composed of (2)5–14 racemes, 12–20 cm. long, erect to patent, 3–5 in a whorl, the superior ones often solitary along a short common axis, sometimes in a second whorl of 2–4 racemes. Rhachis triquetrous, broadly winged, up to 0.8 mm. wide, smooth with scabrous margins. Pedicels 2-nate, 0.5–3.0 mm. long, triangular, scabrous, scarcely broadened at the apex. Spikelets 2.8–3.7 mm. long, oblong-lanceolate. Inferior glume oblong triangular, sometimes slightly bifid. Superior glume c. $\frac{1}{2}$ of the spikelet, oblong triangular, 3-nerved, subglabrous to appressed hairy, hairs very fine, white, acute. Inferior lemma as long as the spikelet, oblong, with a flattened, curved acuminate apex, 7-nerved, appressed hairy, sometimes with a few bulbous based bristles. Superior lemma conspicuously shorter than the spikelet, oblong-lanceolate, acuminate, pale yellow to brown.

Botswana. N: Nokaneng, Taokhne Flood Plain, 21.iii.1961, *Vesey-FitzGerald* 3310 (SRGH). **Zambia**. B: Mongu, 11.iv.1966, *Robinson* 6931 (K; P; SRGH). N: Mbala, 1700 m., 6.iii.1959, *McCallum-Webster* A214 (LISC; K; SRGH). C: Kafue R., 11.i.1963, *van Rensburg* 1198 (K; PRE; SRGH). E: Luangwa R., i.1971, *Abel* 302 (SRGH). S: Kafue Gorge, 14.iii.1960, *Vesey-FitzGerald* 2706 (SRGH). **Zimbabwe**. N: Chirundu, 400 m., 15.iii.1966, *Simon* 720 (BM; K; P; PRE; SRGH). W: Gwampa Forest Reserve, i.1955, *Goldsmith* 133/55 (K; SRGH). E: 15 km. NE. of Chipinge, 10.ii.1977, *Izzett* in GHS 250901 (SRGH). **Mozambique**. N: 2 km. from Namapa, 300 m., 12.xii.1963, *Torre & Paiva* 9512 (COI; LISC; PRE; SRGH). T: 6 km. from Tete on Changara Rd., 200 m., 19.iii.1966, *Torre & Correia* 15250 (LISC). GI: Macia, Tuane, Uagunumbo R., 11.iii.1970, *Balsinhas* 1632 (LMA).
Also in tropical Africa. River banks, flood plains and wet grassland.
Extreme forms of this species are easily recognised. However the presence of so many intermediate specimens makes the delimitation versus *D. sanguinalis* rather arbitrary.

47. **Digitaria ciliaris** (Retz.) Koel., Descr. Gram.: 27 (1802). —Fig. & De Not., Agrost. Aegypt. Fragm. **2**: 46, t. 26 (1853). —Henrard, Monogr. Digit.: 129 (1950). —Blake in Proc. Roy. Soc. Queensl. **81**: 10 (1969). —Bor in Fl. Iran. **70**: 490 (1970). —Simon in Kirkia **8**: 31, 65 (1971). —Clayton in F.W.T.A. ed. 2, **3**: 453 (1972). —Veldkamp in Blumea **21**: 32, fig. 5a, t. 1 (1973). —Clayton in Kew Bull. **29**: 520 (1974). —Rugolo de Agrasar in Hickenia **1**: 27 (1976). —Rose Innes, Man. Ghana Grass.: 134 (1977). —Salem da Silva, Hoehnea **8**: 59, fig. 6–11 (1979). —Bennett in Kirkia **11**: 234 (1980). —Clayton in F.T.E.A., Gramineae: 653 (1982). TAB. **41**, fig. 47. Type from China.
Panicum ciliare Retz, Obs. **4**: 16 (1786). —Willdenow, Sp. Pl. **1**: 344 (1797). Type as above.
Panicum adscendens Kunth in H.B.K., Nov. Gen. Sp. **1**: 80 (1815). —Kunth, Enum. Pl. **1**: 83 (1833). Type from Mexico.
Digitaria marginata Link, Enum. Hort. Berol. **1**: 102 (1821). —Stapf in F.T.A. **9**: 439 (1919). —Stent in Bothalia **1**: 269 (1921). —Robyns, Esp. Cong. Digit.: 25 (1931); in Fl. Agrost. Congo Belge **2**: 24 (1934). Type from Brasil.
Digitaria fimbriata Link, Enum. Hort. Berol. **1**: 226 (1827). Type from Brasil.
Digitaria chrysoblephara Fig. & De Not., Agrost. Aegypt. Fragm. **2**: 48, fig. 27 (1853). Type from N. Africa.
Digitaria marginata var. *fimbriata* (Link) Stapf in F.T.A. **9**: 440 (1919). —Chiov. in Webbia **8**: 69 (1951). Type as for *Digitaria fimbriata*.
Digitaria marginata var. *nubica* Stapf in F.T.A. **9**: 441 (1919). —Chiov. in Webbia **8**: 69 (1951). Type from Sudan.
Digitaria adscendens (Kunth) Henrard in Blumea **1**: 92 (1934). —Eggeling, Annot. List Grass. Uganda: 13 (1947). —Henrard, Monogr. Digit.: 9 (1950). —Sturgeon in Rhod. Agric. Journ. **50**: 284 (1953). —Chippindall in Meredith, Grasses & Pastures of S. Afr.: 399 (1955). —Andrews, Fl. Pl. Sudan **3**: 438 (1956). —Bor in Webbia **11**: 350 (1956); Grass. B.C.I. & P.: 298 (1960). —Harker & Napper, Illustr. Guide Grass. Uganda: 25, t. 59 (1960). —Ebinger in Brittonia **14**: 248 (1962). —Gould in Brittonia **15**: 241 (1963). —Napper, Grass. Tangan.: 76 (1965). —Rugolo de Agrasar in Darwiniana **19**: 153 (1974). Type as for *Panicum adscendens*.
Digitaria adscendens subsp. *chrysoblephara* (Fig. & De Not.) Henrard, Monogr. Digit.: 125

(1950). —Andrews, Fl. Pl. Sudan **3**: 439 (1956). —Bor in Webbia **11**: 352 (1956); Grass. B.C.I. & P.: 299 (1960). Type as for *Digitaria chrysoblephara*.
　　Digitaria adscendens subsp. *nubica* (Stapf) Henrard, Monogr. Digit.: 431 (1950). —Andrews, Fl. Pl. Sudan **3**: 439 (1956). Type as for *Digitaria marginata* var. *nubica*.
　　Digitaria ciliaris subsp. *chrysoblephara* (Fig. & De Not.) Blake in Proc. Roy. Soc. Queensl. **81**: 12 (1969). Type as for *Digitaria chrysoblephara*.
　　Digitaria ciliaris subsp. *nubica* (Stapf) Blake in Proc. Roy. Soc. Queensl. **81**: 12 (1969). Type as for *Digitaria marginata* .
　　Digitaria adscendens var. *fimbriata* (Link) Cufod. in Bull. Jard. Bot. Nat. Belg. **39**, suppl.: 1327 (1969). Type as for *Digitaria fimbriata*.

A loosely caespitose or solitary growing annual. Culms 20–60 cm., decumbent to ascending, glabrous, nodes dark and (sub)glabrous. Leaf sheaths scaberulous, sometimes with a few bulbous based bristles. Ligule up to 2.5 mm. long, truncate, erose. Leaf laminae 4–13 × 0.3–0.8 mm., linear, flat, scaberulous on both sides, with a few bulbous based bristles near the base, scabrous along the margin. Inlforescence composed of 2–9 racemes, 6–12 cm. long, in 1 or 2-more superposed whorls, sometimes a few solitary along the short common axis. Rhachis triquetrous, winged, up to 1.0 mm. broad, smooth with scabrous margins. Pedicels 2-nate, 0.5–2.5 mm. long, triangular, scabrous, scarcely broadened at the apex. Spikelets 2.7–3.4 mm. long, oblonglanceolate. Inferior glume c. 0.5 mm. long, ovate to oblong triangular, often somewhat truncate. Superior glume $\frac{2}{3}$–$\frac{3}{4}$ of the spikelet, oblong triangular, 3-nerved, appressed hairy, hairs very fine, smooth, acute. Inferior lemma as long as the spikelet, oblong-lanceolate, 7-nerved, nerves smooth or nearly smooth, central zone often very broad, appressed hairy, sometimes also with bristle-hairs, rarely glabrous. Superior glume as long as to somewhat shorter than the spikelet, oblong-lanceolate, acute, pale yellow to pale brown, often purplish tinged.

Zimbabwe. N: Karoi, Glen Ellen Farm, 1200 m., 20.vii.1982, *Sheppard* in GHS 280652 (GENT; SRGH). C: Harare, 1700 m., 26.i.1971, *Davies* 3127 (P; SRGH). E: Mutare, 12.ii.1920, *Eyles* 3055 (SRGH). S: Masvingo (Fort Victoria), Glyntor, 27.xii.1947, *Robinson* 63 (K; SRGH). **Mozambique**. N: Erati, Namapa Exp. Stat., 2.iii.1960, *Lemos & Macuácua* 10 (COI; K; LISC; LMA; PRE). MS: Gondola, 7.vi.1956, *Myre & Balsinhas* 2381 (LMA).
Pantropical, but rather rare and scattered in Africa. Weed, on disturbed sites.

48. **Digitaria nuda** Schumach., Beskr. Guin. Pl.: 65 (1827). —Stapf in F.T.A. **9**: 479 (1919). —Henrard, Monogr. Digit.: 500 (1950). —Robyns & Tournay, Fl. Parc Nat. Alb. **3**: 78 (1955). —Veldkamp in Blumea **21**: 41, fig. 6b (1973). —Rose Innes, Man. Ghana Grass.: 138 (1977). —Bennett in Kirkia **11**: 236 (1980). —Clayton in F.T.E.A., Gramineae: 654 (1982). TAB. **41**, fig. 48. Type from W. Africa.
　　Digitaria borbonica Desv., Opusc.: 63 (1831). —Henrard, Monogr. Digit.: 82 (1950). — Chippindall in Meredith, Grasses & Pastures of S. Afr.: 400 (1955). —Jackson & Wiehe, Annot. Check-list Nyasal. Grass.: 35 (1958). Type from Bourbon Is.
　　Panicum diamesum Steud., Syn. Pl. Glum. **1**: 42 (1853). —Durand & Schinz in Consp. Fl. Afr. **5**: 747 (1894). Type from Senegal.
　　Digitaria diamesa (Steud.) A. Chev., Rev. Bot. Appl. **27**: 284 (1947). —Henrard, Monogr. Digit.: 177 (1950). Type as above.
　　Digitaria nuda subsp. *schumacheriana* Henrard, Monogr. Digit.: 952 (1950) nom. illeg. Type as for *Digitaria nuda*.
　　Digitaria nuda subsp. *senegalensis* Henrard, Monogr. Digit.: 831 (1950). Type from Senegal.

A creeping or decumbent, sometimes mat-forming annual. Culms 10–50 cm., glabrous, nodes dark and (sub)glabrous. Leaf sheaths scaberulous, sometimes loosely hairy. Ligule 1–2 mm. long, truncate, erose. Leaf laminae (2)5–12 × 0.3–0.6(0.8) cm., linear, flat, scaberulous on both surfaces, scaberulous along the margin. Inflorescence composed of 2–10 racemes, (3)7–12 cm. long, in one or two superposed whorls of 2–5 racemes each. Rhachis triquetrous, winged, up to 0.7 mm. broad, smooth with scabrous margins. Pedicels 2-nate, 0.5–2 mm. long, triangular, scaberulous, slightly broadened at the apex. Spikelets 2–2.8 mm. long, oblong to lanceolate. Inferior glume shorter than 0.2 mm., poorly developed or even absent. Superior glume $\frac{1}{2}$–$\frac{2}{3}$ of the spikelet, oblong triangular, 3-nerved, appressed hairy, hairs fine, smooth, acute, hyaline. Inferior lemma as long as the spikelet, oblonglanceolate, 5–7-nerved, nerves smooth, mostly (but not always) equidistant, appressed hairy. Superior lemma as long as to slightly shorter than the spikelet, oblong-lanceolate, acute, pale yellow to reddish brown, sometimes bluish grey.

Zambia. B: 3 km. N. of Kabompo, 1020 m., 31.iii.1982, *Drummond & Vernon* 11053 (GENT; SRGH). N: near Bulaya, 1100 m., 13.viii.1962, *Tyrer* 440 (BM; SRGH). W: Kasempa Distr., Kashima Settlement Scheme, 1175 m., 1.iv.1982, *Drummond & Vernon* 11114 (GENT; SRGH). C: Firala, head of Mungala, iii.1941, *Stohr* 566 (SRGH). E: 12 km. from Chipata turnoff on Rd. to Lundazi, 23.ii.1971, *Anton-Smith* 808 (SRGH). S: 22 km. W. of Livingstone, 925 m., 23.iii.1982, *Drummond & Vernon* 10944 (GENT; SRGH). **Zimbabwe**. N: Kerry Farm, 1100 m., 16.vii.1969, *Mogg* 34382 (SRGH). C: Darwendale Dam, 20.iv.1983, *Sheppard* in GHS 281375 (GENT; SRGH). E: 18 km. SW. of Chimanimani (Melsetter), 23.vi.1953, *Crook* 491 (K; PRE; SRGH). **Malawi**. C: Salima Distr., iv.1972, *Williamson* 2174B (SRGH). S: Thornwood Estate, 25.v.1950, *Wiehe* 552 (K; SRGH). **Mozambique**. N: 20.iii.1970, *Macêdo* 744 (LMA). Z: Jagarra, 8.x.1946, *Pedro* 2165 (LMA). T: Zóbuè, 11.vii.1940, *Hornby* 3035 (LISC). MS: Mavita, 28.iv.1948, *Barbosa* 1639 (LISC). M: Bela Vista, 11.iv.1949, *Myre & Balsinhas* 541 (LMA; SRGH).

Tropical Africa. Weed on ruderal land.

49. **Digitaria velutina** (Forssk.) Beauv., Ess. Agrost.: 51 (1812). —Eggeling, Annot. List Grass. Uganda: 78 (1947). —Henrard, Monogr. Digit.: 776 (1950). —Chiovenda in Webbia **8**: 70 (1951). —Sturgeon in Rhod. Agric. Journ. **50**: 284 (1953). —Chippindall in Meredith, Grasses & Pastures of S. Afr.: 400 (1955). —Robyns & Tournay, Fl. Parc Nat. Alb. **3**: 78 (1955). —Andrews, Fl. Pl. Sudan **3**: 435, fig. 105 (1956). —Bogdan, Rev. List Kenya Grass.: 49 (1958). —Jackson & Wiehe, Annot. Check-list Nyasal. Grass.: 36 (1958). —Harker & Napper, Illustr. Guide Grass. Uganda: 26, t. 59 (1960). —Napper, Grass. Tangan.: 78 (1965). —Simon in Kirkia **8**: 31 (1971). —Bennett in Kirkia **11**: 237 (1980). —Clayton in F.T.E.A., Gramineae: 652 (1982). TAB. **41**, fig. 49. Type from Yemen.

Phalaris velutina Forssk., Fl. Aegypt.-Arab. **1**: 17 (1775). Type as above.

Panicum zeyheri Nees, Fl. Afr. Austr.: 25 (1841). —Steudel in Syn. Pl. Glum. **1**: 40 (1853). —Durand & Schinz in Consp. Fl. Afr. **5**: 767 (1894). Type from S. Africa.

Panicum fenestratum Hochst. ex A. Rich., Tent. Fl. Abyss. **2**: 361 (1851). Steudel, Syn. Pl. Glum. **1**: 40 (1853). —Durand & Schinz, Consp. Fl. Afr. **5**: 749 (1894). Type from Ethiopia.

Digitaria fenestrata (Hochst. ex A. Rich.) Rendle, Cat. Afr. Pl. Welw. **2**: 163 (1899). —Henrard, Monogr. Digit.: 248 (1950). Type as above.

Digitaria divaricata Henrard in Blumea **1**: 96 (1934); in Monogr. Digit.: 195 (1950). —Napper, Grass. Tangan.: 78 (1965). Type from Tanzania.

Digitaria zeyheri (Nees) Henrard in Blumea **1**: 105 (1934); in Monogr. Digit.: 806 (1950). —Chippindall in Meredith, Grasses & Pastures of S. Afr.: 400, fig. 336 (1955). —Napper, Grass. Tangan.: 78 (1965). —Launert in Merxm., Prodr. Fl. SW. Afr. **160**: 66 (1970). —Simon in Kirkia **8**: 31 (1971). Type as for *Panicum zeyheri*.

Digitaria ulugurensis Pilg., Notizbl. Bot. Gart. Berlin **15**: 709 (1942). —Napper, Grass. Tangan.: 78 (1965). Type from Tanzania.

A creeping or decumbent annual. Culms creeping to ascending, 20–50 cm., glabrous, nodes dark and glabrous. Leaf sheaths loosely to densely covered with bulbous based bristles, smooth to scaberulous. Ligule up to 2.5 mm. long, truncate, entire. Leaf laminae 3–13 × 0.3–0.8 cm., linear, flat, loosely to densely hairy and sometimes scaberulous on both surfaces, often with a few bulbous based bristles near the base, scabrous along the margin. Inflorescence composed of (2)3–14 racemes, (3)5–10 cm. long, sometimes in a basal whorl of 3–5 racemes, the other racemes solitary along the well developed common axis. Rhachis triquetrous, winged, up to 0.5 mm. broad, smooth to scaberulous, with scabrous margins. Pedicels 2-nate, 0.5–3 mm. long, triangular, scabrous, slightly broadened at the apex. Spikelets 1.7–2.1 mm. (subsessile spikelets) and 2.0–2.3 mm. long (pedicelled spikelets), oblong-lanceolate. Inferior glume short, truncate, nerveless, often poorly developed. Superior glume $\frac{2}{3}$–$\frac{3}{4}$ of the spikelet, sometimes nearly as long, 3-nerved, appressed hairy, hairs fine, smooth, acute, hyaline. Inferior lemma as long as the spikelet, oblong, 7-nerved, appressed hairy. Superior lemma as long as to somewhat shorter than the spikelet, acute, pale to dark brown, often purplish tinged.

Botswana. N: Maun, 27.ii.1976, *Smith* 1644 (GENT; SRGH). SW: Okwa Valley, 5.iii.1978, *Skarpe* 272 (SRGH). SE: Palapye, Malede Village, 1000 m., 17.i.1958, *de Beer* 565 (LISC; SRGH). **Zambia**. C: Mwembeshi, 1300 m., 27.iii.1963, *Vesey-FitzGerald* 4036 (BM; COI). S: Lusitu R., 12.iii.1960, *Vesey-FitzGerald* 2704 (SRGH). **Zimbabwe**. N: Mkumbura, 500 m., 29.i.1970, *Simon* 2107 (BM; K; SRGH). W: Insiza, 6.iii.1967, *Cleghorn* 1490 (SRGH). C: Harare, Marlborough, ii.1954, *Kerr* in GHS 48552 (K; PRE; SRGH). E: Maranke Reserve, 19.iii.1957, *Cleghorn* 237 (SRGH). S: Mrore R. bridge, 28.ii.1967, *Cleghorn* 1472 (K; SRGH). **Malawi**. C: Kamuzu Academy, 1.v.1979, *Banda* 1419 (SRGH). S: Makanga, 90 m., 19.iii.1960, *Phipps* 2539 (BR; COI; K; PRE; SRGH). **Mozambique**. T: 3 km. from Tete on Changara Rd., 13.ii.1968, *Torre & Correia* 17544 (LISC).

From Egypt to S. Africa. Weed of disturbed land.

50. **Digitaria pearsonii** Stapf in F.T.A. **9**: 434 (1919). —Henrard, Monogr. Digit.: 530 (1950). —Bennett in Kirkia **11**: 236 (1980). —Clayton in F.T.E.A., Gramineae: 643 (1982). TAB. **41**, fig. 50. Type from Angola.

Digitaria lancifolia Henrard in Blumea **1**: 102 (1934); in Monogr. Digit.: 372 (1950). —Jackson & Wiehe, Annot. Check-list Nyasal. Grass.: 35 (1958). —Napper, Grass. Tangan.: 78 (1965). —Simon in Kirkia **8**: 31 (1971). Type from Tanzania.

A loosely caespitose perennial on a short branched rhizome, base surrounded by a few cataphylls. Culms 40–70 cm., creeping to ascending, glabrous, nodes dark and glabrous. Leaf sheaths smooth to scaberulous, loosely to densely hairy with bulbous based bristles. Ligule 1–2 mm. long, truncate, erose. Leaf laminae 4–20 × 0.5–1.5 cm., lanceolate to linear, flat, scaberulous, loosely hairy on both surfaces, scabrous along the margin. Inflorescence composed of 7–20 racemes, 4–12 cm. long, patent, often branched, solitary along a well developed common axis. Rhachis triquetrous, scarcely winged, smooth to minutely scaberulous with scabrous margins. Pedicels 2(3)-nate, 0.5–2.5 mm. long, triangular, scabrous, scarcely broadened at the apex, spikeletgroups scattered along the rhachis, often patent. Spikelets 1.8–2.5 mm. long, lanceolate. Inferior glume small, ovate to triangular, sometimes truncate to bifid. Superior glume $\frac{1}{2}$–$\frac{3}{4}$ of the spikelet, oblong triangular, 3-nerved, appressed hairy, hairs fine, hyaline, smooth, acute. Inferior lemma as long as the spikelet, lanceolate, 7-nerved, nerves equidistant, smooth, appressed hairy. Superior lemma somewhat smaller than the spikelet, lanceolate, acute, pale yellow to pale brown, sometimes dark brown or purplish tinged.

Zimbabwe. E: Nyanga (Inyanga) Nat. Park, 2000 m., 5.ii.1966, *Crook* 777 (BM; K; PRE; SRGH). **Malawi**. N: Viphya, Mzuzu, 11.viii.1949, *Wiehe* 192 (K; PRE; SRGH). C: Chipata Mt., 2000 m., 4.v.1963, *Verboom* 982 (SRGH). S: Misuku Hills, 11.i.1959, *Robinson* 3157 (BM; K; PRE; SRGH). **Mozambique**. N: Maniamba, Serra Geci, 29.v.1948, *Pedro & Pedrogão* 4107 (LMA).

Also in Zaire, Rwanda, Burundi, Uganda, Tanzania, and Angola. Woodland clearings and forest edges.

This perennial species is often -and easily- confused with the annual *D. velutina*. Specimens with unbranched racemes and without basal parts cannot be named with certainty.

51. **Digitaria eriantha** Steud. in Flora **12**: 468 (1829). —Stapf in F.C. **7**: 375 (1898). —Stapf in F.T.A. **9**: 429 (1919). —Stent in Bothalia **1**: 266 (1924). —Henrard, Monogr. Digit.: 227, 319, 971 (1950). —Sturgeon in Rhod. Agric. Journ. **50**: 288 (1953). —Chippindall in Meredith, Grasses & Pastures of S. Afr.: 403, fig. 338 (1955). —Bor in Fl. Iran. **70**: 488 (1970). —Launert in Merxm., Prodr. Fl. SW. Afr. **160**: 63 (1970). —Simon in Kirkia **8**: 31 (1971). —Bennett in Kirkia **11**: 235 (1980). —Kok in S. Afr. Journ. Bot. **3**: 184–185 (1984). TAB. **41**, fig. 51. Type from S. Africa.

Digitaria livida Henrard in Blumea **1**: 101 (1934); in Monogr. Digit.: 406, 968 (1950). Type: Zimbabwe, Gold Fields SE. of Bulawayo, 11.ii.1932, *Pottensy* 5503 (SRGH, isotype).

Digitaria nemoralis Henrard, Monogr. Digit.: 488 (1950). —Sturgeon in Rhod. Agric. Journ. **50**: 283 (1953). —Simon in Kirkia **8**: 31, 66 (1971). —Bennett in Kirkia **11**: 236 (1980). Type: Zimbabwe, Gwaai Forest Reserve, 18.iii.1931, *Pardy* 4065 (K, isotype; L, holotype).

A more complete synonymy is published by Kok loc. cit.

A tightly caespitose perennial, often provided with well developed runners; base surrounded by hairy cataphylls and old leaf sheath remnants. Culms 40–120 cm., ascending to erect, glabrous, nodes dark and glabrous. Leaf sheaths scaberulous, glabrous or loosely hairy. Ligule 2–4 mm. long, subtriangular, shortly ciliate. Leaf laminae 5–20 × 0.2–0.4(0.7) mm., linear, flat to involute, minutely scaberulous on both surfaces, often with a few bulbous based bristles near the base, scabrous along the margins. Inflorescence composed of 3–10 racemes, 5–20 cm. long, erect, (2)3–6 together in a inferior whorl, sometimes a few solitary along a short common axis, and mostly 2–4 together in a second, superior whorl. Rhachis triquetrous, narrowly winged, up to 0.5 mm. broad, smooth to scaberulous, with scabrous margins. Pedicels 2-nate, 0.5–2.5 mm. long, subterete to subtriangular, scabrous, broadened at the apex. Spikelets 2.2–3.5 mm. long, oblong. Inferior glume up to 0.5 mm. long, ovate to triangular, sometimes acuminate and ciliate. Superior glume $\frac{1}{2}$–$\frac{2}{3}$ of the spikelet, oblong triangular, 3-nerved, appressed hairy, hairs fine, smooth, acute. Inferior lemma as long as the spikelet, oblong, 7-nerved, nerves smooth or slightly scaberulous, appressed hairy, with or without bristle-hairs. Superior lemma somewhat shorter than the spikelet, oblong, acute, yellowish green to pale brown.

Caprivi Strip. Linyanti, 1000 m., 28.xii.1958, *Killick & Leistner* 3162 (SRGH). **Botswana**. N: 15 km. SW. of Maun, 900 m., 24.i.1972, *Biegel & Gibbs-Russell* 3777 (GENT; K; SRGH). SW: 70 km. from Kang on Ghanzi Rd., 19.ii.1960, *de Winter* 7377 (K; SRGH). SE: Khutse Game Reserve, 18.i.1979, *Timberlake* 1957 (SRGH). **Zimbabwe**. N: Sengwa Res. Stat., 11.i.1974, *Guy* 2214 (SRGH). W: Matopos Res. Stat., 12.ii.1954, *Rattray* 1597 (K; SRGH). C: Kwe Kwe (Que Que), Mlezu School, ii.1960, *Davies* 2696 (BM; PRE; SRGH). E: Sabi Valley, ii.1948, *Rattray* 1283 (K; SRGH). S: Gwanda Distr., Makwe Dam, 1100 m., 8.iii.1967, *Cleghorn* 1528 (SRGH). **Mozambique**. N: 2 km. from Metuge on Quissanga Rd.,

19.xii.1963, *Torre & Paiva* 9608 (COI; K; LISC; PRE). T: Changara, 230 m., 25.iii.1966, *Torre &
Correia* 15351 (LISC). MS: Chemba, Chiou, 13.iv.1960, *Lemos & Macuácua* 96 (COI; LISC; LMA; PRE;
SRGH). GI: San Sebastian Peninsula, 10.xi.1958, *Mogg* 29141 (BM; K; LMU; PRE; SRGH). M:
Inhaca Is., 15.vii.1959, *Mogg* 29368 (K; PRE; SRGH).
    Also in Angola, Namibia and S. Africa. Miombo woodland, degraded savanna, flood plains and
saline marshes.
    Species 51–57 are all closely related and intergrade nearly completely. They are often defined by
rather trifling vegetative characters (except for 54–56) of unascertained taxonomical value. Many
specimens, lacking subterranean parts, cannot be named with certainty. It is clear that herbarium
taxonomy alone is not able to unravel this extremely complex pattern.

52. **Digitaria swazilandensis** Stent in Bothalia **3**: 150 (1930). —Henrard, Monogr. Digit.: 724 (1950).
    —Chase in Hitchcock, Man. Grass. U.S.: 584 (1951). —Chippindall in Meredith, Grasses &
    Pastures of S. Afr.: 397, fig. 334 (1955). —Bennett in Kirkia **11**: 236 (1980). TAB. **41**, fig. 52. Type
    from S. Africa.

A loosely caespitose perennial with a knotty much branched rhizome. Stems prostrate
or creeping and then sometimes branching, nodes dark and glabrous. Leaf sheaths
glabrous, smooth to scaberulous. Ligule c. 0.8 mm. long, rimlike, glabrous. Leaf lamina
3–6(10) × 0.3–0.6 cm., linear, flat, (sub)glabrous and smooth on both surfaces, often
glaucous, scaberulous along the margins. Inflorescence composed of 2–3(4) racemes,
3–10 cm. long, digitately on top of the peduncle. Rhachis triquetrous, winged, up to 0.6
mm. wide, smooth with scaberulous margins. Pedicels 2-nate, 0.5–2.5 mm. long, subterete
to triangular, scaberulous, slightly widened at the apex. Spikelets (2.0)2.2–2.7 mm. long,
oblong to lanceolate. Inferior glume short, ovate to triangular, often truncate. Superior
glume $\frac{1}{2}$–$\frac{1}{3}$ of the spikelet, triangular, 3-nerved, appressed hairy, hairs fine, whitish. Inferior
lemma as long as the spikelet, lanceolate, 7-nerved, nerves smooth to scaberulous,
appressed hairy. Superior lemma as long as the spikelet, lanceolate, acute, pale yellow to
pale brown, sometimes purplish tinged.

Malawi. S: Ndirande, New Industrial Site, 11.i.1968, *Banda* 966 (SRGH). **Mozambique**. MS:
Gorongoza Nat. Park, 9.vi.1966, *Macêdo* 2143 (LMA). GI: Govuro, 11.iii.1970, *Myre, Duarte & Rosa*
5541 (LMA). M: Namaacha, 8.xii.1954, *Myre & Carvalho* 1950 (LMA; P).
    Also in S. Africa. Grassland in open woodland clearings and on alluvial soil, disturbed areas and
roadsides.
    It is not clear whether this species is sufficiently distinct from *D. didactyla* Willd. to keep it separate.
Clayton (F.T.E.A., Gramineae: 649 (1982)) and Kok (Journ. S. Afr. Bot. **3**: 184 (1984)) hold diverging
opinions. Some slender creeping specimens of *D. eriantha* may be difficult to distinguish from this
species.

53. **Digitaria polyphylla** Henrard, Monogr. Digit.: 583 (1950). —Chippindall in Meredith, Grasses &
    Pastures of S. Afr.: 414, fig. 345 (1955). TAB. **41**, fig. 53. Type from S. Africa.
    Digitaria foliosa Stent in Bothalia **3**: 154 (1930) non Lagasca (1816). Type as above.

A rhizomatous perennial, rhizome branched, covered by hairy cataphylls. Culms 20–40
cm., erect, nodes dark and glabrous. Inferior leaves with reduced lamina, soon withering;
median and superior leaves crowded. Leaf sheaths glabrous, smooth. Ligule c. 1 mm.
long, truncate. Leaf lamina 4–10 × 0.2–0.4 cm., linear, flat, smooth and glabrous, with
scaberulous margins. Inflorescence composed of 3–4 racemes, 2–8 cm. long, erect,
whorled. Rhachis winged, 0.5 mm. wide, smooth with scabrous margins. Pedicels 2-nate,
0.5–2 mm. long, triangular, scaberulous, slightly widened at the top. Spikelets 2.1-2.5 mm.
long, oblong-lanceolate. Inferior glume short, rounded triangular, nerveless, hyaline.
Superior glume c. $\frac{2}{3}$ of the spikelet, triangular, 3-nerved, hyaline, appressed hairy, hairs
fine, white. Inferior lemma as long as the spikelet, oblong-lanceolate, with recurved
margins, 7-nerved, nerves about equidistant, appressed hairy. Superior lemma as long as
the spikelet, oblong-lanceolate, acuminate, pale brown.

Mozambique. GI: Maputo to Macia, 29.viii.1949, *Myre & Balsinhas* 767 (SRGH).
    Also in S. Africa. Grassland on sandy soil in low rainfall regions.

54. **Digitaria argyrograpta** (Nees) Stapf in F.C. **7**: 374 (1898). —Stent in Bothalia **1**: 266 (1924).
    —Henrard, Monogr. Digit.: 46 (1950). —Chippindall in Meredith, Grasses & Pastures of S. Afr.:
    414, fig. 346 (1955). TAB. **41**, fig. 54. Type from S. Africa.
    Panicum argyrograptum Nees, Fl. Afr. Austr.: 27 (1841). —Steudel in Syn. Pl. Glum. **1**: 40 (1853).
    Type as above.

A loosely caespitose perennial on a short rhizome. Culms 20–50 cm., ascending, glabrous, nodes dark, glabrous. Leaf sheaths glabrous to loosely hairy. Ligule c. 0.5 mm. long, truncate. Leaf laminae 3–5 × 0.1–0.2 cm., linear, flat, scaberulous on both surfaces, scabrous along the margins. Inflorescence composed of 2–3 racemes, 5–10 cm. long, stiffly erect, close together, 1–2 subsessile, 1 pedunculate. Rhachis triquetrous, scarcely winged, up to 0.5 mm. broad, scabrous with scabrous margins. Pedicels 2-nate, 0.5–3 mm. long, triangular, scabrous, broadened at the apex. Spikelets 3.5–3.8 mm. long, lanceolate. Inferior glume c. 0.5 mm. long, triangular, nerveless, hyaline. Superior glume c. ⅘ of the spikelet, oblong triangular, 3-nerved, appressed hairy, hairs very fine, white, acute. Inferior lemma as long as the spikelet, oblong-lanceolate, 7-nerved, the central and 2 subcentral nerves always close together, appressed hairy. Superior lemma somewhat shorter than the spikelet, lanceolate, acute, pale yellow.

Mozambique. M: Maputo, near Mazeminhama, 18.xi.1952, *Myre* 1277 (LISC; LMA; SRGH). Also in S. Africa. Grassland clearings in open woodland.

55. **Digitaria natalensis** Stent in Bothalia 3: 152 (1930). —Henrard, Monogr. Digit.: 480 (1950). —Chippindall in Meredith, Grasses & Pastures of S. Afr.: 407 (1955). TAB. **41**, fig. 55. Type from S. Africa.
    Panicum commutatum var. *fluviatile* Nees, Fl. Afr. Austr.: 25 (1841). Type from S. Africa.
    Digitaria littoralis Stent in Bothalia 3: 153 (1930) nom. illeg. non *Digitaria littoralis* Salisbury (1796). Type from S. Africa.
    Digitaria rigida Stent in Bothalia 3: 151 (1930). —Henrard, Monogr. Digit.: 625 (1950). —Chippindall in Meredith, Grasses & Pastures of S. Afr.: 407 (1955). Type from S. Africa.
    Digitaria littoralis var. *prostrata* Stent in Bothalia 3: 153 (1930). Type from S. Africa.
    Digitaria macroglossa Henrard, Monogr. Digit.: 404 (1950). Type as for *Digitaria littoralis*.
    Digitaria macroglossa var. *prostrata* (Stent) Henrard, Monogr. Digit.: 419 (1950). Type as for *Digitaria littoralis* var. *prostrata*.
    Digitaria natalensis subsp. *stentiana* Henrard, Monogr. Digit.: 482 (1950). Type from S. Africa.
    Digitaria natalensis var. *paludicola* Henrard, Monogr. Digit.: 483 (1950). Type from S. Africa.

A tightly caespitose perennial, growing on a short, robust rhizome, base surrounded by conspicuously rusty-brown, persistent leaf sheaths, sometimes also by a few densely hairy cataphylls. Culm 70–160 cm., erect to ascending, glabrous, nodes dark and glabrous. Leaf sheaths scaberulous. Ligule 3–12 mm. long, oblong triangular, sometimes truncate and loosely hairy. Leaf laminae 10–40 × 0.3–0.7 cm., linear, flat, scaberulous and glabrous on both surfaces. Inflorescence composed of 5–13 racemes, 15–20 cm. long, erect, in 2 inferior whorls of 3–5 racemes, sometimes a few racemes solitary, and a terminal whorl of 3–5 racemes. Rhachis triquetrous, winged, up to 0.8 mm. broad, loosely to densely scabrous, with scabrous margins. Pedicels 2-nate, 0.5–3.5 mm. long, triangular, scabrous, somewhat asymmetrically and slightly broadened at the apex. Spikelets 2.8–4.2 mm. long, oblong. Inferior glume up to 1 mm., ovate to triangular, sometimes ciliate (exceptionally up to 2 mm., 3-nerved and hairy). Superior glume c. ⅔ of the spikelet, oblong triangular, 3-nerved, appressed hairy, hairs fine, smooth, acute. Inferior lemma as long as the spikelet, oblong, 7-nerved, nerves and interspaces scabrous or scaberulous, appressed hairy, sometimes with bristle-hairs. Superior lemma somewhat shorter than the spikelet, oblong, slightly acuminate, pale yellow to pale brown, sometimes purplish tinged.

Mozambique. GI: 27 km. from Caniçado on Nalazi Rd., 10.iii.1951, *Barbosa & Myre* 64 (LMA; K). M: Ponta do Ouro-Salamanga, 19.xi.1944, *Mendonça* 2950 (K; LISC). Also in S. Africa. Littoral sand dunes, savanna grassland and weed of fallow land.

56. **Digitaria megasthenes** P. Goetghebeur in Bull. Nat. Plantentuin Belg. **45**: 413, fig. 15 (1975). TAB. **41**, fig. 56. Type: Mozambique, Zambezia, Gúruè, 1300 m., 21.ii.1966, *Torre & Correia* 14749 (LISC, holotype).

A caespitose perennial on a very short rhizome, base surrounded by cataphylls. Culms 70–110 cm., erect or ascending, glabrous, nodes dark and glabrous. Leaf sheaths scabrous, glabrous. Ligule up to 3 mm. long, triangular or truncate, shortly ciliate. Leaf laminae 4–15 × 0.2–0.7 mm. long, linear, flat, scabrous and glabrous on both surfaces, scabrous along the margins. Inflorescence composed of 5–11 racemes, 4–12 cm. long, erect, whorled or sometimes partly solitary along a short common axis. Rhachis triquetrous, winged, up to 0.5 mm. broad, smooth to scaberulous, scabrous at the margins. Pedicels 2-nate, 0.2–1.5 mm. long, triangular, scabrous, somewhat broadened at the apex. Spikelets 1.9–2.5 mm. long, oblong-lanceolate. Inferior glume very short, ovate to triangular, often truncate. Superior glume $\frac{1}{5}$–$\frac{1}{3}(\frac{1}{2})$ of the spikelet, ovate to triangular, often bifid, 0(3)-nerved, scabrous. Inferior lemma as long as the spikelet, oblong-lanceolate, 7(9)-nerved, scabrous, central zone often broad and sulcate, glabrous or with scattered bristles. Superior lemma as long as the spikelet, oblong-lanceolate, acute to slightly acuminate, pale yellow, often purplish tinged.

Mozambique. N: Amaramba, 20 km. S. of Nova Freixo, 700 m., 15.ii.1964, *Torre & Paiva* 10577 (LISC). Z: Gúruè, Currarre Mt., 750 m., 11.ii.1964, *Torre & Paiva* 10533 (LISC).
Not known elsewhere. Rocky outcrops in miombo or savanna.

57. **Digitaria milanjiana** (Rendle) Stapf in F.T.A. **9**: 430 (1919). —Robyns, Esp. Congol. Digit.: 21 (1931); in Fl. Agrost. Congo Belge **2**: 20 (1934). —Henrard, Monogr. Digit.: 456, 962 (1950). —Sturgeon in Rhod. Agric. Journ. **50**: 283 (1953). —Chippindall, in Meredith, Grass. Past. S. Afr.: 407 (1955). —Bogdan, Rev. List Kenya Grass.: 29 (1958). —Jackson & Wiehe, Ann. Check List Nyasal. Grass.: 35 (1958). —Harker & Napper, Illustr. Guide Grass. Uganda: 26, pl. 58 (1960). —Napper, Grass. Tangan.: 77 (1965). —Launert in Merxm., Prodr. Fl. S.W. Afr. **160**: 64 (1970). —Simon in Kirkia **8**: 31, 65 (1971). —Bennett in Kirkia **11**: 235 (1980). —Clayton in F.T.E.A., Gramineae: 647 (1982). TAB. **41**, fig. 57. Type: Malawi, Mt. Mulanje, x.1891, *Whyte* s.n. (BM, holotype).
Panicum milanjianum Rendle, Trans. Linn. Soc., Ser. 2, **4**: 56 (1894). Type as above.
Digitaria swynnertonii Rendle, Journ. Linn. Soc., Bot., **40**: 227 (1911). —Henrard, Monogr. Digit.: 725, 961 (1950). Type: Mozambique, Zinyumbo Hills, 200 m., 9.i.1907, *Swynnerton* 1554 (BM, holotype).
Both a more complete synonymy and interesting taxonomic discussion are published already by Clayton, loc. cit.

A rhizomatous perennial, occasionally forming stolons, rhizomes extensively branched, covered by hairy cataphylls; base covered by glabrous or pubescent leaf sheaths, rarely villous or slightly swollen. Culms 50–200 cm., ascending to erect, glabrous, nodes dark and glabrous. Leaf sheaths glabrous to hairy. Ligule up to 2 mm. long, truncate. Leaf lamina 8–30 × 0.3–1.2 cm., linear, glabrous to loosely hairy. Inflorescence composed of 2–18 digitate or subdigitate racemes, 5–20 cm. long, erect to patent. Rhachis triquetrous, winged, up to 0.7 mm. wide, smooth to scabrous with scabrous margins. Pedicels 2-nate, 0.5–2 mm. long, subterete to triangular, scabrous, somewhat widened at the apex. Spikelets (2.1)2.5–3.2(3.5) mm. long, lanceolate. Inferior glume short, triangular or truncate. Superior glume $\frac{1}{2}$–$\frac{2}{3}$ of the spikelet, oblong-triangular, 3-nerved, appressed hairy, hairs fine, whitish. Inferior lemma as long as the spikelet, oblong, 7-nerved, nerves scabrous or nearly smooth, appressed hairy, with or without bristle hairs. Superior lemma nearly as long as the spikelet, oblong, acute, pale brown, greyish or purplish tinged.

Botswana. N: 10 km. S. of Chadum Valley, 14.iii.1965, *Wild & Drummond* 7006 (BM; K; P; SRGH). SW: 60 km. W. of Kang, 19.ii.1960, *Yalala* 74 (BM; K; SRGH). **Zambia**. B: Mongu, 20.i.1966, *Robinson* 6804 (K; SRGH). N: edge of Bwela Flats, 16.ii.1958, *Savory* 276 (PRE; SRGH). W: Kitwe, 1.ii.1958, *Fanshawe* 4215 (P; SRGH). C: Chipapa, 1150 m., 24.ii.1982, *Vernon* 775 (SRGH). E: Senegallia Farm, 30.i.1953, *Grout* 117 (K; P; SRGH). S: 23 km. N. of Livingstone, 23.iii.1982, *Drummond & Vernon* 10996 (SRGH). **Zimbabwe**. N: Mana Pools Floodplain, 31.iii.1981, *Dunham* 71 (GENT; SRGH). W: Matopos Res. Stat., 12.i.1952, *Plowes* 1390 (K; SRGH). C: Gweru (Gwelo), 1500 m., 26.xii.1966, *Biegel* 1572 (BM; K; P; PRE; SRGH). E: Lower Sabi, 400 m., 12.iii.1957, *Phipps* 593 (K; SRGH). S: Sabi-Lundi junction, 3.ii.1971, *Sherry* 100/71 (SRGH). **Malawi**. C: Kasungu Nat. Park, 1100 m., 5.ii.1971, *Hall-Martin* 1632 (K; SRGH). S: Zomba, 10.ii.1949, *Wiehe* 30 (K; SRGH). **Mozambique**. N: Nassapo Mt., 23 km. from Nampula on Meconte Rd., 400 m., 13.i.1964, *Torre & Paiva* 9916 (COI; K; LISC; SRGH). Z: Ile Mt., 3 km. from Errego, 900 m., 3.iii.1966, *Torre & Correia* 15011 (LISC). T: 37 km. from Chicoa on Magoe Rd., 300 m., 16.ii.1970, *Torre & Correia* 18002, (LISC). MS: Chimoio, Nhamissanguere R., 3.ii.1948, *Garcia* 34 (K; LISC). GI: Govuro, 11 km. from Covane on Luido Rd., 11.i.1970, *Myre, Duarte & Rosa* 5285 (LMA). M: near Maputo, 5.iii.1944, *Torre* 6397 (K; LISC).
From Ethiopia, Uganda and Zaire south to Namibia and S. Africa. Miombo woodland, disturbed savanna, saline grassland, weed in plantations and roadsides.
Although in east tropical Africa the scabrid lemma nerves seem to separate this species from its allies, this character is not reliable in the Flora Zambesiaca area. We have used the rhizomatous versus densely caespitose base to distinguish *D. milanjiana* (with mostly scabrid nerves) from *D. eriantha* (with mostly smooth nerves). We are aware that this character may occasionally break down, e.g. in specimens formerly identified as *D. seriata* Stapf (here included in *D. milanjiana*) underground parts consist of much thickened, short branching rhizomes, giving a loosely caespitose habit to the plant.

58. **Digitaria gymnostachys** Pilg., Notizbl. Bot. Gart. Berl. **15**: 709 (1942). —Henrard, Monogr. Digit.: 306 (1950). —Chippindall in Meredith, Grasses & Pastures of S. Afr.: 393 (1955). —Van der Veken in Bull. Jard. Bot. Brux. **32**: 125 (1962). —Napper, Grass. Tangan.: 77 (1965). —Clayton in F.T.E.A., Gramineae: 636 (1982). TAB. **41**, fig. 58. Type from Tanzania.

A tightly caespitose perennial on a short rhizome, base surrounded by cataphylls and firm, pale leaf sheath remnants. Culms 60–120 cm., erect, glabrous, sometimes scaberulous under the dark, hairy nodes. Leaf sheaths densely scaberulous, often loosely hairy. Ligule up to 3.5 mm. long, acute, entire. Leaf laminae 6–25 × 0.7–0.9 cm., linear, flat, densely scaberulous on both surfaces, densely hairy near the ligule, very scabrous along the crisped margin. Inflorescence composed of 6–15 racemes, (10)15–25 cm. long, erect to patent, few together in the inferior part, solitary along the superior part of the well developed common axis. Rhachis triquetrous, scarcely winged, up to 0.3 mm. broad, scaberulous with scabrous margins. Pedicels 2-nate, 1–5 mm. long, terete to subtriangular, scabrous, somewhat broadened at the apex. Spikelets 3.0–3.5 mm. long, lanceolate. Inferior glume 0.3 mm. long, truncate, nerveless, hyaline. Superior glume 0.7 mm. long, truncate, nerveless, hyaline. Both glumes together separated from the lemmata by a short stipe of c. 0.2 mm. long. Inferior lemma as long as the spikelet, lanceolate with recurved margins, 7-nerved, appressed hairy along the margins, hairs fine, wiry, blunt or acute, hyaline or purplish. Superior lemma as long as the spikelet, lanceolate, acuminate, pale to dark brown.

Mozambique. N: Cabo Delgado, Mocimboa da Praia, 180 m., 14.iv.1964, *Torre & Paiva* 11973 (LISC). GI: Govuno, 22.xii.1969, *Myre, Duarte & Rosa* 5252 (LMA). M: Inhaca Is., 4.iii.1958, *Mogg* 30371 (K; LMA; LMU; PRE; SRGH).
Also in Tanzania and S. Africa. Coastal open woodland and savanna, on sandy soil.

59. **Digitaria perrottetii** (Kunth) Stapf in F.T.A. **9**: 435 (1919). —Henrard, Monogr. Digit.: 547 (1950). —Sturgeon in Rhod. Agric. Journ. **50**: 283 (1953). —Chippindall in Meredith, Grasses & Pastures of S. Afr.: 400 (1955). —Bogdan, Rev. List Kenya Grass.: 49 (1958). —Jackson & Wiehe, Annot. Check-list Nyasal. Grass.: 36 (1958). —Napper, Grass. Tangan.: 78 (1965). —Launert in Merxm., Prodr. Fl. SW. Afr. **160**: 65 (1970). —Simon in Kirkia **8**: 31, 66 (1971). —Clayton in F.W.T.A. ed. 2, **3**: 452 (1972); in Kew Bull. **29**: 518 (1974). —Bennett in Kirkia **11**: 236 (1980). —Clayton in F.T.E.A., Gramineae: 649 (1982). TAB. **41**, fig. 59; TAB. **46**. Type from Senegal.
   Panicum perrottetii Kunth, Rev. Gram. **2**: 395 (1829). —Durand & Schinz in Consp. Fl. Afr. **5**: 759 (1894). Type as above.
   Milium minutiflorum Trin., Mem. Acad. St. Petersb. **3**: 121 (1834). Type from ? Senegal.
   Panicum cristatum Andersson in Peters, Naturwiss. Reise Mossamb. Bot. **2**: 548 (1964). Type: Mozambique, Querimba Isl., *Peters* s.n. (B, holotype).
   Digitaria perrottetii var. *gondaensis* Henrard, Monogr. Digit.: 550 (1950). Type from Tanzania.

A loosely caespitose or solitary growing annual. Culms 40–100 cm., ascending, glabrous, nodes dark and glabrous. Leaf sheaths smooth and glabrous, rarely loosely hairy. Ligule c. 1 mm. long, truncate, erose. Leaf laminae 2–10 × 0.5–1.4 cm., lanceolate to linear, flat, glabrous on both surfaces, often with a few bulbous based bristles near the base, scabrous along the margin. Inflorescence composed of 40–100 racemes, (2)3–8 cm. long, the inferior 20–50 in 4–6 whorls of 5–8 racemes, the superior 20–30 racemes solitary along a tall common axis, the racemes in the inferior whorls are conspicuously peduncled. Rhachis triquetrous, unwinged, scabrous, loosely hairy, at the base with a few bulbous based bristles, scabrous at the margins. Pedicels 2-nate, 0.2–2.8 mm. long, triangular, scabrous, slightly broadened at the apex. Spikelets 1.7–2.2 mm. long, oblong-lanceolate. Inferior glume absent. Superior glume somewhat shorter than the spikelet, oblong-lanceolate, 3-nerved, appressed hairy, hairs fine, smooth, acute, slightly overtopping the glume. Inferior lemma as long as the spikelet, oblong-lanceolate, 5-nerved, nerves equidistant, smooth, appressed hairy, hairs slightly overtopping the lemma. Superior lemma shorter than the spikelet, oblong-lanceolate, acute, pale to dark brown.

Caprivi Strip. Ngoma Area, 300–500 m., 19.ii.1969, *de Winter* 9244 (SRGH). **Botswana.** N: 8 km. N. of Aha Hills, 13.iii.1965, *Wild & Drummond* 6968 (BM; P; SRGH). **Zambia.** B: Mongu, 6.i.1966, *Robinson* 6787 (K; P; SRGH). N: Kasaba Bay, 800 m., 16.ii.1959, *McCallum-Webster* A 72 (K; LISC; SRGH). W: 8 km. from Kabompo on Zambezi (Balovale) Rd., 28.i.1971, *Anton-Smith* 741 (SRGH). C: Mwembeshi, 1300 m., 27.iii.1963, *Vesey-FitzGerald* 4035 (BM; COI). S: Kafue Nat. Park, Ngoma Camp, 14.ii.1963, *Mitchell* 17/71 (BM; SRGH). **Zimbabwe.** N: Mkumbura, 500 m., 29.i.1970, *Simon* 2096 (BM; K; SRGH). W: Hwange Nat. Park, 1100 m., 24.i.1981, *Crook* 2367 (SRGH). C: Chegutu (Hartley), Poole Farm, 5.ii.1945, *Hornby* 2364 (K; SRGH). E: Sabi valley, ii.1948, *Rattray* 1274 (K; SRGH). S: Beitbridge, 16.ii.1955, *Exell, Mendonça & Wild* 442 (BM; LISC; SRGH).

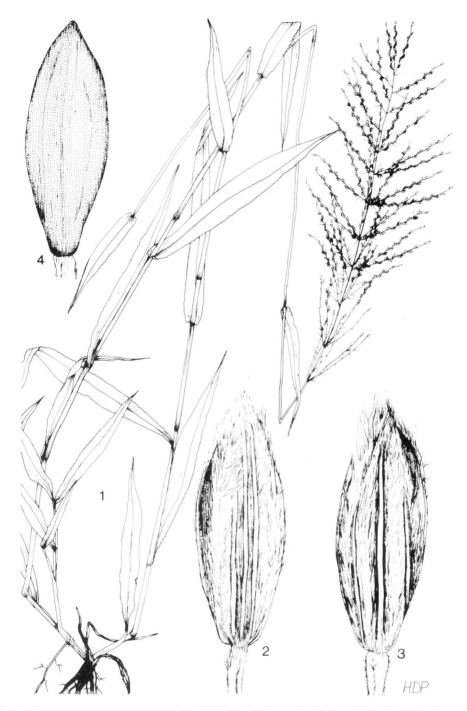

Tab. 46. DIGITARIA PERROTTETII. 1, habit (×$\frac{1}{2}$); spikelet in abaxial view (× 33); 3, spikelet in adaxial view (×33); 4, fertile lemma (× 33), all from *Simon* 2096.

Malawi. C: Salima, 470 m., 16.ii.1959, *Robson* 1631 (BM; K; LISC; PRE; SRGH). S: Makanga, 90 m., 19.iii.1960, *Phipps* 2548 (COI; K; PRE; SRGH). **Mozambique**. Z: 29.5 km. from Mocuba on Manganja da Costa Rd., 17.v.1949, *Barbosa & Carvalho* 2689 (LMA; SRGH). N: Nampula, 7.iv.1961, *Balsinhas & Marrime* 353 (COI; K; LISC; LMA; PRE; SRGH). MS: Chimoio, 2.ii.1948, *Garcia* 13 (LISC). GI: Gaza, Chipenhe, 27.v.1965, *Pereira, Marques & Balsinhas* 516 (LMU). M: Tinonganine, 18.iii.1957, *Barbosa & Lemos* 7574 (COI; LISC; LMA).

Also in Kenya, Zaire, Burundi, Tanzania, Angola, Namibia, S. Africa, Madagascar, and in Senegal (introduced?). Clearings and disturbed sites in miombo woodland, flood plains, weed on fallow land and roadsides.

60. **Digitaria floribunda** P. Goetghebeur in Bull. Nat. Plantentuin Belg. **45**: 415, fig. 14 (1975). TAB. **41**, fig. 60.Type from Tanzania.

A loosely caespitose annual. Culms 40–100 cm., erect or sometimes ascending, glabrous, nodes dark and glabrous. Leaf sheaths loosely to densely hairy with bulbous based bristles. Ligule up to 4 mm. long, triangular or truncate, entire. Leaf laminae 5–25 × 0.2–0.5 cm., linear, flat, glabrous or with scattered bulbous based bristles on both surfaces, scabrous along the margin. Inflorescence composed of 20–80 very slender racemes, 3–8 cm. long, erect, with spikelets to the base, the inferior racemes in one whorl, the other racemes solitary along the well developed common axis. Rhachis triquetrous, unwinged, smooth or somewhat scaberulous, scabrous at the margins. Pedicels 2-nate, 0.2–1.5 mm. long, triangular, scabrous, broadened at the ciliolate apex. Spikelets 1.3–1.6 mm. long, oblong. Inferior glume absent. Superior glume $\frac{3}{4}$–$\frac{4}{5}$ of the spikelet, oblong triangular, 3-nerved, appressed hairy, hairs very fine with an obpyriform apex. Inferior lemma as long as the spikelet, oblong, 7-nerved, appressed hairy. Superior lemma as long as the spikelet, oblong, acute to slightly acuminate, pale to dark brown.

Zambia. N: Mfuwe, Mwamba R., 700 m., 6.iii.1967, *Astle* 5058 (K; P; SRGH). C: Serenje Distr., 16.iii.1970, *Abel* 18 (SRGH). E: Chipata Distr., Mkhania, 700 m., 27.ii.1969, *Astle* 5542 (K; SRGH). Also in Tanzania. Miombo woodland, on sandy soil.

28. PENNISETUM Rich.

Pennisetum Rich. in Pers., Syn. Pl. **1**: 72 (1805).
Beckeropsis Fig. & De Not., Agrost. Aegypt. **2**: 49 (1853).

Inflorescence a cylindrical to subglobose spiciform panicle, terminal or sometimes axillary and then often aggregated into a leafy false panicle, bearing deciduous clusters of 1–several spikelets subtended by an involucre of bristles, these flexuous, filiform and free to the base. Spikelets narrowly lanceolate to oblong. Glumes and inferior lemma variable, absent to as long as spikelet. Superior lemma cartilaginous to thinly coriaceous with flat thin margins.

A genus of c. 80 species. Tropics.
The involucral bristles are apparently derived from the much modified branches of a panicle (Sohns in Journ. Wash. Acad. Sci. **45**: 135–143, 1955. —Butzin in Willdenowia **8**: 67–79, 1977).
The lower spikelet scales are often extremely variable and easily confused with one another, so that spikelet parts are best elucidated by counting back from the superior lemma. Habit, though difficult to describe, is of taxonomic importance in the genus and often seems more reliable than some of the spikelet characters used for diagnosis.

1. Inflorescence reduced to a cluster of 2–4 subsessile spikelets enclosed in the uppermost leaf-sheath, with long protruding filaments and stigmas   -   -   -   - 1. *clandestinum*
 – Inflorescence a spiciform panicle, conspicuously exserted   -   -   -   -   - 2
2. Clusters persistent, usually stipitate; lemmas usually pubescent on the margins; cultivated   -   -   -   -   -   -   -   -   -   -   -   -   - 3. *glaucum*
 – Clusters readily deciduous; lemmas glabrous or almost so; spontaneous   -   -   - 3
3. Involucre borne upon a terete pubescent stipe 1–3 mm. long and falling with it at maturity; bristles plumose   -   -   -   -   -   -   -   -   -   -   -   - 2. *setaceum*
 – Involucre without a stipe (though sometimes with a short glabrous oblong callus)   -   -   -   -   -   -   -   -   -   -   -   -   - 4
4. Rhachis angular, with sharp edged decurrent ribs below the involucre scars   -   - 5
 – Rhachis cylindrical or with rounded ribs   -   -   -   -   -   -   -   - 7
5. Superior lemma firmly membranous, dull, resembling the inferior   -   -   -   - 10
 – Superior lemma coriaceous, shining, readily deciduous   -   -   -   -   - 6

6. Spikelets solitary and sessile within the involucre - - - - - 5. *polystachion*
- Spikelets in clusters of 1–5 within the involucre, at least one of them
  pedicelled - - - - - - - - - - - - 6. *pedicellatum*
7. Panicles mainly axillary, and aggregated into a leafy false inflorescence - - - 8
- Panicles terminal on culms and major branches; bristles numerous - - - 9
8. Bristles single below each spikelet - - - - - - - 7. *unisetum*
- Bristles several below each spikelet - - - - - - - 8. *trisetum*
9. Rhachis pubescent; spikelets in groups of 1–5, 1 sessile the others pedicelled; superior lemma
  subcoriaceous and shining in the inferior half; anther tips minutely hairy; robust plant with
  stout culms and broad leaf laminae - - - - - - - 4. *purpureum*
- Rhachis scaberulous, occasionally pubescent but then the spikelets solitary and the lemmas
  scarcely different in texture - - - - - - - - - - - 10
10. Superior glume ⅔ length of spikelet or more; culm scaberulous below panicle; plant shrubby,
  much branched - - - - - - - - - - - - 9. *massaicum*
- Superior glume up to ½ length of spikelet, sometimes more but then culm pubescent below
  panicle - - - - - - - - - - - - - - - 11
11. Inferior lemma cuspidate, up to ¾ length of spikelet (excluding
  mucro) - - - - - - - - - - - 10. *thunbergii*
- Inferior lemma gradually tapering, ⅘ to as long as spikelet (but sometimes shorter in *P.
  sphacelatum* from southern Africa) - - - - - - - - - 12
12. Plant densely tufted with narrow convolute leaf laminae - - - - 11. *sphacelatum*
- Plant reed-like from a creeping rhizome, often robust with an elongated
  panicle - - - - - - - - - - - - - 12. *macrourum*

1. **Pennisetum clandestinum** Chiov. in Ann. Ist. Bot. Roma **8**: 41 (1903). —Stapf & Hubbard in F.T.A.
    9: 1009 (1934). —Sturgeon in Rhod. Agric. Journ. **50**: 512 (1953). —Chippindall in Meredith,
    Grasses & Pastures of S. Afr.: 444 (1955). —Bor, Grasses of B.C.I. & P.: 344 (1960). —Clayton in
    F.W.T.A. **3**: 459 (1972). —Clayton & Renvoize in F.T.E.A., Gramineae: 675 (1982). Type from
    Ethiopia.
    Pennisetum longistylum var. *clandestinum* (Chiov.) Leeke in Zeitschr. Naturwiss. **79**: 23 (1907).
    Type as above.

Sward-forming perennial with slender rhizomes, and stout rampant stolons amply
clothed with pale sub-inflated leaf-sheaths. Culms 3–15(45) cm. high. Leaf laminae 1–15
cm. long, flat or folded. Inflorescence reduced to a cluster of (1)2–4(6) subsessile spikelets
concealed within the uppermost sheath; involucre sparse; bristles delicate, ⅓–¾ length of
spikelet, scaberulous to ciliolate. Spikelets 10–20 mm. long. Inferior glume absent;
superior 1–3 mm. long, sometimes suppressed. Inferior lemma as long as spikelet.
Superior lemma resembling inferior; stigma up to 3 cm. long, subplumose; anthers
exserted on fine silvery filaments up to 5 cm. long.

Zimbabwe. E: Nyanga (Inyanga) Distr., Pungwe Drift, 25.i.1973, *Simon* 2332 (K; SRGH).
Malawi. N: Nyika Plateau, Chelinda, Old Salt Lick, 17.ii.1976, *Phillips* 1223 (K; MO).
East African highlands, introduced to most tropical and subtropical regions. Upland grassland on
fertile soils subject to grazing, probably introduced. Cultivated for pasture, soil stabilization and
lawns under the name Kikuyu grass; 1300–2500 m.
P. longistylum Hochst. from Ethiopia links the peculiar inflorescence of *P. clandestinum* to the rest
of the genus, for it differs by little more than its short spiciform partially exserted inflorescence.

2. **Pennisetum setaceum** (Forssk.) Chiov. in Bull. Soc. Bot. Ital. **1923**: 113 (1923). —Stapf & Hubbard
    in F.T.A. **9**: 1013 (1934). —Chippindall in Meredith, Grasses & Pastures of S. Afr.: 447 (1955).
    —Clayton in F.W.T.A. **3**: 459 (1972). —Clayton & Renvoize in F.T.E.A., Gramineae: 675 (1982).
    Type from Egypt.
    Phalaris setacea Forssk., Fl. Aegypt.-Arab.: 17 (1775). Type as above.

Densely tufted perennial. Culms 20–130 cm. high. Leaf laminae convolute with midrib
noticeably thickened on upper surface, rigid, harsh, glaucous. Panicle 6–30 cm. long,
linear; rhachis cylindrical with shallow angular ribs below the stumpless scars, glabrous
to pilose; involucre borne upon a slender pubescent stipe 1–3 mm. long, enclosing 1–3
spikelets, one of them sessile the others shortly pedicelled; bristles, at least the inner,
shortly plumose, the longest 16–40 mm. Spikelets 4.5–6.5 mm. long. Inferior glume up to ⅓
length of spikelet, usually subrotund, sometimes suppressed; superior glume ¼–⅔ length of
spikelet. Inferior lemma as long as spikelet. Superior lemma similar to inferior,
occasionally the rhachilla prolonged as a tiny bristle.

Zambia. C: Mt. Makulu, 4.ii.1965, *Lawton* 1199 (K). S: Mazabuka Distr., Kafue Gorge, 1–3 km.
below hydro-electric St., 9.ix.1972, *Strid* 2083 (K). **Zimbabwe.** W: Hwange Distr., Matetsi Safari Area,
15.i.1985, *Gonde* 420 (K; SRGH). E: Mutare, 27.x.1952, *Chase* 4681 (K; SRGH).

Also from Tanzania to Syria; often introduced as an ornamental. Stony slopes and arid places; 1000 m.

Related to *P. foemerianum* Leeke from Namibia, which has shorter panicles and involucres (longest bristle 7–15(20) mm.).

3. **Pennisetum glaucum** (L.) R.Br., Prodr. Fl. Nov. Holl. **1**: 195 (1810). —Stuntz in U.S. Dept. Agric. Bur. Pl. Ind., Inventory **13**: 84 (1914). —Clayton & Renvoize in F.T.E.A., Gramineae: 672 (1982). Type from Sri Lanka.

  Panicum glaucum L., Sp. Pl.: 56 (1753). Type as above.
  Panicum americanum L., Sp. Pl.: 56 (1753). Type from Spain.
  Alopecurus typhoides Burm., Fl. Ind.: 27 (1768). Type from India.
  Panicum lutescens Weigel, Obs. Bot.: 20 (1772) nom. superfl. Based on *Panicum glaucum*.
  Setaria glauca (L.) Beauv., Ess. Agrost.: 51, 178 (1812). Type as for *Panicum glaucum*.
  Pennisetum americanum (L.) Leeke in Zeitschr. Naturwiss. **79**: 52 (1907). —Clayton in F.W.T.A. **3**: 460 (1972). Type as for *Panicum americanum*.
  Setaria lutescens (Weigel) F.T. Hubbard in Rhodora **18**: 232 (1916). Type as for *Panicum glaucum*.
  Pennisetum typhoides (Burm.) Stapf & Hubbard in Kew Bull. **1933**: 271 (1933); in F.T.A. **9**: 1050 (1934). —Sturgeon in Rhod. Agric. Journ. **50**: 512 (1953). —Chippindall in Meredith, Grasses & Pastures of S. Afr.: 447 (1955). —Bor, Grasses of B.C.I. & P.: 350 (1960). Type as for *Alopecurus typhoides*.

Cultivated annual. Culms stout, up to 3 m. high. Leaf laminae up to 1 m. long and 7 cm. wide. Panicle 4 cm.–2 m. long, subglobose to linear; rhachis cylindrical, villous; involucre persistent, borne upon a stipe 1–25 mm. long, enclosing 1–9 spikelets; bristles glabrous or plumose. Spikelets 3–6 mm. long. Both lemmas usually pubescent on the margins.

  Zambia. E: Munkanya, 31.iii.1968, *Phiri* 122 (K). **Malawi**. N: Likoma Is., 28.vii.1900, *Johnston* 65 (K). S: Nsanje, 16.iii.1933, *Laurence* 3 (K). **Mozambique**. Z: Morrumbala, *Scott* (K). MS: Maringua, 25.vi.1950, *Chase* 2501 (K; SRGH). GI: Vilanculos, vii.1938, *Gomes e Sousa* 2154 (K). M: Inhaca Is., 15.vii.1959, *Mogg* 29364 (K).

  From Africa and India. The most drought tolerant of the tropical cereal crops.

  Panicum glaucum was originally based on three discordant elements, generating interminable argument over its proper typification — essentially Bulrush Millet versus *Setaria pumila* (see Terrell in Taxon **25**: 297–304, 1976; Kerguelen in Bull. Soc. Bot. Fr. **124**: 341, 1977). It was published in the same work as *P. americanum*, a misleadingly named Spanish cultivar, and owes its precedence to Stuntz's choice when uniting the two names.

  The cultivated plants were formerly divided into numerous species (Stapf & Hubbard in F.T.A. **9**: 1029–1053, 1934), but modern authors prefer to treat them as cultivars (Brunken in Econ. Bot. **31**: 163–174, 1977; Bono in Agron. Trop. **28**: 229–355, 1973). They belong to a complex of three species which hybridize freely, and which some authors (Brunken in Amer. Journ. Bot. **64**: 161–176, 1977, including full synonymy) prefer to treat as subspecies.

1. *P. glaucum*. The crop plant. Involucres persistent, stipitate.
2. *P. sieberianum* (Schlecht.) Stapf & Hubbard (= *P. americanum* subsp. *stenostachyum* (A.Br. & Bouché) Brunken). Mimics the crop and thus survives as a weed within it, but seems not to persist for more than a year or so in abandoned fields. Involucres deciduous, stipitate.
3. *P. violaceum* (Lam.) Rich. (= *P. americanum* subsp. *monodii* (Maire) Brunken). A weed of inhabited places, native to West Africa. Involucres deciduous, sessile.

4. **Pennisetum purpureum** Schumach., Beskr. Guin. Pl.: 44 (1827). —Stapf & Hubbard in F.T.A. **9**: 1016 (1934). —Sturgeon in Rhod. Agric. Journ. **50**: 513 (1953). —Chippindall in Meredith, Grasses & Pastures of S. Afr.: 443 (1955). —Jackson & Wiehe, Check List Nyasal. Grass.: 53 (1958). —Bor, Grasses of B.C.I. & P.: 347 (1960). —Launert in Merxm., Prodr. Fl. SW. Afr. **160**: 147 (1970). —Clayton in F.W.T.A. **3**: 460 (1972). —Brunken in Amer. Journ. Bot. **64**: 162 (1977). —Clayton & Renvoize in F.T.E.A., Gramineae: 677 (1982). Type from Ghana.

  Gymnothrix nitens Anderss. in Peters, Reise Mossamb. Bot. **2**: 552 (1864). Type: Mozambique, Cabaçeira, *Peters* (whereabouts not known).
  Pennisetum nitens (Anderss.) Hackel in Bol. Soc. Brot. **6**: 142 (1888). Type as above.
  Pennisetum benthamii var. *sambesiense* Hackel in Denkschr. Akad. Wiss., Wien, Math.-Nat. Kl. **78**: 400 (1905). Type: Mozambique, Boroma, *Menyhart* 1118 (Z, isotype).
  Pennisetum benthamii var. *nudum* Hackel in Denkschr. Akad. Wiss., Wien, Math.-Nat. Kl. **78**: 400 (1905). Type: Mozambique, Nhaondue, *Menyhart* 1118a (W, holotype).
  Pennisetum benthamii var. *ternatum* Hackel in Denkschr. Akad. Wiss., Wien, Math.-Nat. Kl. **78**: 400 (1905). Type: Mozambique, Boroma, *Menyhart* 559 (Z, isotype).
  Pennisetum flavicomum Leeke in Zeitschr. Naturwiss. **79**: 45 (1907). Type from Tanzania.
  Pennisetum pruinosum Leeke in Zeitschr. Naturwiss. **79**: 46 (1907). Type from Tanzania.
  Pennisetum lachnorrhachis Peter, Fl. Deutsch Ost-Afr. **1**, Anh.: 70 (1930). Type from Tanzania.
  Pennisetum gossweileri Stapf & Hubbard in Kew Bull. **1933**: 274 (1933); in F.T.A. **9**: 982 (1934). Type from Angola.

Pennisetum blepharideum Gilli in Ann. Naturhist. Mus. Wien **69**: 41 (1966). Type from Tanzania.

Robust perennial, often forming bamboo-like clumps. Culms 1–6 m. high. Leaf laminae up to 120 cm. long and 4 cm. wide. Panicle 7–30 cm. long, linear; rhachis cylindrical, with or without short peduncle stumps, pubescent; involucre enclosing 1–5 spikelets, one of them sessile and bisexual, the others shortly pedicelled and ♂ (but scarcely different in appearance); bristles glabrous or obscurely ciliate, the longest 10–40 mm. Spikelets 4.5–7 mm. long. Inferior glume suppressed, the superior $\frac{1}{4}$–$\frac{1}{2}$ length of spikelet or rarely suppressed. Inferior lemma $\frac{2}{3}$ to as long as spikelet (rarely less). Superior lemma subcoriaceous and shining in the inferior half; anther tips bearing a tiny tuft of hairs (very rarely glabrous).

Zambia. N: Mbala Distr., Kawimbe, 23.iii.1959, *McCallum-Webster* A238 (K; SRGH). W: Muwozi stream, 67 km. S. of Mwinilunga on Kabompo R., 31.v.1963, *Loveridge* 707 (K; SRGH). C: Lusaka Distr., Mwambula, 5.xi.1972, *Strid* 2460 (K). E: Lupande, Munkanya, 2.iii.1968, *Phiri* 62 (K). S: Kalomo, Siantambo, 23.ii.1962, *Mitchell* 13/29 (K). **Zimbabwe**. N: Darwin Distr., Nyarandi R., 27.i.1960, *Phipps* 2428 (K; SRGH). W: Hwange, Denda Farm, 20.iii.1974, *Gonde* 79/74 (K; SRGH). C: Harare, 15.v.1930, *Brain* 1565 (K; SRGH). E: Mutare, Fairbridge Park, 6.v.1975, *Crook* 2079 (K; SRGH). S: Sabi-Lundi Junction, Chitsa's Kraal, 6.vi.1950, *Wild* 3367 (K; SRGH). **Malawi**. N: Nkhata Bay Distr., Chikale Beach, 10.v.1970, *Brummitt* 10553 (K; SRGH). C: Lilongwe, Nsaru Distr., Maone Estate, 2 km. N. of Limbe, 10.iii.1970, *Brummitt* 9016 (K; SRGH). **Mozambique**. N: Erati, between Namapa & Ocúa, 9.iii.1960, *Lemos & Macúaca* 23 (K). Z: Mocuba, 21.ii.1943, *Torre* 4801 (COI; K). T: Massengena, 28.ix.1948, *Wild* 2656 (K; SRGH). MS: Chimoio, 4.vi.1941, *Torre* 2792 (COI; K).

Tropical Africa; introduced to most other tropical countries, and cultivated for fodder under the names 'Elephant' and 'Napier' grass. Riverine sites, valley bottoms and forest margins, with a preference for rich soils; 300–1800 m.

P. purpureum belongs to the same section of the genus as *P. glaucum* and its weedy allies.

5. **Pennisetum polystachion** (L.) Schult., Syst. Veg. Mant. **2**: 146 (1824). —Stapf & Hubbard in F.T.A. **9**: 1057 (1934). —Sturgeon in Rhod. Agric. Journ. **50**: 512 (1953). —Jackson & Wiehe, Annot. Check List Nyasal. Grass.: 53 (1958). —Bor, Grasses of B.C.I. & P.: 346 (1960). —Clayton in F.W.T.A. **3**: 459 (1972). —Clayton & Renvoize in F.T.E.A., Gramineae: 679 (1982). Type from India.

Panicum polystachion L., Syst. Nat., ed. 10, **2**: 870 (1759). Type as above.
Cenchrus setosus Swartz, Prodr. Veg. Ind. Occ.; **26** (1788). Type from Jamaica.
Pennisetum setosum (Swartz) Rich. in Pers., Syn. Pl. **1**: 72 (1805). Type as above.
Pennisetum stenostachyum Peter, Fl. Deutsch Ost-Afr. **1**, Anh.: 70 (1930). Type from Tanzania.
Pennisetum polystachion subsp. *setosum* (Swartz) Brunken in Journ. Linn. Soc., Bot. **79**: 63 (1979). Type as for *Cenchrus setosus*.

Annual or perennial. Culms 30–200 cm. high, copiously branched. Leaf laminae 3–16 mm. wide. Panicle 3–25 cm. long, linear; rhachis angular with sharp decurrent wings below the involucral scars, glabrous; involucre enclosing 1 sessile spikelet; bristles scaberulous or densely ciliate, tawny or purple, the longest 5–25 mm. Spikelets 2–5 mm. long. Inferior glume suppressed or very small, the superior as long as spikelet. Superior floret coriaceous, shining, subacute, readily deciduous at maturity.

Subsp. **polystachion**

Involucral bristles densely ciliate. Annual or perennial.

Zambia. N: Mbala to Kasanga near Tanzania frontier, 29.iii.1959, *McCallum Webster* A 257 (K; SRGH). W: Mwinilunga Distr., Matonchi Farm, 4.xi.1937, *Milne-Redhead* 3099 (K; SRGH). C: Mkushi Distr., Mulungushi Gorge, 17.iii.1973, *Kornaś* 3490 (K). E: Lundazi, 1.vi.1954, *Robinson* 799 (K). **Zimbabwe**. N: Lomagundi Distr., Nyafuta R., iv.1972, *Davies* 3226 (K; SRGH). W: Victoria Falls, v.1915, *Rogers* 13129 (K). E: Ngorima Reserve, near Haroni-Lusitu confluence, 23.xi.1967, *Simon & Ngoni* 1278 (K; SRGH). **Malawi**. N: Nkhata Bay, Bandawe Point, 8.vi.1974, *Pawek* 8701 (K; MA; MO; SRGH; UC). C: Lilongwe Expt. Sta., 2.vii.1961, *Brown* 13 (K; SRGH). S: Zomba, 16.v.1981, *Livingstone* 81–11 (DUKE; K). **Mozambique**. N: Erati, Namapa, 22.iii.1961, *Balsinhas & Marrime* 306 (K). Z: Mocuba, Namagoa, 1945, *Faulkner* 2 (K; SRGH).

Throughout the tropics. Old farmland and disturbed places; sea level to 1500 m.

A polymorphic weedy species, very variable in colour, stiffness and length of the bristles. Brunken has shown that there is some justification for treating annual (subsp. *polystachion*) and perennial (subsp. *setosum*) plants as separate, though intergrading, taxa. These subspecies have not been formally adopted here as the basal parts are seldom well enough collected to make the distinction practical.

Subsp. **atrichum** (Stapf & Hubbard) Brunken in Journ. Linn. Soc. Bot. **79**: 63 (1979). —Clayton & Renvoize in F.T.E.A., Gramineae: 680 (1982). Type: Malawi, Zomba, *Manning* 4 (K, holotype).
   Pennisetum reversum var. *gymnochaetum* Hackel in Bull. Herb. Boiss., sér. 2, **1**: 767 (1901). Types from Tanzania.
   Pennisetum atrichum Stapf & Hubbard in Kew Bull. **1933**: 282 (1933); in F.T.A. **9**: 1061 (1934). —Jackson & Wiehe, Annot. Check List Nyasal. Grass.: 53 (1958). —Clayton in F.W.T.A. **3**: 459 (1972). Type as for *Pennisetum polystachion* subsp. *atrichum*.

Involucral bristles scaberulous or rarely with a few hairs. Perennial.

Zambia. B: Mongu, 24.iii.1964, *Verboom* 1337 (K). N: Samfya Distr., Luapula, iii.1970, *Chabwela* 52 (K; SRGH). W: Solwezi Rd. to Chifubwa Gorge, 19.xii.1969, *Simon & Williamson* 1857 (K; SRGH). C: 40 km. S. of Kabwe, 4.iii.1961, *Phipps & Vesey-FitzGerald* 2926 (K; SRGH). **Zimbabwe**. N: 16 km. N. of Gokwe, 27.iv.1963, *Bingham* 656 (K; SRGH). **Malawi**. S: Namasi, iii.1899, *Cameron* 19 (K). Northwards to Senegal. Weedy places, favouring damp soils; 1000–1500 m.

6. **Pennisetum pedicellatum** Trin. in Mém. Acad. Sci. Pétersb., sér. 6, **3**: 184 (1934). —Stapf & Hubbard in F.T.A. **9**: 1065 (1934). —Bor, Grasses of B.C.I. & P.: 346 (1960). —Clayton in F.W.T.A. **3**: 460 (1972). —Clayton & Renvoize in F.T.E.A., Gramineae: 680 (1982). Type from Cape Verde Is.

Annual. Like *P. polystachion* except: involucre fluffy, ovate, enclosing 1–5 spikelets, at least one of them on a pedicel 0.5–3.5 mm. long; bristles densely woolly.

Zambia. S: Mazabuka Distr., Kariba valley, 27.iii.1958, *Verboom* 324 (K; SRGH); 64 km. from Choma towards Sinazongwe, 18.iv.1965, *Astle* 3067 (K).
   Senegal, through Ethiopia and Tanzania, to India and Thailand. Sandy soils in wooded grassland; 1300 m.
   Our plants belong to subsp. *pedicellatum*. Subsp. *unispiculum* Brunken, characterized by a less fluffy involucre resembling that of *P. polystachion* and containing a single spikelet, occurs in Tanzania and most other parts of the species' range.

7. **Pennisetum unisetum** (Nees) Benth. in Journ. Linn. Soc., Bot. **19**: 47, 49 (1881). —Clayton & Renvoize in F.T.E.A., Gramineae: 681 (1982). Type from S. Africa.
   Gymnothrix uniseta Nees, Fl. Afr. Austr.: 66 (1841). Type as above.
   Beckeropsis uniseta (Nees) K. Schum. in Pflanzenw. Ost-Afr. **B**: 52 (1895). —Stapf & Hubbard in F.T.A. **9**: 949 (1934). —Sturgeon in Rhod. Agric. Journ. **50**: 512 (1953). —Chippindall in Meredith, Grasses & Pastures of S. Afr.: 448 (1955). —Jackson & Wiehe, Annot. Check List Nyasal. Grass.: 30 (1958). —Clayton in F.W.T.A. **3**: 457 (1972). Type as above.
   Pennisetum kirkii Stapf in Kew Bull. **1897**: 286 (1897). Type: Malawi, Soche Hill, *Kirk* (K, holotype).

Perennial. Culms 0.6–4 m. high, typically robust and 2–15 mm. in diam. at the base. Leaf laminae (5)10–30 cm. wide, broadly linear to linear-lanceolate, often falsely petiolate. Panicles 2–4 cm. long, slender, axillary and gathered into a copious false inflorescence; rhachis with rounded ribs and distinct peduncle stumps, minutely pubescent; involucre reduced to a single bristle subtending each spikelet; bristle (2.5)7–40 mm. long, glabrous. Spikelets 2–3 mm. long. Glumes 0.2–0.5 mm. long, obtuse to emarginate, or rarely the superior up to 0.8 mm. long and subacute. Lemmas as long as spikelet, membranous.

Zambia. N: Mbala, Ndundu, 11.iv.1959, *McCallum-Webster* A303 (K; SRGH). W: Ndola Forest Res., 1950, *Jackson* 17 (K). C: Chilanga, Mt. Makulu Res. St., 27.iii.1962, *Angus* 3082 (FHO; K). E: Chipata, iii.1962, *Verboom* 582 (K). **Zimbabwe**. E: Mutare, Murahwa's Hill, 22.iii.1970, *Simon* 2120 (K; SRGH). **Malawi**. N: Mzimba Distr., Mzuzu, Marymount, 30.v.1971, *Pawek* 4876 (K). C: Dedza Mt., 1.iv.1970, *Brummitt* 9579 (K). S: Zomba, 3.iv.1949, *Wiehe* N/35 (K). **Mozambique**. N: Mandimba, 30.iii.1942, *Hornby* 3345 (K; SRGH). Z: Régulo Guja to Derre, 10.vi.1949, *Barbosa & Carvalho* 3021 (K). MS: Espungabera, 8.vi.1942, *Torre* 4264 (COI; K).
   Also from Tropical and S. Africa; Yemen. Wooded grassland and riverine forest, favouring partial shade; 900–1800 m.
   Owing to its distinctive axillary branching and single subtending bristle the species is often accommodated in a separate genus, *Beckeropsis*; but its similarity to *Pennisetum trisetum* renders this treatment untenable.

8. **Pennisetum trisetum** Leeke in Zeitschr. Naturwiss. **79**: 30 (1907). —Stapf & Hubbard in F.T.A. **9**: 970 (1934). —Clayton & Renvoize in F.T.E.A., Gramineae: 682 (1982). Type from Ethiopia.
   Pennisetum schliebenii Pilger in Not. Bot. Gart. Berl. **11**: 804 (1933). —Stapf & Hubbard in F.T.A. **9**: 1069 (1934). Type from Tanzania.

Perennial. Culms 1–3 m. high. Leaf laminae 3–18 mm. wide, broadly linear. Panicles

2–7 cm. long, slender, axillary and gathered into a scanty false inflorescence; rhachis with rounded ribs and distinct peduncle stumps, scaberulous to puberulous; involucre of 2–5(15) bristles around a single spikelet; bristles glabrous, the longest 3–15 mm., the others nearly always shorter than spikelet and often obscure. Spikelets 2.8–4(5) mm. long. Inferior glume c. 0.5 mm. long, truncate to subacute; superior glume 0.8–1.5 mm. long, usually acute, rarely obtuse. Lemmas as long as spikelet, membranous.

Zimbabwe. C: Goromonzi Distr., Ngomakurira, 3.iv.1960, *Phipps* 2793 (K; SRGH); Wedza Mt., 64 km. S. of Marondera, 22.v.1968, *Simon, Rushworth & Mavi* 1826 (K; SRGH).
Northwards to Ethiopia. Forest margins; 1200–1800 m.

9. **Pennisetum massaicum** Stapf in Kew Bull. **1906**: 82 (1906). —Stapf & Hubbard in F.T.A. **9**: 978 (1934). —Clayton & Renvoize in F.T.E.A., Gramineae: 687 (1982). Type from Kenya.

Perennial from a short woody rhizome. Culms 30–90 cm. high, much branched, wiry or woody, ascending, scaberulous below the panicle. Leaf laminae 2–4 mm. wide, flat or folded. Panicle 2–10 cm. long, linear to oblong; rhachis angular with sharp-edged ribs below the involucral scars, scaberulous; involucre enclosing 1 sessile spikelet, the base with a short oblong callus; bristles glabrous or rarely sparsely pilose, the longest 4–10 mm. Spikelets 3.5–5.5 mm. long. Inferior glume ½, the superior ⅔–¾ as long as spikelet. Superior lemma similar to the inferior, membranous, acute to subulate.

Zimbabwe. S: Gwanda Distr., 64 km. from Beitbridge on road to Bulawayo, 5.i.1956, *Rattray* 1706 (K; SRGH).
North to Somalia. Open savanna woodland on black clay soils; 300–600 m.
Intergrades with *P. mezianum*, a doubtfully distinct east African species differing in little more than a smooth culm just below the panicle.
Specimens with hairy bristles are difficult to separate from *Cenchrus ciliaris*, suggesting introgression from that genus. In *C. ciliaris* the involucral callus is obconical, widening into a broad rim which forms the connate base of the bristles. In *P. massaicum* it is oblong, the bristles being free to the base.

10. **Pennisetum thunbergii** Kunth, Rév. Gram. **1**: 50 (1829). —Stapf & Hubbard in F.T.A. **9**: 997 (1934). —Sturgeon in Rhod. Agric. Journ. **50**: 513 (1953). —Chippindall in Meredith, Grasses & Pastures of S. Afr. : 443 (1955). —Jackson & Wiehe, Annot. Check List Nyasal. Grass.: 54 (1958). —Launert in Merxm., Prodr. Fl. SW. Afr. **160**: 147 (1970). —Clayton & Renvoize in F.T.E.A., Gramineae: 687 (1982). TAB. **47**. Type from S. Africa.
   Cenchrus geniculatus Thunb., Prodr. Fl. Cap.: 24 (1794). Type as above.
   Panicum geniculatum (Thunb.) Thunb., Fl. Cap. **1**: 388 (1813) non Poir. (1798). Type as above.
   Pennisetum thunbergii var. *galpinii* Stapf in Fl. Cap. **7**: 437 (1899). Types from S. Africa.
   Pennisetum geniculatum (Thunb.) Leeke in Zeitschr. Naturwiss. **79**: 43 (1907) *non* (Poir.) Jacq. (1820). Type as for *Cenchrus geniculatus*.
   Pennisetum leekei Mez in Notizbl. Bot. Gart. Berl. **7**: 52 (1917). Type from Rwanda.
   Pennisetum paucisetum Peter in Abh. Königl. Ges. Wiss. Göttingen, Math.-Phys. Kl., n.f., **13**, 2: 46 (1928). Type from Tanzania.
   Pennisetum leekei var. *leucostachys* Peter, Fl. Deutsch Ost-Afr. **1**, Anh.: 74 (1930). Type from Tanzania.

Loosely tufted rhizomatous perennial. Culms 10–150 cm. high, erect or geniculate, glabrous or pubescent below panicle. Leaf laminae 2–8 mm. wide, flat or convolute. Panicle 2–15 cm. long, linear to narrowly oblong; rhachis cylindrical with rounded ribs and distinct peduncle stumps, scaberulous; involucre enclosing 1 sessile spikelet, the base truncate; bristles fine, glabrous, the longest 5–14 mm. Spikelets 2.5–5 mm. long. Inferior glume suppressed or up to 1 mm. long, the superior up to ¼ length of spikelet, acute to emarginate. Inferior lemma (½⅓–¾⅔) length of spikelet, abruptly acuminate to cuspidate, usually with a mucro 0.5–1.5 mm. long. Superior lemma thinly coriaceous, acuminate or with a mucro up to 1 mm. long; anther tips smooth to distinctly penicillate.

Zambia. N: Mbala Distr., Kawimbe, 26.iii.1959, *McCallum-Webster* A237 (K; SRGH). Zimbabwe. C: Harare, 4.iii.1984, *Hyde* 169 (K; SRGH). E: Mutare Distr., Penhalonga, 28.ii.1970, *Crook* 910 (K; SRGH). Malawi. N: Nyika Plateau, 29.vi.1952, *Jackson* 889 (K). C: Lilongwe, Cirinda, Civuwo Dambo, 13.iv.1956, *Jackson* 1839 (K). S: Mperere Mission, Chankalamu Dambo, 2.xi.1950, *Jackson* 252 (K).
Yemen to S. Africa and Sri Lanka. Upland grassland, particularly on damp soils; 1500–2500 m.
The abruptly cuspidate tip of the inferior lemma, appreciably shorter than the spikelet, is the best character distinguishing *P. thunbergii* from *P. sphacelatum*. It is usually combined with flat leaf laminae and a rhizomatous habit. Penicillate anther apices are conclusive, but unfortunately these are not always developed.

Tab. 47. PENNISETUM THUNBERGII. 1, habit (× $\frac{2}{3}$); 2, portion of inflorescence rhachis (× 10); 3, spikelet with involucre (× 10); 4, spikelet showing superior glume (× 10); 5, spikelet showing inferior lemma (× 10); 6, anther (× 10), all from *Crook* 910.

11. **Pennisetum sphacelatum** (Nees) Th. Dur. & Schinz, Consp. Fl. Afr. **5**: 784 (1894). —Chippindall in Meredith, Grasses & Pastures of S. Afr.: 442 (1955). —Clayton & Renvoize in F.T.E.A., Gramineae: 689 (1982). Types from S. Africa.

   Gymnothrix sphacelata Nees, Fl. Afr. Austr.: **68** (1841). Types as above.

   Pennisetum schimperi A. Rich., Tent. Fl. Abyss. **2**: 381 (1851). —Stapf & Hubbard in F.T.A. **9**: 992 (1934). —Jackson & Wiehe, Annot. Check List Nyasal. Grass.: 54 (1958). Types from Ethiopia.

   Pennisetum tenuifolium Hackel in Bull. Herb. Boiss. **3**: 380 (1895). Type from S. Africa.

   Pennisetum sphacelatum var. *tenuifolium* (Hackel) Stapf in F.C. **7**: 436 (1899). Type as above.

   Pennisetum macrourum var. *angustifolium* Hackel in Mém. Herb. Boiss. **20**: 8 (1900). Type from S. Africa.

   Pennisetum merkeri Leeke in Zeitschr. Naturwiss. **79**: 27 (1907). Type from Tanzania.

Densely tufted perennial. Culms 30–150 cm. high, erect, mostly pubescent to villous below the panicle. Leaf laminae 2–4 mm. wide, convolute, hard, prominently nerved. Panicle 4–12 cm. long, linear; rhachis cylindrical with rounded ribs and very short almost accrescent peduncle stumps, scaberulous to pubescent; involucre enclosing 1 sessile spikelet, the base slightly rounded; bristles glabrous to weakly plumose, the longest 4–15 mm. Spikelets 2.5–4.5 mm. long. Inferior glume $\frac{1}{8}$–$\frac{1}{2}$, the superior $(\frac{1}{6})\frac{1}{3}$–$\frac{2}{3}(\frac{3}{4})$ length of spikelet. Inferior lemma $\frac{4}{5}$ to as long as spikelet (rarely $\frac{2}{3}$ as long in South Africa), tapering gradually to an acute, acuminate or caudate tip. Superior lemma firmly membranous, similar to the inferior.

   Malawi. S: Mulanje, Chambe Mt., 14.vi.1962, *Robinson* 5349 (K).

   Also from Ethiopia to S. Africa. Upland grassland; 1900–2400 m.

   Distinguished from *P. thunbergii* by the inferior lemma whose tip narrows gradually and usually covers the superior, but it may be shorter in southern Africa. Supporting characters are the hard convolute leaf laminae and densely caespitose habit. These also serve to separate it from *P. macrourum*, with which it intergrades.

12. **Pennisetum macrourum** Trin., Gram. Pan.: 64 (1826). —Chippindall in Meredith, Grasses & Pastures of S. Afr.: 442 (1955). —Clayton & Renvoize in F.T.E.A., Gramineae: 689 (1982). Type from S. Africa.

   Pennisetum angolense Rendle, Cat. Afr. Pl. Welw. **2**: 189 (1899). —Stapf & Hubbard in F.T.A. **9**: 981 (1934). —Jackson & Wiehe, Annot. Check List Nyasal. Grass.: 53 (1958). Type from Angola.

   Pennisetum angolense var. *laxispicatum* Rendle, Cat. Afr. Pl. Welw. **2**: 189 (1899). Type from Angola.

   Pennisetum natalense Stapf in F.C. **7**: 435 (1899). —Chippindall in Meredith, Grasses & Pastures of S. Afr.: 440 (1955). Type from S. Africa.

   Pennisetum stolzii Mez in Bot. Jahrb. **57**: 190 (1921). —Stapf & Hubbard in F.T.A. **9**: 977 (1934). Type from Tanzania.

   Pennisetum validum Mez in Bot. Jahrb. **57**: 191 (1921). —Stapf & Hubbard in F.T.A. **9**: 978 (1934). Type from Tanzania.

   Pennisetum kisantuense Vanderyst in Bull. Agric. Congo Belge **16**: 685 (1925). Types from Zaire.

   Pennisetum haareri Stapf & Hubbard in Kew Bull. **1933**: 273 (1933); in F.T.A. **9**: 980 (1934). Type from Tanzania.

   Pennisetum macropogon Stapf & Hubbard in Kew Bull. **1933**: 275 (1933); in F.T.A. **9**: 983 (1934). —Sturgeon in Rhod. Agric. Journ. **50**: 513 (1953). Type: Zimbabwe, Makabusi R., *Eyles* 4771 (K, holotype).

   Pennisetum glaucocladum Stapf & Hubbard in Kew Bull. **1933**: 276 (1933); in F.T.A. **9**: 984 (1934). —Sturgeon in Rhod. Agric. Journ. 50: 513 (1953). —Chippindall in Meredith, Grasses & Pastures of S. Afr.: 441 (1955). —Jackson & Wiehe, Annot. Check List Nyasal. Grass.: 53 (1958). —Launert in Merxm., Prodr. Fl. SW. Afr. **160**: 147 (1970). —Clayton in F.W.T.A. **3**: 461 (1972). Type: Zimbabwe, Hunyani R., *Eyles* 4903 (K, holotype; SRGH, isotype).

   Pennisetum exile Stapf & Hubbard in Kew Bull. **1933**: 277 (1933); in F.T.A. **9**: 987 (1934). —Jackson & Wiehe, Annot. Check List Nyasal. Grass.: 53 (1958). Type from Tanzania.

   Pennisetum davyi Stapf & Hubbard in Kew Bull. **1933**: 278 (1933); in F.T.A. **9**: 991 (1934). —Jackson & Wiehe, Annot. Check List Nyasal. Grass.: 53 (1958). Type: Malawi, Mt. Mulanje, *Burtt Davy* 2028/29 (K, holotype).

   Pennisetum scaettae Robyns in Bull. Jard. Bot. Brux. **13**: 3 (1934). Type from Zaire.

Reed-like perennial from a creeping rhizome. Culms 0.6–5 m. high, often robust, rarely pubescent below the panicle. Leaf laminae 2–10(15) mm. wide, hard, flat or sometimes convolute. Panicle 6–40 cm. long, linear; rhachis cylindrical with rounded ribs, with or without peduncle stumps, scaberulous or sometimes pubescent; involucre enclosing 1 sessile spikelet without distinct stipe; bristles glabrous, the longest 5–20 mm. Spikelets 2–6 mm. long. Inferior glume up to 1 mm. long, the superior $\frac{1}{8}$–$\frac{1}{4}$ length of spikelet. Inferior lemma $\frac{3}{4}$ to as long as spikelet, acute. Superior lemma similar to the inferior.

Botswana. N: Dikgathong, Thaoge R., 4.v.1978, *Smith* 2403 (K; SRGH). **Zambia**. B: Mongu Distr., Luena flood plain, 14.vi.1964, *Verboom* 1115 (K). N: L. Bangweulu, Mpanta Point, 24.iv.1964, *Verboom* 2509 (K). S: Choma, 23.iv.1963, *van Rensburg* 2088 (K). **Zimbabwe**. N: Guruve Distr., Nyamunyeche Estate, 11.iv.1979, *Nyariri* 794 (K; SRGH). W: Masvingo Distr., Great Zimbabwe, 12.iv.1973, *Chiparawasha* 702 (K; SRGH). C: Makoni Distr., Chiduku Reserve, Macheke R., iv.1955, *Davies* 1079 (K; SRGH). E: Mutare, 30.iii.1969, *Crook* 863 (K; SRGH). S: Beitbridge Distr., 11 km. downstream from Tuli Police Camp, 3.v.1959, *Drummond* 6068 (K; SRGH). **Malawi**. N: Mzimba Distr., Viphya, vi.1951, *Jackson* 570 (K). C: Dedza, 5.ix.1950, *Jackson* 204 (K). S: Mulanje, above Chisambo Tea Estate, 3.ix.1970, *Müller* 1520 (K).

Also from Sudan and Ethiopia to S. Africa and Yemen. River banks and stream beds; 800–2100 m.

The waterside reeds are here gathered into a single polymorphic species. Their variation is considerable, but apparently continuous and uncorrelated with geographical distribution. Arbitrary morphological subdivisions seem pointless until something is known of the ecological and cytogenetic factors at work.

29. CENCHRUS L.

Cenchrus L., Sp. Pl.: 1049 (1753). —De Lisle in Iowa State Journ. Sci. **37**: 259–351 (1963).

Inflorescence a cylindrical spiciform panicle with angular rhachis, bearing deciduous clusters of 1–several spikelets subtended by an involucre of bristles, these flexuous or spinous, ± flattened and united below (at least the inner whorl), the degree of union varying from a basal disk to a deep cupule. Spikelets lanceolate to ovate. Inferior glume up to $\frac{1}{2}$ length of spikelet, sometimes suppressed. Superior glume and inferior lemma as long as spikelet or a little shorter. Superior lemma chartaceous to thinly coriaceous with flat thin margins.

A genus of 22 species. Tropics.

1. Involucral bristles antrorsely scaberulous - - - - - - - - - - - 2
   – Involucral bristles retrorsely barbellate, spinous, tenaciously prickly - - - - 3
2. Bristles united only at the base to form a shallow disk 0.5–1.5 mm in
   diam. - - - - - - - - - - - - - - - - 1. *ciliaris*
   – Bristles united for $\frac{1}{3}-\frac{2}{3}$ their length to form a cup - - - - - - - 2. *mitis*
3. Inner spines united into a disk, the outer aciculate and borne in a whorl around
   its rim - - - - - - - - - - - - - - - - 3. *biflorus*
   – Inner spines united into a cup, the outer flattened and borne irregularly on its
   surface - - - - - - - - - - - - - - - - 4. *incertus*

1. **Cenchrus ciliaris** L., Mant. Pl. Alt.: 302 (1771).—Stapf & Hubbard in F.T.A. **9**: 1072 (1934). —Sturgeon in Rhod. Agric. Journ. **50**: 514 (1953). —Chippindall in Meredith, Grasses & Pastures of S. Afr.: 451 (1955). —Bor, Grasses of B.C.I. & P.: 289 (1960). —Ramaswamy, Raman & Menon in Journ. Ind. Bot. Soc. **48**: 102–111 (1969). —Clayton in F.W.T.A. **3**: 464 (1972). —Clayton & Renvoize in F.T.E.A., Gramineae: 691 (1982). TAB. **48**. Type from S. Africa.

Pennisetum ciliare (L.) Link, Hort. Reg. Bot. Berol. **1**: 213 (1827). Type as above.
Pennisetum rangei Mez in Bot. Jahrb. **57**: 190 (1921). Type from Namibia.
Pennisetum oxyphyllum Peter, Fl. Deutsch Ost-Afr. **1**, Anh.: 72 (1930). Type from Tanzania.

Perennial, often forming mats or tussocks. Culms 10–150 cm. high, ascending, wiry or sometimes almost woody. Panicle 2–14 cm. long, cylindrical to ovoid, grey, purple or straw-coloured; involucre 6–16 mm. long, connate only at the base to form a disk 0.5–1.5 mm. in diam.; inner bristles flexuous, often wavy, sparsely to densely ciliate below, filiform above, one of them longer and stouter than the rest, this at least somewhat flattened towards the base; outer bristles filiform. Spikelets 1–4 per bur, 2–5.5 mm. long, acutely lanceolate. Glumes distinct, acute, the inferior from $\frac{1}{4}-\frac{1}{2}$, the superior from $\frac{1}{2}$ to as long as spikelet. Superior lemma chartaceous.

Caprivi Strip. 80 km. from Katima Mulilo on Rd. to Linyanti, 27.xii.1958, *Killick & Leistner* 3142 (K; PRE). **Botswana**. N: Chobe Nat. Park, 21.viii.1978, *Smith* 2478 (K; SRGH). SW: Kgalagadi, Mabua Sefhubi Pan, 28.ii.1963, *Leistner* 3101 (K; PRE). SE: NE. of Mopipi, 16.iv.1973, *Tyers* 1 (K). **Zambia**. S: Kalomo, 26.ii.1962, *Mitchell* 13/35 (K). **Zimbabwe**. N: Sengwa Res. St., 21.i.1981, *Mahlangu* 11 (K; SRGH). W: Bulawayo, Khami Ruins, 24.xii.1965, *Simon* 608 (K; SRGH). C: Chegutu, Poole Farm, 24.i.1944, *Hornby* 2286 (K; SRGH). E: Mtanda Range, Achnashie Farm, 3.i.1969, *Crook* 835 (K; SRGH). S: Beitbridge, 30.iii.1973, *Crook* 1070 (K; SRGH). **Mozambique**. MS: Chemba, Chiou, 13.iv.1960, *Lemos & Macuáca* 94 (K). M: Boane, 19.ii.1981, *de Koning & Boane* 8666 (K).

Tab. 48. CENCHRUS CILIARIS. 1, habit (×⅔); 2, spikelet with involucre (×4); 3, inferior glume (×6); 4, superior glume (×6); 5, inferior lemma (×6); 6, superior lemma (×6); 7, superior palea (×6); 8, flower (×8); 9, grian (×8); 10, ligule (×8), all from *Guest et al.* 17628. From Fl. Iraq.

From Africa, extending through Arabia to India. Savanna woodland; 0–1500 m.

The cupule is so weakly developed that the species is easily mistaken for a *Pennisetum*. However, both spikelet shape and intergradation with *C. pennisetiformis* (a northeast African species with well formed cup) indicate that it is better retained in *Cenchrus*.

2. **Cenchrus mitis** Anderss. in Peters, Reise Mossamb. Bot. **2**: 554 (1864). —Stapf & Hubbard in F.T.A. **9**: 1078 (1934). —Clayton & Renvoize in F.T.E.A., Gramineae: 694 (1982). Type: Mozambique, Querimba, *Peters* (K, isotype).

Annual. Culms 20–100 cm. high, decumbent or ascending. Panicle 4–18 cm. long; involucre 6–9 mm. long, connate for $\frac{1}{3}-\frac{2}{3}$ its length to form a cup; inner spines flattened, pubescent on the face, ciliate on the margins, aciculate at the apex; outer bristles filiform, rarely much reduced. Spikelets 2(3) per bur, 4–6 mm. long. Superior lemma coriaceous.

Mozambique. N: Querimba, *Peters* (K).
Northwards to Kenya. Coastal bushland.

3. **Cenchrus biflorus** Roxb., Fl. Indica **1**: 238 (1820). —Bor, Grasses of B.C.I. & P.: 289 (1960). —Clayton in F.W.T.A. **3**: 463 (1972). —Clayton & Renvoize in F.T.E.A., Gramineae: 695 (1982). Type from India.
    Cenchrus annularis Anderss. in Peters, Reise Mossamb., Bot. **2**: 553 (1864). Type: Mozambique, Querimba Is., *Peters* (S, holotype).
    Cenchrus perinvolucratus Stapf & Hubbard in Kew Bull. **1933**: 299 (1933); in F.T.A. **9**: 1081 (1934). Type from Zanzibar.

Annual. Culms 5–90 cm. high, ascending. Panicle 2–15 cm. long; involucre 4–11 mm. long, connate at the base to form an ovoid or diamond-shaped disk 2–4 mm. across; inner spines flattened, ciliate below, retrorsely barbellate and mostly pungent at the apex; outer spines numerous, aciculate, borne in a whorl around the rim of the disk, rarely suppressed. Spikelets 1–3 per bur, 3.5–6 mm. long. Superior lemma thinly coriaceous.

Caprivi Strip. Katima Mulilo, 15.ii.1969, *de Winter* 9194 (K; PRE). **Botswana**. N: Dautsa Flats, 22 km. WSW. of Sehithwa, 20.ii.1966, *Drummond* 8816 (K; SRGH). **Zambia**. B: 14 km. N. of Lukulu, 2.iv.1982, *Drummond & Vernon* 11157 (K). C: Luangwa Valley, Kapamba R., 28.vi.1967, *Astle* 5075 (K). S: Livingstone, 3.iv.1956, *Robinson* 1386 (K). **Zimbabwe**. W: Hwange Distr., Malindi, 7.ii.1964, *Clark* 357 (K; SRGH). **Mozambique**. N: Umparapara, Messalo R., 20.iii.1912, *Allen* 126 (K). GI: Bazaruto Is., 20.x.1958, *Mogg* 28522 (K). M: Maputo, 3.i.1958, *Macêdo* 95 (K).
From Tropical Africa, extending through Arabia to India. A noxious weed of old farmland and waste places, usually on sandy soils; sea level to 1300 m.

4. **Cenchrus incertus** M.A. Curtis in Bost. Journ. Nat. Hist. **1**: 135 (1837). Type from USA.

Annual. Culms 5–80 cm. high, erect or ascending. Panicle 2–8.5 cm. long; involucre 5.5–10 mm. long, connate for $\frac{1}{3}-\frac{2}{3}$ its length to form a cup cleft on 2 sides; inner spines flattened, ± pubescent, retrorsely barbellate, pungent; outer spines triangular, arising irregularly from the surface of the bur. Spikelets 2–4 per bur, 3.5-6 mm. long. Superior lemma thinly coriaceous.

Mozambique. M: Costa de Sol, 1.iv.1948, *Schweickerdt* 1913 (K; PRU).
Also from Southern USA to Argentina; introduced as a weed elsewhere. Sand dunes; sea level.

30. PARATHERIA Griseb.

Paratheria Griseb., Cat. Pl. Cub.: 236 (1866).

Inflorescence loosely spiciform, bearing deciduous racemelets appressed to the axis, each racemelet composed of a single spikelet supported on a pungent stipe and subtended by a stout bristle. Spikelets lanceolate. Glumes tiny. Inferior lemma as long as spikelet. Superior lemma cartilaginous with flat thin margins.

A genus of 2 species. Africa and South America.

Paratheria prostrata Griseb., Cat. Pl. Cub.: 236 (1866). —Stapf & Hubbard in F.T.A. **9**: 1085 (1934). —Clayton in F.W.T.A. **3**: 457 (1972). TAB. **49**. Type from Cuba.

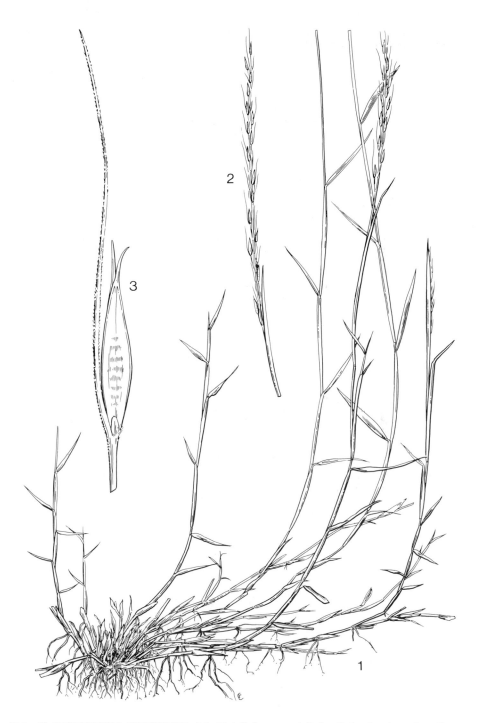

Tab. 49. PARATHERIA PROSTRATA. 1, habit (×⅔); 2, raceme (×⅔); 3, spikelet showing subtending bristle and inferior glume (× 6), all from *Robinson* 4311.

Prostrate perennial. Inflorescence 5–15 cm. long; stipe 5–10 mm. long, attached obliquely to axis; bristle 2–4 cm. long. Spikelets (6)10–12 mm. long.

Zambia. B: Mongu, 6.i.1966, *Robinson* 6784 (K). N: Chambeshi R. on Mpika to Kasama Rd., 29.xii.1967, *Simon & Williamson* 1610 (K; SRGH). S: Livingstone Distr., Katambora, 5–9.iii.1956, *Rattray* 1795 (K; SRGH).

Also from Madagascar, western Africa and South America. Marshy soils and in water; 1000–1200 m.

31. **ANTHEPHORA** Schreb.

Anthephora Schreb., Beschr. Gräser **2**: 105 (1779).

Inflorescence a cylindrical spiciform panicle, bearing deciduous clusters of several spikelets surrounded by an involucre of stiffly coriaceous bracts, these narrowly elliptic, several-nerved and free to the base (except *A. pungens*). Spikelets as long as the bracts, lanceolate. Inferior glume absent; superior glume subulate from a broad base, its back facing inwards. Superior lemma cartilaginous with flat thin margins.

A genus of 12 species. Africa and Arabia; 1 species in tropical America.

The homology of the bracts is contentious, but they seem best explained as derived from sterile spikelets.

1. Base of involucre prolonged into a pungent stipe -   -   -   -   -   *1. pungens*
 - Base of involucre conical, obtuse -   -   -   -   -   -   -   -   -   -   - 2
2. Plant annual -   -   -   -   -   -   -   -   -   -   -   -   - 3
 - Plant perennial -   -   -   -   -   -   -   -   -   -   -   - 4
3. Involucral bracts awned -   -   -   -   -   -   -   -   -   *2. truncata*
 - Involucral bracts caudate -   -   -   -   -   -   -   -   *3. schinzii*
4. Involucral bract scabrid, the margin (see from inside) thickened -   -   *4. elongata*
 - Involucral bract pubescent to villous, the margin not thickened -   -   -   -   - 5
5. Leaf laminae flat, scarcely ribbed, the margins thickened -   -   -   -   *5. pubescens*
 - Leaf laminae folded, rigid, strongly ribbed above, the margins not thickened -   -   -   -   -   -   -   -   -   -   -   -   *6. argentea*

1. **Anthephora pungens** Clayton in Kew Bull. **32**: 2 (1977). Type: Zambia, Cassava Sands, *Richards* 9232 (K, holotype).

Annual. Culms 30–40 cm. high. Panicle 2–6 cm. long; involucre 5–7 mm. long, its base prolonged into a pungent pubescent stipe 4 mm. long, the bracts pubescent with cartilaginous margins, connate for up to ½ their length, each tipped by an awn 6–15 mm. long.

Zambia. N. Mbala Distr., Kasaba Game Reserve, 16.ii.1959, *McCallum-Webster* A62 (K; SRGH). **Malawi**. C: Nkhota Kota, 2.v.1963, *Verboom* 991 (K). **Mozambique**. N: Lichinga (Vila Cabral), 10.i.1961, *Carvalho* 490 (K).

Not known elsewhere. Sand dunes; 800–1100 m.

2. **Anthephora truncata** Robyns in Bull. Jard. Bot. Brux. **9**: 198 (1932). —Stapf & Hubbard in F.T.A. **9**: 947 (1934). —Clayton & Renvoize in F.T.E.A., Gramineae: 662 (1982). Type from Zaire.
    Anthephora gracilis Stapf & Hubbard in Kew Bull. **1933**: 271 (1933). —Stapf & Hubbard in F.T.A. **9**: 946 (1934). Type from Tanzania.

Annual. Culms 25–100 cm. high. Panicle 3–10 cm. long; involucre 3.5–6 mm. long, the bracts appressed pubescent with narrow membranous wings above, acute to emarginate and bearing an awn 1–25 mm. long.

Zambia. N: Kasama, 15.iii.1974, *Davidse & Handlos* 7174 (K; MO). W: 8 km. SE. of Chingola, 15.v.1972, *Kornaś* 1803 (K). E: Chipangali, *Verboom* 575 (K). S: Kalomo, 15.iii.1963, *Astle* 2262 (K). **Zimbabwe**. N: Binga Distr., Lusulu Veterinary Ranch, 28.ii.1965, *Bingham* 1399 (K; SRGH).

Also from Tanzania and Zaire. Open places on poor soils in savanna woodland and thicket; 900–1600 m.

3. **Anthephora schinzii** Hack. in Verh. Bot. Ver. Prov. Brand. **30**: 139 (1888). —Stapf & Hubbard in F.T.A. **9**: 942 (1930). —Chippindall in Meredith, Grasses & Pastures of S. Afr.: 438 (1955). —Launert in Merxm., Prodr. Fl. SW. Afr. **160**: 24 (1970). Type from Namibia.

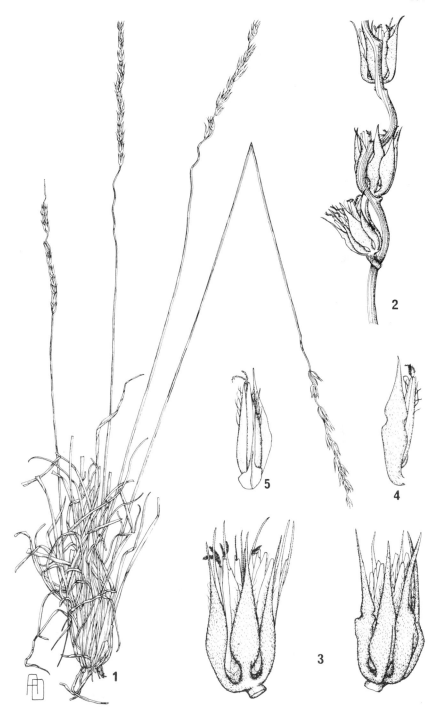

Tab. 50. ANTHEPHORA ELONGATA. 1, habit (× ½); 2, portion of inflorescence (× 3); 3, spikelet clusters (× 5); 4, spikelet broken from cluster to show involucral scale (× 5); 5, spikelet seen from within to show superior glume (× 5), all from *Burtt* 2627. From F.T.E.A.

Anthephora undulatifolia Hack. in Bull. Herb. Boiss. **4**, App. 3: 12 (1896). —Stapf & Hubbard in F.T.A. **9**: 945 (1934). Type from Namibia.

Annual. Culms 12–35 cm. high. Panicle 2.5–6 cm. long. Involucre 4–20 mm. long, pilose at the base, the bracts tapering to a caudate-acuminate apex or the shortest merely acute, commonly reflexed above the middle leaving the spikelets exposed.

Botswana. N: Qangwa R. headwaters, 22.iv.1981, *Smith* 3633 (K; SRGH).
Also from Namibia and Angola. Dry places in savanna woodland.
Length of the bracts varies greatly, even on the same plant.

4. **Anthephora elongata** De Wild. in Ann. Mus. Congo, Bot., sér. 4, **1**: 2 (1902). —Stapf & Hubbard in F.T.A. **9**: 941 (1930). —Clayton & Renvoize in F.T.E.A., Gramineae: 662 (1982). TAB. **50**. Type from Zaire.
    Anthephora elegans var. *acuminata* Rendle, Cat. Afr. Pl. Welw. **2**: 193 (1899). Type from Angola.
    Anthephora hochstetteri var. *glabra* Pilger in Bot. Jahrb. **30**: 269 (1901). Type from Tanzania.
    Anthephora elongata var. *undulata* Chiov. in Nuov. Giorn. Bot. Ital., n.s., **26**: 79 (1919). Type from Zaire.
    Anthephora acuminata (Rendle) Stapf & Hubbard in F.T.A. **9**: 940 (1930). —Jackson & Wiehe, Annot. Check List Nyasal. Grass.: 29 (1958). Type as for *Anthephora elegans* var. *acuminata*.
    Anthephora burttii Stapf & Hubbard in F.T.A. **9**: 941 (1930). Types from Tanzania.

Tufted perennial. Culms 50–100 cm. high. Panicle 8–20 cm. long. Involucre 7–12 mm. long including awn, scabrid, with thick cartilaginous margins (viewed from inside), setaceously acuminate usually with an awn up to 4 mm. long. Inferior lemma glabrous, or rarely pilose at the apex.

Zambia. B: 40 km. W. of Kabompo, 7.iii.1963, *Robinson* 5562 (K). N: Mbala Distr., Kalambo Falls, 25.xii.1967, *Simon & Williamson* 1556 (K; SRGH). W: Solwezi, 23.i.1963, *van Rensburg* 1251 (K). C: Serenje Distr., Kundalila Falls, 4.ii.1973, *Strid* 2865 (K). S: Mukulaikwa, 1954, *Hinds* 194 (K). **Malawi**. N: Mzimba, 9.v.1970, *Brummitt* 10513 (K).
Tanzania, Zaire and Angola. Savanna woodland; 1000–2000 m.
Anthephora ampullacea Stapf & Hubbard from west Africa has awnless involucres 5–6 mm. long, but it intergrades with *A. elongata* and is doubtfully distinct at species level.

5. **Anthephora pubescens** Nees, Fl. Afr. Austr.: 74 (1841). —Chippindall in Meredith, Grasses & Pastures of S. Afr.: 436 (1955). —Launert in Merxm., Prodr. Fl. SW. Afr. **160**: 24 (1970). —Clayton & Renvoize in F.T.E.A., Gramineae: 664 (1982). Type from S. Africa.
    Anthephora ramosa Goossens in Trans. Roy. Soc. S. Afr. **20**: 192 (1932). —Chippindall in Meredith, Grasses & Pastures of S. Afr.: 438 (1955). —Launert in Merxm., Prodr. Fl. SW. Afr. **160**: 24 (1970). Type from S. Africa.

Tufted perennial. Culms 30–100 cm. high. Leaf laminae 2–7 mm. wide, flat, the base as wide as the sheath, the margins thickened and often crinkled. Panicle 5–25 cm. long. Involucre 5–10 mm. long including awn, bearded below, glabrous to villous above, without cartilaginous margin, acute to acuminate and then often attenuate to an awn up to 4 mm. long. Inferior lemma hirsute above.

Botswana. N: Maun to Matabe, 19.x.1978, *Smith* 2498 (K). SW: Western border of Molopo ranches along Lobatse to Werda road, 31.i.1978, *Skarpe* S-224 (K). SE: 6 km. W. of Kanye, 15.iv.1978, *Hansen* (C; GAB; K). **Zimbabwe**. S: Sengwe Tribal Trust land SE. of Mwenezi, 15.ii.1973, *Cleghorn* 2911 (K; SRGH). **Mozambique**. GI: Massingir, 5 km. from Lagoa Nova, 10.iv.1972, *Myre, Lousã & Rosa* 5729 (K).
Also from S. Africa to Mali and eastwards to Iran. Savanna woodland; 600–1300 m.
The size and indumentum of the involucre varies greatly and in the northern part of the range there is intergradation with *A. nigritana* Stapf & Hubbard, a species having glabrous elliptic (rather than narrowly ovate) bracts.

6. **Anthephora argentea** Goossens in Trans. Roy. Soc. S. Afr. **20**: 198 (1932). —Chippindall in Meredith, Grasses & Pastures of S. Afr.: 436 (1955). —Launert in Merxm., Prodr. Fl. SW. Afr. **160**: 24 (1970). Type from S. Africa.
    Anthephora angustifolia Goosens in Trans. Roy. Soc. S. Afr. **20**: 194 (1932). Type from S. Africa.

Tufted perennial. Culms 40–90 cm. high. Leaf laminae 1–2 mm. wide, folded, the base narrower than the sheath, rigid, ribbed above, without thickened margins. Panicle 7–13 cm. long. Involucre 6–8 mm. long, villous, without cartilaginous margins, acute. Inferior lemma hirsute above.

Botswana. SW: 85 km. N. of Kang, 19.ii.1960, *de Winter* 7379 (K; PRE). SE: Morapedi Ranch, 19.i.1978, *Hansen* 3333 (C; GAB; K; PRE; SRGH; UPS; WAG).
Also from S. Africa and Namibia. Savanna bushland and grassland on sandy soils; 1000–1100 m.

XXV. ISACHNEAE Benth.

By W.D. Clayton

Isachneae Benth. in Journ. Linn. Soc., Bot. **19**: 30, 92 (1881). —C.E. Hubbard in Hook., Ic. Pl. **35**: t.3432 (1943).

Inflorescence a panicle or of racemes, terminal. Spikelets all alike, dorsally compressed, 2-flowered without rhachilla extension (rarely 1-flowered), disarticulating above glumes and usually between florets. Glumes 2, tardily deciduous, membranous, shorter than or equalling spikelet. Florets similar or dissimilar, the inferior bisexual (rarely ♂), the superior ♀ or bisexual. Lemmas membranous to coriaceous, the margins inrolled and clasping edges of palea. Stamens 2–3. Stigmas 2. Caryopsis ellipsoid to plano-convex with large embryo; hilum round to oval. Chromosomes small, basic number 10.

A tribe comprising 5 genera. Tropics.

Isachne resembles *Panicum* sect. *Verruculosa* (for example *P. gracilicaule*), and the tribe appears to be a derivative of *Panicum* reverting to the condition of both florets fertile.

1. Inflorescence a panicle - - - - - - - - - - - 2
– Inflorescence of racemes - - - - - - - - - 34. **Heteranthoecia**
2. Superior lemma becoming crisply chartaceous to crustaceous;glumes deciduous 32. **Isachne**
– Superior lemma remaining membranous, thinner than the inferior; glumes persistent - - - - - - - - - - - - 33. **Coelachne**

32. ISACHNE R.Br.

Isachne R.Br., Prodr. Fl. Nov. Holl.: 196 (1810).

Inflorescence a panicle. Florets similar or dissimilar, separated by an abbreviated internode, the lower sometimes ♂. Glumes subequal, $\frac{3}{4}$ to as long as spikelet, 5–9-nerved. Inferior lemma membranous to coriaceous. Superior lemma crisply chartaceous to crustaceous.

A genus of c. 100 species. Throughout the tropics, but mainly Asia.

1. Florets obovoid to subrotund, crustaceous, obtuse; panicle diffuse - - - - 2
– Florets elliptic, thinly coriaceous, acute; panicle contracted about primary branches - - - - - - - - - - - *3. angolensis*
2. Lemmas glabrous - - - - - - - - - - *1. mauritiana*
– Lemmas minutely puberulous - - - - - - - - *2. buettneri*

1. **Isachne mauritiana** Kunth, Rév. Gram. **1**: 243 (1830). —Clayton in F.W.T.A. **3**: 420 (1972). —Clayton, Phillips & Renvoize in F.T.E.A., Gramineae: 434 (1974). Type from Mauritius.

Isachne aethiopica Stapf & Hubbard in Kew Bull. **1933**: 300 (1933); in F.T.A. **9**: 1092 (1934). —Sturgeon in Rhod. Agric. Journ. **51**: 215 (1954). Type from Tanzania.

Rambling perennial. Culms 30–60 cm. high, geniculately ascending from a prostrate base rooting at the nodes. Leaf laminae 5–20 cm. long, 3–17 mm. wide, linear-lanceolate to lanceolate. Panicle 10–20 cm. long, diffuse. Spikelets 1.2–1.8 mm. long, obovoid, the florets bisexual and contiguous. Glumes as long as spikelet, glabrous or with a few hairs towards the apex. Lemmas broadly elliptic to subrotund, the inferior a little larger than the superior, crustaceous, glabrous, obtuse.

Zambia. W: Luanshya Distr., Baluba stream on Fisenge to Kitwe Rd., 4.xi.1967, *Simon & Williamson* 1231 (K; SRGH). E: Nyika Plateau, 7.vi.1962, *Verboom* 713 (K). **Zimbabwe**. E: Mutare Distr., Cloudlands Estate, *Crook* 2016 (K; SRGH). **Malawi**. N: Nkhata Bay, 8 km. E. of Mzuzu, 1.iii.1969, *Pawek* 1780A (K). C: Ntchisi Mt., 6.v.1963, *Verboom* 964 (K). S: Mulanje Distr., Great Ruo Gorge, 17.vi.1962, *Robinson* 5369 (K). **Mozambique**. Z: Milange Mt., 26.ii.1943, *Torre* 4865 (COI; K). MS: Gorongosa Mt., 24.vii.1970, *Müller & Gordon* 1451 (K; SRGH).

Northwards to Kenya and Ghana; also in Mauritius and Madagascar. Forest shade; 1000–2000 m.

There is some variation in the sex of the florets, the inferior being occasionally ♂ and the superior ♀.

Tab. 51. ISACHNE ANGOLENSIS. 1, habit (×⅔); 2, panicle (×⅔); 3, spikelet (× 15); 4, spikelet with glumes removed to show florets (× 15), all from *Simon & Williamson* 1002.

2. **Isachne buettneri** Hackel in Verh. Bot. Ver. Prov. Brand. **31**: 69 (1889). —Stapf & Hubbard in F.T.A. **9**: 1091 (1934). —Clayton in F.W.T.A. **3**: 420 (1972). —Clayton, Phillips & Renvoize in F.T.E.A., Gramineae: 434 (1974). Type from Gabon.

Isachne scandens C.E. Hubbard in Kew Bull. **4**: 360 (1949). Type: Zambia, Mwinilunga, Dobeka Bridge, *Milne-Redhead* 3163 (K, holotype).

Like *Isachne mauritiana* except: Spikelets 0.8–1.4 mm. long, subglobose; glumes a little shorter than spikelet; lemmas minutely puberulous.

Zambia. N: Kawambwa, Chishinga Ranch, 1.vi.1965, *Lawton* 1220 (K). W: Mwinilunga, Dobeka Bridge, 21.xii.1969, *Simon & Williamson* 1924 (K; SRGH).
Also from Angola to Sierra Leone. Forest shade; 1500–1600 m.

3. **Isachne angolensis** Rendle, Cat. Afr. Pl. Welw. **2**: 166 (1899). —Stapf & Hubbard in F.T.A. **9**: 1093 (1934). —Jackson & Wiehe, Annot. Check List Nyasal. Grass.: 33 (1958). —Clayton in F.W.T.A. **3**: 420 (1972). TAB. **51**. Type from Angola.

Rambling perennial. Culms 20–40 cm. high, ascending from a prostrate base. Leaf laminae 3–10 cm. long, 2–4 mm. wide, broadly linear. Panicle 4–15 cm. long, the spikelets condensed about the primary branches. Spikelets 1.5–2 mm. long, elliptic, the florets bisexual and contiguous. Glumes a little shorter than spikelet, glabrous. Lemmas narrowly elliptic, the inferior a little larger than superior, thinly coriaceous, glabrous, acute in side view.

Zambia. N: Mpika Distr., Lwitikila R., 6.x.1969, *Verboom* 2531 (K). C: Serenje Distr., Kundalila Falls, 15.x.1967, *Simon & Williamson* 1002 (K; SRGH). E: Chipata, Mpangwe Hill, 10.vi.1963, *Verboom* 995A (K). **Malawi**. N: Misuku, 28.vi.1951, *Jackson* 565 (K).
Also in Western Africa from Angola to Nigeria. Marshy places in forest; 600–2300 m.

33. COELACHNE R.Br.

Coelachne R.Br., Prodr. Fl. Nov. Holl.: 187 (1810).

Inflorescence a panicle. Florets dissimilar, separated by a slender internode, both fertile. Glumes persistent, subequal, $\frac{1}{3}-\frac{2}{3}$ length of spikelet, obtuse. Inferior lemma firmly cartilaginous. Superior lemma remaining membranous at maturity, $\frac{2}{3}$ to almost as long as the inferior.

A genus of 10 species. Old World tropics.

Coelachne africana Pilger in R.E. Fries, Wiss. Ergebn, Schwed. Rhod.-Kongo-Exped. **1**: 208 (1916). —C.E. Hubbard in F.T.A. **10**: 86 (1937). —Sturgeon in Rhod. Agric. Journ. **51**: 215 (1954). —Jackson & Wiehe, Annot. Check List Nyasal. Grass.: 33 (1958). —Clayton, Phillips & Renvoize in F.T.E.A., Gramineae: 436 (1974). TAB. **52**. Types: Zambia, Luwingu, *Fries* 1105 (B, syntype); Lunzua R., near Mbala, *Fries* 1217 (B, syntype) & from Tanzania.
Coelachne paludosa Peter in Abh. Königl. Ges. Wiss. Göttingen, n.f., **13**, 2 : 40 (1928). Types from Tanzania.

Mat-forming perennial. Culms 5–20 cm. high, ascending from a creeping base and rooting at the nodes. Leaf laminae 0.6–2 cm. long, 1.5–4.5 mm. wide, narrowly lanceolate. Panicle 1.2–6 cm. long, sparse. Spikelets 2–2.5 mm. long, ovate oblong to oblong. Lemmas soft, the inferior glabrous except at the base, the superior a little shorter, thinner and ± pubescent. Stamens 2.

Zambia. N: Mbala Distr., Lunzua Falls, 19 km. S. of Mpulungu, 22.x.1967, *Simon & Williamson* 1162 (K; SRGH). **Zimbabwe**. E: Mutare, Vumba Mt., 29.xii.1967, *Biegel* 2427 (K; SRGH). **Malawi**. N: Nyika Plateau, Nchena Chena stream, 14.viii.1949, *Wiehe* N/198 (K). S: Zomba Plateau, Chagwa Dam, 14.iii.1975, *Seyani* 107 (K; SRGH). **Mozambique**. MS: Gorongosa Mt., 12.iii.1972, *Tinley* 2412 (K; SRGH).
Northwards to Uganda and Ethiopia; also in Madagascar. Upland bogs, stream banks and bordering waterfalls; 1300–1800 m.
C. africana belongs to a complex of interrelated species including *C. simpliciuscula* Benth. from India (spiciform panicles) and *C. infirma* Büse from Indonesia (3 stamens).

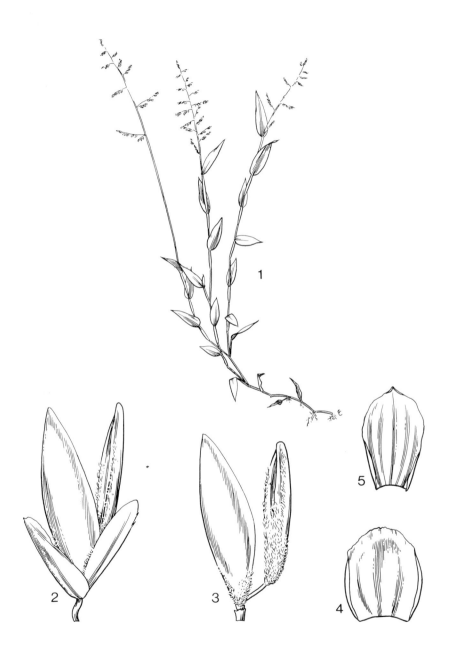

Tab. 52. COELACHNE AFRICANA. 1, habit (×⅔); 2, spikelet (× 25); 3, spikelet with glumes removed (× 25); 4, inferior glume (× 25); 5, superior glume (× 25), all from *Simon & Williamson* 1162.

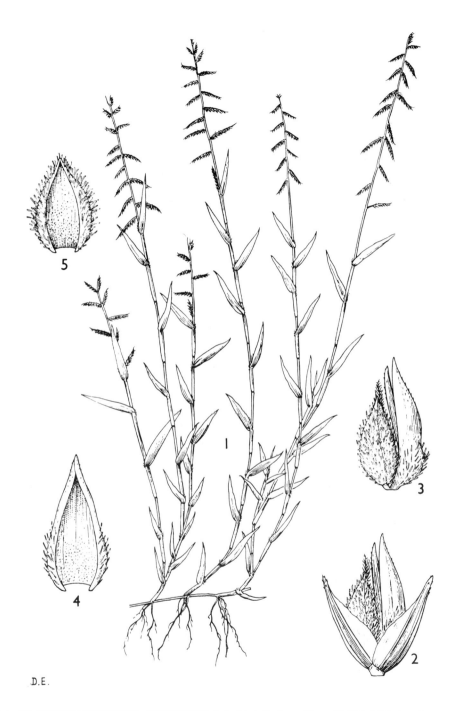

Tab. 53. HETERANTHOECIA GUINEENSIS. 1, habit (× ⅔), from *Haarer* 2065 & 2471; 2, spikelet (× 20); 3, spikelet with glumes removed (× 20); 4, inferior lemma (× 20); 5, superior lemma (× 20), 2–5 from *Haarer* 2471. From F.T.E.A.

34. HETERANTHOECIA Stapf

Heteranthoecia Stapf in Hook., Ic. Pl. **30**: t. 2927 (1911).

Inflorescence of short unilateral racemes along a central axis, the raceme rhachis terminating in a point. Florets dissimilar, both fertile. Glumes persistent, subequal, $\frac{1}{2}$–$\frac{2}{3}$ length of spikelet. Inferior lemma firmly membranous, acuminate. Superior lemma crisply chartaceous smaller, obtuse to subacute.

A monotypic genus. Tropical Africa.

Heteranthoecia guineensis (Franch.) Robyns in Bull. Jard. Bot. Brux **9**: 201 (1932). —Stapf & Hubbard in F.T.A. **9**: 1099 (1934). —Clayton in F.W.T.A. **3**: 421 (1972). —Clayton, Phillips & Renvoize in F.T.E.A., Gramineae: 432 (1974). TAB. **53**. Types from Gabon and Congo (Brazzaville).
> Dinebra guineensis Franch. in Bull. Soc. Hist. Nat. Autun **8**: 376 (1895). Types as above.

Trailing annual. Culms 10–30 cm. high, the base decumbent, rooting at the nodes and forming mats. Leaf laminae 1–3 cm. long, narrowly lanceolate. Inflorescence 2–8 cm. long, the racemes 3–12 mm. long, spreading or eventually reflexed. Spikelets 1.7–2.3 mm. long, gaping. Inferior lemma as long as spikelet, hairy only on base and margins. Superior lemma $\frac{1}{2}$ as long, densely and minutely pubescent all over.

Zambia. B: Mongu, 10.vii.1964, *Robinson* 5414 (K). N: 112 km. NE. of Mpika along road to Isoka, 16.iii.1974, *Davidse & Handlos* 7217 (K; MO). W: Mwinilunga Distr., Matonchi Dambo, 19.xii.1937, *Milne-Redhead* 3739 (K; SRGH).
Western Africa from Angola to Sierra Leone. Swamps and shallow water; 1300–1500 m.

XXVI. ARUNDINELLEAE Stapf

By W.D. Clayton

Arundinelleae Stapf in F.C. **7**: 314 (1898). —Hubbard in Kew Bull. **1936**: 317–322 (1936). —Conert in Bot. Jahrb. **77**: 226–354 (1957). —Phipps in Kirkia **4**: 87–124 (1964); in Blumea **15**: 477–517 (1967). —Clayton in Kew Bull. **21**: 119–124 (1967); **26**: 111–123 (1971). —Phipps in Canad. Journ. Bot. **50**: 1309–1336 (1972).

Inflorescence a panicle. Spikelets all alike, often in triads these sometimes with connate pedicels, lanceolate, slightly laterally compressed, 2-flowered without rhachilla extension, shedding 1 or both florets. Glumes 2, persistent, membranous to coriaceous, the superior as long as spikelet the inferior usually shorter, often brown or beset with tubercle-based hairs, rarely awned. Inferior floret ♂ or barren, its lemma similar to superior glume, often persistent. Superior floret bisexual, subterete; lemma thinly coriaceous, often decorated with hair tufts, bidentate or bilobed, awned from sinus, rarely entire and awnless; awn usually geniculate, often deciduous. Lodicules 2, fleshy. Stamens 2–3. Stigmas 2. Caryopsis ellipsoid with large embryo; hilum linear or punctiform. Chromosomes small, basic number 10 or 12.

A tribe comprising 12 genera. Tropics, mainly Old World.

The tribe suffers from an overabundance of potentially significant generic characters which occur in a bewildering number of different combinations. Consequently there is considerable disagreement over generic limits, and a variety of alternative treatments may be found among the references cited above.

The tribe appears to be a precursor of Andropogoneae, and can usually be recognised by its lanceolate spikelets with papery brown or purplish glumes. The callus at the base of the superior floret is of diagnostic value, and can readily be examined if a mature floret is gently pulled out by its awn. A peculiar feature is the emergence of the panicle in an immature condition, and juvenile spikelets with undeveloped awns frequently puzzle the unwary.

1. Lemma of superior floret scaberulous; ligule a short membrane   35. **Arundinella**
 – Lemma of superior floret hairy or quite smooth; ligule a line of hairs -   -   -   2
2. Spikelets in triads with connate pedicels   -   -   -   -   -   -   -   -   -   3
 – Spikelets with free pedicels   -   -   -   -   -   -   -   -   -   -   -   4

3. Callus of superior floret conical, pungent - - - - - - - 38. **Tristachya**
– Callus of superior floret short, broadly rounded - - - - - - 39. **Zonotriche**
4. Superior lemma with tufts of hair across the back - - - - - - 5
– Superior lemma glabrous or hairy, but the hair not in tufts - - - - - 7
5. Palea of inferior floret woody at maturity; inferior glume awned 37. **Gilgiochloa**
– Palea of inferior floret membranous; inferior glume obtuse to acuminate - - - 6
6. Hair tufts on superior lemma 6–8 - - - - - - - - 36. **Danthoniopsis**
– Hair tufts on superior lemma 2 - - - - - - - - 40. **Trichopteryx**
7. Inferior lemma 3-nerved; spikelets 6–18 mm. long - - .- - - 41. **Loudetia**
– Inferior lemma 5–7 nerved - - - - - - - - - - - - 8
8. Callus of superior floret pungent; spikelets 10–35 mm. long - - - 38. **Tristachya**
– Callus of superior floret short, broadly rounded; leaves falsely petiolate 36. **Danthoniopsis**

35. ARUNDINELLA Raddi

Arundinella Raddi, Agrost. Bras.: 36(1823).

Panicle open or contracted, often with simple raceme-like primary branches, the spikelets usually paired. Spikelets purplish, disarticulating beneath superior floret. Inferior lemma 3–7 nerved. Superior lemma scaberulous, entire to bilobed, with or without a straight or geniculate awn; callus short, rounded; palea wingless, sometimes the margins auriculate below. Hilum punctiform.

A genus of c. 50 species. Tropics and subtropics but mainly in Asia.

Arundinella nepalensis Trin., Gram. Pan.: 62 (1826). —Sturgeon in Rhod. Agric. Journ. **51**: 215 (1954). —Chippindall in Meredith, Grasses and Pastures of S. Afr.: 275 (1955). —Jackson & Wiehe, Annot. Check List Nyasal. Grass.: 30 (1958). —Phipps in Kirkia **4**: 97 (1964). —Clayton in F.W.T.A. **3**: 413 (1972). —Clayton, Phillips & Renvoize in F.T.E.A., Gamineae: 409 (1974). TAB. **54**. Type from Nepal.
    Arundinella ecklonii Nees, Fl. Afr. Austr.: 80 (1841). —Hubbard in F.T.A. **10**: 2 (1937). Type from S. Africa.
    Arundinella rigida Nees, Fl. Afr. Austr.: 80 (1841). Type from S. Africa.

Tufted perennial with short scaly rhizomes. Culms 60–180 cm. high. Panicle 10–40 cm. long, open or contracted, the branches subsecund. Spikelets 4–6 mm. long. Glumes acuminate. Inferior lemma obtuse. Superior lemma 2–2.5 mm. long, emarginate, with a deciduous geniculate awn 4–6 mm. long; palea margins auriculate below.

Zambia. N: Mbala Distr., Kawimbe, 26.ii.1959, *McCallum-Webster* A170 (K; SRGH). **Zimbabwe**. W: Matopos, 7.iii.1931, *Brain* 2761 (K). C: Harare Distr., Cleveland Dam, 14.iii.1969, *Biegel* 2885 (K; SRGH). E: Nyanga, 15.v.1930, *Allen* 1448 (K). S: Masvingo Distr., Great Zimbabwe Nat. Park, 16.v.1973, *Vernon* 35 (K; SRGH). **Malawi**. N: Kasuni, Chindi Jere, 23.vi.1951, *Jackson* 546 (K).
    Also in tropical and South Africa, eastwards to China and Australia. Streamsides and marshy places; 1000–1900 m.
    Closely allied to *A. hispida* (Willd.) Kuntze of tropical America, which has slightly smaller spikelets 3–4 mm. long.

36. DANTHONIOPSIS Stapf

Danthoniopsis Stapf in Ic. Pl. Hook. **31**: t 3075 (1916).
    Rattraya Phipps in Kirkia **4**: 100 (1964).
    Jacquesfelixia Phipps in Kirkia **4**: 115 (1964).
    Gazachloa Phipps in Kirkia **4**: 116 (1964).
    Petrina Phipps in Kirkia **4**: 117 (1964).

Panicle open or contracted, bearing spikelets in groups of 2–3. Spikelets mostly purplish, disarticulating beneath each floret. Glumes usually glabrous. Inferior lemma 5–9-nerved (*D. pruinosa* 3-nerved). Superior lemma usually with 2–8 transversely arranged tufts of hair, sometimes glabrous, bilobed, geniculately awned, the awn usually deciduous; callus square to oblong, obtuse; palea keels winged, the wings clasped by inrolled margins of lemma and often terminating in an auricle or clavate swelling.

A genus of 12 species. Tropical Africa and eastward to Pakistan.

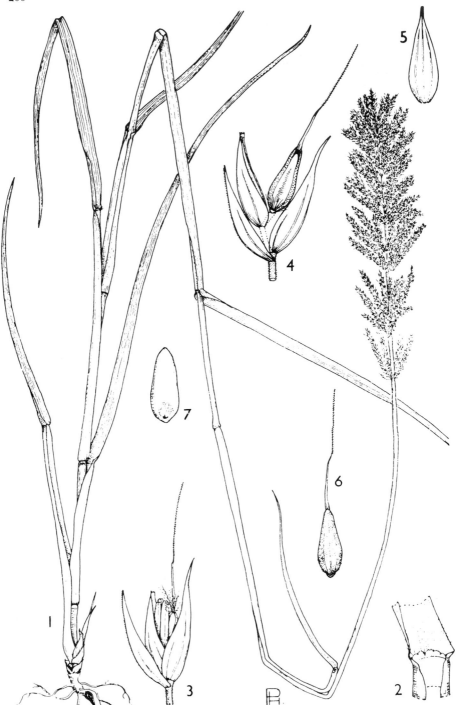

Tab. 54. ARUNDINELLA NEPALENSIS. 1, habit (× ⅓); 2, ligule (× 1); 3, spikelet (×8); 4, spikelet, exploded view (× 8), 5, inferior lemma (× 8); 6, superior lemma (× 8); 7, caryopsis (× 12), all from *Procter* 3312. From F.T.E.A.

1. Leaf lamina with neither margin crinkled; palea of superior floret narrowly winged without
   auricles - - - - - - - - - - - - - - - 2
   − Leaf lamina with one margin crinkled, the other not; palea of superior floret auriculate 3
2. Leaf lamina flat; inferior lemma usually 3-nerved - - - - - 1. *pruinosa*
   − Leaf lamina setaceous; inferior lemma 5-nerved - - - - - 2. *chimanimaniensis*
3. Plants annual - - - - - - - - - - - - - - 6. *dinteri*
   − Plants perennial - - - - - - - - - - - - - - 4
4. Leaf lamina, or at least the inferior, falsely petiolate; superior lemma pilose but the hairs
   not in tufts - - - - - - - - - - - - - - 5. *petiolata*
   − Leaf lamina not falsely petiolate; superior lemma bearing tufts of hairs in a row across
   the back - - - - - - - - - - - - - - - 5
5. Inferior glume obtuse to subacute, ½ of spikelet - - - - - - 3. *viridis*
   − Inferior glume acuminate, ⅔ length of spikelet. - - - - - - 4. *acutigluma*

1. **Danthoniopsis pruinosa** C.E. Hubbard in Kew Bull. **1934**: 436 (1934); in F.T.A. **10**: 80 (1937).
   —Sturgeon in Rhod. Agric. Journ. **51**: 221 (1954). —Chippindall in Meredith, Grasses and
   Pastures of S.Afr.: 285(1955). Type: Zambia, Pemba, *Trapnell* 997 (K, holotype).
   Danthoniopsis pruinosa var. *gracilis* C.E. Hubbard in Kew Bull. **1935**: 310 (1935); in F.T.A. **10**:
   81 (1937). Type: Zambia, Musha Hills, *St. Clair-Thompson* 1292 (K, holotype).
   Petrina pruinosa (C.E. Hubbard) Phipps in Kirkia **4**: 118 (1964). Type as for *Danthoniopsis
   pruinosa*.

Perennial with swollen rootstock. Culms 60–150 cm. high, woody. Leaf laminae 4–12
mm. wide, flat, glaucous, the margins not crinkly. Panicle 10–35 cm. long. Spikelets 5–9
mm. long. Inferior glume ½–¾ length of spikelet, acute to setaceously acuminate. Inferior
lemma 3(5)-nerved. Superior lemma 5-nerved, pilose and with a row of 6 tufts across the
back, setaceously bilobed (setae up to 1 mm. long), with an awn 6–12 mm. long; palea
keels narrowly winged, papillose between keels.

Zambia. C: Mumbwa, 13km. N. of Chombwa Cultivation Project, 25.iii.1964, *van Rensburg* 2882
(K). S: Choma, Sikalonga, 14.vii.1955, *Verboom* (K). **Zimbabwe**. N: Binga Distr., Chizarira Game
Res., 25.iii.1974, *Thomson* 1043 (K; SRGH). W: Matopos Hills, 26.ii.1954, *Rattray* 1668 (K;
SRGH). C: Makoni Distr., 15 km. N. of Rusape, 12.ii.1974, *Davidse, Simon & Pope* 6506 (K;
MO). E: Mutare, Chitsanza, 12.iii.1973, *Crook* 1064 (K; SRGH). S: Mberengwa Distr., Mt. Buhwa,
2.v.1973, *Simon, Pope & Biegel* 2423 (K; SRGH).
   Also from S. Tanzania and S. Africa (Transvaal). Rocky places in savanna woodland and thicket;
550–1600 m.
   D. parva (Phipps) Clayton from the Transvaal is distinguished by its more delicate habit. A
specimen of similar stature from Zambia (Kanona, 1600 m., *Phipps & Vesey-FitzGerald* 2967) seems
merely to be an upland variant of *D. pruinosa*.
   D. lignosa C.E. Hubbard from rocky steamsides in Angola and Namibia also resembles *D. pruinosa*,
but has a 7–9-nerved inferior lemma and 8 hair tufts on the superior lemma.

2. **Danthoniopsis chimanimaniensis** (Phipps) Clayton in Kew Bull. **21**: 123 (1967). Type: Zimbabwe,
   Chimanimani Mts., *Plowes* 1242 (K, isotype; SRGH, holotype).
   Gazachloa chimanimaniensis Phipps in Kirkia **4**: 116 (1964). Type as above.

Tufted perennial. Culms 60–100 cm. high. Leaf laminae setaceous, the margins not
crinkly. Panicle 6–15 cm. long. Spikelets 8–10 mm. long. Inferior glume ⅘ length of
spikelet, acuminate. Inferior lemma 5-nerved. Superior lemma 7-nerved, pilose and with
a row of 8 tufts across the back, setaceously bilobed (setae 1–2 mm. long), with a persistent
awn 12–18 mm. long; palea keels narrowly winged to apex; callus cuneate, narrowly
obtuse.

Zimbabwe. E: Chimanimani Distr., Bundi R. just N. of Southern Lakes, 7.xi.1972, *Simon* 2278 (K;
SRGH). **Mozambique**. MS: Chimanimani Mts., Mukurupini Falls, 25.xii.1967, *Simon & Ngoni* 1322
(K; SRGH).
   Not known elsewhere. Rocky places along streams; 300–1600 m.

3. **Danthoniopsis viridis** (Rendle) C.E. Hubbard in Kew Bull. **1935**: 310 (1935); in F.T.A. **10**: 77 (1935).
   —Phipps in Kirkia **4**: 115 (1964). Type from Angola.
   Trichopteryx viridis Rendle, Cat. Afr. Pl. Welw. **2**: 216 (1899). Type as above.
   Danthoniopsis gossweileri Stapf in Ic. Pl. Hook. **31**: t. 3075 (1916). Type from Angola.
   Danthoniopsis minor Stapf & Hubbard in Kew Bull. **1927**: 269 (1927). —C.E. Hubbard in F.T.A.
   10: 78 (1937). —Sturgeon in Rhod. Agric. Journ. **51**: 220 (1954). —Phipps in Kirkia **4**: 114 (1964).
   Type: Zimbabwe, Harare, *Eyles* 2968 (K, holotype; SRGH, isotype).
   Danthoniopsis intermedia C.E. Hubbard in Kew Bull. **1936**: 500 (1936); in F.T.A. **10**: 78 (1937).
   —Phipps in Kirkia **4**: 114 (1964). Type: Zambia, Kabwe, *Broken Hill Govt. School* 12 (K, holotype).

Danthoniopsis westii Phipps in Kirkia **4**: 114 (1964). Type: Zambia, Kitwe, *West* 4016 (K, isotype; SRGH, holotype).

Tufted perennial. Culms 30–140 cm. high. Leaf laminae 2–15 mm. wide, one margin crinkly. Panicle 5–30 cm. long, contracted, the branches glabrous to villous. Spikelets 7–14 mm. long. Inferior glume ½ as long as spikelet, glabrous to pubescent, obtuse to acute. Inferior lemma 5–7-nerved. Superior lemma 9-nerved, pilose at base with a row of 8 tufts across the back, acutely bilobed, with an awn 7–15 mm. long; palea keels winged to midway, the wing terminating in an auricle.

Zambia. B: Mongu Distr., 48 km. along Kaoma (Mankoya) Rd., 2. iv. 1964, *Verboom* 1373 (K). N: Misamfu, 6 km. N. of Kasama, 5. iv. 1961, *Angus* 2704 (FHO; K). W: Kalulushi, 2. iv. 1964, *Griffin* (K). C: Kabwe Distr., Chibombo, 25.ii.1973, *Kornaś* 3284 (K). S: Mochipapa, 13. iii. 1962, *Astle* 1526 (K). **Zimbabwe**. N: Gokwe Distr., Water Pump II, 14. iii. 1984, *Mahlangu* 951 (K; SRGH). C: Kwekwe, 25 km. along Gokwe road, 16. iii. 1966, *Wild* 7545 (K; SRGH).
Also from Zaire and Angola. Rocky places in *Brachystegia* woodland; 1200–1600 m.
Very similar to *D. wasaensis* (Vanderyst) C.E. Hubbard from Zaire, whose superior floret has a palea with almost obsolete wings and no auricle.

4. **Danthoniopsis acutigluma** Chippindall in Blumea, Suppl. **3**: 27 (1946). —Phipps in Kirkia **4**: 114 (1964). Type: Zambia, Mwendafye, *Stohr* 786 (K, isotype; PRE, holotype).

Tufted perennial. Culms 60–120 cm. high. Leaf laminae 3–15 mm. wide, one margin crinkly. Panicle 5–25 cm. long, contracted, the branches pubescent to pilose. Spikelets 10–12 mm. long. Inferior glume ⅔ as long as spikelet, glabrous to pilose, acuminate. Inferior lemma 5–7-nerved. Superior lemma 9-nerved, pilose at base and with a row of 8 tufts across the back, acutely bilobed, with an awn 10–17 mm. long; palea keels winged to midway, the wing terminating in an auricle.

Zambia. C: Mkushi Distr., Kashitu R. near Kapiri Mposhi, 5. iv. 1961, *Phipps & Vesey-FitzGerald* 2941 (K; SRGH).
Also in Zaire (Katanga). Rocky slopes; 600–1300 m.

5. **Danthoniopsis petiolata** (Phipps) W.D. Clayton in Kew Bull. **21**: 123 (1967). Type: Zimbabwe, between Hwange and Victoria Falls, *Rattray* 1784 (K, isotype; SRGH holotype).
Rattraya petiolata Phipps in Kirkia **4**: 101 (1964). Type as above.

Perennial. Culms 1–2.5 m. high. Leaf laminae 10–35 mm. wide, one margin crinkly, at least the inferior with a false petiole 1–15 cm. long. Panicle 20–35 cm. long, loose. Spikelets 9–13 mm. long. Inferior glume ½ length of spikelet, acuminate. Inferior lemma 7-nerved. Superior lemma 9-nerved, pilose but the hairs not in tufts, bidentate, with an awn 10–15 mm. long; palea keels narrowly winged, without auricles or appendages.

Zambia. S: Livingstone, Zambezi R. gorges, 10.v.1969, *van Rensburg* 3130 (K; SRGH). **Zimbabwe**. W: Victoria Falls Nat. Park, Masuwe R., 22.iv.1970, *Simon & Hill* 2133 (K; SRGH).
Not known elsewhere. Cliffs and shallow soils; 600–1100 m.

6. **Danthoniopsis dinteri** (Pilger) C.E. Hubbard in Kew Bull. **1934**: 436 (1934); in F.T.A. **10**: 81 (1937); in Ic. Pl. Hook. **35**: t. 3447 (1943). —Chippindall in Meredith, Grasses & Pastures of S. Afr.: 285 (1955). —Launert in Merxm., Prodr. Fl. SW. Afr. **160**: 55 (1970). TAB. **55**. Type from Namibia.
Trichopteryx dinteri Pilger in Bot. Jahrb. **51**: 414 (1914). Type as above.
Jacquesfelixia dinteri (Pilger) Phipps in Kirkia **4**: 115 (1964). Type as above.

Robust annual. Culms 0.5–3 m. high. Leaf laminae 5–20 mm. wide, one margin crinkly. Panicle 25–50 cm. long, loose. Spikelets 14–20 mm. long. Inferior glume ⅓ length of spikelet, acute. Inferior lemma 7-nerved. Superior lemma 7-nerved, pilose on the body and with a row of 6–8 tufts across the back, bilobed, with an awn 2–3.5 cm. long; palea keels narrowly winged to the apex, the wings terminating in a small clavate appendage, papillose between keels.

Botswana. SE: Zhilo Hill, 2. v. 1963, *Drummond* 8021 (K; SRGH). **Zimbabwe**. S: Ndanga Distr., Rundi R. gorge near Chipinda Pools, 11.iv.1961, *Goodier* 1065 (K; SRGH). **Mozambique**. GI: Massingir, 19. iv. 1972, *Lousa & Rosa* 242 (K).
Also in Angola, Namibia and S. Africa (Transvaal). Rocky slopes; 400–500 m.

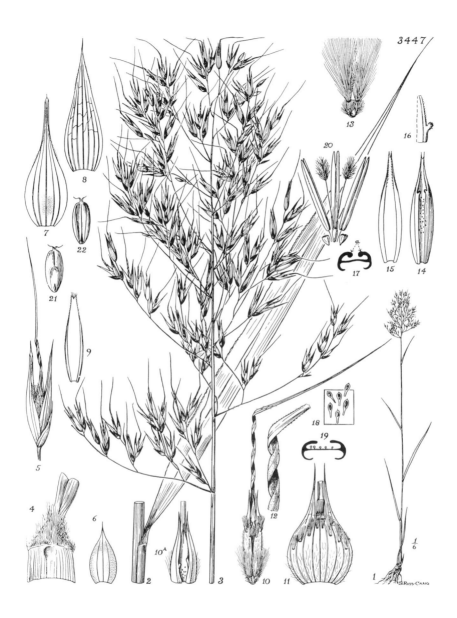

Tab. 55. **DANTHONIOPSIS DINTERI**. 1, habit (× $\frac{1}{8}$); 2, leaf (× 1); 3, panicle (× $\frac{2}{3}$); 4, ligule (× 4); 5, spikelet (× 3); 6, inferior glume (× 4); 7, superior glume (× 4); 8, inferior lemma (× 4); 9, inferior floret palea (× 4); 10, superior floret, adaxial view (× 4); 10A, superior floret, abaxial view (× 6); 11, superior lemma (× 6); 12, portion of awn (× 12); 13, superior floret callus (× 6); 14 & 15, superior floret palea, adaxial and abaxial (× 6); 16, appendage on palea keel (× 12); 17, palea T/S in region of appendages (a) (× 12); 18, papillae between palea keels (× 94); 19, palea T/S in region of papillae (× 12); 20, flower (× 6); 21 & 22, caryopsis, back and front views (× 4). From Ic. Pl. **35**: t.3447.

37. GILGIOCHLOA Pilger

Gilgiochloa Pilger in Bot. Jahrb. **51**: 415 (1914).

Panicle spiciform, the pedicels ± accrescent to the axis. Spikelets purplish, disarticulating beneath each floret. Inferior lemma 5–7-nerved, its palea becoming thickened and hardened like a chip of wood. Superior lemma with hair tufts on margin and at base of awn, bilobed, with a deciduous geniculate awn; callus oblong, truncate; palea keels like *Danthoniopsis*.

A monotypic genus. Central Africa.

Gilgiochloa indurata Pilger in Bot. Jahrb. **51**: 416 (1914). —Hubbard in F.T.A. **10**: 83 (1937). —Jackson & Wiehe, Annot. Check List Nyasal. Grass.: 42 (1958). —Phipps in Kirkia **4**: 113 (1964). —Clayton, Phillips & Renvoize in F.T.E.A., Gramineae: 413 (1974). TAB. **56**. Type from Tanzania.
  Gilgiochloa alopecuroides Peter, Fl. Deutsch Ost-Afr. **1**, Anh.: 92 (1930). Type from Tanzania.

Annual. Culms 30–90 cm. high. Panicle 3–15 cm. long. Spikelets 7–8 mm. long. Inferior glume 3–4 mm. long with an awn 4–7 mm. long, the superior glume acuminate. Inferior lemma acuminate, its palea 3.5–5 mm. long, elliptic, woody. Superior lemma with aristulate lobes, the central awn 14–20 mm. long; palea keels terminating in a small swelling.

Zambia. N: Mporokoso Distr., near Muzombwe, 15.iv.1961, *Phipps & Vesey-FitzGerald* 3208 (K; SRGH). C: 40 km. S. of Lusaka, 6.ii.1954, *Hinds* 148 (K). S: Pemba, 19. vii. 1963, *Astle* 2601 (K). **Malawi**. N: Livingstonia Hills, 25.iii.1954, *Jackson* 1277 (K). **Mozambique.** N: Cabo Delgado, Palma, 18.iv.1964, *Torre & Paiva* 12143 (COI; K).
Also from Tanzania and Burundi. Open places in deciduous thicket; 200–1300 m.

38. TRISTACHYA Nees

Tristachya Nees, Agrost. Bras.: 458 (1829).
Apochaete (C.E. Hubbard) Phipps in Kirkia **4**: 105 (1964).
Dolichochaete (C.E. Hubbard) Phipps in Kirkia **4**: 109 (1964).
Muantijamvella Phipps in Kirkia **4**: 106 (1964).
Veseyochloa Phipps in Kirkia **4**: 106 (1964).

Panicle open or contracted (*T. thollonii* racemose), sometimes bearing the spikelets in groups of 2 or 3 on long unequal pedicels, but more often bearing them in triads whose pedicels are usually connate. Spikelets narrowly lanceolate (except *T. viridiaristata*), brown, often large, disarticulating beneath each floret. Inferior lemma 5–7-nerved. Superior lemma usually pubescent, rarely with 2 or more tufts of hair, bilobed, geniculately awned, the awn usually deciduous; callus conical, pungent (*T. huillensis* linear and bluntly pointed); palea keels thickened but not winged.

A genus of 18 spp. Tropical Africa and America.

1. Pedicels free - - - - - - - - - - - - - - - 2
 – Pedicels connate - - - - - - - - - - - - - - 4
2. Panicle broadly ovate, lax - - - - - - - - - 1. *hubbardiana*
 – Panicle linear, contracted - - - - - - - - - - - - 3
3. Spikelets 25–35 mm. long - - - - - - - - - - - 2. *superba*
 – Spikelets 10–18 mm. long - - - - - - - - - - 3. *lualabaensis*
4. Inflorescence racemose, the triads subsessile - - - - - - 4. *thollonii*
 – Inflorescence paniculate, the triads peduncled - - - - - - - 5
5. Peduncle with a hook-like bend; annual; superior lemma with longitudinal lines
   of hairs - - - - - - - - - - - - - 11. *viridiaristata*
 – Peduncle straight or flexuous; perennial - - - - - - - - 6
6. Superior lemma with a tuft of hairs at the base of each lobe - - - 7. *bicrinita*
 – Superior lemma without hair tufts - - - - - - - - - - 7
7. Lobes of superior lemma acute to acuminate, rarely with a bristle up to 3 mm. long; back of
   lemma usually glabrous - - - - - - - - - - - - 8
 – Lobes of superior lemma tapering to an awn 7–20 mm. long; back of lemma usually
   pubescent - - - - - - - - - - - - - - - 9

Tab. 56. GILGIOCHLOA INDURATA. 1, habit (× ⅔), from *Burtt* 2023; 2, spikelet (× 6); 3, spikelet, exploded view (× 6); 4, spikelet, with glumes removed (× 6); 5, inferior lemma (× 6); 6, palea of inferior floret (× 8); 7, superior lemma (× 8); 2–7 from *van Rensburg* 320.

8. Spikelets 20–45 mm. long - - - - - - - - - - - 5. *leucothrix*
– Spikelets 8–10 mm. long - - - - - - - - - - - 6. *huillensis*
9. Panicle comprising 1–6 triads; leaf laminae 1–3 mm. wide; culms slender, 20–90 cm. high;
   spikelets nearly always glabrous - - - - - - - - - - 8. *rehmannii*
– Panicle comprising (5)8–70 triads; leaf laminae (2)3–13 mm. wide; culms robust, 60–200 cm.
   high; spikelets usually hairy - - - - - - - - - - - 10
10. Inferior glume a little shorter than spikelet; basal leaf-sheaths silky villous   9. *nodiglumis*
– Inferior glume as long as spikelet; basal leaf-sheaths woolly tomentose   10. *bequaertii*

1. **Tristachya hubbardiana** Conert in Bot. Jahrb. **77**: 299 (1957). —Phipps in Kirkia **4**: 103 (1964).
—Clayton, Phillips & Renvoize in F.T.E.A., Gramineae: 424 (1974). Type from Zaire.
    Trichopteryx bequaertii De Wild. in Ann. Soc. Sci. Brux. **39**, Mém.: 146 (1920) non *Tristachya
bequaertii* De Wild. Type as above.
    Loudetia bequaertii (De Wild.) C.E. Hubbard in Kew Bull. **1934**: 431 (1934); in F.T.A. **10**: 49
(1937). Type as above.
    Tristachya pseudoligulata Phipps in Kirkia **4**: 103 (1964). Type: Zambia, Mwinilunga, *Robinson*
3576 (K, isotype; SRGH, holotype).

Tufted perennial. Culms 1–2.7 m. high, the base bulbous and clad in silky pubescent
leaf-sheaths. Panicle 15–45 cm. long, broadly ovate, lax, with slender flexuous branches,
the spikelets in groups of 2(3) on filiform pedicels. Spikelets 20–27 mm. long, brown,
glabrous. Inferior glume ½–⅔ length of spikelet. Superior lemma sparsely pubescent, the
lobes acute, with a central awn 3–10 cm. long.

Zambia. N: Mbala, Kellet's Farm, 14.iv.1959, *McCallum-Webster* A316 (K). W: Ndola, 7.ii.1960,
Robinson 3332 (K).
    Also in Zaire, Tanzania and Angola. *Brachystegia* woodland; 1000–1600 m.

2. **Tristachya superba** (De Not.) Schweinf. & Aschers., Beitr. Fl. Aeth.: 302 (1867). —Phipps in Kirkia
4: 102 (1964). —Launert in Merxm., Prodr. Fl. SW. Afr. **160**: 214 (1970). —Clayton in F.W.T.A. **3**:
413 (1972). —Clayton, Phillips & Renvoize in F.T.E.A., Gramineae: 422 (1974). Type from
Sudan.
    Loudetia superba De Not. in Ind. Sem. Hort. Genuens.: 24 (1852); Ann. Sci. Nat., sér. 3, **19**: 369
(1853). —Hubbard in F.T.A. **10**: 47 (1937). —Sturgeon in Rhod. Agric. Journ. **51**: 218 (1954).
—Chippindall in Meredith, Grasses & Pastures of S. Afr.: 281 (1955). —Jackson & Weihe,
Annot. Check List Nyasal. Grass.: 47 (1958). Type as above.
    Trichopteryx gigantea Stapf in Kew Bull. **1897**: 295 (1897). Types: Zimbabwe, Victoria Falls,
Baines (K, syntype); Deka R., *Holub* (K, syntype) and another syntype from Sudan.
    Trichopteryx gigantea var. *gracilis* Rendle in Cat. Afr. Pl. Welw. **2**: 215 (1899). Types from
Angola.
    Trichopteryx gigantea var. *spiciformis* Pilg. in Fries, Wiss. Ergebn. Schwed. Rhod.-Kongo-Exped.
1: 209 (1916). Type: Zambia, Kamindas, *Fries* 942 (UPS, holotype).
    Trichopteryx superba (De Not.) Chiov. in Nuov. Giorn. Bot. Ital., n.s., **26**: 79 (1919).
    Trichopteryx hockii De Wild in Ann. Soc. Sci. Brux. **39**, Mém.: 154 (1920). Type from Zaire.
    Trichopteryx homblei De Wild. in Ann. Soc. Sci. Brux. **39**, Mém.: 155 (1920). Type from Zaire.
    Trichopteryx gigantea var. *phalacrotes* Peter, Fl. Deutsch Ost-Afr. **1**, Anh.: 95 (1930). Type from
Tanzania.
    Tristachya augusta Phipps in Kirkia **4**: 102 (1964). Type: Zambia, Mporokoso, *Phipps &
Vesey-FitzGerald* 3136 (K, isotype; SRGH, holotype).

Tufted perennial from a woody rootstock. Culms 1.2–2.4 m. high, the base bulbous and
clad in pubescent or tomentose leaf-sheaths. Panicle 20–40 cm. long, linear, contracted,
occasionally the branches villous, the spikelets in groups of (2)3 on short pedicels.
Spikelets 25–35 mm. long, light brown, glabrous (rarely the inferior glume pubescent).
Inferior glume ⅓–½ length of spikelet. Inferior lemma 5(7)-nerved. Superior lemma
sparsely pubescent, the lobes acuminate to a setiform apex, with a central awn 4–12 cm.
long.

Caprivi Strip. Katima Mulilo, 26.xii.1958, *Killick & Leistner* 3109 (K; PRE). **Botswana**.
N: Movombi, 15.ii.1983, *Smith* 4059 (K; SRGH). **Zambia**. N: Mbala Distr., Lucheche R., Muswilo,
4.ii.1965, *Richards* 19595 (K). W: Mufulira, 19.v.1934, *Eyles* 8395 (K; SRGH). C: Serenje Distr., 8 km.
S. of Kanona, 16.ii.1970, *Drummond & Williamson* 9617 (K; SRGH). S: Choma, 14.i.1963, *van
Rensburg* 1207 (K). **Zimbabwe**. N: Miami, 7.iii.1947, *Wild* 1718 (K; SRGH). W: Hwange Distr.,
Lupani to Victoria Falls, 5–9.iii.1956, *Rattray* 1770 (K; SRGH). **Malawi**. C: Nkhota Kota, Sani road,
22.ii.1953, *Jackson* 1097 (K). S: Zomba, 1936, *Cormack* 323 (K). **Mozambique**. N: Milanje to Mocuba,
3.iii.1943, *Torre* 4878 (COI; K). Z: 30 km. from Imala towards Mocuburi, 16.i.1964, *Torre & Paiva*
10007 (COI; K).
    Tropical Africa. Deciduous woodland; 450–1900 m.

T. pedicellata Stent is a segregate from S. Africa (Transvaal), doubtfully worthy of species rank. It has spikelets 20–28 mm. long, the inferior lemma usually 7-nerved and no bulbous base.

3. **Tristachya lualabaensis** (De Wild.) Phipps in Kirkia **4**: 104 (1964). —Clayton, Phillips & Renvoize in F.T.E.A., Gramineae: 423 (1974). Type from Zaire.
  Trichopteryx lualabaensis De Wild., Notes Fl. Kat. **5**: 36 (1920). Type as above.
  Loudetia lualabaensis (De Wild.) C.E. Hubbard in Kew Bull. **1935**: 309 (1935); in F.T.A. **10**: 51 (1937). Type as above.
  Loudetia hitchcockii C.E. Hubbard in Kew Bull. **1935**: 309 (1935); in F.T.A. **10**: 50 (1937). —Sturgeon in Rhod. Agric. Journ. **51**: 218 (1954). Type: Zimbabwe, Victoria Falls, *Hitchcock* 24229 (K, holotype).
  Tristachya hitchcockii (C.E. Hubbard) Conert in Bot. Jahrb. **77**: 301 (1957). —Phipps in Kirkia **4**: 104 (1964). Type as above.

Tufted perennial. Culms 70–140 cm. high, the base not bulbous, clad in glabrous or pubescent leaf-sheaths. Panicle 8–30 cm. long, linear, loosely contracted, the spikelets in groups of (2)3 on short pedicels. Spikelets 10–18 mm. long, yellowish brown, glabrous. Inferior glume $\frac{1}{2}$–$\frac{1}{3}$ length of spikelet. Inferior lemma 5-nerved. Superior lemma pubescent, the lobes usually tipped by a seta 0.5–2 mm. long, with a central awn 1.5–3.5 cm. long.

Botswana. N: Ngamiland, i.1931, *Curzon* 669 (K; PRE). **Zambia**. B: Kabompo Distr., Manyinga, 17.xii.1969, *Simon & Williamson* 2060 (K; SRGH). N: Luwingu Distr., Nsombo, 14.x.1947, *Greenway & Brenan* 8206 (EA; K). S: Namwala, 14.i.1964, *van Rensburg* 2747 (K). **Zimbabwe**. W: Hwange Distr., Kazungula, 8.iii.1981, *Chivandire* 4 (K; SRGH).

Also in Zaire and Tanzania. River flood plains; 1000–1300 m.

4. **Tristachya thollonii** Franch. in Bull. Soc. Hist. Nat. Autun **8**: 374 (1895). —C.E. Hubbard in F.T.A. **10**: 62 (1937). —Jackson & Wiehe, Annot. Check List Nyasal. Grass.: 64 (1958). —Clayton in F.W.T.A. **3**: 413 (1972). —Clayton, Phillips & Renvoize in F.T.E.A., Gramineae: 424 (1974). Type from Gabon.
  Tristachya elymoides Chiov. in Ann. Bot. Roma **13**: 51 (1914). Type from Zaire.
  Tristachya elymoides var. *laevis* Chiov. in Nuov. Giorn. Bot. Ital., n.s., **26**: 69 (1919). Type from Zaire.
  Tristachya homblei De Wild. in Bull. Jard. Bot. Brux. **6**: 49 (1919). Type from Zaire.
  Tristachya spiculata Peter, Fl. Deutsch Ost-Afr. **1**, Anh.: 89 (1930). Type from Tanzania.
  Apochaete thollonii (Franch.) Phipps in Kirkia **4**: 105 (1964). Type as for *Tristachya thollonii*.

Densely tufted perennial. Culms 30–150 cm. high, the basal leaf-sheaths silky pubescent. Inflorescence 4–20 cm. long, racemose, bearing 1–8 triads of spikelets with connate pedicels, the triads subsessile (peduncles 2–3 mm. long) and appressed to the rhachis. Spikelets 22–37 mm. long, pale yellow or yellowish brown, pilose from rows of black tubercles. Inferior glume $\frac{2}{3}$ length of spikelet. Superior lemma glabrous, the lobes acute, with a central awn 5–16 cm. long.

Zambia. N: Chambeshi Flats, Nkweto, 10.i.1962, *Astle* 1244 (K). W: Mwinilunga Distr., Dobeka Bridge, 8.xi. 1937, *Milne-Redhead* 3162 (K; SRGH). C: Mkushi R., 16.xii.1967, *Simon & Williamson* 1397 (K; SRGH). **Malawi**. N: Mzimba Distr., Mbawa, 20.i.1951, *Jackson* 373 (K). C: Nkhota Kota, 21.ii.1953, *Jackson* 1081 (K). **Mozambique**. N: Nampula, Namaita, 26.iii.1964, *Torre & Paiva* 11396 (COI; K).

Also in Nigeria, Cameroon, Gabon, Zaire and Angola. Seasonally wet grassland; 350–1600 m.

5. **Tristachya leucothrix** Nees, Agrost. Bras.: 460 (1829). —Clayton, Phillips & Renvoize in F.T.E.A., Gramineae: 424 (1974). Type from S. Africa.
  Avena hispida Thunb., Prodr. Pl. Cap.: 22 (1794) non *A. hispida* L.f. (1781). Type from S. Africa.
  Tristachya monocephala Hochst. in Flora **29**: 120 (1846). Type from S. Africa.
  Tristachya hispida K. Schum., Pflanzenw. Ost-Afr. **C**: 109 (1895). —C.E. Hubbard in F.T.A. **10**: 63 (1937). —Sturgeon in Rhod. Agric. Journ. **51**: 219 (1954). —Chippindall in Meredith, Grasses & Pastures of S. Afr. (1955). —Jackson & Wiehe, Annot. Check List Nyasal. Grass.: 64 (1958). Type as for *A. hispida* Thunb.
  Tristachya leucothrix var. *sapinii* De Wild. in Bull Jard. Bot. Brux. **3**: 278 (1911). Type from Zaire.
  Tristachya granulosa Chiov. in Ann. Bot. Roma **13**: 52 (1914). Type from Zaire.
  Tristachya leucothrix var. *bolusii* De Wild. in Bull. Jard. Bot. Brux. **6**: 51 (1919). Type from S. Africa.
  Tristachya leucothrix var. *longiaristata* De Wild. in Bull. Jard. Bot. Brux. **6**: 52 (1919). Type from S. Africa.

Tufted perennial. Culms 30–120 cm. high, the basal sheaths silky pubescent to fulvously tomentose. Panicle 7–20 cm. long, linear, nodding, bearing 1–7 triads of spikelets with connate pedicels. Spikelets 20–45 mm. long, yellow or yellowish brown, villous from rows of black tubercles. Inferior glume $\frac{3}{4}$ length of spikelet. Superior lemma glabrous or rarely obscurely pubescent near apex, the lobes acute to acuminate, rarely prolonged into setae up to 3 mm. long, with a central awn 6–10 cm. long.

Zambia. B: Kaoma, 20.xi.1959, *Drummond & Cookson* 6679 (K; SRGH). N: Mbala Distr., Zombe Plain, Sumbawanga Rd., 29.xii.1964, *Richards* 19408 (K). W: Mwinilunga Distr., SW. of Dobeka Bridge, 11.x.1937, *Milne-Redhead* 2702 (K). S: 9 km. E. of Choma, 18.xii.1956, *Robinson* 1985 (K; SRGH). **Zimbabwe**. C: Beatrice, 15.xii.1947, *Rattray* 1542 (K; SRGH). E: Nyanga Distr., Pungwe Gorge, 9.vii.1955, *Cleghorn* 3055 (K; SRGH). **Malawi**. N: Viphya, Luwawa, 12.vii.1952, *Jackson* 968 (K). S: Thyolo, 11.i.1950, *Wiehe* N/406 (K). **Mozambique**. MS: Chimanimani Mts., Mevumozi R. near Pillar Rock, 28.ix.1966, *Simon* 879 (K; SRGH).

Also from Zaire and S. Africa. Damp grassland; 1100–2200 m.

The nomenclature of the species has been confused by combinations — such as *Apochaete hispida* (L.f.) Phipps in Kirkia **4**: 105 (1964) — inadvertently based on *Avena hispida* L.f. The latter is a superfluous name for *Avena capensis* Burman, whose identity is uncertain though commonly equated with *Themeda triandra*.

6. **Tristachya huillensis** Rendle, Cat. Afr. Pl. Welw. **2**: 217 (1899). —Hubbard in F.T.A. **10**: 61 (1937). —Jackson & Wiehe, Annot. Check List Nyasal. Grass.: 64 (1958). Type from Angola.
  Muantizambella huillensis (Rendle) Phipps in Kirkia **4**: 106 (1964). Type as above.

Tufted perennial. Culms 60–90 cm. high. Panicle 5–12 cm. long, ovate, bearing 5–40 triads of spikelets with connate pedicels. Spikelets 8–10 mm. long, golden brown, pilose from rows of black tubercles. Inferior glume almost as long as spikelet. Superior lemma glabrous, the lobes acute, with a persistent central awn 5–12 mm. long, its column very short; callus linear, bluntly pointed.

Zambia. N: Nkali Dambo by Lumi stream, 4. xi. 1965, *Richards* 20659 (K). W: Chingola, 13.ix.1964, *van Rensburg* 2983 (K). C: Kabwe, 33.ix.1947, *Brenan & Greenway* 7947 (FHO; K). **Malawi**. N: Katoto, 5 km. W. of Mzuzu, 7.xi.1970, *Pawek* 3950 (K; MAL). C: Mchinji, Nyoka, 11.x.1951, *Jackson* 616 (K). S: Kirk Range, Daviko, x.1950, *Briars* N/670 (K).

Also in Zaire and Angola. Wet grassland; 1300–1900 m.

7. **Tristachya bicrinita** (Phipps) W.D. Clayton in Kew Bull. **21**: 124 (1967). —Clayton, Phillips & Renvoize in F.T.E.A., Gramineae: 425 (1974). Type from Tanzania.
  Dolichochaete bicrinita Phipps in Kirkia **4**: 112 (1964). Type as above.

Densely tufted perennial. Culms 60–100 cm. high, the basal sheaths pubescent. Panicle 5–17 cm. long, lanceolate, bearing 5–30 triads of spikelets with connate pedicels. Spikelets 14–23 mm. long, straw-coloured, ciliate from rows of black tubercles. Inferior glume as long as spikelet. Superior lemma pubescent with a tuft of hair below each lobe, the lobes tapering to a bristle 3–7 mm. long, with a persistent central awn 1.5–2.5 cm. long.

Zambia. N: Mbala Distr., Kalambo Falls, 21. iv. 1959, *McCallum-Webster* A345 (K; SRGH).
Also in Tanzania. *Brachystegia* woodland; 1300–1700 m.

8. **Tristachya rehmannii** Hackel in Bull. Herb. Boiss. **3**: 384 (1895). —Chippindall in Meredith, Grasses & Pastures of S. Afr.: 279 (1955). —Clayton, Phillips & Renvoize in F.T.E.A., Gramineae: 425 (1974). Type from S. Africa.
  Tristachya glabra Stapf in Kew Bull. **1897**: 294 (1897). Type: Malawi, Blantyre, *Scott* (K, syntype) and three S. African syntypes including an isotype of *T. rehmannii*.
  Tristachya helenae Buscal. & Muschl. in Bot. Jahrb. **49**: 485 (1913). Type: Zambia, Kabwe to Bwana Mkubwa, *Aosta* 287 (B, holotype).
  Tristachya hockii De Wild. in Bull. Jard. Bot. Brux. **6**: 47 (1919). —C.E. Hubbard in F.T.A. **10**: 67 (1937). Type from Zaire.
  Tristachya rehmannii var. *helenae* (Buscal. & Muschl.) C.E. Hubbard in Kew Bull. **1934**: 434 (1934); in F.T.A. **10**: 70 (1937). —Sturgeon in Rhod. Agric. Journ. **51**: 220 (1954). —Jackson & Wiehe, Annot. Check List Nyasal. Grass.: 64 (1958). Type as for *Tristachya helenae*.
  Tristachya rehmannii var. *pilosa* C.E. Hubbard in Kew Bull. **1934**: 435 (1934); in F.T.A. **10**: 71 (1937). —Sturgeon in Rhod. Agric. Journ. **51**: 220 (1954). Type from Angola.
  Dolichochaete rehmannii (Hack.) Phipps in Kirkia **4**: 110 (1964). Type as for *Tristachya rehmannii*.
  Dolichochaete rehmannii subsp. *mosambicensis* Phipps in Kirkia **4**: 111 (1964). Type: Mozambique, Pessene to Vundica, *Myre & Carvalho* 1689 (SRGH, holotype; LMU, isotype).
  Dolichochaete rehmannii var. *helenae* (Buscal. & Muschl.) Phipps in Kirkia **4**: 111 (1964). Type as for *Tristachya helenae*.

Dolichochaete rehmannii var. *pilosa* (C.E. Hubbard) Phipps in Kirkia **4**: 111 (1964).
Type as for *Tristachya rehmannii* var. *pilosa* C.E. Hubbard.

Compactly tufted perennial. Culms 20–90 cm. high, slender, the basal sheaths silky villous. Leaf laminae 1–3(6) mm. wide, flat. Panicle up to 20 cm. long, narrowly lanceolate, nodding, bearing 1–6(8) triads of spikelets with connate pedicels. Spikelets 18–33 mm. long, yellowish with the inferior glume purple-brown, glabrous or very rarely tuberculate-ciliate. Inferior glume $\frac{1}{2}$–$\frac{3}{4}$ length of spikelet. Superior lemma pubescent, the lobes tapering into bristles 10–20 mm. long, these deciduous with the central awn which is 3–9 cm. long.

Botswana. N: Gwiligwa, 18.x.1979, *Smith* 2850 (K; SRGH). **Zambia**. N: Mbala Distr., Nakatali, Lumi Marsh, 25.i.1957, *Richards* 8026 (K). W: Mwinilunga Distr., Matonchi Dambo, 16.xii.1937, *Milne-Redhead* 3737 (K). C: Mkushi R. Hotel, 1.xi.1965, *Lawton* 1314 (K). E: Chipata to Katete, 9.i.1959, *Robson* 1120 (K). S: Mazabuka, 26.i.1960, *White* 6396 (FHO; K). **Zimbabwe**. N: Guruve Distr., Nyamnyetsi Estate, 4.i.1979, *Nyariri* 615 (K; SRGH). C: Charter Estate, 25.i.1969, *Thomas* 739 (K; SRGH). **Malawi**. N: Karonga Distr., 9 km. N. of Chilumba, 24.ii.1978, *Pawek* 13888 (K; MA; MO; SRGH; UC). C: Mehinje, Guilleme, 31.i.1952, *Jackson* 724 (K). S: Thyolo, Tung Station, 6.iii.1951, *Jackson* 405 (K). **Mozambique**. Z: Molumbo, 3.xii.1970, *Bowbrick* BA 900 A (K; NU).

Also in Zaire, Tanzania, Angola and South Africa. Margins of seasonally wet drainage tracts; 600–1700 m.
The culm is scabrid below the inflorescence (var. *rehmannii*) in South Africa, but usually smooth (var. *helenae*) in our area. Angolan specimens have the penultimate culm internode pilose (var. *pilosa*). None of these distinctions seem to be of much taxonomic significance.
Tristachya biseriata Stapf, from South Africa, is similar but has filiform leaf laminae and the inferior glume equalling the spikelet.

9. **Tristachya nodiglumis** K. Schum. in Bot. Jahrb. **24**: 334 (1897). —C.E. Hubbard in F.T.A. **10**: 64 (1937). —Clayton, Phillips & Renvoize in F.T.E.A., Gramineae: 426 (1974). TAB. **57**. Type from Angola.
   Tristachya nodiglumis var. *laeviglumis* K. Schum. in Bot. Jahrb. **24**: 334 (1897). Type from Angola.
   Tristachya welwitschii Rendle, Cat. Afr. Pl. Welw. **2**: 217 (1899). —C.E. Hubbard in F.T.A. **10**: 68 (1937). —Sturgeon in Rhod. Agric. Journ. **51**: 219 (1954). —Jackson & Wiehe, Annot. Check List Nyasal. Grass.: 64 (1958). Type from Angola.
   Tristachya superbiens Pilger in Bot. Jahrb. **40**: 82 (1907). Type from Angola.
   Tristachya vanderystii De Wild. in Bull. Jard. Bot. Brux. **6**: 56 (1919). Type from Zaire.
   Tristachya bequaertii var. *vanderystii* De Wild. in Bull. Jard. Bot. Brux **6**: 46 (1919). Type from Zaire.
   Tristachya welwitschii var. *atricha* Chiov. in Bull. Soc. Bot. Ital. **1924**: 45 (1924). Types from Angola.
   Tristachya welwitschii var. *trichophora* Chiov. in Bull. Soc. Bot. Ital. **1924**: 45 (1924). Type from Angola.
   Tristachya atricha Peter, Fl. Deutsch Ost-Afr. **1**, Anh.: 91 (1930). Type from Tanzania.
   Tristachya eylesii Stent & Rattray in Proc. Rhod. Sci. Ass. **32**: 41 (1933). —C.E. Hubbard in F.T.A. **10**: 65 (1937). —Sturgeon in Rhod. Agric. Journ. **51**: 219 (1954). Type: Zimbabwe, Chinamora Reserve, *Eyles* 3403A (SRGH, holotype).
   Tristachya longispiculata C.E. Hubbard in Kew Bull. **1934**: 263 (1934). —Launert in Merxm., Prodr. Fl. SW. Afr. **160**: 214 (1970). Type from Angola.
   Tristachya welwitschii var. *superbiens* (Pilger) C.E. Hubbard in Kew Bull. **1934**: 434 (1934); in F.T.A. **10**: 69 (1937). Type as for *T. superbiens*.
   Tristachya welwitschii var. *major* C.E. Hubbard in Kew Bull. **1934**: 434 (1934); in F.T.A. **10**: 69 (1937). —Sturgeon in Rhod. Agric. Journ. **51**: 219 (1954). Type: Zimbabwe, Harare, *Eyles* 3403 (K, holotype; SRGH, isotype).
   Dolichochaete nodiglumis (K. Schum.) Phipps in Kirkia **4**: 111 (1964). Type as for *Tristachya nodiglumis*.
   Dolichochaete longispiculata (C.E. Hubbard) Phipps in Kirkia **4**: 111 (1964). Type as for *Tristachya longispiculata*.

Densely tufted perennial. Culms 60–200 cm. high, robust, the basal sheaths silky villous. Leaf laminae 3–13 mm. wide, flat. Panicle 12–30 cm. long, narrowly ovate, open or contracted, bearing 8–70 triads with connate pedicels. Spikelets 18–26(34) mm. long, straw-coloured to brown, glabrous or ciliate from black tubercles. Inferior glume $\frac{1}{2}$–$\frac{4}{5}$ length of spikelet. Superior lemma pubescent, the lobes tapering into bristles 7–20 mm. long, these deciduous with the central awn which is 3–6 cm. long.

Caprivi Strip. 11 km. S. of Katima Mulilo, 20.xii. 1958, *Killick & Leistner* 3040 (K; PRE). **Botswana**. N: Kwando R. Plain, 24.i.1978, *Smith* 2230 (K; SRGH). **Zambia**. B: Mongu, 25.iii.1964, *Verboom* 1326 (K). N: 48 km. W. of Mporokoso, 13.iv.1961, *Phipps & Vesey-FitzGerald* 3110 (K;

Tab. 57. TRISTACHYA NODIGLUMIS. 1, habit (× ⅔); 2, triad of spikelets (× 2); 3, single spikelet (× 3); 4, superior lemma (× 6), all from *Milne-Redhead & Taylor* 9412.

SRGH). W: Mwinilunga Distr., Ikelenge, 16.iv.1965, *Robinson* 6595 (K). C: Kabwe Distr., Muka Mwanje Hills, 25.11.1973, *Kornaś* 3305 (K). E: Chipata Distr., Sinde Misale, xii.1962, *Verboom* 554 (K). S: Choma, 17.ii.1963, *Astle* 2140 (K). **Zimbabwe**. N: Guruve Distr., Nyamnyetsi Estate, 19.iii.1979, *Nyariri* 769 (K; SRGH). W: Bulawayo, *Jeffreys* in GHS 39914 (K; SRGH). C: Makoni Distr., Rusape, 11.ii.1969, *Cleghorn* 1852 (K; SRGH). E: Chimanimani, 6.v.1951, *Crook* 403 (K; SRGH). S: Masvingo, Glenlivet Hotel, 15.ii.1974, *Davidse, Simon & Pope* 6647 (K; MO; SRGH). **Malawi**. C: Lilongwe, Nsam, 27.i.1953, *Jackson* 1039 (K). **Mozambique**. N: Ribáuè, 1.ii.1964, *Torre & Paiva* 10364 (COI; K). Z: Massingire, Serra da Marrumbala, 1.v.1943, *Torre* 5275 (K). MS: Chimoio, iii.1952, *Schweickerdt* 2358 (K; PRE).

Also in Zaire, Tanzania, Angola and Namibia. *Brachystegia* woodland; 500–2000 m.

T. nodiglumis, T. rehmannii and *T. bequaertii* are closely related, and intermediate specimens can be difficult to identify.

10. **Tristachya bequaertii** De Wild. in Bull Jard. Bot. Brux. **6**: 45 (1919). —C.E. Hubbard in F.T.A. **10**: 66 (1937). —Jackson & Wiehe, Annot. Check List Nyasal. Grass.: 64 (1958). —Clayton, Phillips & Renvoize in F.T.E.A., Gramineae: 428 (1974). Type from Zaire.
    Dolichochaete bequaertii (De Wild.) Phipps in Kirkia **4**: 112 (1964). Type as above.

Like *T. nodiglumis* except; basal sheaths woolly tomentose, spikelets 15–25 mm. long, inferior glume equalling spikelet.

Zambia. N: 27 km. E. of Kawambwa on Mushota Rd., 7.iv.1961, *Angus* 2743 (FHO; K). W: Mwinilunga Distr., Luao R., 27.xii.1937, *Milne-Redhead* 3845 (K). C: Kundalila Falls, 17.iii.1974, *Davidse & Handlos* 7258 (K; MO). **Malawi**. N: Mzimba Distr., Mtanga-Mtanga Forest Reserve, 4.ii.1968, *Simon, Williamson & Ball* 1634 (K; SRGH). C: Nkhota Kota Game Reserve, 28.ii.1957, *Jackson* 1130 (K).
Also in Zaire and Tanzania. *Brachystegia* woodland; 1300–2000 m.

11. **Tristachya viridiaristata** (Phipps) Clayton in Kew Bull. **21**: 124 (1967). —Clayton, Phillips & Renvoize in F.T.E.A., Gramineae: 429 (1974). Type: Zambia, Mporokoso, *Phipps & Vesey-FitzGerald* 3153 (SRGH holotype).
    Veseyochloa viridiaristata Phipps in Kirkia **4**: 107 (1964). Type as above.

Annual. Culms 50–100 cm. long, slender. Panicle 5–15 cm. long, linear, bearing 3–15 triads of spikelets with connate pedicels, the peduncle of each triad hook-like and fracturing at maturity. Spikelets 28–33 mm. long, yellowish, pilose from dark tubercles. Inferior glume $\frac{1}{2}$–$\frac{2}{3}$ length of spikelet. Superior lemma turbinate at maturity, densely pilose from longitudinal lines of rufescent hairs, the lobes tapering to bristles 5–10 mm. long, with a central awn 3.5–6.5 cm. long.

Zambia. N: Mbala Distr., Sanzyo Falls, 19.iv.1969, *Sanane* 594 (K).
Also in Burundi and Tanzania. *Brachystegia* woodland; 1600 m.

39. ZONOTRICHE (C.E. Hubbard) Phipps

Zonotriche (C.E. Hubbard) Phipps in Kirkia **4**: 113 (1964).
Piptostachya (C.E. Hubbard) Phipps in Kirkia **4**: 108 (1964).

Panicle bearing spikelets in triads with connate pedicels, the peduncle developing a fragile hook-like flexure at maturity and the triads falling entire. Spikelets brown, the superior florets eventually disarticulating. Inferior lemma 5-nerved. Superior lemma with or without hair tufts, bilobed, with a persistent geniculate awn; callus short, broadly rounded; palea keels with or without narrow wings, but these not clasped by the lemma.

A genus of 3 species. Central Africa.

Triads clad in golden brown hairs; superior lemma glabrous or villous but without tufts
    of hair - - - - - - - - - - - - - - - - - 1. *inamoena*
Triads clad in white hairs; superior lemma with a row of 6–8 tufts of hair - - 2. *decora*

1. **Zonotriche inamoena** (K. Schum.) W.D. Clayton in Kew Bull. **21**: 124 (1967). —Clayton, Phillips & Renvoize in F.T.E.A., Gramineae: 429 (1974). Type from Malawi, Shire Highlands, *Buchanan* 49 (B, holotype; K, isotype).

Tab. 58. ZONOTRICHE DECORA. 1, habit (× $\frac{2}{3}$); 2, rootstock (× $\frac{2}{3}$); 3, triad of spikelets (× 2); 4, superior lemma, adaxial view (× 6); 5, superior lemma, abaxial view (× 3), all from *McCallum-Webster* T201.

Tristachya inamoena K. Schum. in Pflanzenw. Ost-Afr. **C:** 109 (1895). —C.E. Hubbard in F.T.A.
10: 71 (1937). —Sturgeon in Rhod. Agric. Journ. **51:** 220 (1959). —Jackson & Wiehe, Annot.
Check List Nyasal. Grass.: 64 (1958). Type as above.
   Tristachya ringoetii De Wild. in Bull. Jard. Bot. Brux. **6:** 55 (1919). Type from Zaire.
   Tristachya aurea Chiov. in Nuov. Giorn. Bot. Ital., n.s., **26:** 69 (1919). Type from Zaire.
   Piptostachya inamoena (K. Schum.) Phipps in Kirkia **4:** 109 (1964). Type as for *Tristachya inamoena*.

Tufted perennial. Culms 60–150 cm. high. Panicle 10–25 cm. long, linear to lanceolate.
Spikelets 10–15(19) mm. long, narrowly lanceolate. Inferior glume $\frac{2}{3}$ to as long as spikelet,
villous with tubercle-based golden brown hairs, acuminate. Superior lemma villous all
over or only on margins, the lateral lobes usually attenuate into setae 1–3(6) mm. long, the
central awn 9–18 mm. long.

Zambia. B: Kaoma, 20.xi.1959, *Drummond & Cookson* 6658 (K; SRGH). N: Mbala Distr.,
Kawimbe, 25.ii.1959, *McCallum-Webster* A 152 (K). W: Ndola, v.1961, *Wilberforce* A 69 (K).
C: Chakwenga Headwaters, 14.ii.1965, *Robinson* 6378 (K). E: Kachalola, 17.iii.1959, *Robson* 1734
(K). S: Mazabuka Distr., Kanchale, 40 km. N. of Pemba, 11.ii.1960, *White* 6947 (FHO; K).
Zimbabwe. N: Gokwe Distr., Sanyati Tsetse Camp, 26.iv.1963, *Bingham* 652 (K; SRGH).
Malawi. N: Rumphi Distr., foothills to Nyika Plateau, 12.ii.1968, *Simon & Williamson* 1770 (K;
SRGH). S: Blantyre, 13.iv.1970, *Brummitt* 9823 (K). **Mozambique**. N: Marrupa, 15 km. on Rd. to
Nungo, 22.ii.1982, *Jansen & Boane* 8061 (K). Z: Gúrùe, 22 km. from Nintulo to Lioma, 10.ii.1964, *Torre
& Paiva* 10519 (COI; K).
   Also in Zaire, Tanzania and Angola. *Brachystegia* woodland; 500–2000 m.

2. **Zonotriche decora** (Stapf) Phipps in Kirkia **4:** 113 (1964). —Clayton, Phillips & Renvoize in
   F.T.E.A., Gramineae: 431 (1974). TAB. **58**. Type: Zambia, Fwambo, *Carson* 56 (K, holotype).
      Tristachya decora Stapf in Kew Bull. **1895:** 75 (1895). —C.E. Hubbard in F.T.A. **10:** 72 (1937); Ic.
   Pl. Hook. **35:** t. 3446 (1943). Type as above.
      Tristachya pilgeriana Buscal. & Muschl. in Bot. Jahrb. **49:** 459 (1913). Type: Zambia, L.
   Bangweulu to L. Tanganyika, *Aosta* (B, holotype).

Perennial from a knotty rootstock. Culms 60–120 cm. high. Panicle 12–20 cm. long,
narrowly elliptic. Spikelets 15–28 mm. long, narrowly lanceolate. Inferior glume as long
as spikelet, hirsute with tubercle-based white hairs, attenuate to a filiform apex. Superior
lemma glabrous on the back except for a row of 6–8 tufts of hair, the lateral lobes often
attenuate into setae 1–2 mm. long, the central awn 15–25 mm. long.

Zambia. N: Kaputa Distr., 17 km. N. of Mporokoso, 1.iv.1984, *Brummitt, Chisumpa & Nshingo*
17080 (K; SRGH). W: Mwinilunga Distr., Ikelenge, 16.iv.1965, *Robinson* 6646 (K).
   Also in Zaire and Tanzania. *Brachystegia* woodland; 1400–1600 m.

40. TRICHOPTERYX Nees

Trichopteryx Nees in Lindley, Nat. Syst. ed. 2: 449 (1836) as 'Trichopteria', corr. Nees in
Fl. Afr. Austr.: 339 (1841).

Panicle open or contracted, bearing spikelets singly or in pairs. Spikelets brown,
disarticulating beneath each floret. Inferior lemma 3-nerved. Superior lemma bilobed,
the lobes finely awned, with a tuft of hair below each lobe and a persistent awn from the
sinus; callus semicircular (except *T. stolziana*), c. 0.1 mm. long; palea not winged.

A genus of 5 species. Tropical Africa and Madagascar.

1. Plants annual - - - - - - - - - - - - - - - 2
- Plants perennial - - - - - - - - - - - - - - 3
2. Callus of superior floret broadly rounded; spikelets 2.5–3.5 mm. long   1. *elegantula*
- Callus of superior floret truncate or emarginate; spikelets 3.5–5.5 mm. long   2. *stolziana*
3. Culms stiff, the nodes generally hidden by overlapping leaf-sheaths; spikelets glabrous or rarely
   softly pilose; superior lemma with hair tufts 1–2 mm. long; leaf laminae
   linear-lanceolate - - - - - - - - - - - - - 3. *fruticulosa*
- Culms trailing, the nodes exposed - - - - - - - - - - 4
4. Spikelets glabrous or with a few hairs at apex of inferior glume; superior lemma with hair tufts
   2–13 mm. long; leaf laminae linear - - - - - - - - 4. *dregeana*
- Spikelets minutely pubescent; superior lemma with hair tufts 0.4–1 mm. long; leaf laminae
   lanceolate - - - - - - - - - - - - - - 5. *marungensis*

1. **Trichopteryx elegantula** (Hook.f.) Stapf in Ic. Pl. Hook. **24**: t. 2394 (1895). —C.E. Hubbard in F.T.A. **10**: 11 (1937). —Clayton in F.W.T.A. **3**: 420 (1972). —Clayton, Phillips & Renvoize in F.T.E.A., Gramineae: 410 (1974). Type from Cameroon.
  Arundinella elegantula Hook.f. in Journ. Linn. Soc. **7**: 233 (1864). Type as above.
  Trichopteryx elegantula var. *katangensis* Chippindall in Blumea, Suppl. **3**: 29 (1946). Type from Zaire.
  Trichopteryx delicatissima Phipps in Kirkia **4**: 119 (1964). Type: Zambia, Kalambo Falls, *Exell, Mendonça & Wild* 1284 (SRGH, holotype; BM, isotype).

Annual. Culms 2–30 cm. high, erect or ascending, usually slender and wiry. Leaf laminae 0.5–2.5 cm. long, 1–6 mm. wide, lanceolate to ovate. Panicle 0.5–7 cm. long, ovate, open or compact. Spikelets 2.5–3.5 mm. long, glabrous or tuberculate setose. Superior lemma with short lateral tufts; central awn 7–15 mm. long, the laterals 3–6 mm.; callus broadly rounded.

Zambia. N: Mbala Distr., Kalambo Falls, 11.iv.1961, *Phipps & Vesey-FitzGerald* 3079 (K; SRGH). **Malawi.** N: Mzimba Distr., Viphya, 21.vi.1974, *Salubeni* 1948 (K; SRGH). C: Ntchisi For. Res., 26.iii.1970, *Brummitt* 9424 (K; SRGH). S: Zomba Plateau, 2.vi.1946, *Brass* 16153 (K; NY).
Also in Zaire and Tanzania, and westwards through Uganda to Sierra Leone. Damp slopes; 1200–2300 m.

2. **Trichopteryx stolziana** Henr. in Fedde, Repert. **18**: 242 (1922). —C.E. Hubbard in F.T.A. **10**: 10 (1937). —Jackson & Wiehe, Annot. Check List Nyasal. Grass.: 63 (1958). —Clayton, Phillips & Renvoize in F.T.E.A., Gramineae: 410 (1974). TAB. **59**. Type from Tanzania.
  Loudetia pusilla Chiov. in Atti Reale Accad. It., Mem. Cl. Sci. Fis. Math. Nat. **11**: 64 (1940). Type from Tanzania.
  Trichopteryx elegantula subsp. *stolziana* Phipps in Kirkia **4**: 120 (1964). Type as for *T. stolziana*.

Like *T. elegantula* except; panicle usually compact; spikelets 3.5–5-5 mm. long; callus truncate to emarginate.

Zambia. N: Mbala Distr., Ndundu, 3.iv.1959, *McCallum-Webster* A262 (K). W: Kasempa Distr., Chifubwe Gorge, near Solwezi, 20.iii.1961, *Drummond* 7109 (K; SRGH). C: Serenje Distr., 50 km. NE. of Kanona, 6.iv.1961, *Phipps & Vesey-FitzGerald* 2969 (K; SRGH). **Zimbabwe.** E: Mutare Distr., Sheba Forest Estates, 1.v.1970, *Crook* 922 (K; SRGH). **Malawi.** N: Misuku Hills, 28.iv.1951, *Jackson* 557 (K). **Mozambique.** N: Vila Cabral, 25.ii.1964, *Torre & Paiva* 10830 (COI; K). Z: Gúrùe, near Pico Namuli, 4. iv. 1943, *Torre* 5156 (COI; K).
Also in Zaire and Tanzania. Wet flushes on banks and among rocks; 1300–2500 m.
Closely related to *T. elegantula*, and sometimes intergrading with it.

3. **Trichopteryx fruticulosa** Chiov. in Ann. Bot. Roma **13**: 50 (1914). —C.E. Hubbard in F.T.A. **10**: 6 (1937). —Phipps in Kirkia **4**: 120 (1964). —Clayton, Phillips & Renvoize in F.T.E.A., Gramineae: 411 (1974). Type from Zaire.
  Trichopteryx katangensis Chiov. in Nuov. Giorn. Bot. Ital, n.s., **26**: 68 (1919). Types from Zaire.
  Trichopteryx katangensis De Wild., Notes Fl. Katanga **5**: 35 (1920) non Chiov. Type from Zaire.
  Trichopteryx fruticulosa var. *whytei* C.E. Hubbard in Kew Bull. **1934**: 425 (1934); in F.T.A. **10**: 7 (1937). —Jackson & Wiehe, Annot. Check List Nyasal. Grass.: 63 (1958). Type: Malawi, Kondowe to Karonga, *Whyte* (K, holotype).
  Trichopteryx perlaxa Pilger in Not. Bot. Gart. Berlin **12**: 383 (1935). —Phipps in Kirkia **4**: 120 (1964). Type from Tanzania.
  Trichopteryx fruticulosa var. *perlaxa* (Pilger) C.E. Hubbard in F.T.A. **10**: 7 (1937). Type as above.

Perennial. Culms 30–80 cm. high, slender, somewhat woody, stiffly erect or ascending, the internodes shorter than the leaf-sheaths which cover the nodes. Leaf laminae 3–8 cm. long, 3–10 mm. wide, linear-lanceolate, ± reflexed. Panicle 6–30 cm. long, lanceolate; pedicel hairs 1–2 mm. long. Spikelets 3–6 mm. long, glabrous, occasionally with a few hairs at apex of inferior glume, rarely softly pilose. Superior lemma glabrous or sparsely pilose, with lateral tufts 1–2 mm. long; central awn 8–11 mm. long, the laterals 1.5–3 mm.

Zambia. N: Mbala, Sunzu Mt., 24.iii.1966, *Richards* 21391 (K). W: Mwinilunga Distr., 15 km. SW. of Kalene Hill, 14.vi.1963, *Drummond* 8334 (K; SRGH). C: Kundalila Falls, 17.iii.1974, *Davidse & Handlos* 7263 (K; MO; SRGH). E: Ntumbacushi Falls, 22.vi.1957, *Robinson* 2352 (K; SRGH). **Malawi.** N: Nkhata Bay, Chikale Beach, 15.iv.1977, *Pawek* 12590 (K; MA; MO; SRGH; UC). S: Mulanje, Chambe plateau, 18.xi.1949, *Wiehe* N/354 (K).
Also in Zaire, Tanzania and Angola. Rocky slopes in *Brachystegia* woodland; 500–2300 m.

4. **Trichopteryx dregeana** Nees in Lindl., Nat. Syst. ed. 2: 449 (1836). —C.E. Hubbard in F.T.A. **10**: 7 (1937). —Sturgeon in Rhod. Agric. Journ. **51**: 216 (1954). —Chippindall in Meredith, Grasses & Pastures of S. Afr.: 287 (1955). —Jackson & Wiehe, Annot. Check List Nyasal. Grass.: 63 (1958). —Phipps in Kirkia **4**: 120 (1964). Type from S. Africa.

Tab. 59. TRICHOPTERYX STOLZIANA. 1, habit (× ½), from *Richards* 8827; 2, spikelet (× 9); 3, superior lemma (× 12), 2–3 from *Milne-Redhead & Taylor* 10432.

Perennial. Culms 30–100 cm. long, wiry, trailing, the internodes longer than the leaf-sheaths. Leaf laminae 1–3 cm. long, 1–3 mm. wide, linear, ± reflexed. Panicle 3–15 cm. long, narrowly ovate; pedicel hairs usually conspicuous, 0.5–2 mm. long. Spikelets 4–5 mm. long, glabrous or with a few hairs at the apex of the inferior glume. Superior lemma glabrous with conspicuous lateral tufts 2–3 mm. long; central awn 4–6 mm. long, the laterals 1.5–4 mm.

Zambia. B: Mongu, 24.iii.1964, *Verboom* 1332 (K). N: Mbala Distr., Lunzua R., 30 km. from Mbala on Kambole road, 5.iv.1959, *McCallum-Webster* A274 (K; SRGH). C: Serenje Distr., Kundalila Falls, 5.v.1972, *Kornaś* 1686 (K). **Zimbabwe**. E: Chimanimani, Cashel Scenic Rd., 22.iv.1973, *Crook* 1082 (K; SRGH). S: Great Zimbabwe, 8.iv.1973, *Chiparawasha* 688 (K; SRGH). **Malawi**. S: Zomba Mt., 29.vi.1955, *Jackson* 1693 (K).
Also in Zaire, Angola, S. Africa and Madagascar. Wet flushes and marshy places; 1100–2100 m.

5. **Trichopteryx marungensis** Chiov. in Nuov. Giorn. Bot. Ital., n.s., **26**: 67 (1919). —C.E. Hubbard in F.T.A. **10**: 9 (1937). —Clayton in F.W.T.A. **3**: 420 (1972). —Clayton, Phillips & Renvoize in F.T.E.A., Gramineae: 411 (1974). Type from Zaire.
    Arundinella marungensis (Chiov.) Chiov. in Nuov. Giorn. Bot. Ital., n.s., **29**: 113 (1923). Type as above.
    Trichopteryx gracillima C.E. Hubbard in Kew Bull. **1934**: 425 (1934); in F.T.A. **10**: 8 (1937). —Jackson & Wiehe, Annot. Check List Nyasal. Grass.: 63 (1958). —Phipps in Kirkia **4**: 120 (1964). Type from Uganda.
    Trichopteryx decumbens C.E. Hubbard in Kew Bull. **4**: 350 (1949). Type: Zambia, Mwinilunga, *Milne-Redhead* 3705 (K, holotype).

Perennial. Culms 20–60 cm. long, wiry, trailing or geniculately ascending, the internodes longer than the leaf-sheath. Leaf laminae 1–4 cm. long, 2–6 mm. wide, lanceolate to narrowly ovate, ± reflexed. Panicle 5–12 cm. long, lanceolate to ovate; pedicel hairs inconspicuous, up to 1 mm. long. Spikelets 3.5–5 mm. long, minutely pubescent, sometimes also tuberculate-setose. Superior lemma pubescent, with lateral tufts 0.4–1 mm. long; central awn 6–12 mm. long, the laterals 1.5–3.5 mm.

Zambia. N: Shiwa Ngandu, 1.vi.1956, *Robinson* 1518 (K). W: Mwinilunga Distr., Ikelenge, 16.iv.1965, *Robinson* 6607 (K; SRGH). C: Serenje Distr., Kundalila Falls, 15.x.1967, *Simon & Williamson* 1004 (K; SRGH). **Zimbabwe**. E: Nyanga, Pungwe Gorge, 7.ix.1954, *Wild* 4612 (K; SRGH). **Malawi**. N: Mzimba Distr., 16 km. SW. of Mzuzu, 30.vi.1974, *Pawek* 8775 (K; MA; MO; SRGH; UC). C: Lilongwe Res. Sta., 14.iv.1951, *Jackson* 458 (K).

41. LOUDETIA Steud.

Loudetia Steud., Syn. Pl. Glum. **1**: 238 (1854) nom. conserv.

Panicle open, contracted or spiciform, bearing spikelets singly or in pairs. Spikelets brown, disarticulating beneath the superior floret and tardily so beneath the inferior. Inferior lemma 3-nerved. Superior lemma bidentate, with a deciduous (except *L. phragmitoides*) awn from the sinus; callus oblong to linear, truncate, 2-toothed or obliquely pungent; palea not winged.

A genus of 23 species. Mainly in tropical Africa, but also in Madagascar and South America.

1. Superior lemma loosely hirsute with silky white hairs 1–2 mm. long; inferior floret barren - - - - - - - - - - - 1. *phragmitoides*
 – Superior lemma pubescent; inferior floret ♂ - - - - - - - - 2
2. Panicle spiciform - - - - - - - - - - - - - - - 3
 – Panicle open or contracted, but not spiciform - - - - - - - - 4
3. Callus emarginate, oblong, 0.4–0.7 mm. long; culm nodes distinct, dark 2. *coarctata*
 – Callus pungent, obliquely 2-toothed, linear, 1–1.5 mm. long; culm nodes indistinct - - - - - - - - - - - - - - 3. *densispica*
4. Inferior lemma and superior glume acuminate; stamens 3; callus linear, narrowly truncate - - - - - - - - - - - - - - 4. *flavida*
 – Inferior lemma and superior glume obtuse to acute; stamens 2 - - - - 5
5. Callus truncate - - - - - - - - - - - - - - - 6
 – Callus 1- or 2-toothed - - - - - - - - - - - - - - 7
6. Panicle loose, open; callus oblong - - - - - - - - 5. *arundinacea*
 – Panicle contracted, linear; callus linear - - - - - - - 6. *angolensis*
7. Callus symmetrically 2-toothed - - - - - - - - - 7. *simplex*
 – Callus pungent or obliquely and unequally 2-toothed - - - - - 8. *lanata*

1. **Loudetia phragmitoides** (Peter) C.E. Hubbard in Kew Bull. **1934**: 428 (1934); in F.T.A. **10**: 18 (1937). —Jackson & Wiehe, Annot. Check List Nyasal. Grass.: 47 (1958). —Phipps in Kirkia **4**: 99 (1964). —Clayton in F.W.T.A. **3**: 417 (1972). —Clayton, Phillips & Renvoize in F.T.E.A., Gramineae: 415 (1974). Type from Burundi.
    Trichopteryx phragmitoides Peter, Fl. Deutsch. Ost-Afr. **1**, Anh.: 96 (1930). Type as above.

Robust perennial forming large tussocks. Culms 2–4 m. high. Panicle 40–60 cm. long, narrowly oblong, dense, contracted. Spikelets 6–13 mm. long, glabrous or tuberculate-setose. Glumes obtuse to subacute. Inferior floret barren. Superior lemma loosely hirsute with spreading silky hairs 1–2 mm. long; awn persistent with a short (1.5–3 mm.) column and falcate limb 9–16 mm. long; stamens 2; callus 0.2 mm. long, oblong, truncate.

Zambia. N: Fisaka R. on Mporokoso to Kawambwa Rd., 10.v.1954, *Hinds* 264 (K). W: Mwinilunga Distr., Kalene Hill Mission, 10.vi.1963, *Drummond* 8282 (K; SRGH). **Malawi**. S: Thyolo, Nehima Estate, 19.ix.1950, *Wiehe* N/652 (K). **Mozambique**. N: Malema, Murralelo, 19.iii.1964, *Torre & Paiva* 11270 (COI; K). Z: Nhamarroi, 10.vi.1943, *Torre* 5468 (COI; K). MS: Cheringoma Plateau, Chinizuia Drainage, vi.1972, *Tinley* 2631 (K; SRGH).
Also in Angola and northwards to Senegal and Sudan. Marshy places; 650–1300 m.

2. **Loudetia coarctata** (A. Camus) C.E. Hubbard in Kew Bull. **1934**: 428 (1934); in F.T.A. **10**: 37 (1937). —Clayton in F.W.T.A. **3**: 417 (1972). Type from Guinée.
    Tristachya coarctata A. Camus in Bull. Soc. Bot. Fr. **53**: 774 (1933). Type as above.

Tufted perennial, the basal sheaths becoming fibrous. Culms 30–75 cm. high, with distinct dark nodes. Panicle 4–12 cm. long, spiciform. Spikelets 9–11 mm. long. Inferior glume fringed with a line of tubercle-based setae on either flank, acute. Superior lemma bidenticulate; awn 2.5–4 cm. long; stamens 2; callus 0.4–0.7 mm. long, oblong, emarginate.

Zambia. N: Mansa, 24.iv.1969, *Verboom* 2502 (K).
Also in Zaire and northwards to Guinée and Sudan. Seasonally wet pans and drainage tracts; 1400 m.
It is related to *L. vanderystii* (De Wild.) C.E. Hubbard from Zaire, a more delicate species with shorter spikelets (7–9 mm.) and awns (1–1.5 cm.).

3. **Loudetia densispica** (Rendle) C.E. Hubbard in Kew Bull. **1934**: 428 (1934); in F.T.A. **10**: 38 (1937). —Chippindall in Meredith, Grasses & Pastures of S. Afr.: 284 (1955). Type from Angola.
    Trichopteryx densispica Rendle, Cat. Afr. Pl. Welw. **2**: 214 (1899). Type as above.
    Loudetia gossweileri C.E. Hubbard in Kew Bull. **1934**: 428 (1934); in F.T.A. **10**: 36 (1937). —Phipps in Kirkia **4**: 99 (1964). Type from Angola.

Tufted perennial, the basal sheaths sometimes fibrous. Culms 50–120 cm. high, juncoid, the nodes scarcely differentiated. Panicle 4–14 cm. long, spiciform. Spikelets 12–18 mm. long. Inferior glume glabrous or with a line of tubercle-based setae on either flank, acute. Superior lemma bidenticulate; awn 3–7 cm. long; stamens 2; callus 1–1.5 mm. long, linear, pungent or obliquely 2-toothed.

Zambia. B: Lukalanya R. resettlement area, 80 km. E. of Mongu, 8.ii.1955, *Hinds* 284 (K); Mata School area, Senanga, 3.ii.1987, *Jeanes* 333 (K; SRGH).
Also in Zaire, Angola, S. Africa (Transvaal and Swaziland). Woodland margins on freely drained soil.

4. **Loudetia flavida** (Stapf) C.E. Hubbard in Kew Bull. **1934**: 429 (1934); in F.T.A. **10**: 34 (1937). —Sturgeon in Rhod. Agric. Journ. **51**: 217 (1954). —Chippindall in Meredith, Grasses & Pastures of S. Afr.: 282 (1955). —Jackson & Wiehe, Annot. Check List Nyasal. Grass.: 47 (1958). —Phipps in Kirkia **4**: 100 (1964). —Clayton in F.W.T.A. **3**: 417 (1972). —Clayton, Phillips & Renvoize in F.T.E.A., Gramineae: 416 (1982). Types from S. Africa.
    Trichopteryx flavida Stapf in Kew Bull. **1897**: 298 (Sept. 1897). Type as above.
    Trichopteryx pennata Chiov. in Ann. Ist. Bot. Roma **7**: 69 (Nov. 1897). Type from Ethiopia.
    Trichopyteryx glabra Hackel in Bull. Herb. Boiss. sér. 2, **1**: 770 (1901). Type: Mozambique, Boroma, *Menyharth* 916 (W, holotype).
    Loudetia pennata (Chiov.) C.E. Hubbard in Kew Bull. **1934**: 428 (1934); in F.T.A. **10**: 32 (1937). —Phipps in Kirkia **4**: 100 (1964). Type as for *Trichopteryx pennata*.
    Trichopteryx cuspidata Gilli in Ann. Naturhist. Mus. Wien **69**: 43 (1966). Type from Tanzania.

Tufted perennial, the basal sheaths pubescent to woolly tomentose. Culms 30–150 cm. high, unbranched. Leaf blades up to 5 mm. wide, flat or filiform. Panicle 10–30 cm. long, linear to narrowly ovate, open or dense, the branches glabrous to pubescent or rarely villous. Spikelets 6–12 mm. long, glabrous or rarely tuberculate setose. Inferior glume acute to acuminate; superior glume and inferior lemma acuminate. Superior lemma 9-nerved, bidentate; awn 1.5–4.5 mm. long; stamens 3; callus 0.5–0.7 mm. long, linear, narrowly truncate.

Botswana. N: Pandamatenga, 18.iii.1966, *Rains* 55 (K). **Zambia**. N: Mbala Distr., Mpulungu, 12.iv.1959, *McCallum-Webster* A310 (K). C: Serenje, 5.iv.1961, *Phipps & Vesey-FitzGerald* 2948 (K; SRGH). E: Chipata, Katete R., 3.ii.1968, *Simon & Williamson* 1633 (K; SRGH). S: Kalomo, Senkobo, 15.i.1963, *Astle* 2877 (K). **Zimbabwe**. N: Bindura Distr., Kingston Hill, 7.v.1969, *Wild* 7765 (K; SRGH). W: Hwange Distr., Kazuma Forest, 24.i.1974, *Gonde* 16/74 (K; SRGH). C: Harare Distr., Lake McIlwaine Nat. Park, 19.xii.1965, *Crook* 762 (K; SRGH). E: Nyanga Distr., Lawley's Concession, 21.ii.1954, *West* 3398 (K; SRGH). S: 16 km. S. of Umzingwane R. on Beitbridge Rd., 4.i.1956, *Rattray* 1759 (K; SRGH). **Malawi**. N: Mzimba Distr., Mbawa, 19.vi.1952, *Jackson* 843 (K). C: Nkhota Kota Distr., Matichi, 24.ii.1953, *Jackson* 1110 (K). **Mozambique**. N: Mandimba, 28.iii.1942, *Hornby* 3362 (K; SRGH).

Also recorded from tropical and S. Africa (Transvaal). Dry sandy or rocky soils in bushland and woodland: 600–2000 m.

Two closely related species, distinguished mainly by their wiry culms which branch towards the base, are *L. filifolia* Schweick. from South Africa and *L. cuanzensis* Lubke & Phipps from Angola. The former also differs in its shorter oblong callus, and the latter in its 5–7-nerved superior lemma.

5. **Loudetia arundinacea** (A. Rich.) Steud., Syn. Pl. Glum. **1**: 238 (1854). —C.E. Hubbard in F.T.A. **10**: 20 (1937). —Phipps in Kirkia **4**: 99 (1964). —Clayton in F.W.T.A. **5**: 417 (1972). —Clayton, Phillips & Renvoize in F.T.E.A., Gramineae: 417 (1974). Type from Ethiopia.
   Tristachya arundinacea A. Rich., Tent. Fl. Abyss. **2**: 417 (1851). Type as above.
   Trichopteryx arundinacea (A. Rich.) Engl., Hochgebirgsfl. Trop. Afr.: 129 (1892). Type as above.
   Trichopteryx convoluta De Wild., Notes Fl. Katanga **5**: 25 (1920). Types from Zaire.
   Trichopteryx dobbelaerei De Wild., Notes Fl. Katanga **5**: 28 (1920). Type from Zaire.
   Trichopteryx elegans var. *hensii* De Wild., Notes Fl. Katanga **5**: 30 (1920). Type from Zaire.
   Trichopteryx verticillata De Wild., Notes Fl. Katanga **5**: 38 (1920). Type from Zaire.
   Trichopteryx arundinacea var. *trichantha* Peter, Fl. Deutsch Ost-Afr. **1**, Anh.: 96 (1930). Type from Tanzania.
   Loudetia arundinacea var. *trichantha* (Peter) Hutch. in F.T.A. **2**: 544 (1936). —C.E. Hubbard in F.T.A. **10**: 22 (1937). Type as above.
   Loudetia arundinacea var. *hensii* (De Wild.) Pic.-Serm., Miss. Studio Lago Tana **7**: 180 (1951). Type as for *Trichopteryx elegans* var. *hensii*.

Tufted perennial, the basal leaf sheaths ± silky pubescent. Culms 60–300 cm. high, the nodes bearded or not. Panicle 18–70 cm. long, narrowly oblong to ovate, the branches spreading and conspicuously whorled. Spikelets 6–13 mm. long, glabrous or tuberculate setose. Inferior glume obtuse; superior glume and inferior lemma obtuse to subacute. Superior lemma bidentate with teeth 0.7–1 mm. long; awn 2–4 cm. long; stamens 2; callus 0.5–1 mm. long, truncate or slightly emarginate.

Zambia. N: Mbala to Mpulungu, 10.iv.1961, *Phipps & Vesey-FitzGerald* 3008 (K; SRGH). W: Mwinilunga Distr., Matonchi, 6.xi.1937, *Milne-Redhead* 3133 (K; SRGH). **Mozambique**. N: Malema, Murralelo, 19.iii.1964, *Torre & Paiva* 11264 (COI; K). Z: Quelimane to Milange, 11.iii.1943, *Torre* 4924 (COI; K).

Also in Angola and northwards to Senegal and Ethiopia. *Brachystegia* woodland, sometimes on the margins of drainage tracts; 700–1900 m.

6. **Loudetia angolensis** C.E. Hubbard in Kew Bull. **1934**: 426 (1934); in F.T.A. **10**: 19 (1937). —Phipps in Kirkia **4**: 99 (1964). Type from Angola.

Like *L. arundinacea* except; panicle 20–35 cm. long, contracted, linear; spikelets 6–7 mm. long; inferior glume obtuse to subacute; callus linear, narrowly truncate.

Zambia. B: Kalabo, 16.xi.1959, *Drummond & Cookson* 6872 (K; SRGH). W: Mwinilunga Distr., Dobeka Bridge, 8.ii.1938, *Milne-Redhead* 4489 (K; SRGH).

Also in Angola. Marshy places and margins of drainage tracts; 1100–1500 m.

Very closely allied to *L. arundinacea*, being distinguished by the narrow panicle and slender callus.

7. **Loudetia simplex** (Nees) C.E. Hubbard in Kew Bull. **1934**: 431 (1934); in F.T.A. **10**: 25 (1937). —Sturgeon in Rhod. Agric. Journ. **51**: 216 (1954). —Chippindall in Meredith, Grasses & Pastures of S. Afr.: 282 (1955). —Jackson & Wiehe, Annot. Check List Nyasal. Grass.: 47 (1958). —Phipps in Kirkia **4**: 99 (1964). —Clayton in F.W.T.A. **3**: 419 (1972). —Clayton, Phillips & Renvoize in F.T.E.A., Gramineae: 418 (1974). TAB. **60**. Type from S. Africa.

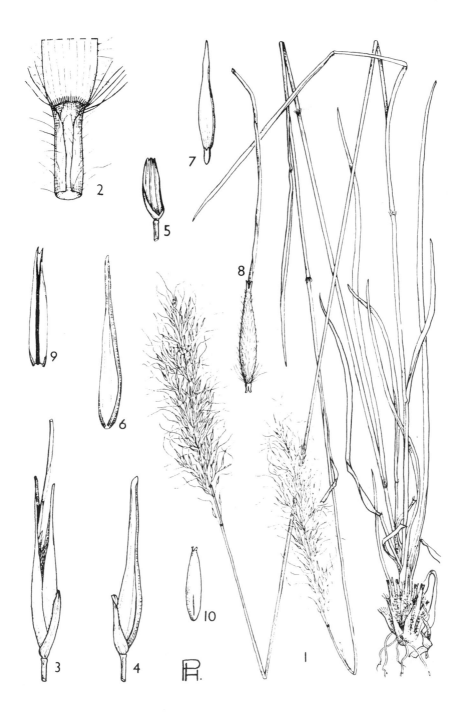

Tab. 60. LOUDETIA SIMPLEX. 1, habit (× ½); 2, ligule (× 8); 3, spikelet (× 4); 4, glumes (× 4); 5, inferior glume (× 4); 6, inferior lemma (× 4); 7, palea of inferior floret (× 4); 8, superior lemma (× 4); 9, palea of superior floret (× 6); 10, caryopsis (× 6), all from *Boaler* 835.

Tristachya simplex Nees, Fl. Afr. Austr.: 269 (1841). Type as above.
Trichopteryx simplex (Nees) Engl., Hochgebirgsfl. Trop. Afr.: 129 (1892). Type as above.
Trichopteryx simplex var. *gracilis* Rendle, Cat. Afr. Pl. Welw. **2**: 214 (1899). Types from Angola.
Trichopteryx simplex var. *crinita* Stapf in F.C. **7**: 450 (1899). Types from S. Africa.
Trichopteryx simplex var. *minor* Stapf in F.C. **7**: 450 (1899). Type from S. Africa.
Trichopteryx simplex var. *sericea* Stapf in F.C. **7**: 450 (1899). Types from S. Africa.
Trichopteryx stipoides var. *natalensis* Hackel in Mem. Herb. Boiss. **20**: 9 (1900). Types from S. Africa.
Trichopteryx kapirensis De Wild., Notes Fl. Katanga **5**: 34 (1920). Type from Zaire.
Trichopteryx elisabethvilleana De Wild., Notes Fl. Katanga **5**: 31 (1920). Type from Zaire.
Trichopteryx gracilis Peter, Fl. Deutsch Ost-Afr. **1**, Anh.: 96 (1930). Type from Tanzania.
Trichopteryx simplex var. *longifolia* Peter, Fl. Deutsch Ost-Afr. **1**, Anh.: 96 (1930). Type from Tanzania.
Arundinella simplex (Nees) Roberty in Bull. Inst. Fond. Afr. Noire **17**: 56 (1955). Type as for *Tristachya simplex*.

Tufted perennial, the basal leaf-sheaths usually becoming fibrous, typically woolly tomentose but sometimes silky pubescent to glabrecent. Culms 30–150 cm. high. Panicle 5–30 cm. long, linear to narrowly ovate, the branches contracted or loose but seldom markedly whorled. Spikelets 8–14 mm. long, glabrous or tuberculate-setose. Inferior glume obtuse; superior glume and inferior lemma obtuse to acute. Superior lemma bidentate with teeth 0.2–1 mm. long; stamens 2; callus 0.8–1 mm. long, narrowly oblong, 2-toothed.

Zambia. B: Mongu, 22.iii.1964, *Verboom* 1323 (K). N: Mbala Distr., Ndundu, 23.ii.1959, *McCallum-Webster* A126 (K; SRGH). W: Mwinilunga Distr., 6 km. N. of Kalene Hill, 12.xii.1963, *Robinson* 5898 (K). C: Kabwe Distr., Mpunde Mission, 20.i.1973, *Kornaś* 3048 (K). E: Chipata, i.1963, *Verboom* 902 (K). S: Choma, 17.ii.1963, *Astle* 2135 (K). **Zimbabwe**. N: Makonde (Lomagundi) Distr., Maryland, 9.ii.1961, *Phipps* 2850 (K; SRGH). W: Matopos Research Station, 19.ii.1954, *Rattray* 1627 (K; SRGH). C: Chegutu Distr., Msengezi Purchase Land, ii.1972, *Davies* 3360 (K; SRGH). E: Mutare, 28.ii.1969, *Crook* 854 (K; SRGH). S: Mberengwa Distr., Mt. Bukwa, 1.v.1973, *Simon, Pope & Biegel* 2403 (K; SRGH). **Malawi**. N: Viphya Plateau, 60 km. SW. of Mzuzu, 23.ii.1975, *Pawek* 9091 (K; MO). C: Mchinje, Bua R., 29.i.1952, *Jackson* 718A (K). S: Zomba Plateau, 29.iii.1937, *Laurence* 351 (K). **Mozambique**. N: Marrupa, Naboina, 17.ii.1981, *Nuvunga* 557 (K; SRGH). Z: Pico Namuli, 9.iv.1943, *Torre* 5132 (COI; K). MS: Beira to Dondo, 31.xii.1943, *Torre* 6334 (COI; K).
Also recorded from and tropical and S. Africa, and from Madagascar. Wooded grassland, on both stony and seasonally waterlogged soils; 1000–2500 m.
Loudetia kagerensis (K. Schum.) Hutch. is very similar. It occurs to the north of the Flora Zambesiaca area and is distinguished by the emarginate (lobes less than 0.2 mm.) apex of the superior lemma.

8. **Loudetia lanata** (Stent & Rattray) C.E. Hubbard in Kew Bull. **1934**: 429 (1934); in F.T.A. **10**: 30 (1937). —Sturgeon in Rhod. Agric. Journ. **51**: 217 (1954). —Phipps in Kirkia **4**: 100 (1964). —Launert in Merxm., Prodr. Fl. SW. Afr. **160**: 127 (1970). Type: Zimbabwe, Harare, *Eyles* 2955 (SRGH, holotype; K, isotype).
  Trichopteryx lanata Stent & Rattray in Proc. Rhod. Sci. Ass. **32**: 39 (1933). Type as above.
  Loudetia crassipes C.E. Hubbard in Kew Bull. **1934**: 430 (1934); in F.T.A. **10**: 32 (1937). Type from Angola.
  Loudetia longipes C.E. Hubbard in Kew Bull. **1934**: 430 (1934); in F.T.A. **10**: 29 (1937). Type from Angola.

Tufted perennial, the basal leaf-sheaths woolly and usually becoming fibrous. Culms 40–100 cm. high. Panicle 5–35 cm. long, ovate, loose or somewhat contracted. Spikelets 8–14 mm. long, tuberculate setose, rarely glabrous. Inferior glume obtuse to subacute; superior glume and inferior lemma obtuse to acute. Superior lemma bidentate with teeth 0.5–1 mm. long; awn 4–6 cm. long, including a column 0.5–1 cm. long; stamens 2; callus 1–1.5 mm. long, linear, obliquely pungent or asymmetrically 2-toothed.

Caprivi Strip. Katima Mulilo, 26.xii.1958, *Killick & Leistner* 3105 (K; PRE). **Zambia**. B: 56 km. W. of Kabompo, 26.xii.1969, *Simon & Williamson* 2047 (K; SRGH). N: 25 km. N. of Chipili Mission on Kawambwa to Mansa road, 27.ii.1970, *Drummond & Williamson* 10065 (K; SRGH). **Zimbabwe**. C: Harare, S. of Mukuvisi R., 19.i.1985, *Hyde* 318 (K; SRGH). S: Masvingo, 5.i.1948, *Robinson* 173 (K; SRGH).
Also in Angola, Namibia and S. Africa (Transvaal). Savanna woodland, often on damp soils; 1000–1600 m.
Intergrades with *L. demeusei* (De Wild.) C.E. Hubbard from Zaire. This is a taller (100–200 cm.) plant with a contracted panicle, but the only significant difference seems to be the longer (1.2–2 cm.) column of its awn.

INDEX TO BOTANICAL NAMES